CAMBRIDGE LIBRARY COLLECTION

Books of enduring scholarly value

Physical Sciences

From ancient times, humans have tried to understand the workings of the world around them. The roots of modern physical science go back to the very earliest mechanical devices such as levers and rollers, the mixing of paints and dyes, and the importance of the heavenly bodies in early religious observance and navigation. The physical sciences as we know them today began to emerge as independent academic subjects during the early modern period, in the work of Newton and other 'natural philosophers', and numerous sub-disciplines developed during the centuries that followed. This part of the Cambridge Library Collection is devoted to landmark publications in this area which will be of interest to historians of science concerned with individual scientists, particular discoveries, and advances in scientific method, or with the establishment and development of scientific institutions around the world.

The Scientific Papers of James Clerk Maxwell

The publication in 1890 of the two-volume *Scientific Papers of James Clerk Maxwell*, edited by W.D. Niven, was one of the two objects of a committee formed 'for the purpose of securing a fitting memorial of him' (the other object being the commissioning of a marble bust for the Cavendish Laboratory). Before his death in 1879 at the age of 48, Clerk Maxwell had made major contributions to many areas of theoretical physics and mathematics, not least his discoveries in the fields of electromagnetism and of the kinetic theory of gases, which have been regarded as laying the foundations of all modern physics. He is generally considered the third most important physicist of all time, after Newton and Einstein. These collected shorter works, beginning with a paper written at the age of 15, show the wide range of Clerk Maxwell's interests across mathematics, physics and chemistry.

Cambridge University Press has long been a pioneer in the reissuing of out-of-print titles from its own backlist, producing digital reprints of books that are still sought after by scholars and students but could not be reprinted economically using traditional technology. The Cambridge Library Collection extends this activity to a wider range of books which are still of importance to researchers and professionals, either for the source material they contain, or as landmarks in the history of their academic discipline.

Drawing from the world-renowned collections in the Cambridge University Library, and guided by the advice of experts in each subject area, Cambridge University Press is using state-of-the-art scanning machines in its own Printing House to capture the content of each book selected for inclusion. The files are processed to give a consistently clear, crisp image, and the books finished to the high quality standard for which the Press is recognised around the world. The latest print-on-demand technology ensures that the books will remain available indefinitely, and that orders for single or multiple copies can quickly be supplied.

The Cambridge Library Collection will bring back to life books of enduring scholarly value (including out-of-copyright works originally issued by other publishers) across a wide range of disciplines in the humanities and social sciences and in science and technology.

The Scientific Papers of James Clerk Maxwell

VOLUME 1

JAMES CLERK MAXWELL
EDITED BY W. D. NIVEN

CAMBRIDGE
UNIVERSITY PRESS

CAMBRIDGE UNIVERSITY PRESS

Cambridge, New York, Melbourne, Madrid, Cape Town, Singapore,
São Paolo, Delhi, Dubai, Tokyo, Mexico City

Published in the United States of America by Cambridge University Press, New York

www.cambridge.org
Information on this title: www.cambridge.org/9781108012256

© in this compilation Cambridge University Press 2010

This edition first published 1890
This digitally printed version 2010

ISBN 978-1-108-01225-6 Paperback

THE

SCIENTIFIC PAPERS

OF

JAMES CLERK MAXWELL.

London: C. J. CLAY & SONS,
CAMBRIDGE UNIVERSITY PRESS WAREHOUSE,
AVE MARIA LANE.

Cambridge: DEIGHTON, BELL AND CO.
Leipzig: F. A. BROCKHAUS.

James Clerk Maxwell

Engraved by G. J. Stodart from a Photograph by Fergus of Greenock

THE

SCIENTIFIC PAPERS

OF

JAMES CLERK MAXWELL,

M.A., LL.D. EDIN., D.C.L., F.R.SS. LONDON AND EDINBURGH,

HONORARY FELLOW OF TRINITY COLLEGE,

CAVENDISH PROFESSOR OF EXPERIMENTAL PHYSICS IN THE UNIVERSITY
OF CAMBRIDGE.

EDITED BY

W. D. NIVEN, M.A., F.R.S.,

DIRECTOR OF STUDIES AT THE ROYAL NAVAL COLLEGE, GREENWICH ;
FORMERLY FELLOW OF TRINITY COLLEGE.

VOL. I.

CAMBRIDGE :
AT THE UNIVERSITY PRESS.
1890

CAMBRIDGE:

PRINTED BY C. J. CLAY, M.A. AND SONS,
AT THE UNIVERSITY PRESS.

TO HIS GRACE

THE DUKE OF DEVONSHIRE K.G.

CHANCELLOR OF THE UNIVERSITY OF CAMBRIDGE

FOUNDER OF THE CAVENDISH LABORATORY

THIS MEMORIAL EDITION

OF

THE SCIENTIFIC PAPERS

OF

THE FIRST CAVENDISH PROFESSOR OF EXPERIMENTAL PHYSICS

IS

BY HIS GRACE'S PERMISSION

RESPECTFULLY AND GRATEFULLY DEDICATED

SHORTLY after the death of Professor James Clerk Maxwell a Committee was formed, consisting of graduate members of the University of Cambridge and of other friends and admirers, for the purpose of securing a fitting memorial of him.

The Committee had in view two objects: to obtain a likeness of Professor Clerk Maxwell, which should be placed in some public building of the University; and to collect and publish his scattered scientific writings, copies of which, so far as the funds at the disposal of the Committee would allow, should be presented to learned Societies and Libraries at home and abroad.

It was decided that the likeness should take the form of a marble bust. This was executed by Sir J. E. Boehm, R.A., and is now placed in the apparatus room of the Cavendish Laboratory.

In carrying out the second part of their programme the Committee obtained the cordial assistance of the Syndics of the University Press, who willingly consented to publish the present work. At the request of the Syndics, Mr W. D. Niven, M.A., Fellow and Assistant Tutor of Trinity College and now Director of Studies at the Royal Naval College, Greenwich, undertook the duties of Editor.

The Committee and the Syndics desire to take this opportunity of acknowledging their obligation to Messrs Adam and Charles Black, Publishers of the ninth Edition of the *Encyclopædia Britannica*, to Messrs Taylor and Francis, Publishers of the *London, Edinburgh, and Dublin Philosophical Magazine and Journal of Science*, to Messrs Macmillan and Co., Publishers of *Nature* and of the *Cambridge and Dublin Mathematical Journal*, to Messrs Metcalfe and Co., Publishers of the *Quarterly Journal of Pure and Applied Mathematics*, and to the Lords of the Committee of Council on Education, Proprietors of the *Handbooks* of the South Kensington Museum, for their courteous consent to allow the articles which Clerk Maxwell had contributed to these publications to be included in the present work; to Mr Norman Lockyer for the assistance which he rendered in the selection of the articles re-printed from *Nature*; and their further obligation to Messrs Macmillan and Co. for permission to use in this work the steel engravings of Faraday, Clerk Maxwell, and Helmholtz from the *Nature* Series of Portraits.

Numerous and important Papers, contributed by Clerk Maxwell to the *Transactions* or *Proceedings* of the Royal Societies of London and of Edinburgh, of the Cambridge Philosophical Society, of the Royal Scottish Society of Arts, and of the London Mathematical Society; Lectures delivered by Clerk Maxwell at the Royal Institution of Great Britain published in its *Proceedings*; as well as Communications and Addresses to the British Association published in its *Reports,* are also included in the present work with the sanction of the above mentioned learned bodies.

The Essay which gained the Adams Prize for the year 1856 in the University of Cambridge, the introductory Lecture on the Study of Experimental Physics delivered in the Cavendish Laboratory, and the Rede Lecture delivered before the University in 1878, complete this collection of Clerk Maxwell's scientific writings.

The diagrams in this work have been re-produced by a photographic process from the original diagrams in Clerk Maxwell's Papers by the Cambridge Scientific Instrument Company.

It only remains to add that the footnotes inserted by the Editor are enclosed between square brackets.

CAMBRIDGE, *August,* 1890.

PREFACE.

CLERK MAXWELL'S biography has been written by Professors Lewis Campbell and Wm. Garnett with so much skill and appreciation of their subject that nothing further remains to be told. It would therefore be presumption on the part of the editor of his papers to attempt any lengthened narrative of a biographical character. At the same time a memorial edition of an author's collected writings would hardly be complete without some account however slight of his life and works. Accordingly the principal events of Clerk Maxwell's career will be recounted in the following brief sketch, and the reader who wishes to obtain further and more detailed information or to study his character in its social relations may consult the interesting work to which reference has been made.

James Clerk Maxwell was descended from the Clerks of Penicuick in Midlothian, a well-known Scottish family whose history can be traced back to the 16th century. The first baronet served in the parliament of Scotland. His eldest son, a man of learning, was a Baron of the Exchequer in Scotland. In later times John Clerk of Eldin a member of the family claimed the credit of having invented a new method of breaking the enemy's line in naval warfare, an invention said to have been adopted by Lord Rodney in the battle which he gained over the French in 1782. Another John Clerk, son of the naval tactitian, was a lawyer of much acumen and became a Lord of the Court of Session. He was distinguished among his Edinburgh contemporaries by his ready and sarcastic wit.

The father of the subject of this memoir was John, brother to Sir George Clerk of Penicuick. He adopted the surname of Maxwell on succeeding to an estate in Kirkcudbrightshire which came into the Clerk family through marriage with a Miss Maxwell. It cannot be said that he was possessed of the energy and activity of mind which lead to distinction. He was in truth a somewhat easy-going but shrewd and intelligent man, whose most notable characteristics were his perfect sincerity and extreme benevolence. He took an enlightened interest in mechanical and scientific pursuits and was of an essentially practical turn of mind. On leaving the University he had devoted himself to law and was called to the Scottish Bar. It does not appear however that he met with any great success in that profession. At all events, a quiet life in the country

         b

presented so many attractions to his wife as well as to himself that he was easily induced to relinquish his prospects at the bar. He had been married to Frances, daughter of Robert Cay of N. Charlton, Northumberland, a lady of strong good sense and resolute character.

The country house which was their home after they left Edinburgh was designed by John Clerk Maxwell himself and was built on his estate. The house, which was named Glenlair, was surrounded by fine scenery, of which the water of Urr with its rocky and wooded banks formed the principal charm.

James was born at Edinburgh on the 13th of June, 1831, but it was at Glenlair that the greater part of his childhood was passed. In that pleasant spot under healthful influences of all kinds the child developed into a hardy and courageous boy. Not precociously clever at books he was yet not without some signs of future intellectual strength, being remarkable for a spirit of inquiry into the causes and connections of the phenomena around him. It was remembered afterwards when he had become distinguished, that the questions he put as a child shewed an amount of thoughtfulness which for his years was very unusual.

At the age of ten, James, who had lost his mother, was placed under the charge of relatives in Edinburgh that he might attend the Edinburgh Academy. A charming account of his school days is given in the narrative of Professor Campbell who was Maxwell's schoolfellow and in after life an intimate friend and constant correspondent. The child is father to the man, and those who were privileged to know the man Maxwell will easily recognise Mr Campbell's picture of the boy on his first appearance at school,—the home-made garments more serviceable than fashionable, the rustic speech and curiously quaint but often humorous manner of conveying his meaning, his bewilderment on first undergoing the routine of schoolwork, and his Spartan conduct under various trials at the hands of his schoolfellows. They will further feel how accurate is the sketch of the boy become accustomed to his surroundings and rapidly assuming the place at school to which his mental powers entitled him, while his superfluous energy finds vent privately in carrying out mechanical contrivances and geometrical constructions, in reading and even trying his hand at composing ballads, and in sending to his father letters richly embellished with grotesquely elaborate borders and drawings.

An event of his school-days, worth recording, was his invention of a mechanical method of drawing certain classes of Ovals. An account of this method was printed in the Proceedings of the Royal Society of Edinburgh and forms the first of his writings collected in the present work. The subject was introduced to the notice of the Society by the celebrated Professor James Forbes, who from the first took the greatest possible interest in Maxwell's progress. Professor Tait, another schoolfellow, mentions that at the time when the paper on the Ovals was written, Maxwell had received no instruction in Mathematics beyond a little Euclid and Algebra.

In 1847 Maxwell entered the University of Edinburgh where he remained for three sessions. He attended the lectures of Kelland in Mathematics, Forbes in Natural Philosophy, Gregory in Chemistry, Sir W. Hamilton in Mental Philosophy, Wilson (Christopher North) in Moral Philosophy. The lectures of Sir W. Hamilton made a strong impression upon him, in stimulating the love of speculation to which his mind was prone, but, as might have been expected, it was the Professor of Natural Philosophy who obtained the chief share of his devotion. The enthusiasm which so distinguished a man as Forbes naturally inspired in young and ardent disciples, evoked a feeling of personal attachment, and the Professor, on his part, took special interest in his pupil and gave to him the altogether unusual privilege of working with his fine apparatus.

What was the nature of this experimental work we may conjecture from a perusal of his paper on Elastic Solids, written at that time, in which he describes some experiments made with the view of verifying the deductions of his theory in its application to Optics. Maxwell would seem to have been led to the study of this subject by the following cir cumstance. He was taken by his uncle John Cay to see William Nicol, the inventor of the polarising prism which bears his name, and was shewn by Nicol the colours of unannealed glass in the polariscope. This incited Maxwell to study the laws of polarised light and to construct a rough polariscope in which the polariser and analyser were simple glass reflectors. By means of this instrument he was able to obtain the colour bands of unannealed glass. These he copied on paper in water colours and sent to Nicol. It is gratifying to find that this spirited attempt at experimenting on the part of a mere boy was duly appreciated by Nicol, who at once encouraged and delighted him by a present of a couple of his prisms.

The paper alluded to, viz. that entitled "On the Equilibrium of Elastic Solids," was read to the Royal Society of Edinburgh in 1850. It forms the third paper which Maxwell addressed to that Society. The first in 1846 on Ovals has been already mentioned. The second, under the title "The Theory of Rolling Curves," was presented by Kelland in 1849.

It is obvious that a youth of nineteen years who had been capable of these efforts must have been gifted with rare originality and with great power of sustained exertion. But his singular self-concentration led him into habits of solitude and seclusion, the tendency of which was to confirm his peculiarities of speech and of manner. He was shy and reserved with strangers, and his utterances were often obscure both in substance and in his manner of expressing himself, so many remote and unexpected allusions perpetually obtruding themselves. Though really most sociable and even fond of society he was essentially reticent and reserved. Mr Campbell thinks it is to be regretted that Maxwell did not begin his Cambridge career earlier for the sake of the social intercourse which he would have found it difficult to avoid there. It is a question, however, whether in losing the opportunity of using Professor Forbes' apparatus he would not thereby have lost what was perhaps the most valuable part of his early scientific training.

It was originally intended that Maxwell should follow his father's profession of advocate, but this intention was abandoned as soon as it became obvious that his tastes lay in a direction so decidedly scientific. It was at length determined to send him to Cambridge and accordingly in October, 1850, he commenced residence in Peterhouse, where however he resided during the Michaelmas Term only. On December 14 of the same year he migrated to Trinity College.

It may readily be supposed that his preparatory training for the Cambridge course was far removed from the ordinary type. There had indeed for some time been practically no restraint upon his plan of study and his mind had been allowed to follow its natural bent towards science, though not to an extent so absorbing as to withdraw him from other pursuits. Though he was not a sportsman,—indeed sport so called was always repugnant to him—he was yet exceedingly fond of a country life. He was a good horseman and a good swimmer. Whence however he derived his chief enjoyment may be gathered from the account which Mr Campbell gives of the zest with which he quoted on one occasion the lines of Burns which describe the poet finding inspiration while wandering along the banks of a stream in the free indulgence of his fancies. Maxwell was not only a lover of poetry but himself a poet, as the fine pieces gathered together by Mr Campbell abundantly testify. He saw however that his true calling was Science and never regarded these poetical efforts as other than mere pastime. Devotion to science, already stimulated by successful endeavour, a tendency to ponder over philosophical problems and an attachment to English literature, particularly to English poetry,—these tastes, implanted in a mind of singular strength and purity, may be said to have been the endowments with which young Maxwell began his Cambridge career. Besides this, his scientific reading, as we may gather from his papers to the Royal Society of Edinburgh referred to above, was already extensive and varied. He brought with him, says Professor Tait, a mass of knowledge which was really immense for so young a man but in a state of disorder appalling to his methodical private tutor.

Maxwell's undergraduate career was not marked by any specially notable feature. His private speculations had in some measure to be laid aside in favour of more systematic study. Yet his mind was steadily ripening for the work of his later years. Among those with whom he was brought into daily contact by his position, as a Scholar of Trinity College, were some of the brightest and most cultivated young men in the University. In the genial fellowship of the Scholars' table Maxwell's kindly humour found ready play, while in the more select coterie of the Apostle Club, formed for mutual cultivation, he found a field for the exercise of his love of speculation in essays on subjects beyond the lines of the ordinary University course. The composition of these essays doubtless laid the foundation of that literary finish which is one of the characteristics of Maxwell's scientific writings. His biographers have preserved several extracts on a variety of subjects chiefly of a speculative character. They are remarkable mainly for the weight of thought contained in them but occasionally also for smart epigrams and for a vein of dry and sarcastic humour.

These glimpses into Maxwell's character may prepare us to believe that, with all his shyness, he was not without confidence in his own powers, as also appears from the account which was given by the late Master of Trinity College, Dr Thompson, who was Tutor when Maxwell personally applied to him for permission to migrate to that College. He appeared to be a shy and diffident youth, but presently surprised Dr Thompson by producing a bundle of papers, doubtless copies of those we have already mentioned, remarking "Perhaps these may shew you that I am not unfit to enter at your College."

He became a pupil of the celebrated William Hopkins of Peterhouse, under whom his course of study became more systematic. One striking characteristic was remarked by his contemporaries. Whenever the subject admitted of it he had recourse to diagrams, though his fellow students might solve the question more easily by a train of analysis. Many illustrations of this manner of proceeding might be taken from his writings, but in truth it was only one phase of his mental attitude towards scientific questions, which led him to proceed from one distinct idea to another instead of trusting to symbols and equations.

Maxwell's published contributions to Mathematical Science during his undergraduate career were few and of no great importance. He found time however to carry his investigations into regions outside the prescribed Cambridge course. At the lectures of Professor Stokes* he was regular in his attendance. Indeed it appears from the paper on Elastic Solids, mentioned above, that he was acquainted with some of the writings of Stokes before he entered Cambridge. Before 1850, Stokes had published some of his most important contributions to Hydromechanics and Optics; and Sir W. Thomson, who was nine years' Maxwell's senior in University standing, had, among other remarkable investigations, called special attention to the mathematical analogy between Heat-conduction and Statical Electricity. There is no doubt that these authors as well as Faraday, of whose experimental researches he had made a careful study, exercised a powerful directive influence on his mind.

In January, 1854, Maxwell's undergraduate career closed. He was second wrangler, but shared with Dr Routh, who was senior wrangler, the honours of the First Smith's Prize. In due course he was elected Fellow of Trinity and placed on the staff of College Lecturers.

No sooner was he released from the restraints imposed by the Trinity Fellowship Examination than he plunged headlong into original work. There were several questions he was anxious to deal with, and first of all he completed an investigation on the Transformation of Surfaces by Bending, a purely geometrical problem. This memoir he presented to the Cambridge Philosophical Society in the following March. At this period he also set about an enquiry into the quantitative measurement of mixtures of colours and the causes of colour-blindness. During his undergraduateship he had, as we have seen, found time for the study of Electricity. This had already borne fruit and now resulted in the first of his important memoirs on that subject,—the memoir on Faraday's Lines of Force.

* Now Sir George Gabriel Stokes, Bart., M.P. for the University.

The number and importance of his papers, published in 1855—6, bear witness to his assiduity during this period. With these labours, and in the preparation of his College lectures, on which he entered with much enthusiasm, his mind was fully occupied and the work was congenial. He had formed a number of valued friendships, and he had a variety of interests, scientific and literary, attaching him to the University. Nevertheless, when the chair of Natural Philosophy in Marischal College, Aberdeen, fell vacant, Maxwell became a candidate. This step was probably taken in deference to his father's wishes, as the long summer vacation of the Scottish College would enable him to reside with his father at Glenlair for half the year continuously. He obtained the professorship, but unhappily the kind intentions which prompted him to apply for it were frustrated by the death of his father, which took place in April, 1856.

It is doubtful whether the change from the Trinity lectureship to the Aberdeen professorship was altogether prudent. The advantages were the possession of a laboratory and the long uninterrupted summer vacation. But the labour of drilling classes composed chiefly of comparatively young and untrained lads, in the elements of mechanics and physics, was not the work for which Maxwell was specially fitted. On the other hand, in a large college like Trinity there could not fail to have been among its undergraduate members, some of the most promising young mathematicians of the University, capable of appreciating his original genius and immense knowledge, by instructing whom he would himself have derived advantage.

In 1856 Maxwell entered upon his duties as Professor of Natural Philosophy at Marischal College, and two years afterwards he married Katharine Mary Dewar, daughter of the Principal of the College. He in consequence ceased to be a Fellow of Trinity College, but was afterwards elected an honorary Fellow, at the same time as Professor Cayley.

During the years 1856—60 he was still actively employed upon the subject of colour sensation, to which he contributed a new method of measurement in the ingenious instrument known as the colour-box. The most serious demands upon his powers and upon his time were made by his investigations on the Stability of Saturn's Rings. This was the subject chosen by the Examiners for the Adams Prize Essay to be adjudged in 1857, and was advertised in the following terms:—

> "The Problem may be treated on the supposition that the system of Rings is exactly or very approximately concentric with Saturn and symmetrically disposed about the plane of his equator and different hypotheses may be made respecting the physical constitution of the Rings. It may be supposed (1) that they are rigid; (2) that they are fluid and in part aeriform; (3) that they consist of masses of matter not materially coherent. The question will be considered to be answered by ascertaining on these hypotheses severally whether the conditions of mechanical stability are satisfied by the mutual attractions and motions of the Planet and the Rings."

"It is desirable that an attempt should also be made to determine on which of the above hypotheses the appearances both of the bright rings and the recently discovered dark ring may be most satisfactorily explained; and to indicate any causes to which a change of form such as is supposed from a comparison of modern with the earlier observations to have taken place, may be attributed."

It is sufficient to mention here that Maxwell bestowed an immense amount of labour in working out the theory as proposed, and that he arrived at the conclusion that "the only system of rings which can exist is one composed of an indefinite number of unconnected particles revolving round the planet with different velocities according to their respective distances. These particles may be arranged in a series of narrow rings, or they may move about through each other irregularly. In the first case the destruction of the system will be very slow, in the second case it will be more rapid, but there may be a tendency towards an arrangement in narrow rings which may retard the process."

Part of the work, dealing with the oscillatory waves set up in a ring of satellites, was illustrated by an ingenious mechanical contrivance which was greatly admired when exhibited before the Royal Society of Edinburgh.

This essay, besides securing the prize, obtained for its author great credit among scientific men. It was characterized by Sir George Airy as one of the most remarkable applications of Mathematics to Physics that he had ever seen.

The suggestion has been made that it was the irregular motions of the particles which compose the Rings of Saturn resulting on the whole in apparent regularity and uniformity, which led Maxwell to the investigation of the Kinetic Theory of Gases, his first contribution to which was read to the British Association in 1859. This is not unlikely, but it must also be borne in mind that Bernoulli's Theory had recently been revived by Herapath, Joule and Clausius whose writings may have drawn Maxwell's attention to the subject.

In 1860 King's College and Marischal College were joined together as one institution, now known as the University of Aberdeen. The new chair of Natural Philosophy thus created was filled up by the appointment of David Thomson, formerly Professor at King's College and Maxwell's senior. Professor Thomson, though not comparable to Maxwell as a physicist, was nevertheless a remarkable man. He was distinguished by singular force of character and great administrative faculty and he had been prominent in bringing about the fusion of the Colleges. He was also an admirable lecturer and teacher and had done much to raise the standard of scientific education in the north of Scotland. Thus the choice made by the Commissioners, though almost inevitable, had the effect of making it appear that Maxwell failed as a teacher. There seems however to be no evidence to support such an inference. On the contrary, if we may judge from the number of voluntary students attending his classes in his last College session, he would seem to have been as popular as a professor as he was personally estimable.

This is also borne out by the fact that he was soon afterwards elected Professor of Natural Philosophy and Astronomy in King's College, London. The new appointment had the advantage of bringing him much more into contact with men in his own department of science, especially with Faraday, with whose electrical work his own was so intimately connected. In 1862—63 he took a prominent part in the experiments organised by a Committee of the British Association for the determination of electrical resistance in absolute measure and for placing electrical measurements on a satisfactory basis. In the experiments which were conducted in the laboratory of King's College upon a plan due to Sir W. Thomson, two long series of measurements were taken in successive years. In the first year, the working members were Maxwell, Balfour Stewart and Fleeming Jenkin; in the second, Charles Hockin took the place of Balfour Stewart. The work of this Committee was communicated in the form of reports to the British Association and was afterwards republished in one volume by Fleeming Jenkin.

Maxwell was a professor in King's College from 1860 to 1865, and this period of his life is distinguished by the production of his most important papers. The second memoir on Colours made its appearance in 1860. In the same year his first papers on the Kinetic Theory of Gases were published. In 1861 came his papers on Physical Lines of Force and in 1864 his greatest memoir on Electricity,—a Dynamical Theory of the Electromagnetic Field. He must have been occupied with the Dynamical Theory of Gases in 1865, as two important papers appeared in the following year, first the Bakerian lecture on the Viscosity of Gases, and next the memoir on the Dynamical Theory of Gases.

The mental strain involved in the production of so much valuable work, combined with the duties of his professorship which required his attention during nine months of the year, seems to have influenced him in a resolution which in 1865 he at length adopted of resigning his chair and retiring to his country seat. Shortly after this he had a severe illness. On his recovery he continued his work on the Dynamical Theory of Gases, to which reference has just been made. For the next few years he led a quiet and secluded life at Glenlair, varied by annual visits to London, attendances at the British Association meetings and by a tour in Italy in 1867. He was also Moderator or Examiner in the Mathematical Tripos at Cambridge on several occasions, offices which entailed a few weeks' residence at the University in winter. His chief employment during those years was the preparation of his now celebrated treatise on Electricity and Magnetism which, however, was not published till 1873. He also wrote a treatise on Heat which was published in 1871.

In 1871 Maxwell was, with some reluctance, induced to quit his retreat in the country and to enter upon a new career. The University of Cambridge had recently resolved to found a professorship of physical science, especially for the cultivation and teaching of the subjects of Heat, Electricity and Magnetism. In furtherance of this object her Chancellor, the Duke of Devonshire, had most generously undertaken to build a laboratory and furnish it with the necessary apparatus. Maxwell was invited to fill the new chair thus formed and to superintend the erection of the laboratory. In October, 1871, he delivered his inaugural lecture.

The Cavendish Laboratory, so called after its founder, the present venerable chief of the family which produced the great physicist of the same name, was not completed for practical work until 1874. In June of that year it was formally presented to the University by the Chancellor. The building itself and the fittings of the several rooms were admirably contrived mainly by Maxwell himself, but the stock of apparatus was smaller than accorded with the generous intentions of the Chancellor. This defect must be attributed to the anxiety of the Professor to procure only instruments by the best makers and with such improvements as he could himself suggest. Such a defect therefore required time for its removal and afterwards in great measure disappeared, apparatus being constantly added to the stock as occasion demanded.

One of the chief tasks which Maxwell undertook was that of superintending and directing the energies of such young Bachelors of Arts as became his pupils after having acquired good positions in the University examinations. Several pupils, who have since acquired distinction, carried out valuable experiments under the guidance of the Professor. It must be admitted, however, that the numbers were at first small, but perhaps this was only to be expected from the traditions of so many years. The Professor was singularly kind and helpful to these pupils. He would hold long conversations with them, opening up to them the stores of his mind, giving them hints as to what they might try and what avoid, and was always ready with some ingenious remedy for the experimental troubles which beset them. These conversations, always delightful and instructive, were, according to the account of one of his pupils, a liberal education in themselves, and were repaid in the minds of the pupils by a grateful affection rarely accorded to any teacher.

Besides discharging the duties of his chair, Maxwell took an active part in conducting the general business of the University and more particularly in regulating the courses of study in Mathematics and Physics.

For some years previous to 1866 when Maxwell returned to Cambridge as Moderator in the Mathematical Tripos, the studies in the University had lost touch with the great scientific movements going on outside her walls. It was said that some of the subjects most in vogue had but little interest for the present generation, and loud complaints began to be heard that while such branches of knowledge as Heat, Electricity and Magnetism, were left out of the Tripos examination, the candidates were wasting their time and energy upon mathematical trifles barren of scientific interest and of practical results. Into the movement for reform Maxwell entered warmly. By his questions in 1866 and subsequent years he infused new life into the examination; he took an active part in drafting the new scheme introduced in 1873; but most of all by his writings he exerted a powerful influence on the younger members of the University, and was largely instrumental in bringing about the change which has been now effected.

VOL. I. c

In the first few years at Cambridge Maxwell was busy in giving the final touches to his great work on Electricity and Magnetism and in passing it through the press. This work was published in 1873, and it seems to have occupied the most of his attention for the two previous years, as the few papers published by him during that period relate chiefly to subjects forming part of the contents. After this publication his contributions to scientific journals became more numerous, those on the Dynamical Theory of Gases being perhaps the most important. He also wrote a great many short articles and reviews which made their appearance in *Nature* and the *Encyclopædia Britannica*. Some of these essays are charming expositions of scientific subjects, some are general criticisms of the works of contemporary writers and others are brief and appreciative biographies of fellow workers in the same fields of research.

An undertaking in which he was long engaged and which, though it proved exceedingly interesting, entailed much labour, was the editing of the "Electrical Researches" of the Hon. Henry Cavendish. This work, published in 1879, has had the effect of increasing the reputation of Cavendish, disclosing as it does the unsuspected advances which that acute physicist had made in the Theory of Electricity, especially in the measurement of electrical quantities. The work is enriched by a variety of valuable notes in which Cavendish's views and results are examined by the light of modern theory and methods. Especially valuable are the methods applied to the determination of the electrical capacities of conductors and condensers, a subject in which Cavendish himself shewed considerable skill both of a mathematical and experimental character.

The importance of the task undertaken by Maxwell in connection with Cavendish's papers will be understood from the following extract from his introduction to them.

"It is somewhat difficult to account for the fact that though Cavendish had prepared a complete description of his experiments on the charges of bodies, and had even taken the trouble to write out a fair copy, and though all this seems to have been done before 1774 and he continued to make experiments in Electricity till 1781 and lived on till 1810, he kept his manuscript by him and never published it."

"Cavendish cared more for investigation than for publication. He would undertake the most laborious researches in order to clear up a difficulty which no one but himself could appreciate or was even aware of, and we cannot doubt that the result of his enquiries, when successful, gave him a certain degree of satisfaction. But it did not excite in him that desire to communicate the discovery to others which in the case of ordinary men of science, generally ensures the publication of their results. How completely these researches of Cavendish remained unknown to other men of science is shewn by the external history of electricity."

It will probably be thought a matter of some difficulty to place oneself in the position of a physicist of a century ago and to ascertain the exact bearing of his experiments. But Maxwell entered upon this undertaking with the utmost enthusiasm and succeeded in completely identifying himself with Cavendish's methods. He shewed that Cavendish had really anticipated several of the discoveries in electrical science which have been made since his time. Cavendish was the first to form the conception of and to measure Electrostatic Capacity and Specific Inductive Capacity; he also anticipated Ohm's law.

The Cavendish papers were no sooner disposed of than Maxwell set about preparing a new edition of his work on Electricity and Magnetism; but unhappily in the summer term of 1879 his health gave way. Hopes were however entertained that when he returned to the bracing air of his country home he would soon recover. But he lingered through the summer months with no signs of improvement and his spirits gradually sank. He was finally informed by his old fellow-student, Professor Sanders, that he could not live more than a few weeks. As a last resort he was brought back to Cambridge in October that he might be under the charge of his favourite physician, Dr Paget*. Nothing however could be done for his malady, and, after a painful illness, he died on the 5th of November, 1879, in his 49th year.

Maxwell was thus cut off in the prime of his powers, and at a time when the departments of science, which he had contributed so much to develop, were being every day extended by fresh discoveries. His death was deplored as an irreparable loss to science and to the University, in which his amiable disposition was as universally esteemed as his genius was admired.

It is not intended in this preface to enter at length into a discussion of the relation which Maxwell's work bears historically to that of his predecessors, or to attempt to estimate the effect which it has had on the scientific thought of the present day. In some of his papers he has given more than usually copious references to the works of those by whom he had been influenced; and in his later papers, especially those of a more popular nature which appeared in the *Encyclopædia Britannica*, he has given full historical outlines of some of the most prominent fields in which he laboured. Nor does it appear to the present editor that the time has yet arrived when the quickening influence of Maxwell's mind on modern scientific thought can be duly estimated. He therefore proposes to himself the duty of recalling briefly, according to subjects, the most important speculations in which Maxwell engaged.

His works have been arranged as far as possible in chronological order but they fall naturally under a few leading heads; and perhaps we shall not be far wrong if we place first in importance his work in Electricity.

His first paper on this subject bearing the title "On Faraday's Lines of Force" was read before the Cambridge Philosophical Society on Dec. 11th, 1855. He had been previously attracted by Faraday's method of expressing electrical laws, and he here set before himself the task of shewing that the ideas which had guided Faraday's researches were not inconsistent with the mathematical formulæ in which Poisson and others had cast the laws of

* Now Sir George Edward Paget, K.C.B.

Electricity.   His object, he says, is to find a physical analogy which shall help the mind to grasp the results of previous investigations "without being committed to any theory founded on the physical science from which that conception is borrowed, so that it is neither drawn aside from the subject in the pursuit of analytical subtleties nor carried beyond the truth by a favorite hypothesis."

The laws of electricity are therefore compared with the properties of an incompressible fluid the motion of which is retarded by a force proportional to the velocity, and the fluid is supposed to possess no inertia.   He shews the analogy which the lines of flow of such a fluid would have with the lines of force, and deduces not merely the laws of Statical Electricity in a single medium but also a method of representing what takes place when the action passes from one dielectric into another.

In the latter part of the paper he proceeds to consider the phenomena of Electromagnetism and shews how the laws discovered by Ampère lead to conclusions identical with those of Faraday.   In this paper three expressions are introduced which he identifies with the components of Faraday's electrotonic state, though the author admits that he has not been able to frame a physical theory which would give a clear mental picture of the various connections expressed by the equations.

Altogether this paper is most important for the light which it throws on the principles which guided Maxwell at the outset of his electrical work.   The idea of the electrotonic state had already taken a firm hold of his mind though as yet he had formed no physical explanation of it.   In the paper "On Physical Lines of Force" printed in the *Philosophical Magazine*, Vol. XXI. he resumes his speculations.   He explains that in his former paper he had found the geometrical significance of the Electrotonic state but that he now proposes "to examine magnetic phenomena from a mechanical point of view."   Accordingly he propounds his remarkable speculation as to the magnetic field being occupied by molecular vortices, the axes of which coincide with the lines of force.   The cells within which these vortices rotate are supposed to be separated by layers of particles which serve the double purpose of transmitting motion from one cell to another and by their own motions constituting an electric current.   This theory, the parent of several working models which have been devised to represent the motions of the dielectric, is remarkable for the detail with which it is worked out and made to explain the various laws not only of magnetic and electromagnetic action, but also the various forms of electrostatic action.   As Maxwell subsequently gave a more general theory of the Electromagnetic Field, it may be inferred that he did not desire it to be supposed that he adhered to the views set forth in this paper in every particular; but there is no doubt that in some of its main features, especially the existence of rotation round the lines of magnetic force, it expressed his permanent convictions.   In his treatise on "Electricity and Magnetism," Vol. II. p. 416, (2nd edition 427) after quoting from Sir W. Thomson on the explanation of the magnetic rotation of the plane of the polarisation of light, he goes on to say of the present paper,

"A theory of molecular vortices which I worked out at considerable length was published in the *Phil. Mag.* for March, April and May, 1861, Jan. and Feb. 1862."

"I think we have good evidence for the opinion that some phenomenon of rotation is going on in the magnetic field, that this rotation is performed by a great number of very small portions of matter, each rotating on its own axis, that axis being parallel to the direction of the magnetic force, and that the rotations of these various vortices are made to depend on one another by means of some mechanism between them."

"The attempt which I then made to imagine a working model of this mechanism must be taken for no more than it really is, a demonstration that mechanism may be imagined capable of producing a connection mechanically equivalent to the actual connection of the parts of the Electromagnetic Field."

This paper is also important as containing the first hint of the Electromagnetic Theory of Light which was to be more fully developed afterwards in his third great memoir "On the Dynamical Theory of the Electromagnetic Field." This memoir, which was presented to the Royal Society on the 27th October, 1864, contains Maxwell's mature thoughts on a subject which had so long occupied his mind. It was afterwards reproduced in his Treatise with trifling modifications in the treatment of its parts, but without substantial changes in its main features. In this paper Maxwell reverses the mode of treating electrical phenomena adopted by previous mathematical writers; for while they had sought to build up the laws of the subject by starting from the principles discovered by Ampère, and deducing the induction of currents from the conservation of energy, Maxwell adopts the method of first arriving at the laws of induction and then deducing the mechanical attractions and repulsions.

After recalling the general phenomena of the mutual action of currents and magnets and the induction produced in a circuit by any variation of the strength of the field in which it lies, the propagation of light through a luminiferous medium, the properties of dielectrics and other phenomena which point to a medium capable of transmitting force and motion, he proceeds.—

"Thus then we are led to the conception of a complicated mechanism capable of a vast variety of motions but at the same time so connected that the motion of one part depends, according to definite relations, on the motion of other parts, these motions being communicated by forces arising from the relative displacement of their connected parts, in virtue of their elasticity. Such a mechanism must be subject to the laws of Dynamics."

On applying dynamical principles to such a connected system he attains certain general propositions which, on being compared with the laws of induced currents, enable him to identify certain features of the mechanism with properties of currents. The induction of currents and their electromagnetic attraction are thus explained and connected.

In a subsequent part of the memoir he proceeds to establish from these premises the general equations of the Field and obtains the usual formulæ for the mechanical force on currents, magnets and bodies possessing an electrostatic charge.

He also returns to and elaborates more fully the electromagnetic Theory of Light. His equations shew that dielectrics can transmit only transverse vibrations, the speed of propagation of which in air as deduced from electrical data comes out practically identical with the known velocity of light. For other dielectrics the index of refraction is equal to the square root of the product of the specific inductive capacity by the coefficient of magnetic induction, which last factor is for most bodies practically unity. Various comparisons have been made with the view of testing this deduction. In the case of paraffin wax and some of the hydrocarbons, theory and experiment agree, but this is not the case with glass and some other substances. Maxwell has also applied his theory to media which are not perfect insulators, and finds an expression for the loss of light in passing through a stratum of given thickness. He remarks in confirmation of his result that most good conductors are opaque while insulators are transparent, but he also adds that electrolytes which transmit a current freely are often transparent, while a piece of gold leaf whose resistance was determined by Mr Hockin allowed far too great an amount of light to pass. He observes however that it is possible "there is less loss of energy when the electromotive forces are reversed with the rapidity of light than when they act for sensible times as in our experiments." A similar explanation may be given of the discordance between the calculated and observed values of the specific inductive capacity. Prof. J. J. Thomson in the *Proceedings of the Royal Society*, Vol. 46, has described an experiment by which he has obtained the specific inductive capacities of various dielectrics when acted on by alternating electric forces whose frequency is 25,000,000 per second. He finds that under these conditions the specific inductive capacity of glass is very nearly the same as the square of the refractive index, and very much less than the value for slow rates of reversals. In illustration of these remarks may be quoted the observations of Prof. Hertz who has shewn that vulcanite and pitch are transparent for waves, whose periods of vibration are about three hundred millionths of a second. The investigations of Hertz have shewn that electro-dynamic radiations are transmitted in waves with a velocity, which, if not equal to, is comparable with that of light, and have thus given conclusive proof that a satisfactory theory of Electricity must take into account in some form or other the action of the dielectric. But this does not prove that Maxwell's theory is to be accepted in every particular. A peculiarity of his theory is, as he himself points out in his treatise, that the variation of the electric displacement is to be treated as part of the current as well as the current of conduction, and that it is the total amount due to the sum of these which flows as if electricity were an incompressible fluid, and which determines external electrodynamic actions. In this respect it differs from the theory of Helmholtz which also takes into account the action of the dielectric. Professor J. J. Thomson in his Review of Electric Theories has entered into a full discussion of the points at issue between the two above mentioned theories, and the reader is referred to his paper for further information *. Maxwell in the memoir before us has also applied his theory to the passage of light through crystals, and gets rid at once of the wave of normal vibrations which has hitherto proved the stumbling block in other theories of light.

The electromagnetic Theory of Light has received numerous developments at the hands of Lord Rayleigh, Mr Glazebrook, Professor J. J. Thomson and others. These volumes also contain various shorter papers on Electrical Science, though perhaps the most complete record of Maxwell's work in this department is to be found in his Treatise on Electricity and Magnetism in which they were afterwards embodied.

Another series of papers of hardly less importance than those on Electricity are the various memoirs on the Dynamical Theory of Gases. The idea that the properties of matter might be explained by the motions and impacts of their ultimate atoms is as old as the time of the Greeks, and Maxwell has given in his paper on "Atoms" a full sketch of the ancient controversies to which it gave rise. The mathematical difficulties of the speculation however were so great that it made little real progress till it was taken up by Clausius and shortly afterwards by Maxwell. The first paper by Maxwell on the subject is entitled "Illustrations of the Dynamical Theory of Gases" and was published in the *Philosophical Magazine* for January and July, 1860, having been read at a meeting of the British Association of the previous year. Although the methods developed in this paper were afterwards abandoned for others, the paper itself is most interesting, as it indicates clearly the problems in the theory which Maxwell proposed to himself for solution, and so far contains the germs of much that was treated of in his next memoir. It is also epoch-making, inasmuch as it for the first time enumerates various propositions which are characteristic of Maxwell's work in this subject. It contains the first statement of the distribution of velocities according to the law of errors. It also foreshadows the theorem that when two gases are in thermal equilibrium the mean kinetic energy of the molecules of each system is the same; and for the first time the question of the viscosity of gases is treated dynamically.

In his great memoir "On the Dynamical Theory of Gases" published in the *Philosophical Transactions of the Royal Society* and read before the Society in May, 1866, he returns to this subject and lays down for the first time the general dynamical methods appropriate for its treatment. Though to some extent the same ground is traversed as in his former paper, the methods are widely different. He here abandons his former hypothesis that the molecules are hard elastic spheres, and supposes them to repel each other with forces varying inversely as the fifth power of the distance. His chief reason for assuming this law of action appears to be that it simplifies considerably the calculation of the collisions between the molecules, and it leads to the conclusion that the coefficient of viscosity is directly proportional to the absolute temperature. He himself undertook an experimental enquiry for the purpose of verifying this conclusion, and, in his paper on the Viscosity of Gases, he satisfied himself of its correctness. A re-examination of the numerical

---

* *British Association Report*, 1885.

reductions made in the course of his work discloses however an inaccuracy which materially affects the values of the coefficient of viscosity obtained. Subsequent experiments also seem to shew that the concise relation he endeavoured to establish is by no means so near the truth as he supposed, and it is more than doubtful whether the action between two molecules can be represented by any law of so simple a character.

In the same memoir he gives a fresh demonstration of the law of distribution of velocities, but though the method is of permanent value, it labours under the defect of assuming that the distribution of velocities in the neighbourhood of a point is the same in every direction, whatever actions may be taking place within the gas. This flaw in the argument, first pointed out by Boltzmann, seems to have been recognised by Maxwell, who in his next paper "On the Stresses in Rarefied Gases arising from inequalities of Temperature," published in the *Philosophical Transactions* for 1879, Part I., adopts a form of the distribution function of a somewhat different shape. The object of this paper was to arrive at a theory of the effects observed in Crookes's Radiometer. The results of the investigation are stated by Maxwell in the introduction to the paper, from which it would appear that the observed motion cannot be explained on the Dynamical Theory, unless it be supposed that the gas in contact with a solid can slide along the surface with a finite velocity between places whose temperatures are different. In an appendix to the paper he shews that on certain assumptions regarding the nature of the contact of the solid and gas, there will be, when the pressure is constant, a flow of gas along the surface from the colder to the hotter parts. The last of his longer papers on this subject is one on Boltzmann's Theorem. Throughout these volumes will be found numerous shorter essays on kindred subjects, published chiefly in *Nature* and in the *Encyclopædia Britannica*. Some of these contain more or less popular expositions of this subject which Maxwell had himself in great part created, while others deal with the work of other writers in the same field. They are profoundly suggestive in almost every page, and abound in acute criticisms of speculations which he could not accept. They are always interesting; for although the larger papers are sometimes difficult to follow, Maxwell's more popular writings are characterized by extreme lucidity and simplicity of style.

The first of Maxwell's papers on Colour Perception is taken from the Transactions of the Royal Scottish Society of Arts and is in the form of a letter to Dr G. Wilson dated Jan. 4, 1855. It was followed directly afterwards by a communication to the Royal Society of Edinburgh, and the subject occupied his attention for some years. The most important of his subsequent work is to be found in the papers entitled "An account of Experiments on the Perception of Colour" published in the *Philosophical Magazine*, Vol. XIV. and "On the Theory of Compound Colours and its relation to the colours of the spectrum" in the *Philosophical Transactions* for the year 1860. We may also refer to two lectures delivered at the Royal Institution, in which he recapitulates and enforces his main positions in his usual luminous style. Maxwell from the first adopts Young's Theory of Colour Sensation, according to which all colours may ultimately be reduced to three, a red, a green and a violet. This theory had been revived by Helmholtz who endeavoured to find for it a physiological basis. Maxwell however devoted himself chiefly to the invention of accurate methods for combining and recording mixtures of colours. His first method of obtaining mixtures, that of the Colour Top, is an adaptation of one formerly employed, but in Maxwell's hands it became an instrument capable of giving precise numerical results by means which he added of varying and measuring the amounts of colour which were blended in the eye. In the representation of colours diagrammatically he followed Young in employing an equilateral triangle at the angles of which the fundamental colours were placed. All colours, white included, which may be obtained by mixing the fundamental colours in any proportions will then be represented by points lying within the triangle. Points without the triangle represent colours which must be mixed with one of the fundamental tints to produce a mixture of the other two, or with which two of them must be mixed to produce the third.

In his later papers, notably in that printed in the *Philosophical Transactions*, he adopts the method of the Colour Box, by which different parts of the spectrum may be mixed in different proportions and matched with white, the intensity of which has been suitably diminished. In this way a series of colour equations are obtained which can be used to evaluate any colour in terms of the three fundamental colours. These observations on which Maxwell expended great care and labour, constitute by far the most important data regarding the combinations of colour sensations which have been yet obtained, and are of permanent value whatever theory may ultimately be adopted of the physiology of the perception of colour.

In connection with these researches into the sensations of the normal eye, may be mentioned the subject of colour-blindness, which also engaged Maxwell's attention, and is discussed at considerable length in several of his papers.

Geometrical Optics was another subject in which Maxwell took much interest. At an early period of his career he commenced a treatise on Optics, which however was never completed. His first paper "On the general laws of optical instruments," appeared in 1858, but a brief account of the first part of it had been previously communicated to the Cambridge Philosophical Society. He therein lays down the conditions which a perfect optical instrument must fulfil, and shews that if an instrument produce perfect images of an object, i.e. images free from astigmatism, curvature and distortion, for two different positions of the object, it will give perfect images at all distances. On this result as a basis, he finds the relations between the foci of the incident and emergent pencils, the magnifying power and other characteristic quantities. The subject of refraction through optical combinations was afterwards treated by him in a different manner, in three papers communicated to the London Mathematical Society. In the first (1873), "On the focal lines of a refracted pencil," he applies Hamilton's characteristic function to determine the focal lines of a thin pencil refracted from one isotropic medium into another at any surface of separation. In the second (1874), "On

Hamilton's characteristic function for a narrow beam of light," he considers the more general question of the passage of a ray from one isotropic medium into another, the two media being separated by a third which may be of a heterogeneous character. He finds the most general form of Hamilton's characteristic function from one point to another, the first being in the medium in which the pencil is incident and the second in the medium in which it is emergent, and both points near the principal ray of the pencil. This result is then applied in two particular cases, viz. to determine the emergent pencil (1) from a spectroscope, (2) from an optical instrument symmetrical about its axis. In the third paper (1875) he resumes the last-mentioned application, discussing this case more fully under a somewhat simplified analysis.

It may be remarked that all these papers are connected by the same idea, which was— first to study the optical effects of the entire instrument without examining the mechanism by which these effects are produced, and then, as in the paper in 1858, to supply whatever data may be necessary by experiments upon the instrument itself.

Connected to some extent with the above papers is an investigation which was published in 1868 "On the cyclide." As the name imports, this paper deals chiefly with the geometrical properties of the surface named, but other matters are touched on, such as its conjugate isothermal functions. Primarily however the investigation is on the orthogonal surfaces to a system of rays passing accurately through two lines. In a footnote to this paper Maxwell describes the stereoscope which he invented and which is now in the Cavendish Laboratory.

In 1868 was also published a short but important article entitled "On the best arrangement for producing a pure spectrum on a screen."

The various papers relating to the stresses experienced by a system of pieces joined together so as to form a frame and acted on by forces form an important group connected with one another. The first in order was "On reciprocal figures and diagrams of forces," published in 1864. It was immediately followed by a paper on a kindred subject, "On the calculation of the equilibrium and stiffness of frames." In the first of these Maxwell demonstrates certain reciprocal properties in the geometry of two polygons which are related to one another in a particular way, and establishes his well-known theorem in Graphical Statics on the stresses in frames. In the second he employs the principle of work to problems connected with the stresses in frames and structures and with the deflections arising from extensions in any of the connecting pieces.

A third paper "On the equilibrium of a spherical envelope," published in 1867, may here be referred to. The author therein considers the stresses set up in the envelope by a system of forces applied at its surface, and ultimately solves the problem for two normal forces applied at any two points. The solution, in which he makes use of the principle of inversion as it is applied in various electrical questions, turns ultimately on the determination of a certain function first introduced by Sir George Airy, and called by Maxwell

Airy's Function of Stress. The methods which in this paper were attended with so much success, seem to have suggested to Maxwell a reconsideration of his former work, with the view of extending the character of the reciprocity therein established. Accordingly in 1870 there appeared his fourth contribution to the subject, "On reciprocal figures, frames and diagrams of forces." This important memoir was published in the Transactions of the Royal Society of Edinburgh, and its author received for it the Keith Prize. He begins with a remarkably beautiful construction for drawing plane reciprocal diagrams, and then proceeds to discuss the geometry and the degrees of freedom and constraint of polyhedral frames, his object being to lead up to the limiting case when the faces of the polyhedron become infinitely small and form parts of a continuous surface. In the course of this work he obtains certain results of a general character relating to inextensible surfaces and certain others of practical utility relating to loaded frames. He then attacks the general problem of representing graphically the internal stress of a body and by an extension of the meaning of "Diagram of Stress," he gives a construction for finding a diagram which has mechanical as well as geometrical reciprocal properties with the figure supposed to be under stress. It is impossible with brevity to give an account of this reciprocity, the development of which in Maxwell's hands forms a very beautiful example of analysis. It will be sufficient to state that under restricted conditions this diagram of stress leads to a solution for the components of stress in terms of a single function analogous to Airy's Function of Stress. In the remaining parts of the memoir there is a discussion of the equations of stress, and it is shewn that the general solution may be expressed in terms of three functions analogous to Airy's single function in two dimensions. These results are then applied to special cases, and in particular the stresses in a horizontal beam with a uniform load on its upper surface are fully investigated.

On the subjects in which Maxwell's investigations were the most numerous it has been thought necessary, in the observations which have been made, to sketch out briefly the connections of the various papers on each subject with one another. It is not however intended to enter into an account of the contents of his other contributions to science, and this is the less necessary as the reader may readily obtain the information he may require in Maxwell's own language. It was usually his habit to explain by way of introduction to any paper his exact position with regard to the subject matter and to give a brief account of the nature of the work he was contributing. There are however several memoirs which though unconnected with others are exceedingly interesting in themselves. Of these the essay on Saturn's Rings will probably be thought the most important as containing the solution of a difficult cosmical problem; there are also various papers on Dynamics, Hydromechanics and subjects of pure mathematics, which are most useful contributions on the subjects of which they treat.

The remaining miscellaneous papers may be classified under the following heads: (a) Lectures and Addresses, (b) Essays or Short Treatises, (c) Biographical Sketches, (d) Criticisms and Reviews.

d 2

Class (*a*) comprises his addresses to the British Association, to the London Mathematical Society, the Rede Lecture at Cambridge, his address at the opening of the Cavendish Laboratory and his Lectures at the Royal Institution and to the Chemical Society.

Class (*b*) includes all but one of the articles which he contributed to the *Encyclopædia Britannica* and several others of a kindred character to *Nature*.

Class (*c*) contains such articles as "Faraday" in the *Encyclopædia Britannica* and "Helmholtz" in *Nature*.

Class (*d*) is chiefly occupied with the reviews of scientific books as they were published. These appeared in *Nature* and the most important have been reprinted in these pages.

In some of these writings, particularly those in class (*b*), the author allowed himself a greater latitude in the use of mathematical symbols and processes than in others, as for instance in the article "Capillary Attraction," which is in fact a treatise on that subject treated mathematically. The lectures were upon one or other of the three departments of Physics with which he had mainly occupied himself;—Colour Perception, Action through a Medium, Molecular Physics; and on this account they are the more valuable. In the whole series of these more popular sketches we find the same clear, graceful delineation of principles, the same beauty in arrangement of subject, the same force and precision in expounding proofs and illustrations. The style is simple and singularly free from any kind of haze or obscurity, rising occasionally, as in his lectures, to a strain of subdued eloquence when the emotional aspects of the subject overcome the purely speculative.

The books which were written or edited by Maxwell and published in his lifetime but which are not included in this collection were the "Theory of Heat" (1st edition, 1871); "Electricity and Magnetism" (1st edition, 1873); "The Electrical Researches of the Honourable Henry Cavendish, F.R.S., written between 1771 and 1781, edited from the original manuscripts in the possession of the Duke of Devonshire, K.G." (1879). To these may be added a graceful little introductory treatise on Dynamics entitled "Matter and Motion" (published in 1876 by the Society for promoting Christian Knowledge). Maxwell also contributed part of the British Association Report on Electrical Units which was afterwards published in book form by Fleeming Jenkin.

The "Theory of Heat" appeared in the Text Books of Science series published by Longmans, Green and Co., and was at once hailed as a beautiful exposition of a subject, part of which, and that the most interesting part, the mechanical theory, had as yet but commenced the existence which it owed to the genius and labours of Rankine, Thomson and Clausius. There is a certain charm in Maxwell's treatise, due to the freshness and originality of its expositions which has rendered it a great favourite with students of Heat.

After his death an "Elementary Treatise on Electricity," the greater part of which he had written, was completed by Professor Garnett and published in 1881. The aim of this treatise and its position relatively to his larger work may be gathered from the following extract from Maxwell's preface.

"In this smaller book I have endeavoured to present, in as compact a form as I can, those phenomena which appear to throw light on the theory of electricity and to use them, each in its place, for the development of electrical ideas in the mind of the reader."

"In the larger treatise I sometimes made use of methods which I do not think the best in themselves, but without which the student cannot follow the investigations of the founders of the Mathematical Theory of Electricity. I have since become more convinced of the superiority of methods akin to those of Faraday, and have therefore adopted them from the first."

Of the "Electricity and Magnetism" it is difficult to predict the future, but there is no doubt that since its publication it has given direction and colour to the study of Electrical Science. It was the master's last word upon a subject to which he had devoted several years of his life, and most of what he wrote found its proper place in the treatise. Several of the chapters, notably those on Electromagnetism, are practically reproductions of his memoirs in a modified or improved form. The treatise is also remarkable for the handling of the mathematical details no less than for the exposition of physical principles, and is enriched incidentally by chapters of much originality on mathematical subjects touched on in the course of the work. Among these may be mentioned the dissertations on Spherical Harmonics and Lagrange's Equations in Dynamics.

The origin and growth of Maxwell's ideas and conceptions of electrical action, culminating in his treatise where all these ideas are arranged in due connection, form an interesting chapter not only in the history of an individual mind but in the history of electrical science. The importance of Faraday's discoveries and speculations can hardly be overrated in their influence on Maxwell, who tells us that before he began the study of electricity he resolved to read none of the mathematics of the subject till he had first mastered the "Experimental Researches." He was also at first under deep obligations to the ideas contained in the exceedingly important papers of Sir W. Thomson on the analogy between Heat-Conduction and Statical Electricity and on the Mathematical Theory of Electricity in Equilibrium. In his subsequent efforts we must perceive in Maxwell, possessed of Faraday's views and embued with his spirit, a vigorous intellect bringing to bear on a subject still full of obscurity the steady light of patient thought and expending upon it all the resources of a never failing ingenuity.

ROYAL NAVAL COLLEGE,
GREENWICH,
August, 1890.

TABLE OF CONTENTS.

ERRATA.

Page 40. In the first of equations (12), second group of terms, read

$$\frac{d^2}{dx^2}\,\delta x + \frac{d^2}{dy^2}\,\delta x + \frac{d^2}{dz^2}\,\delta x$$

instead of

$$\frac{d^2}{dx^2}\,\delta x + \frac{d^2}{dy^2}\,\delta y + \frac{d^2}{dz^2}\,\delta z$$

with corresponding changes in the other two equations.

Page 153, five lines from bottom of page, read 127 instead of 276.

Page 591, four lines from bottom of page the equation should be

$$\frac{d^2M}{da^2} + \frac{d^2M}{db^2} - \frac{1}{a}\frac{dM}{da} = 0.$$

Page 592, in the first line of the expression for L change

$$-\frac{\pi}{3}\cos 2\theta \quad \text{into} \quad -\frac{\pi}{3}\operatorname{cosec} 2\theta.$$

[From the *Proceedings of the Royal Society of Edinburgh*, Vol. II. April, 1846.]

I. *On the Description of Oval Curves, and those having a plurality of Foci; with remarks by Professor Forbes.* Communicated by PROFESSOR FORBES.

MR CLERK MAXWELL ingeniously suggests the extension of the common theory of the foci of the conic sections to curves of a higher degree of complication in the following manner :—

(1) As in the ellipse and hyperbola, any point in the curve has the *sum* or *difference* of two lines drawn from two points or *foci* = a constant quantity, so the author infers, that curves to a certain degree analogous, may be described and determined by the condition that the simple distance from one focus *plus* a multiple distance from the other, may be = a constant quantity; or more generally, *m* times the one distance + *n* times the other = constant.

(2) The author devised a simple mechanical means, by the wrapping of a thread round pins, for producing these curves. See Figs. 1 and 2. He

Fig. 1. Two Foci. Ratios 1, 2.   Fig. 2. Two Foci. Ratios 2, 3.

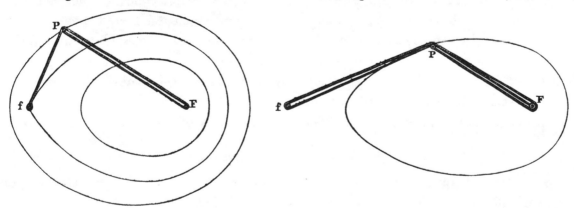

then thought of extending the principle to other curves, whose property should be, that the sum of the simple or multiple distances of any point of

the curve from three or more points or foci, should be = a constant quantity;
and this, too, he has effected mechanically, by a very simple arrangement of
a string of given length passing round three or more fixed pins, and con-
straining a tracing point, *P*. See Fig. 3. Farther, the author regards curves

Fig. 3.   Three Foci.   Ratios of Equality.

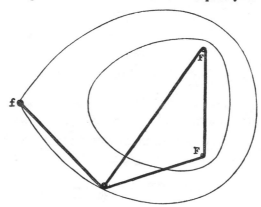

of the first kind as constituting a particular class of curves of the second
kind, two or more foci coinciding in one, a focus in which two strings meet
being considered a double focus; when three strings meet a treble focus, &c.

Professor Forbes observed that the equation to curves of the first class is
easily found, having the form

$$\sqrt{x^2+y^2} = a + b\sqrt{(x-c)^2+y^2},$$

which is that of the curve known under the name of the First Oval of
Descartes*. Mr Maxwell had already observed that when one of the foci was
at an infinite distance (or the thread moved parallel to itself, and was confined
in respect of length by the edge of a board), a curve resembling an ellipse
was traced; from which property Professor Forbes was led first to infer the
identity of the oval with the Cartesian oval, which is well known to have this
property. But the simplest analogy of all is that derived from the method of
description, *r* and *r'* being the radients to any point of the curve from the two
foci;

$$mr + nr' = \text{constant},$$

which in fact at once expresses on the undulatory theory of light the optical
character of the surface in question, namely, that light diverging from one
focus *F* without the medium, shall be correctly convergent at another point *f*

* Herschel, *On Light*, Art. 232; Lloyd, *On Light and Vision*, Chap. VII.

within it; and in this case the ratio $\dfrac{n}{m}$ expresses the index of refraction of the medium*.

If we denote by *the power of either focus* the number of strings leading to it by Mr Maxwell's construction, and if one of the foci be removed to an infinite distance, if the powers of the two foci be *equal* the curve is a parabola; if the power of the nearer focus be *greater* than the other, the curve is an ellipse; if the power of the infinitely distant focus be the greater, the curve is a hyperbola. The first case evidently corresponds to the case of the reflection of parallel rays to a focus, the velocity being unchanged after reflection; the second, to the refraction of parallel rays to a focus in a dense medium (in which light moves slower); the third case to refraction into a rarer medium.

The ovals of Descartes were described in his *Geometry*, where he has also given a mechanical method of describing one of them†, but only in a particular case, and the method is less simple than Mr Maxwell's. The *demonstration* of the optical properties was given by Newton in the *Principia*, Book I., prop. 97, by the law of the sines; and by Huyghens in 1690, on the Theory of Undulations in his *Traité de la Lumière*. It probably has not been suspected that so easy and elegant a method exists of describing these curves by the use of a thread and pins whenever the powers of the foci are commensurable. For instance, the curve, Fig. 2, drawn with powers 3 and 2 respectively, give the proper form for a refracting surface of a glass, whose index of refraction is 1·50, in order that rays diverging from f may be refracted to F.

As to the higher classes of curves with three or more focal points, we cannot at present invest them with equally clear and curious physical properties, but the method of drawing a curve by so simple a contrivance, which shall satisfy the condition

$$mr + nr' + pr'' + \&c. = \text{constant},$$

is in itself not a little interesting; and if we regard, with Mr Maxwell, the ovals above described, as the limiting case of the others by the coalescence of two or more foci, we have a farther generalization of the same kind as that so highly recommended by Montucla‡, by which Descartes elucidated the conic sections as particular cases of his oval curves.

---

* This was perfectly well shewn by Huyghens in his *Traité de la Lumière*, p. 111. (1690.)

† Edit. 1683. *Geometria*, Lib. II. p. 54.

‡ *Histoire des Mathématiques*. First Edit. II. 102.

[From the *Transactions of the Royal Society of Edinburgh,* Vol. XVI. Part V.]

II. *On the Theory of Rolling Curves.* Communicated by the Rev. Professor KELLAND.

THERE is an important geometrical problem which proposes to find a curve having a given relation to a series of curves described according to a given law. This is the problem of Trajectories in its general form.

The series of curves is obtained from the general equation to a curve by the variation of its parameters. In the general case, this variation may change the form of the curve, but, in the case which we are about to consider, the curve is changed only in position.

This change of position takes place partly by rotation, and partly by transference through space. The rolling of one curve on another is an example of this compound motion.

As examples of the way in which the new curve may be related to the series of curves, we may take the following :—

1. The new curve may cut the series of curves at a given angle. When this angle becomes zero, the curve is the envelope of the series of curves.

2. It may pass through corresponding points in the series of curves. There are many other relations which may be imagined, but we shall confine our attention to this, partly because it affords the means of tracing various curves, and partly on account of the connection which it has with many geometrical problems.

Therefore the subject of this paper will be the consideration of the relations of three curves, one of which is fixed, while the second rolls upon it and traces the third. The subject of rolling curves is by no means a new one. The first idea of the cycloid is attributed to Aristotle, and involutes and evolutes have been long known.

In the *History of the Royal Academy of Sciences* for 1704, page 97, there is a memoir entitled "Nouvelle formation des Spirales," by M. Varignon, in which he shews how to construct a polar curve from a curve referred to rectangular co-ordinates by substituting the radius vector for the abscissa, and a circular arc for the ordinate. After each curve, he gives the curve into which it is "unrolled," by which he means the curve which the spiral must be rolled upon in order that its pole may trace a straight line; but as this is not the principal subject of his paper, he does not discuss it very fully.

There is also a memoir by M. de la Hire, in the volume for 1706, Part II., page 489, entitled "Methode generale pour réduire toutes les Lignes courbes à des Roulettes, leur generatrice ou leur base étant donnée telle qu'on voudra."

M. de la Hire treats curves as if they were polygons, and gives geometrical constructions for finding the fixed curve or the rolling curve, the other two being given; but he does not work any examples.

In the volume for 1707, page 79, there is a paper entitled, "Methode generale pour déterminer la nature des Courbes formées par le roulement de toutes sortes de Courbes sur une autre Courbe quelconque." Par M. Nicole.

M. Nicole takes the equations of the three curves referred to rectangular co-ordinates, and finds three general equations to connect them. He takes the tracing-point either at the origin of the co-ordinates of the rolled curve or not. He then shews how these equations may be simplified in several particular cases. These cases are—

    (1) When the tracing-point is the origin of the rolled curve.
    (2) When the fixed curve is the same as the rolling curve.
    (3) When both of these conditions are satisfied.
    (4) When the fixed line is straight.

He then says, that if we roll a geometric curve on itself, we obtain a new geometric curve, and that we may thus obtain an infinite number of geometric curves.

The examples which he gives of the application of his method are all taken from the cycloid and epicycloid, except one which relates to a parabola, rolling on itself, and tracing a cissoid with its vertex. The reason of so small a number of examples being worked may be, that it is not easy to eliminate the co-ordinates of the fixed and rolling curves from his equations.

The case in which one curve rolling on another produces a circle is treated of in Willis's *Principles of Mechanism.* Class C. *Rolling Contact.*

He employs the same method of finding the one curve from the other which is used here, and he attributes it to Euler (see the *Acta Petropolitana*, Vol. v.).

Thus, nearly all the simple cases have been treated of by different authors; but the subject is still far from being exhausted, for the equations have been applied to very few curves, and we may easily obtain new and elegant properties from any curve we please.

Almost all the more notable curves may be thus linked together in a great variety of ways, so that there are scarcely two curves, however dissimilar, between which we cannot form a chain of connected curves.

This will appear in the list of examples given at the end of this paper.

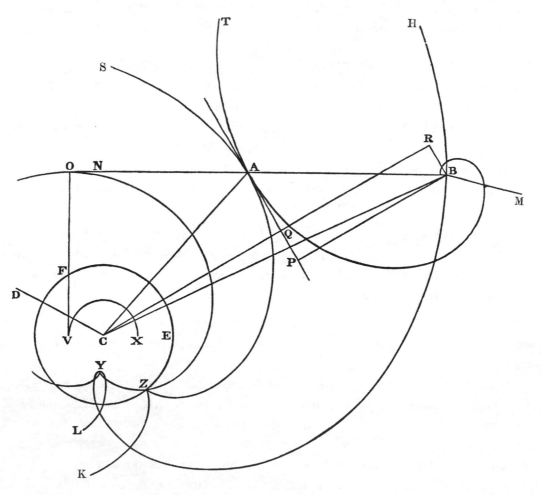

Let there be a curve *KAS*, whose pole is at *C*.

Let the angle $DCA = \theta_1$ and $CA = r_1$ and let

$$\theta_1 = \phi_1(r_1).$$

Let this curve remain fixed to the paper.

Let there be another curve BAT, whose pole is B.

Let the angle $MBA = \theta_2$, and $BA = r_2$, and let

$$\theta_2 = \phi_2(r_2).$$

Let this curve roll along the curve KAS without slipping.

Then the pole B will describe a third curve, whose pole is C.

Let the angle $DCB = \theta_3$, and $CB = r_3$, and let

$$\theta_3 = \phi_3(r_3).$$

We have here six unknown quantities $\theta_1\theta_2\theta_3r_1r_2r_3$; but we have only three equations given to connect them, therefore the other three must be sought for in the enunciation.

But before proceeding to the investigation of these three equations, we must premise that the three curves will be denominated as follows :—

The Fixed Curve, Equation, $\theta_1 = \phi_1(r_1)$.

The Rolled Curve, Equation, $\theta_2 = \phi_2(r_2)$.

The Traced Curve, Equation, $\theta_3 = \phi_3(r_3)$.

When it is more convenient to make use of equations between rectangular co-ordinates, we shall use the letters x_1y_1, x_2y_2, x_3y_3. We shall always employ the letters $s_1s_2s_3$ to denote the length of the curve from the pole, $p_1p_2p_3$ for the perpendiculars from the pole on the tangent, and $q_1q_2q_3$ for the intercepted part of the tangent.

Between these quantities, we have the following equations :—

$$r = \sqrt{x^2 + y^2}, \qquad\qquad \theta = \tan^{-1}\frac{y}{x},$$

$$x = r\cos\theta, \qquad\qquad y = r\sin\theta,$$

$$s = \int \sqrt{r^2 + \left(\frac{dr}{d\theta}\right)^2}\, d\theta, \qquad\qquad s = \int \sqrt{1 + \left(\frac{dy}{dx}\right)^2}\, dx,$$

$$p = \frac{r^2}{\sqrt{r^2 + \left(\frac{dr}{d\theta}\right)^2}}, \qquad\qquad p = \frac{ydx - xdy}{\sqrt{(dx)^2 + (dy)^2}},$$

$$q = \frac{r\frac{dr}{d\theta}}{\sqrt{r^2 + \left(\frac{dr}{d\theta}\right)^2}}, \qquad\qquad q = \frac{x\,dx + y\,dy}{\sqrt{(dx)^2 + (dy)^2}},$$

$$R = \frac{\left\{r^2 + \left(\frac{dr}{d\theta}\right)^2\right\}^{\frac{3}{2}}}{r^2 + 2\left(\frac{dr}{d\theta}\right)^2 - r\frac{d^2r}{d\theta^2}}, \qquad R = \frac{\left\{1 + \left(\frac{dy}{dx}\right)^2\right\}^{\frac{3}{2}}}{\frac{d^2y}{dx^2}}.$$

We come now to consider the three equations of rolling which are involved in the enunciation. Since the second curve rolls upon the first *without slipping*, the length of the fixed curve at the point of contact is the measure of the length of the rolled curve, therefore we have the following equation to connect the fixed curve and the rolled curve—

$$s_1 = s_2.$$

Now, by combining this equation with the two equations

$$\begin{Bmatrix} \theta_1 = \phi_1(r_1) \\ \theta_2 = \phi_2(r_2) \end{Bmatrix} \text{ or } \begin{Bmatrix} x_1 = \psi_1(y_1) \\ x_2 = \psi_2(y_2) \end{Bmatrix},$$

it is evident that from any of the four quantities $\theta_1 r_1 \theta_2 r_2$ or $x_1 y_1 x_2 y_2$, we can obtain the other three, therefore we may consider these quantities as known functions of each other.

Since the curve *rolls* on the fixed curve, they must have a common tangent.

Let PA be this tangent, draw BP, CQ perpendicular to PA, produce CQ, and draw BR perpendicular to it, then we have $CA = r_1$, $BA = r_2$, and $CB = r_3$; $CQ = p_1$, $PB = p_2$, and $BN = p_3$; $AQ = q_1$, $AP = q_2$, and $CN = q_3$.

Also
$$r_3^2 = CB^2 = CR^2 + RB^2 = (CQ + PB)^2 + (AP - AQ)^2$$
$$= (p_1 + p_2)^2 + (q_2 - q_1)^2$$
$$= p_1^2 + 2p_1 p_2 + p_2^2 + r_2^2 - p_2^2 - 2q_1 q_2 + r_1^2 - p_1^2$$
$$r_3^2 = r_1^2 + r_2^2 + 2p_1 p_2 - 2q_1 q_2.$$

Since the first curve is fixed to the paper, we may find the angle θ_3.

Thus
$$\theta_3 = DCB = DCA + ACQ + RCB$$
$$= \theta_1 + \tan^{-1}\frac{q_1}{p_1} + \tan^{-1}\frac{RB}{RC}$$
$$\theta_3 = \theta_1 + \tan^{-1}\frac{dr_1}{r_1 d\theta_1} + \tan^{-1}\frac{q_2 - q_1}{p_2 + p_1}.$$

THE THEORY OF ROLLING CURVES.

Thus we have found three independent equations, which, together with the equations of the curves, make up six equations, of which each may be deduced from the others. There is an equation connecting the radii of curvature of the three curves which is sometimes of use.

The angle through which the rolled curve revolves during the description of the element ds_3, is equal to the angle of contact of the fixed curve and the rolling curve, or to the sum of their curvatures,

$$\therefore \frac{ds_3}{r_2} = \frac{ds_1}{R_1} + \frac{ds_2}{R_2}.$$

But the radius of the rolled curve has revolved in the opposite direction through an angle equal to $d\theta_2$, therefore the angle between two successive positions of r_2 is equal to $\frac{ds_3}{r_2} - d\theta_2$. Now this angle is the angle between two successive positions of the normal to the traced curve, therefore, if O be the centre of curvature of the traced curve, it is the angle which ds_3 or ds_1 subtends at O. Let $OA = T$, then

$$\frac{ds_3}{R_3} = \frac{r_2 d\theta_2}{T} = \frac{ds_3}{r_2} - d\theta_2 = \frac{ds_2}{R_1} + \frac{ds_2}{R_2} - d\theta_2,$$

$$\therefore r_2 \frac{d\theta_2}{ds_2} \frac{1}{T} = \frac{1}{R_1} + \frac{1}{R_2} - \frac{d\theta_2}{ds_2},$$

$$\therefore \frac{p_2}{r_2} \left(\frac{1}{T} + \frac{1}{r_2} \right) = \frac{1}{R_1} + \frac{1}{R_2}.$$

As an example of the use of this equation, we may examine a property of the logarithmic spiral.

In this curve, $p = mr$, and $R = \frac{r}{m}$, therefore if the rolled curve be the logarithmic spiral

$$m \left(\frac{1}{T} + \frac{1}{r_2} \right) = \frac{1}{R_1} + \frac{m}{r_2},$$

$$\frac{m}{T} = \frac{1}{R_1},$$

therefore AO in the figure $= mR_1$, and $\frac{AO}{R_1} = m$.

Let the locus of O, or the evolute of the traced curve $LYBH$, be the curve OZY, and let the evolute of the fixed curve $KZAS$ be FEZ, and let us consider FEZ as the fixed curve, and OZY as the traced curve.

Then in the triangles BPA, AOF, we have $OAF = PBA$, and $\dfrac{OA}{AF} = m = \dfrac{BP}{AB}$, therefore the triangles are similar, and $FOA = APB = \dfrac{\pi}{2}$, therefore OF is perpendicular to OA, the tangent to the curve OZY, therefore OF is the radius of the curve which when rolled on FEZ traces OZY, and the angle which the curve makes with this radius is $OFA = PAB = \sin^{-1} m$, which is constant, therefore the curve, which, when rolled on FEZ, traces OZY, is the logarithmic spiral. Thus we have proved the following proposition: "The involute of the curve traced by the pole of a logarithmic spiral which rolls upon any curve, is the curve traced by the pole of the same logarithmic spiral when rolled on the involute of the primary curve."

It follows from this, that if we roll on any curve a curve having the property $p_1 = m_1 r_1$, and roll another curve having $p_2 = m_2 r_2$ on the curve traced, and so on, it is immaterial in what order we roll these curves. Thus, if we roll a logarithmic spiral, in which $p = mr$, on the nth involute of a circle whose radius is a, the curve traced is the $n+1$th involute of a circle whose radius is $\sqrt{1 - m^2}$.

Or, if we roll successively m logarithmic spirals, the resulting curve is the $n+m$th involute of a circle, whose radius is

$$a \sqrt{1 - m_1^2} \ \sqrt{1 - m_2^2}, \ \sqrt{\&c.}$$

We now proceed to the cases in which the solution of the problem may be simplified. This simplification is generally effected by the consideration that the radius vector of the rolled curve is the normal drawn from the traced curve to the fixed curve.

In the case in which the curve is rolled on a straight line, the perpendicular on the tangent of the rolled curve is the distance of the tracing point from the straight line; therefore, if the traced curve be defined by an equation in x_3 and y_3,

$$x_3 = p_2 = \frac{r_2^2}{\sqrt{r_2^2 + \left(\dfrac{dr_2}{d\theta_2}\right)^2}} \quad \dotfill \quad (1),$$

and

$$r_2 = x_3 \sqrt{\left(\dfrac{dx_3}{dy_3}\right)^2 + 1} \quad \dotfill \quad (2).$$

By substituting for r_2 in the first equation, its value, as derived from the second, we obtain

$$x_3^2\left(\frac{dx_3}{dy_3}\right)^2\left[\left(\frac{dx_3}{dy_3}\right)^2+1\right]=\left(\frac{dr_2}{d\theta_2}\right)^2.$$

If we know the equation to the rolled curve, we may find $\left(\frac{dr_2}{d\theta_2}\right)^2$ in terms of r_2, then by substituting for r_2 its value in the second equation, we have an equation containing x_3 and $\frac{dx_3}{dy_3}$, from which we find the value of $\frac{dx_3}{dy_3}$ in terms of x_3; the integration of this gives the equation of the traced curve.

As an example, we may find the curve traced by the pole of a hyperbolic spiral which rolls on a straight line.

The equation of the rolled curve is $\theta_2=\frac{a}{r_2}$,

$$\therefore\left(\frac{dr_2}{d\theta_2}\right)^2=\frac{r_2^4}{a^2}$$

$$=x_3^2\left(\frac{dx_3}{dy_3}\right)^2\left[\left(\frac{dx_3}{dy_3}\right)^2+1\right]=\frac{x_3^4}{a^2}\left[\left(\frac{dx_3}{dy_3}\right)^2+1\right]^2,$$

$$\therefore\ a^2\left(\frac{dx_3}{dy_3}\right)^2=x_3^2\left[\left(\frac{dx_3}{dy_3}\right)^2+1\right],$$

$$\therefore\ \frac{dx_3}{dy_3}=\frac{x_3}{\sqrt{a^2-x_3^2}}.$$

This is the differential equation of the tractory of the straight line, which is the curve traced by the pole of the hyperbolic spiral.

By eliminating x_3 in the two equations, we obtain

$$\frac{dr_2}{d\theta_2}=r_2\left(\frac{dx_3}{dy_3}\right).$$

This equation serves to determine the rolled curve when the traced curve is given.

As an example we shall find the curve, which being rolled on a straight line, traces a common catenary.

Let the equation to the catenary be

$$x=\frac{a}{2}\left(e^{\frac{y}{a}}+e^{-\frac{y}{a}}\right).$$

2—2

Then
$$\frac{dx_3}{dy_3} = \sqrt{\frac{x_3^2}{a^2} - 1},$$

$$\therefore \left(\frac{dr_2}{d\theta_2}\right)^2 = \frac{r_2^2}{a^2} \frac{r_2^4}{\left(\frac{dr_2}{d\theta_2}\right)^2 + r_2^2} - r_2^2,$$

$$\therefore \left[\left(\frac{dr_2}{d\theta_2}\right)^2 + r_2^2\right]^2 = \left(\frac{r_2^3}{a}\right)^2,$$

$$\therefore \left(\frac{dr_2}{d\theta_2}\right)^2 = \frac{r_2^2}{a}(r_2 - a),$$

$$\therefore \frac{d\theta}{dr} = \frac{1}{r\sqrt{\frac{r}{a} - 1}},$$

then by integration
$$\theta = \cos^{-1}\left(\frac{2a}{r} - 1\right),$$

$$r = \frac{2a}{1 + \cos\theta}.$$

This is the polar equation of the parabola, the focus being the pole; therefore, if we roll a parabola on a straight line, its focus will trace a catenary.

The rectangular equation of this parabola is $x^2 = 4ay$, and we shall now consider what curve must be rolled along the axis of y to trace the parabola.

By the second equation (2),

$$r_2 = x_3 \sqrt{\frac{4a^2}{x_3^2} + 1}, \quad \text{but } x_3 = p_2,$$

$$\therefore r_2 = \sqrt{4a^2 + p_2^2},$$

$$\therefore r_2^2 = p_2^2 + 4a^2,$$

$$\therefore 2a = \sqrt{r_2^2 - p_2^2} = q_2,$$

but q_2 is the perpendicular on the normal, therefore the normal to the curve always touches a circle whose radius is $2a$, therefore the curve is the involute of this circle.

Therefore we have the following method of describing a catenary by continued motion.

Describe a circle whose radius is twice the parameter of the catenary; roll a straight line on this circle, then any point in the line will describe an involute of the circle; roll this curve on a straight line, and the centre of the circle will describe a parabola; roll this parabola on a straight line, and its focus will trace the catenary required.

We come now to the case in which a straight line rolls on a curve.

When the tracing-point is in the straight line, the problem becomes that of involutes and evolutes, which we need not enter upon; and when the tracing-point is not in the straight line, the calculation is somewhat complex; we shall therefore consider only the relations between the curves described in the first and second cases.

Definition.—The curve which cuts at a given angle all the circles of a given radius whose centres are in a given curve, is called a tractory of the given curve.

Let a straight line roll on a curve A, and let a point in the straight line describe a curve B, and let another point, whose distance from the first point is b, and from the straight line a, describe a curve C, then it is evident that the curve B cuts the circle whose centre is in C, and whose radius is b, at an angle whose sine is equal to $\dfrac{a}{b}$, therefore the curve B is a tractory of the curve C.

When $a = b$, the curve B is the orthogonal tractory of the curve C. If tangents equal to a be drawn to the curve B, they will be terminated in the curve C; and if one end of a thread be carried along the curve C, the other end will trace the curve B.

When $a = 0$, the curves B and C are both involutes of the curve A, they are always equidistant from each other, and if a circle, whose radius is b, be rolled on the one, its centre will trace the other.

If the curve A is such that, if the distance between two points measured along the curve is equal to b, the two points are similarly situate, then the curve B is the same with the curve C. Thus, the curve A may be a re-entrant curve, the circumference of which is equal to b.

When the curve A is a circle, the curves B and C are always the same.

The equations between the radii of curvature become

$$\frac{1}{T} + \frac{1}{r_2} = \frac{r}{a R_1}.$$

When $a=0$, $T=0$, or the centre of curvature of the curve B is at the point of contact. Now, the normal to the curve C passes through this point, therefore—

"The normal to any curve passes through the centre of curvature of its tractory."

In the next case, one curve, by rolling on another, produces a straight line. Let this straight line be the axis of y, then, since the radius of the rolled curve is perpendicular to it, and terminates in the fixed curve, and since these curves have a common tangent, we have this equation,

$$x_1 \frac{dy_1}{dx_1} = r_2^2 \frac{d\theta_2}{dr_2}.$$

If the equation of the rolled curve be given, find $\dfrac{d\theta_2}{dr_2}$ in terms of r_2, substitute x_1 for r_2, and multiply by x_1, equate the result to $\dfrac{dy_1}{dx_1}$, and integrate.

Thus, if the equation of the rolled curve be

$$\theta = Ar^{-n} + \&c. + Kr^{-2} + Lr^{-1} + M \log r + Nr + \&c. + Zr^n,$$

$$\frac{d\theta}{dr} = -nAr^{-(n+1)} - \&c. - 2Kr^{-3} - Lr^{-2} + Mr^{-1} + N + \&c. + nZr^{n-1},$$

$$\frac{dy}{dx} = -nAx^{-n} - \&c. - 2Kx^{-2} - Lx^{-1} + M + Nx + \&c. + nZx^n,$$

$$y = \frac{n}{n-1} Ax^{1-n} + \&c. + 2Kx^{-1} - L \log x + Mx + \tfrac{1}{2}Nx^2 + \&c. + \frac{n}{n+1} Zx^{n+1},$$

which is the equation of the fixed curve.

If the equation of the fixed curve be given, find $\dfrac{dy}{dx}$ in terms of x, substitute r for x, and divide by r, equate the result to $\dfrac{d\theta}{dr}$, and integrate.

Thus, if the fixed curve be the orthogonal tractory of the straight line, whose equation is

$$y = a \log \frac{x}{a + \sqrt{a^2 - x^2}} + \sqrt{a^2 - x^2},$$

$$\frac{dy}{dx} = \frac{\sqrt{a^2 - x^2}}{x},$$

$$\frac{d\theta}{dr} = \frac{\sqrt{a^2 - r^2}}{r^2},$$

$$\theta = \cos^{-1}\frac{r}{a} - \sqrt{\frac{a^2}{r^2} - 1},$$

this is the equation to the orthogonal tractory of a circle whose diameter is equal to the constant tangent of the fixed curve, and its constant tangent equal to half that of the fixed curve.

This property of the tractory of the circle may be proved geometrically, thus—Let P be the centre of a circle whose radius is PD, and let CD be a line constantly equal to the radius. Let BCP be the curve described by the point C when the point D is moved along the circumference of the circle, then if tangents equal to CD be drawn to the curve, their extremities will be in the circle. Let ACH be the curve on which BCP rolls, and let OPE be the straight line traced by the pole, let CDE be the common tangent, let it cut the circle in D, and the straight line in E.

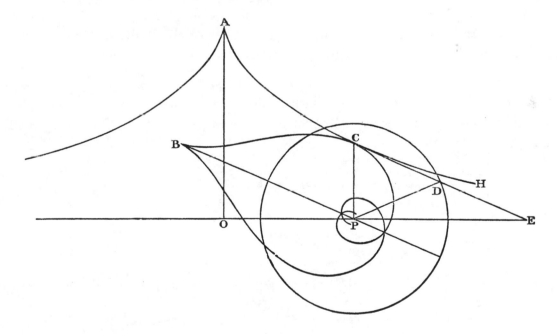

Then $CD = PD$, $\therefore$ $\angle DCP = \angle DPC$, and CP is perpendicular to OE, $\therefore$ $\angle CPE = \angle DCP + \angle DEP$. Take away $\angle DCP = \angle DPC$, and there remains $DPE = DEP$, $\therefore$ $PD = DE$, $\therefore$ $CE = 2PD$.

Therefore the curve ACH has a constant tangent equal to the diameter of the circle, therefore ACH is the orthogonal tractory of the straight line, which is the tractrix or equitangential curve.

The operation of finding the fixed curve from the rolled curve is what Sir John Leslie calls "divesting a curve of its radiated structure."

The method of finding the curve which must be rolled on a circle to trace a given curve is mentioned here because it generally leads to a double result, for the normal to the traced curve cuts the circle in two points, either of which may be a point in the rolled curve.

Thus, if the traced curve be the involute of a circle concentric with the given circle, the rolled curve is one of two similar logarithmic spirals.

If the curve traced be the spiral of Archimedes, the rolled curve may be either the hyperbolic spiral or the straight line.

In the next case, one curve rolls on another and traces a circle.

Since the curve traced is a circle, the distance between the poles of the fixed curve and the rolled curve is always the same; therefore, if we fix the rolled curve and roll the fixed curve, the curve traced will still be a circle, and, if we fix the poles of both the curves, we may roll them on each other without friction.

Let a be the radius of the traced circle, then the sum or difference of the radii of the other curves is equal to a, and the angles which they make with the radius at the point of contact are equal,

$$\therefore \ r_1 = \pm(a \pm r_2) \text{ and } r_1 \frac{d\theta_1}{dr_1} = r_2 \frac{d\theta_2}{dr_2}.$$

$$\therefore \ \frac{d\theta_2}{dr_2} = \frac{\pm(a \pm r_2)}{r_2} \frac{d\theta_1}{dr_1}.$$

If we know the equation between θ_1 and r_1, we may find $\frac{d\theta_1}{dr_1}$ in terms of r_1, substitute $\pm(a \pm r_2)$ for r_1, multiply by $\frac{\pm(a \pm r_2)}{r_2}$, and integrate.

Thus, if the equation between θ_1 and r_1 be

$$r_1 = a \sec \theta_1,$$

which is the polar equation of a straight line touching the traced circle whose equation is $r = a$, then

$$\frac{d\theta}{dr_1} = \frac{a}{r_1\sqrt{r_1{}^2 - a^2}}$$

$$= \frac{a}{(r_2 \pm a)\sqrt{r_2{}^2 \pm 2r_2 a}},$$

$$\frac{d\theta_2}{dr_2} = \frac{r_2 \pm a}{r_2} \frac{a}{(r_2 \pm a)\sqrt{r_2{}^2 \pm 2r_2 a}}$$

$$= \frac{a}{r_2\sqrt{r_2{}^2 \pm 2ar_2}},$$

$$\theta_2 = \pm\sqrt{1 \pm 2\frac{a}{r_2}},$$

$$r_2 = \frac{2a}{\theta_2{}^2 - 1} = \frac{2a}{\theta^2 - 1}.$$

Now, since the rolling curve is a straight line, and the tracing point is not in its direction, we may apply to this example the observations which have been made upon tractories.

Let, therefore, the curve $r = \dfrac{2a}{\theta^2 - 1}$ be denoted by A, its involute by B, and the circle traced by C, then B is the tractory of C; therefore the involute of the curve $r = \dfrac{2a}{\theta^2 - 1}$ is the tractory of the circle, the equation of which is $\theta = \cos^{-1}\dfrac{r}{a} - \sqrt{\dfrac{a^2}{r^2} - 1}$. The curve whose equation is $r = \dfrac{2a}{\theta^2 - 1}$ seems to be among spirals what the catenary is among curves whose equations are between rectangular co-ordinates; for, if we represent the vertical direction by the radius vector, the tangent of the angle which the curve makes with this line is proportional to the length of the curve reckoned from the origin; the point at the distance a from a straight line rolled on this curve generates a circle, and when rolled on the catenary produces a straight line; the involute of this curve is the tractory of the circle, and that of the catenary is the tractory of the straight line, and the tractory of the circle rolled on that of the straight line traces the straight line; if this curve is rolled on the catenary, it produces the straight line touching the catenary at its vertex; the method of drawing tangents is the same as in the catenary, namely, by describing a circle whose radius is a on the production of the radius vector, and drawing a tangent to the circle from the given point.

In the next case the rolled curve is the same as the fixed curve. It is evident that the traced curve will be similar to the locus of the intersection of the tangent with the perpendicular from the pole; the magnitude, however, of the traced curve will be double that of the other curve; therefore, if we call $r_0 = \phi_0 \theta_0$ the equation to the fixed curve, $r_1 = \phi_1 \theta_1$ that of the traced curve, we have

$$r_1 = 2p_0, \qquad \theta_1 = \theta_0 - \cos^{-1} \frac{p_0}{r_0} = \theta_0 - \frac{\pi}{2} + \sin^{-1} \frac{p_0}{r_0},$$

also,
$$\frac{p_1}{r_1} = \frac{p_0}{r_0}.$$

Similarly, $\quad r_2 = 2p_1 = 2r_1 \frac{p_0}{r_0} = 4 \frac{p_0^2}{r_0} \, 4r_0 \left(\frac{p_0}{r_0}\right)^2, \quad \theta_2 = \theta_0 - 2\cos^{-1} \frac{p_0}{r_0}.$

Similarly, $\quad r_n = 2p_{n-1} = 2r_{n-1} \frac{p_0}{r_0}$ &c. $= 2^n r_0 \left(\frac{p_0}{r_0}\right)^n,$

and
$$\frac{p_n}{r_n} = \frac{p_0}{r_0},$$

$$\theta_n = \theta_0 - n \cos^{-1} \frac{p_0}{r_0},$$

$$\theta_n = \theta_0 - n \cos^{-1} \frac{p_n}{r_n}.$$

Let θ_n become $\theta_n{}^1$; θ_0, $\theta_0{}^1$ and $\frac{p_0}{r_0}$, $\frac{p_0{}^1}{r_0{}^1}$. Let $\theta_n{}^1 - \theta_n = a,$

$$\theta_n{}^1 = \theta_0{}^1 - n \cos^{-1} \frac{p^1}{r_n{}^1},$$

$$a = \theta_n{}^1 - \theta_n = \theta_0{}^1 - \theta_0 - n \cos^{-1} \frac{p_n{}^1}{r_n{}^1} + n \cos^{-1} \frac{p_n}{r_n};$$

$$\therefore \cos^{-1} \frac{p_n}{r_n} - \cos^{-1} \frac{p_n{}^1}{r_n{}^1} = \frac{a}{n} + \frac{\theta_0 - \theta_0{}^1}{n}.$$

Now, $\cos^{-1} \dfrac{p_n}{r_n}$ is the complement of the angle at which the curve cuts the radius vector, and $\cos^{-1} \dfrac{p_n}{r_n} - \cos^{-1} \dfrac{p_n^{\;1}}{r_n^{\;1}}$ is the variation of this angle when θ_n varies by an angle equal to α. Let this variation $= \phi$; then if $\theta_0 - \theta_0^{\;1} = \beta$,

$$\phi = \frac{\alpha}{n} + \frac{\beta}{n}.$$

Now, if n increases, ϕ will diminish; and if n becomes infinite,

$$\phi = \frac{\alpha}{\infty} + \frac{\beta}{\infty} = 0 \text{ when } \alpha \text{ and } \beta \text{ are finite.}$$

Therefore, when n is infinite, ϕ vanishes; therefore the curve cuts the radius vector at a constant angle; therefore the curve is the logarithmic spiral.

Therefore, if any curve be rolled on itself, and the operation repeated an infinite number of times, the resulting curve is the logarithmic spiral.

Hence we may find, analytically, the curve which, being rolled on itself, traces itself.

For the curve which has this property, if rolled on itself, and the operation repeated an infinite number of times, will still trace itself.

But, by this proposition, the resulting curve is the logarithmic spiral; therefore the curve required is the logarithmic spiral. As an example of a curve rolling on itself, we will take the curve whose equation is

$$r_0 = 2^n a \left(\cos \frac{\theta_0}{n} \right)^n.$$

Here
$$-\frac{dr_0}{d\theta_0} = 2^n a \left(\sin \frac{\theta_0}{n} \right) \left(\cos \frac{\theta_0}{n} \right)^{n-1};$$

$$\therefore r_1 = 2p_0 = 2 \; \frac{2^{2n} a^2 \left(\cos \dfrac{\theta_0}{n} \right)^{2n}}{\sqrt{2^{2n} a^2 \left(\cos \dfrac{\theta_0}{n} \right)^{2n} + 2^{2n} a^2 \left(\sin \dfrac{\theta_0}{n} \right) \left(\cos \dfrac{\theta_0}{n} \right)^{2n-2}}},$$

$$r_1 = 2 \; \frac{2^n a \left(\cos \dfrac{\theta_0}{n} \right)^{n+1}}{\sqrt{\left(\cos \dfrac{\theta_0}{n} \right)^2 + \left(\sin \dfrac{\theta_0}{n} \right)^2}} = 2^{n+1} a \left(\cos \frac{\theta_0}{n} \right)^{n+1}.$$

Now $\theta_1 - \theta_0 = -\cos^{-1}\dfrac{p_0}{r_0} = -\cos^{-1}\cos\dfrac{\theta_0}{n} = \dfrac{\theta_0}{n}$,

$$\therefore \; \theta_0 = \theta_1 \frac{n}{n+1} \, ;$$

substituting this value of θ_0 in the expression for r_1,

$$r_1 = 2^{n+1} a \left(\cos \frac{\theta_1}{n+1} \right)^{n+1},$$

similarly, if the operation be repeated m times, the resulting curve is

$$r_m = 2^{n+m} a \left(\cos \frac{\theta_m}{n+m} \right)^{n+m}.$$

When $n = 1$, the curve is

$$r = 2a \cos \theta,$$

the equation to a circle, the pole being in the circumference.

When $n = 2$, it is the equation to the cardioid

$$r = 4a \left(\cos \frac{\theta}{2} \right)^2.$$

In order to obtain the cardioid from the circle, we roll the circle upon itself, and thus obtain it by one operation; but there is an operation which, being performed on a circle, and again on the resulting curve, will produce a cardioid, and the intermediate curve between the circle and cardioid is

$$r = 2^{\frac{3}{2}} a \left(\cos \frac{2\theta}{3} \right)^{\frac{3}{2}}.$$

As the operation of rolling a curve on itself is represented by changing n into $(n+1)$ in the equation, so this operation may be represented by changing n into $(n+\frac{1}{2})$.

Similarly there may be many other fractional operations performed upon the curves comprehended under the equation

$$r = 2^n a \left(\cos \frac{\theta}{n} \right)^n.$$

We may also find the curve, which, being rolled on itself, will produce a given curve, by making $n = -1$.

We may likewise prove by the same method as before, that the result of performing this inverse operation an infinite number of times is the logarithmic spiral.

As an example of the inverse method, let the traced line be straight, let its equation be

$$r_0 = 2a \sec \theta_0,$$

then

$$\frac{p_{-1}}{r_{-1}} = \frac{p_0}{r_0} = \frac{2a}{r_0} = \frac{2a}{2p_{-1}},$$

$$\therefore \; p^2_{-1} = ar_{-1},$$

therefore suppressing the suffix,

$$\frac{r^4}{r^2 + \dfrac{dr^2}{d\theta^2}} = ar,$$

$$\therefore \; \left(\frac{dr}{d\theta}\right)^2 = \frac{r^3}{a} - r^2,$$

$$\therefore \; \frac{d\theta}{dr} = \frac{1}{r\sqrt{\dfrac{r}{a} - 1}},$$

$$\therefore \; \theta = \cos^{-1}\left(\frac{2a}{r} - 1\right),$$

$$r = \frac{2a}{1 - \cos\theta},$$

the polar equation of the parabola whose parameter is $4a$.

The last case which we shall here consider affords the means of constructing two wheels whose centres are fixed, and which shall roll on each other, so that the angle described by the first shall be a given function of the angle described by the second.

Let $\theta_2 = \phi\theta_1$, then $r_1 + r_2 = a$, and $\dfrac{d\theta_2}{d\theta_1} = \dfrac{r_1}{r_2}$;

$$\therefore \; \frac{d\,(\phi\theta_1)}{d\theta_1} = \frac{r_1}{a - r_1}.$$

Let us take as an example, the pair of wheels which will represent the angular motion of a comet in a parabola.

Here $\qquad \theta_2 = \tan\dfrac{\theta_1}{2}, \quad \therefore \dfrac{d\theta_2}{d\theta_1} = \dfrac{1}{2\cos^2\dfrac{\theta_1}{2}} = \dfrac{r_1}{a - r_1},$

$$\therefore \frac{r_1}{a} = \frac{1}{2 + \cos\theta_1},$$

therefore the first wheel is an ellipse, whose major axis is equal to $\frac{4}{3}$ of the distance between the centres of the wheels, and in which the distance between the foci is half the major axis.

Now since $\qquad \theta_1 = 2\tan^{-1}\theta_2$ and $r_1 = a - r_2,$

$$\frac{r}{a} = 1 + \frac{1}{2\,(2 - \theta^4)},$$

$$\theta^4 = 2 - \frac{1}{\dfrac{2r}{a} - 2},$$

which is the equation to the wheel which revolves with constant angular velocity.

Before proceeding to give a list of examples of rolling curves, we shall state a theorem which is almost self-evident after what has been shewn previously.

Let there be three curves, A, B, and C. Let the curve A, when rolled on itself, produce the curve B, and when rolled on a straight line let it produce the curve C, then, if the dimensions of C be doubled, and B be rolled on it, it will trace a straight line.

A Collection of Examples of Rolling Curves.

First. Examples of a curve rolling on a straight line.

Ex. 1. When the rolling curve is a circle whose tracing-point is in the circumference, the curve traced is a cycloid, and when the point is not in the circumference, the cycloid becomes a trochoid.

Ex. 2. When the rolling curve is the involute of the circle whose radius is $2a$, the traced curve is a parabola whose parameter is $4a$.

Ex. 3. When the rolled curve is the parabola whose parameter is $4a$, the traced curve is a catenary whose parameter is a, and whose vertex is distant a from the straight line.

Ex. 4. When the rolled curve is a logarithmic spiral, the pole traces a straight line which cuts the fixed line at the same angle as the spiral cuts the radius vector.

Ex. 5. When the rolled curve is the hyperbolic spiral, the traced curve is the tractory of the straight line.

Ex. 6. When the rolled curve is the polar catenary

$$\theta = \pm \sqrt{1 \pm \frac{2a}{r}},$$

the traced curve is a circle whose radius is a, and which touches the straight line.

Ex. 7. When the equation of the rolled curve is

$$\theta = \log\left(\sqrt{\frac{r^4}{a^4} - 1} + \frac{r^2}{a^2}\right) - \log\left(\sqrt{\frac{a^4}{r^4} + 1} - \frac{a^2}{r^2}\right),$$

the traced curve is the hyperbola whose equation is

$$y^2 = a^2 + x^2.$$

Second. In the examples of a straight line rolling on a curve, we shall use the letters A, B, and C to denote the three curves treated of in page 22.

Ex. 1. When the curve A is a circle whose radius is a, then the curve B is the involute of that circle, and the curve C is the spiral of Archimedes, $r = a\theta$.

Ex. 2. When the curve A is a catenary whose equation is

$$x = \frac{a}{2}\left(e^{\frac{y}{a}} + e^{-\frac{y}{a}}\right),$$

the curve B is the tractory of the straight line, whose equation is

$$y = a \log \frac{x}{a + \sqrt{a^2 - x^2}} + \sqrt{a^2 - x^2},$$

and C is a straight line at a distance a from the vertex of the catenary.

Ex. 3.   When the curve A is the polar catenary

$$\theta = \pm \sqrt{1 \pm \frac{2a}{r}},$$

the curve B is the tractory of the circle

$$\theta = \cos^{-1}\frac{r}{a} - \sqrt{\frac{a^2}{r^2} - 1},$$

and the curve C is a circle of which the radius is $\frac{a}{2}$.

Third.   Examples of one curve rolling on another, and tracing a straight line.

Ex. 1.   The curve whose equation is

$$\theta = Ar^{-n} + \&c. + Kr^{-2} + Lr^{-1} + M\log r + Nr + \&c. + Zr^r,$$

when rolled on the curve whose equation is

$$y = \frac{n}{n-1}Ax^{1-n} + \&c. + 2Kx^{-1} - L\log x + Mx + \tfrac{1}{2}Nx^2 + \&c. + \frac{n}{n+1}Zx^{n+1},$$

traces the axis of y.

Ex. 2.   The circle whose equation is $r = a\cos\theta$ rolled on the circle whose radius is a traces a diameter of the circle.

Ex. 3.   The curve whose equation is

$$\theta = \sqrt{\frac{2a}{r} - 1} - \text{versin}^{-1}\frac{r}{a},$$

rolled on the circle whose radius is a, traces the tangent to the circle.

Ex. 4.   If the fixed curve be a parabola whose parameter is $4a$, and if we roll on it the spiral of Archimedes $r = a\theta$, the pole will trace the axis of the parabola.

Ex. 5.   If we roll an equal parabola on it, the focus will trace the directrix of the first parabola.

Ex. 6.   If we roll on it the curve $r = \dfrac{a}{4\theta^2}$ the pole will trace the tangent at the vertex of the parabola.

Ex. 7. If we roll the curve whose equation is

$$r = a \cos\left(\frac{a}{b}\theta\right)$$

on the ellipse whose equation is

$$\frac{x^2}{a^2} + \frac{y^2}{b^2} = 1,$$

the pole will trace the axis b.

Ex. 8. If we roll the curve whose equation is

$$r = \frac{a}{2}\left(e^{\frac{a\theta}{b}} - e^{-\frac{a\theta}{b}}\right)$$

on the hyperbola whose equation is

$$\frac{y^2}{b^2} - \frac{x^2}{a^2} = 1,$$

the pole will trace the axis b.

Ex. 9. If we roll the lituus, whose equation is

$$r^2 = \frac{a^2}{3\theta},$$

on the hyperbola whose equation is

$$xy = a^2,$$

the pole will trace the asymptote.

Ex. 10. The cardioid whose equation is

$$r = a\,(1 + \cos\theta),$$

rolled on the cycloid whose equation is

$$y = a\operatorname{versin}^{-1}\frac{x}{a} + \sqrt{2ax - x^2},$$

traces the base of the cycloid.

Ex. 11. The curve whose equation is

$$\theta = \operatorname{versin}^{-1}\frac{r}{a} + 2\sqrt{\frac{2a}{r} - 1},$$

rolled on the cycloid, traces the tangent at the vertex.

Ex. 12. The straight line whose equation is

$$r = a \sec \theta,$$

rolled on a catenary whose parameter is a, traces a line whose distance from the vertex is a.

Ex. 13. The part of the polar catenary whose equation is

$$\theta = \pm \sqrt{1 + \frac{2a}{r}},$$

rolled on the catenary, traces the tangent at the vertex.

Ex. 14. The other part of the polar catenary whose equation is

$$\theta = \pm \sqrt{1 - \frac{2a}{r}},$$

rolled on the catenary, traces a line whose distance from the vertex is equal to $2a$.

Ex. 15. The tractory of the circle whose diameter is a, rolled on the tractory of the straight line whose constant tangent is a, produces the straight line.

Ex. 16. The hyperbolic spiral whose equation is

$$r = \frac{a}{\theta},$$

rolled on the logarithmic curve whose equation is

$$y = a \log \frac{x}{a},$$

traces the axis of y or the asymptote.

Ex. 17. The involute of the circle whose radius is a, rolled on an orthogonal trajectory of the catenary whose equation is

$$y = \frac{x}{2a} \sqrt{x^2 - a^2} + \frac{a}{2} \log \left(\sqrt{\frac{x^2}{a^2} - 1} + \frac{x}{a} \right),$$

traces the axis of y.

Ex. 18. The curve whose equation is

$$\theta = \left(\frac{a}{r} + 1 \right) \sqrt{2\frac{a}{r} + 1},$$

rolled on the witch, whose equation is

$$y = 2a \sqrt{\frac{2a}{x} - 1},$$

traces the asymptote.

Ex. 19. The curve whose equation is

$$r = a \tan \theta,$$

rolled on the curve whose equation is

$$y = \frac{a}{2} \log \left(\frac{x^2}{a^2} - 1 \right),$$

traces the axis of y.

Ex. 20. The curve whose equation is

$$\theta = \frac{2r}{\sqrt{a^2 - r^2}},$$

rolled on the curve whose equation is

$$y = \frac{x^2}{\sqrt{a^2 - x^2}}, \quad \text{or} \quad r = a \tan \theta,$$

traces the axis of y.

Ex. 21. The curve whose equation is

$$r = a \left(\sec \theta - \tan \theta \right),$$

rolled on the curve whose equation is

$$y = a \log \left(\frac{x^2}{a^2} + 1 \right),$$

traces the axis of y.

Fourth. Examples of pairs of rolling curves which have their poles at a fixed distance $= a$.

Ex. 1. $\begin{cases} \text{The straight line whose equation is} \quad \theta = \sec^{-1} \frac{r}{a}. \\ \text{The polar catenary whose equation is} \quad \theta = \pm \sqrt{1 \pm \frac{2a}{r}}. \end{cases}$

Ex. 2. Two equal ellipses or hyperbolas centered at the foci.

Ex. 3. Two equal logarithmic spirals.

Ex. 4. $\begin{cases} \text{Circle whose equation is} \qquad\qquad r = 2a \cos \theta. \\ \text{Curve whose equation is} \qquad\qquad \theta = \sqrt{2 \frac{a}{r} - 1} + \text{versin}^{-1} \frac{r}{a}. \end{cases}$

4—2

Ex. 5. $\begin{cases} \text{Cardioid whose equation is} & r = 2a\,(1 + \cos\theta). \\ \\ \text{Curve whose equation is} & \theta = \sin^{-1}\dfrac{r}{a} + \log\dfrac{r}{\sqrt{a^2 - r^2} + a}\,. \end{cases}$

Ex. 6. $\begin{cases} \text{Conchoid,} & r = a\,(\sec\theta - 1). \\ \\ \text{Curve,} & \theta = \sqrt{1 - \dfrac{a^2}{r^2}} + \sec^{-1}\dfrac{r}{a}\,. \end{cases}$

Ex. 7. $\begin{cases} \text{Spiral of Archimedes,} & r = a\theta. \\ \\ \text{Curve,} & \theta = \dfrac{r}{a} + \log\dfrac{r}{a}\,. \end{cases}$

Ex. 8. $\begin{cases} \text{Hyperbolic spiral,} & r = \dfrac{a}{\theta}\,. \\ \\ \text{Curve,} & r = \dfrac{a}{e^\theta + 1}\,. \end{cases}$

Ex. 9. $\begin{cases} \text{Ellipse whose equation is} & r = a\,\dfrac{1}{2 + \cos\theta}\,. \\ \\ \text{Curve,} & r = a\left(1 + \dfrac{1}{2\,(2 - \theta^4)}\right). \end{cases}$

Ex. 10. $\begin{cases} \text{Involute of circle,} & \theta = \sqrt{\dfrac{r^2}{a^2} - 1}\;\sec^{-1}\dfrac{r}{a}\,. \\ \\ \text{Curve,} & \theta = \sqrt{\dfrac{r^2}{a^2} \pm 2\dfrac{r}{a}} \pm \log\left(\dfrac{r}{a} \pm 1 + \sqrt{\dfrac{r^2}{a^2} \pm 2\dfrac{r}{a}}\right). \end{cases}$

Fifth. Examples of curves rolling on themselves.

Ex. 1. When the curve which rolls on itself is a circle, equation

$$r = a\cos\theta,$$

the traced curve is a cardioid, equation $r = a\,(1 + \cos\theta)$.

Ex. 2. When it is the curve whose equation is

$$r = 2^n a\left(\cos\frac{\theta}{r}\right)^n,$$

the equation of the traced curve is

$$r = 2^{n+1}\,a\left(\cos\frac{\theta}{n+1}\right)^{n+1}.$$

Ex. 3. When it is the involute of the circle, the traced curve is the spiral of Archimedes.

Ex. 4. When it is a parabola, the focus traces the directrix, and the vertex traces the cissoid.

Ex. 5. When it is the hyperbolic spiral, the traced curve is the tractory of the circle.

Ex. 6. When it is the polar catenary, the equation of the traced curve is

$$\theta = \sqrt{\frac{2a}{r} - 1} - \operatorname{versin}^{-1} \frac{r}{a}.$$

Ex. 7. When it is the curve whose equation is

$$\theta = \log \left(\sqrt{\frac{r^4}{a^4} - 1} + \frac{r^2}{a^2} \right) - \log \left(\sqrt{1 + \frac{a^4}{r^4}} - \frac{a^2}{r^2} \right),$$

the equation of the traced curve is $r = a \left(e^\theta - e^{-\theta} \right)$.

This paper commenced with an outline of the nature and history of the problem of rolling curves, and it was shewn that the subject had been discussed previously, by several geometers, amongst whom were De la Hire and Nicolè in the *Mémoires de l'Académie*, Euler, Professor Willis, in his *Principles of Mechanism*, and the Rev. H. Holditch in the *Cambridge Philosophical Transactions*.

None of these authors, however, except the two last, had made any application of their methods; and the principal object of the present communication was to find how far the general equations could be simplified in particular cases, and to apply the results to practice.

Several problems were then worked out, of which some were applicable to the generation of curves, and some to wheelwork; while others were interesting as shewing the relations which exist between different curves; and, finally, a collection of examples was added, as an illustration of the fertility of the methods employed.

[From the *Transactions of the Royal Society of Edinburgh*, Vol. **xx**. Part I.]

III.—*On the Equilibrium of Elastic Solids.*

THERE are few parts of mechanics in which theory has differed more from experiment than in the theory of elastic solids.

Mathematicians, setting out from very plausible assumptions with respect to the constitution of bodies, and the laws of molecular action, came to conclusions which were shewn to be erroneous by the observations of experimental philosophers. The experiments of Œrsted proved to be at variance with the mathematical theories of Navier, Poisson, and Lamé and Clapeyron, and apparently deprived this practically important branch of mechanics of all assistance from mathematics.

The assumption on which these theories were founded may be stated thus:—

Solid bodies are composed of distinct molecules, which are kept at a certain distance from each other by the opposing principles of attraction and heat. When the distance between two molecules is changed, they act on each other with a force whose direction is in the line joining the centres of the molecules, and whose magnitude is equal to the change of distance multiplied into a function of the distance which vanishes when that distance becomes sensible.

The equations of elasticity deduced from this assumption contain only *one* coefficient, which varies with the nature of the substance.

The insufficiency of one coefficient may be proved from the existence of bodies of different degrees of solidity.

No effort is required to retain a liquid in any form, if its volume remain unchanged; but when the form of a solid is changed, a force is called into action which tends to restore its former figure; and this constitutes the difference between elastic solids and fluids. Both tend to recover their *volume*, but fluids do not tend to recover their *shape*.

Now, since there are in nature bodies which are in every intermediate state from perfect solidity to perfect liquidity, these two elastic powers cannot exist in every body in the same proportion, and therefore all theories which assign to them an invariable ratio must be erroneous.

I have therefore substituted for the assumption of Navier the following axioms as the results of experiments.

If three pressures in three rectangular axes be applied at a point in an elastic solid,—

1. *The sum of the three pressures is proportional to the sum of the compressions which they produce.*

2. *The difference between two of the pressures is proportional to the difference of the compressions which they produce.*

The equations deduced from these axioms contain two coefficients, and differ from those of Navier only in not assuming any invariable ratio between the cubical and linear elasticity. They are the same as those obtained by Professor Stokes from his equations of fluid motion, and they agree with all the laws of elasticity which have been deduced from experiments.

In this paper *pressures* are expressed by the number of units of weight to the unit of surface; if in English measure, in pounds to the square inch, or in atmospheres of 15 pounds to the square inch.

Compression is the proportional change of any dimension of the solid caused by pressure, and is expressed by the quotient of the change of dimension divided by the dimension compressed*.

Pressure will be understood to include tension, and compression dilatation; pressure and compression being reckoned positive.

Elasticity is the force which opposes pressure, and the *equations of elasticity* are those which express the relation of pressure to compression†.

Of those who have treated of elastic solids, some have confined themselves to the investigation of the laws of the bending and twisting of rods, without

---

* The laws of pressure and compression may be found in the Memoir of Lamé and Clapeyron. See note A.

† See note B.

considering the relation of the coefficients which occur in these two cases; while others have treated of the general problem of a solid body exposed to any forces.

The investigations of Leibnitz, Bernoulli, Euler, Varignon, Young, La Hire, and Lagrange, are confined to the equilibrium of bent rods; but those of Navier, Poisson, Lamé and Clapeyron, Cauchy, Stokes, and Wertheim, are principally directed to the formation and application of the general equations.

The investigations of Navier are contained in the seventh volume of the *Memoirs of the Institute*, page 373; and in the *Annales de Chimie et de Physique*, 2ᵉ Série, xv. 264, and xxxviii. 435; *L'Application de la Mécanique*, Tom. i.

Those of Poisson in *Mém. de l'Institut*, viii. 429; *Annales de Chimie*, 2ᵉ Série, xxxvi. 334; xxxvii. 337; xxxviii. 338; xlii. *Journal de l'École Polytechnique*, cahier xx., with an abstract in *Annales de Chimie* for 1829.

The memoir of MM. Lamé and Clapeyron is contained in Crelle's *Mathematical Journal*, Vol. vii.; and some observations on elasticity are to be found in Lamé's *Cours de Physique*.

M. Cauchy's investigations are contained in his *Exercices d'Analyse*, Vol. iii. p. 180, published in 1828.

Instead of supposing each pressure proportional to the linear compression which it produces, he supposes it to consist of two parts, one of which is proportional to the linear compression in the direction of the pressure, while the other is proportional to the diminution of volume. As this hypothesis admits two coefficients, it differs from that of this paper only in the values of the coefficients selected. They are denoted by K and k, and $K = \mu - \frac{1}{3} m$, $k = m$.

The theory of Professor Stokes is contained in Vol. viii. Part 3, of the *Cambridge Philosophical Transactions*, and was read April 14, 1845.

He states his general principles thus:—"The capability which solids possess of being put into a state of isochronous vibration, shews that the pressures called into action by small displacements depend on homogeneous functions of those displacements of one dimension. I shall suppose, moreover, according to the general principle of the superposition of small quantities, that the pressures due to different displacements are superimposed, and, consequently, that the pressures are linear functions of the displacements."

Having assumed the proportionality of pressure to compression, he proceeds to define his coefficients.—"Let $-A\delta$ be the pressures corresponding to a uniform linear dilatation δ when the solid is in equilibrium, and suppose that it becomes $mA\delta$, in consequence of the heat developed when the solid is in a state of rapid vibration. Suppose, also, that a displacement of shifting parallel to the plane xy, for which $\delta x = kx$, $\delta y = -ky$, and $\delta z = 0$, calls into action a pressure $-Bk$ on a plane perpendicular to the axis of x, and a pressure Bk on a plane perpendicular to the axis of y; the pressure on these planes being equal and of contrary signs; that on a plane perpendicular to z being zero, and the tangential forces on those planes being zero." The coefficients A and B, thus defined, when expressed as in this paper, are $A = 3\mu$, $B = \dfrac{m}{2}$.

Professor Stokes does not enter into the solution of his equations, but gives their results in some particular cases.

1. A body exposed to a uniform pressure on its whole surface.
2. A rod extended in the direction of its length.
3. A cylinder twisted by a statical couple.

He then points out the method of finding A and B from the last two cases.

While explaining why the equations of motion of the luminiferous ether are the same as those of incompressible elastic solids, he has mentioned the property of *plasticity* or the tendency which a constrained body has to relieve itself from a state of constraint, by its molecules assuming new positions of equilibrium. This property is opposed to linear elasticity; and these two properties exist in all bodies, but in variable ratio.

M. Wertheim, in *Annales de Chimie*, 3ᵉ Série, XXIII., has given the results of some experiments on caoutchouc, from which he finds that $K = k$, or $\mu = \frac{4}{3}m$; and concludes that $k = K$ in all substances. In his equations, μ is therefore made equal to $\frac{4}{3}m$.

The accounts of experimental researches on the values of the coefficients are so numerous that I can mention only a few.

Canton, Perkins, Œrsted, Aimé, Colladon and Sturm, and Regnault, have determined the cubical compressibilities of substances; Coulomb, Duleau, and Giulio, have calculated the linear elasticity from the torsion of wires; and a great many observations have been made on the elongation and bending of beams.

I have found no account of any experiments on the relation between the doubly refracting power communicated to glass and other elastic solids by compression, and the pressure which produces it*; but the phenomena of bent glass seem to prove, that, in homogeneous singly-refracting substances exposed to pressures, the principal axes of pressure coincide with the principal axes of double refraction; and that the difference of pressures in any two axes is proportional to the difference of the velocities of the oppositely polarised rays whose directions are parallel to the third axis. On this principle I have calculated the phenomena seen by polarised light in the cases where the solid is bounded by parallel planes.

In the following pages I have endeavoured to apply a theory identical with that of Stokes to the solution of problems which have been selected on account of the possibility of fulfilling the conditions. I have not attempted to extend the theory to the case of imperfectly elastic bodies, or to the laws of permanent bending and breaking. The solids here considered are supposed not to be compressed beyond the limits of perfect elasticity.

The equations employed in the transformation of co-ordinates may be found in Gregory's *Solid Geometry*.

I have denoted the displacements by δx, δy, δz. They are generally denoted by α, β, γ; but as I had employed these letters to denote the principal axes at any point, and as this had been done throughout the paper, I did not alter a notation which to me appears natural and intelligible.

The laws of elasticity express the relation between the changes of the dimensions of a body and the forces which produce them.

These forces are called Pressures, and their effects Compressions. Pressures are estimated in pounds on the square inch, and compressions in fractions of the dimensions compressed.

Let the position of material points in space be expressed by their co-ordinates x, y, and z, then any change in a system of such points is expressed by giving to these co-ordinates the variations δx, δy, δz, these variations being functions of x, y, z.

* See note C.

The quantities δx, δy, δz, represent the absolute motion of each point in the directions of the three co-ordinates; but as compression depends not on absolute, but on relative displacement, we have to consider only the nine quantities—

$$\frac{d\delta x}{dx}, \quad \frac{d\delta x}{dy}, \quad \frac{d\delta x}{dz},$$

$$\frac{d\delta y}{dx}, \quad \frac{d\delta y}{dy}, \quad \frac{d\delta y}{dz},$$

$$\frac{d\delta z}{dx}, \quad \frac{d\delta z}{dy}, \quad \frac{d\delta z}{dz}.$$

Since the number of these quantities is nine, if nine other independent quantities of the same kind can be found, the one set may be found in terms of the other. The quantities which we shall assume for this purpose are—

1. Three compressions, $\dfrac{\delta a}{a}$, $\dfrac{\delta \beta}{\beta}$, $\dfrac{\delta \gamma}{\gamma}$, in the directions of three principal axes a, β, γ.

2. The nine *direction-cosines* of these axes, with the *six connecting equations*, leaving three independent quantities. (See Gregory's *Solid Geometry*.)

3. The small angles of rotation of this system of axes about the axes of x, y, z.

The cosines of the angles which the axes of x, y, z make with those of a, β, γ are

$$\cos(a0x) = a_1, \quad \cos(\beta 0x) = b_1, \quad \cos(\gamma 0x) = c_1,$$
$$\cos(a0y) = a_2, \quad \cos(\beta 0y) = b_2, \quad \cos(\gamma 0y) = c_2,$$
$$\cos(a0z) = a_3, \quad \cos(\beta 0z) = b_3, \quad \cos(\gamma 0z) = c_3.$$

These *direction-cosines* are connected by the six equations,

$$a_1^2 + b_1^2 + c_1^2 = 1, \qquad a_1 a_2 + b_1 b_2 + c_1 c_2 = 0,$$
$$a_2^2 + b_2^2 + c_2^2 = 1, \qquad a_2 a_3 + b_2 b_3 + c_2 c_3 = 0,$$
$$a_3^2 + b_3^2 + c_3^2 = 1, \qquad a_3 a_1 + b_3 b_1 + c_3 c_1 = 0.$$

The rotation of the system of axes a, β, γ, round the axis of

$$x, \text{ from } y \text{ to } z, = \delta\theta_1,$$
$$y, \text{ from } z \text{ to } x, = \delta\theta_2,$$
$$z, \text{ from } x \text{ to } y, = \delta\theta_3;$$

By resolving the displacements δa, $\delta \beta$, $\delta \gamma$, $\delta \theta_1$, $\delta \theta_2$, $\delta \theta_3$, in the directions of the axes x, y, z, the displacements in these axes are found to be

$$\delta x = a_1 \delta a + b_1 \delta \beta + c_1 \delta \gamma - \delta \theta_2 z + \delta \theta_3 y,$$
$$\delta y = a_2 \delta a + b_2 \delta \beta + c_2 \delta \gamma - \delta \theta_3 x + \delta \theta_1 z,$$
$$\delta z = a_3 \delta a + b_3 \delta \beta + c_3 \delta \gamma - \delta \theta_1 y + \delta \theta_2 x.$$

But

$$\delta a = a \frac{\delta a}{a}, \quad \delta \beta = \beta \frac{\delta \beta}{\beta}, \quad \text{and} \quad \delta \gamma = \gamma \frac{\delta \gamma}{\gamma},$$

and

$$a = a_1 x + a_2 y + a_3 z, \quad \beta = b_1 x + b_2 y + b_3 z, \quad \text{and} \quad \gamma = c_1 x + c_2 y + c_3 z.$$

Substituting these values of δa, $\delta \beta$, and $\delta \gamma$ in the expressions for δx, δy, δz, and differentiating with respect to x, y, and z, in each equation, we obtain the equations

$$\left.\begin{aligned}
\frac{d\delta x}{dx} &= \frac{\delta a}{a} a_1{}^2 + \frac{\delta \beta}{\beta} b_1{}^2 + \frac{\delta \gamma}{\gamma} c_1{}^2 \\[1ex]
\frac{d\delta y}{dy} &= \frac{\delta a}{a} a_2{}^2 + \frac{\delta \beta}{\beta} b_2{}^2 + \frac{\delta \gamma}{\gamma} c_2{}^2 \\[1ex]
\frac{d\delta z}{dz} &= \frac{\delta a}{a} a_3{}^2 + \frac{\delta \beta}{\beta} b_3{}^2 + \frac{\delta \gamma}{\gamma} c_3{}^2
\end{aligned}\right\} \quad \ldots\ldots\ldots\ldots\ldots (1).$$

Equations of compression.

$$\left.\begin{aligned}
\frac{d\delta x}{dy} &= \frac{\delta a}{a} a_1 a_2 + \frac{\delta \beta}{\beta} b_1 b_2 + \frac{\delta \gamma}{\gamma} c_1 c_2 + \delta \theta_3 \\[1ex]
\frac{d\delta x}{dz} &= \frac{\delta a}{a} a_1 a_3 + \frac{\delta \beta}{\beta} b_1 b_3 + \frac{\delta \gamma}{\gamma} c_1 c_3 - \delta \theta_2 \\[1ex]
\frac{d\delta y}{dz} &= \frac{\delta a}{a} a_2 a_3 + \frac{\delta \beta}{\beta} b_2 b_3 + \frac{\delta \gamma}{\gamma} c_2 c_3 + \delta \theta_1 \\[1ex]
\frac{d\delta y}{dx} &= \frac{\delta a}{a} a_2 a_1 + \frac{\delta \beta}{\beta} b_2 b_1 + \frac{\delta \gamma}{\gamma} c_2 c_1 - \delta \theta_3 \\[1ex]
\frac{d\delta z}{dx} &= \frac{\delta a}{a} a_3 a_1 + \frac{\delta \beta}{\beta} b_3 b_1 + \frac{\delta \gamma}{\gamma} c_3 c_1 + \delta \theta_2 \\[1ex]
\frac{d\delta z}{dy} &= \frac{\delta a}{a} a_3 a_2 + \frac{\delta \beta}{\beta} b_3 b_2 + \frac{\delta \gamma}{\gamma} c_3 c_2 - \delta \theta_1
\end{aligned}\right\} \quad \ldots\ldots\ldots\ldots (2).$$

Equations of the equilibrium of an element of the solid.

The forces which may act on a particle of the solid are :—

1.   Three attractions in the direction of the axes, represented by X, Y, Z.

2.   Six pressures on the six faces.

3. Two tangential actions on each face.

Let the six faces of the small parallelopiped be denoted by x_1, y_1, z_1, x_2, y_2, and z_2, then the forces acting on x_1 are:—

1. A normal pressure p_1 acting in the direction of x on the area $dydz$.

2. A tangential force q_3 acting in the direction of y on the same area.

3. A tangential force q_2^1 acting in the direction of z on the same area, and so on for the other five faces, thus:—

Forces which act in the direction of the axes of

		x	y	z
On the face	x_1	$-p_1 dydz$	$-q_3 dydz$	$-q_2^1 dydz$
... ...	x_2	$\left(p_1+\dfrac{dp_1}{dx}dx\right)dydz$	$\left(q_3+\dfrac{dq_3}{dx}dx\right)dydx$	$\left(q_2^1+\dfrac{dq_2^1}{dx}dx\right)dydz$
... ...	y_1	$-q_3^1 dzdx$	$-p_2 dzdx$	$-q_1 dzdx$
... ...	y_2	$\left(q_3^1+\dfrac{dq_3^1}{dy}dy\right)dzdx$	$\left(p_2+\dfrac{dp_2}{dy}dy\right)dzdx$	$\left(q_1+\dfrac{dq_1}{dy}dy\right)dzdx$
... ...	z_1	$-q_2 dxdy$	$-q^1 dxdy$	$-p_3 dxdy$
... ...	z_2	$\left(q_2+\dfrac{dq_2}{dz}dz\right)dxdy$	$\left(q^1+\dfrac{dq^1}{dz}dz\right)dxdy$	$\left(p_3+\dfrac{dp_3}{dz}dz\right)dxdy$
Attractions,		$\rho X dxdydz$	$\rho Y dxdydz$	$\rho Z dxdydz$

Taking the moments of these forces round the axes of the particle, we find

$$q_1^1 = q_1, \quad q_2^1 = q_2, \quad q_3^1 = q_3;$$

and then equating the forces in the directions of the three axes, and dividing by dx, dy, dz, we find the equations of pressures,

$$\left. \begin{aligned} \frac{dp_1}{dx}+\frac{dq_3}{dy}+\frac{dq_2}{dz}+\rho X &= 0 \\[2mm] \frac{dp_2}{dy}+\frac{dq_1}{dz}+\frac{dq_3}{dx}+\rho Y &= 0 \\[2mm] \frac{dp_3}{dz}+\frac{dq_2}{dx}+\frac{dq_1}{dy}+\rho Z &= 0 \end{aligned} \right\} \quad \text{Equations of Pressures.} \quad \ldots\ldots\ldots (3).$$

The resistance which the solid opposes to these pressures is called Elasticity, and is of two kinds, for it opposes either change of *volume* or change of *figure*. These two kinds of elasticity have no necessary connection, for they are possessed in very different ratios by different substances. Thus *jelly* has a cubical elasticity little different from that of water, and a linear elasticity as small as we please; while *cork*, whose cubical elasticity is very small, has a much greater linear elasticity than jelly.

Hooke discovered that the elastic forces are proportional to the changes that excite them, or as he expressed it, "Ut tensio sic vis."

To fix our ideas, let us suppose the compressed body to be a parallelopiped, and let pressures P_1, P_2, P_3 act on its faces in the direction of the axes a, β, γ, which will become the principal axes of compression, and the compressions will be $\dfrac{\delta a}{a}$, $\dfrac{\delta \beta}{\beta}$, $\dfrac{\delta \gamma}{\gamma}$.

The fundamental assumption from which the following equations are deduced is an extension of Hooke's law, and consists of two parts.

I.   The sum of the compressions is proportional to the sum of the pressures.

II.   The difference of the compressions is proportional to the difference of the pressures.

These laws are expressed by the following equations:—

$$\text{I.}\quad (P_1 + P_2 + P_3) = 3\mu\left(\frac{\delta a}{a} + \frac{\delta \beta}{\beta} + \frac{\delta \gamma}{\gamma}\right) \quad\ldots\ldots\ldots\ldots\ldots\ldots (4).$$

$$\text{II.}\quad \left.\begin{aligned}
(P_1 - P_2) &= m\left(\frac{\delta a}{a} - \frac{\delta \beta}{\beta}\right) \\
(P_2 - P_3) &= m\left(\frac{\delta \beta}{\beta} - \frac{\delta \gamma}{\gamma}\right) \\
(P_3 - P_1) &= m\left(\frac{\delta \gamma}{\gamma} - \frac{\delta a}{a}\right)
\end{aligned}\right\} \quad\ldots\ldots\ldots\ldots\ldots\ldots (5).$$

Equations of Elasticity.

The quantity μ is the coefficient of cubical elasticity, and m that of linear elasticity.

By solving these equations, the values of the pressures P_1, P_2, P_3, and the compressions $\frac{\delta a}{a}$, $\frac{\delta \beta}{\beta}$, $\frac{\delta \gamma}{\gamma}$ may be found.

$$
\left.\begin{aligned}
P_1 &= (\mu - \tfrac{1}{3}m)\left(\frac{\delta a}{a} + \frac{\delta \beta}{\beta} + \frac{\delta \gamma}{\gamma}\right) + m\frac{\delta a}{a} \\
P_2 &= (\mu - \tfrac{1}{3}m)\left(\frac{\delta a}{a} + \frac{\delta \beta}{\beta} + \frac{\delta \gamma}{\gamma}\right) + m\frac{\delta \beta}{\beta} \\
P_3 &= (\mu - \tfrac{1}{3}m)\left(\frac{\delta a}{a} + \frac{\delta \beta}{\beta} + \frac{\delta \gamma}{\gamma}\right) + m\frac{\delta \gamma}{\gamma}
\end{aligned}\right\} \quad \ldots\ldots\ldots\ldots\ldots (6).
$$

$$
\left.\begin{aligned}
\frac{\delta a}{a} &= \left(\frac{1}{9\mu} - \frac{1}{3m}\right)(P_1 + P_2 + P_3) + \frac{1}{m}P_1 \\
\frac{\delta \beta}{\beta} &= \left(\frac{1}{9\mu} - \frac{1}{3m}\right)(P_1 + P_2 + P_3) + \frac{1}{m}P_2 \\
\frac{\delta \gamma}{\gamma} &= \left(\frac{1}{9\mu} - \frac{1}{3m}\right)(P_1 + P_2 + P_3) + \frac{1}{m}P_3
\end{aligned}\right\} \quad \ldots\ldots\ldots\ldots\ldots (7).
$$

From these values of the pressures in the axes a, β, γ, may be obtained the equations for the axes x, y, z, by resolutions of pressures and compressions[*].

For
$$p = a^2 P_1 + b^2 P_2 + c^2 P_3$$
and
$$q = aa P_1 + bb P_2 + cc P_3;$$

$$
\left.\begin{aligned}
p_1 &= (\mu - \tfrac{1}{3}m)\left(\frac{d\delta x}{dx} + \frac{d\delta y}{dy} + \frac{d\delta z}{dz}\right) + m\frac{d\delta x}{dx} \\
p_2 &= (\mu - \tfrac{1}{3}m)\left(\frac{d\delta x}{dx} + \frac{d\delta y}{dy} + \frac{d\delta z}{dz}\right) + m\frac{d\delta y}{dy} \\
p_3 &= (\mu - \tfrac{1}{3}m)\left(\frac{d\delta x}{dx} + \frac{d\delta y}{dy} + \frac{d\delta z}{dz}\right) + m\frac{d\delta z}{dz}
\end{aligned}\right\} \quad \ldots\ldots\ldots (8).
$$

$$
\left.\begin{aligned}
q_1 &= \frac{m}{2}\left(\frac{d\delta y}{dz} + \frac{d\delta z}{dy}\right) \\
q_2 &= \frac{m}{2}\left(\frac{d\delta z}{dx} + \frac{d\delta x}{dz}\right) \\
q_3 &= \frac{m}{2}\left(\frac{d\delta x}{dy} + \frac{d\delta y}{dx}\right)
\end{aligned}\right\} \quad \ldots\ldots\ldots\ldots (9).
$$

* See the Memoir of Lamé and Clapeyron, and note A.

$$\frac{d\delta x}{dx} = \left(\frac{1}{9\mu} - \frac{1}{3m}\right)(p_1 + p_2 + p_3) + \frac{1}{m}\,p_1$$

$$\frac{d\delta y}{dy} = \left(\frac{1}{9\mu} - \frac{1}{3m}\right)(p_1 + p_2 + p_3) + \frac{1}{m}\,p_2 \qquad \Bigg\} \quad \dots\dots\dots (10).$$

$$\frac{d\delta z}{dz} = \left(\frac{1}{9\mu} - \frac{1}{3m}\right)(p_1 + p_2 + p_3) + \frac{1}{m}\,p_3$$

$$\frac{d\delta x}{dy} - \delta\theta_3 = \frac{d\delta y}{dx} + \delta\theta_3 = \frac{1}{m}\,q_3$$

$$\frac{d\delta y}{dz} - \delta\theta_1 = \frac{d\delta z}{dy} + \delta\theta_1 = \frac{1}{m}\,q_1 \qquad \Bigg\} \quad \dots\dots\dots\dots(11).$$

$$\frac{d\delta z}{dx} - \delta\theta_2 = \frac{d\delta x}{dz} + \delta\theta_2 = \frac{1}{m}\,q_2$$

By substituting in Equations (3) the values of the forces given in Equations (8) and (9), they become

$$\left(\mu + \frac{1}{6}m\right)\left\{\frac{d}{dx}\left(\frac{d\delta x}{dx} + \frac{d\delta y}{dy} + \frac{d\delta z}{dz}\right)\right\} + \frac{m}{2}\left(\frac{d^2}{dx^2}\,\delta x + \frac{d^2}{dy^2}\,\delta y + \frac{d^2}{dz^2}\,\delta z\right) + \rho X = 0$$

$$\left(\mu + \frac{1}{6}m\right)\left\{\frac{d}{dy}\left(\frac{d\delta x}{dx} + \frac{d\delta y}{dy} + \frac{d\delta z}{dz}\right)\right\} + \frac{m}{2}\left(\frac{d^2}{dx^2}\,\delta x + \frac{d^2}{dy^2}\,\delta y + \frac{d^2}{dz^2}\,\delta z\right) + \rho Y = 0 \quad \Bigg\} \dots (12).$$

$$\left(\mu + \frac{1}{6}m\right)\left\{\frac{d}{dz}\left(\frac{d\delta x}{dx} + \frac{d\delta y}{dy} + \frac{d\delta z}{dz}\right)\right\} + \frac{m}{2}\left(\frac{d^2}{dx^2}\,\delta x + \frac{d^2}{dy^2}\,\delta y + \frac{d^2}{dz^2}\,\delta z\right) + \rho Z = 0$$

These are the general equations of elasticity, and are identical with those of M. Cauchy, in his *Exercices d'Analyse*, Vol. III., p. 180, published in 1828, where k stands for m, and K for $\mu - \dfrac{m}{2}$, and those of Mr Stokes, given in the *Cambridge Philosophical Transactions*, Vol. VIII., part 3, and numbered (30); in his equations $A = 3\mu$, $B = \dfrac{m}{2}$.

If the temperature is variable from one part to another of the elastic solid, the compressions $\dfrac{d\delta x}{dx}$, $\dfrac{d\delta y}{dy}$, $\dfrac{d\delta z}{dz}$, at any point will be diminished by a quantity proportional to the temperature at that point. This principle is applied in Cases X. and XI. Equations (10) then become

$$\left.\begin{aligned}\frac{d\delta x}{dx} &= \left(\frac{1}{9\mu} - \frac{1}{3m}\right)(p_1 + p_2 + p_3) + c_3 v + \frac{1}{m}p_1 \\ \frac{d\delta y}{dy} &= \left(\frac{1}{9\mu} - \frac{1}{3m}\right)(p_1 + p_2 + p_3) + c_3 v + \frac{1}{m}p_2 \\ \frac{d\delta x}{dz} &= \left(\frac{1}{9\mu} - \frac{1}{3m}\right)(p_1 + p_2 + p_3) + c_3 v + \frac{1}{m}p_3\end{aligned}\right\} \dots\dots\dots (13),$$

$c_3 v$ being the linear expansion for the temperature v.

Having found the general equations of the equilibrium of elastic solids, I proceed to work some examples of their application, which afford the means of determining the coefficients μ, m, and ω, and of calculating the stiffness of solid figures. I begin with those cases in which the elastic solid is a hollow cylinder exposed to given forces on the two concentric cylindric surfaces, and the two parallel terminating planes.

In these cases the co-ordinates x, y, z are replaced by the co-ordinates

$x = x$, measured along the axis of the cylinder.

$y = r$, the radius of any point, or the distance from the axis.

$z = r\theta$, the arc of a circle measured from a fixed plane passing through the axis.

$\frac{d\delta x}{dx} = \frac{d\delta x}{dx}$, $p_1 = o$, are the compression and pressure in the direction of the axis at any point.

$\frac{d\delta y}{dy} = \frac{d\delta r}{dr}$, $p_2 = p$, are the compression and pressure in the direction of the radius.

$\frac{d\delta z}{dz} = \frac{d\delta r\theta}{dr\theta} = \frac{\delta r}{r}$, $p_3 = q$, are the compression and pressure in the direction of the tangent.

Equations (9) become, when expressed in terms of these co-ordinates—

$$\left.\begin{aligned}q_1 &= \frac{m}{2}\, r\, \frac{d\delta\theta}{dr} \\ q_2 &= \frac{m}{2}\, r\, \frac{d\delta\theta}{dx} \\ q_3 &= \frac{m}{2} \cdot \frac{d\delta x}{dr}\end{aligned}\right\} \dots\dots\dots\dots\dots\dots\dots\dots (14).$$

The length of the cylinder is b, and the two radii a_1 and a_2 in every case.

CASE I.

The first equation is applicable to the case of a hollow cylinder, of which the outer surface is fixed, while the inner surface is made to turn through a small angle $\delta\theta$, by a couple whose moment is M.

The twisting force M is resisted only by the elasticity of the solid, and therefore the whole resistance, in every concentric cylindric surface, must be equal to M.

The resistance at any point, multiplied into the radius at which it acts, is expressed by

$$rq_1 = \frac{m}{2}\, r^2\, \frac{d\delta\theta}{dr}\,.$$

Therefore for the whole cylindric surface

$$\frac{d\delta\theta}{dr}\, m\pi r^3 b = M.$$

Whence
$$\delta\theta = \frac{M}{2\pi mb}\left(\frac{1}{a_1^2} - \frac{1}{a_2^2}\right),$$

and
$$m = \frac{M}{2\pi b\delta\theta}\left(\frac{1}{a_1^2} - \frac{1}{a_2^2}\right) \dots\dots\dots\dots\dots\dots (16).$$

The optical effect of the pressure of any point is expressed by

$$I = \omega q_1 b = \omega\,.\,\frac{Mb}{2\pi r^2} \dots\dots\dots\dots\dots\dots (15).$$

Therefore, if the solid be viewed by polarized light (transmitted parallel to the axis), the difference of retardation of the oppositely polarized rays at any point in the solid will be inversely proportional to the square of the distance from the axis of the cylinder, and the planes of polarization of these rays will be inclined 45° to the radius at that point.

The general appearance is therefore a system of coloured rings arranged oppositely to the rings in uniaxal crystals, the tints ascending in the scale as they approach the centre, and the distance between the rings decreasing towards the centre. The whole system is crossed by two dark bands inclined 45° to the plane of primitive polarization, when the plane of the analysing plate is perpendicular to that of the first polarizing plate.

A jelly of isinglass poured when hot between two concentric cylinders forms, when cold, a convenient solid for this experiment; and the diameters of the rings may be varied at pleasure by changing the force of torsion applied to the interior cylinder.

By continuing the force of torsion while the jelly is allowed to dry, a hard plate of isinglass is obtained, which still acts in the same way on polarized light, even when the force of torsion is removed.

It seems that this action cannot be accounted for by supposing the interior parts kept in a state of constraint by the exterior parts, as in unannealed and heated glass; for the optical properties of the plate of isinglass are such as would indicate a strain preserving in every part of the plate the direction of the original strain, so that the strain on one part of the plate cannot be maintained by an opposite strain on another part.

Two other uncrystallised substances have the power of retaining the polarizing structure developed by compression. The first is a mixture of wax and resin pressed into a thin plate between two plates of glass, as described by Sir David Brewster, in the *Philosophical Transactions* for 1815 and 1830.

When a compressed plate of this substance is examined with polarized light, it is observed to have no action on light at a perpendicular incidence; but when inclined, it shews the segments of coloured rings. This property does not belong to the plate as a whole, but is possessed by every part of it. It is therefore similar to a plate cut from a uniaxal crystal perpendicular to the axis.

I find that its action on light is like that of a *positive* crystal, while that of a plate of isinglass similarly treated would be *negative*.

The other substance which possesses similar properties is gutta percha. This substance in its ordinary state, when cold, is not transparent even in thin films; but if a thin film be drawn out gradually, it may be extended to more than double its length. It then possesses a powerful double refraction, which it retains so strongly that it has been used for polarizing light*. As one of its refractive indices is nearly the same as that of Canada balsam, while the other is very different, the common surface of the gutta percha and Canada balsam will transmit one set of rays much more readily than the other, so that a film of extended gutta percha placed between two layers of Canada balsam acts like

* By Dr Wright, I believe.

6—2

a plate of nitre treated in the same way. That these films are in a state of constraint may be proved by heating them slightly, when they recover their original dimensions.

As all these permanently compressed substances have passed their limit of perfect elasticity, they do not belong to the class of elastic solids treated of in this paper; and as I cannot explain the method by which an uncrystallised body maintains itself in a state of constraint, I go on to the next case of twisting, which has more practical importance than any other. This is the case of a cylinder fixed at one end, and twisted at the other by a couple whose moment is M.

CASE II.

In this case let $\delta\theta$ be the angle of torsion at any point, then the resistance to torsion in any circular section of the cylinder is equal to the twisting force M.

The resistance at any point in the circular section is given by the second Equation of (14).

$$q_2 = \frac{m}{2}\, r\, \frac{d\delta\theta}{dx}\,.$$

This force acts at the distance r from the axis; therefore its resistance to torsion will be $q_2 r$, and the resistance in a circular annulus will be

$$q_2 r 2\pi r dr = m\pi r^3 \frac{d\delta\theta}{dx}\, dr$$

and the whole resistance for the hollow cylinder will be expressed by

$$M = \frac{m\pi}{4}\, \frac{d\delta\theta}{dx}\, (a_1^4 - a_2^4)\dots\dots\dots\dots\dots\dots\dots(16).$$

$$m = 4M\, \frac{1}{\pi\, \dfrac{\delta\theta}{b}\, (a_1^4 - a_2^4)}\,.$$

$$m = \frac{720}{\pi^2}\, \frac{M}{n}\left(\frac{b}{a_1^4 - a_2^4}\right)\dots\dots\dots\dots\dots\dots\dots(17).$$

In this equation, m is the coefficient of linear elasticity; a_1 and a_2 are the radii of the exterior and interior surfaces of the hollow cylinder in inches; M is the moment of torsion produced by a weight acting on a lever, and is expressed by the product of the number of pounds in the weight into the number of inches in the lever; b is the distance of two points on the cylinder whose angular motion is measured by means of indices, or more accurately by small mirrors attached to the cylinder; n is the difference of the angle of rotation of the two indices in degrees.

This is the most accurate method for the determination of m independently of μ, and it seems to answer best with thick cylinders which cannot be used with the balance of torsion, as the oscillations are too short, and produce a vibration of the whole apparatus.

Case III.

A hollow cylinder exposed to normal pressures only. When the pressures parallel to the axis, radius, and tangent are substituted for p_1, p_2, and p_3, Equations (10) become

$$\frac{d\delta x}{dx} = \left(\frac{1}{9\mu} - \frac{1}{3m}\right)(o + p + q) + \frac{1}{m}o \dots\dots\dots\dots (18).$$

$$\frac{d\delta r}{dr} = \left(\frac{1}{9\mu} - \frac{1}{3m}\right)(o + p + q) + \frac{1}{m}p \dots\dots\dots\dots (19).$$

$$\frac{d\delta(r\theta)}{d(r\theta)} = \frac{\delta r}{r} = \left(\frac{1}{9\mu} - \frac{1}{3m}\right)(o + p + q) + \frac{1}{m}q \dots\dots\dots\dots (20).$$

By multiplying Equation (20) by r, differentiating with respect to r, and comparing this value of $\frac{d\delta r}{dr}$ with that of Equation (19),

$$\frac{p - q}{rm} = \left(\frac{1}{9\mu} - \frac{1}{3m}\right)\left(\frac{do}{dr} + \frac{dp}{dr} + \frac{dq}{dr}\right) - \frac{1}{m}\frac{dq}{dr}.$$

The equation of the equilibrium of an element of the solid is obtained by considering the forces which act on it in the direction of the radius. By equating the forces which press it outwards with those pressing it inwards, we find the equation of the equilibrium of the element,

$$\frac{q - p}{r} = \frac{dp}{dr} \dots\dots\dots\dots\dots\dots\dots\dots (21).$$

By comparing this equation with the last, we find

$$\left(\frac{1}{9\mu}-\frac{1}{3m}\right)\frac{do}{dr}+\left(\frac{1}{9\mu}+\frac{2}{3m}\right)\left(\frac{dp}{dr}+\frac{dq}{dr}\right)=0.$$

Integrating,

$$\left(\frac{1}{9\mu}-\frac{1}{3m}\right)o+\left(\frac{1}{9\mu}+\frac{2}{3m}\right)(p+q)=c_1.$$

Since o, the longitudinal pressure, is supposed constant, we may assume

$$c_2=\frac{c_1-\left(\dfrac{1}{9\mu}-\dfrac{1}{3m}\right)o}{\dfrac{1}{9\mu}+\dfrac{2}{3m}}=(p+q).$$

Therefore $q-p=c_2-2p,$ therefore by (21),

$$\frac{dp}{dr}+\frac{2p}{r}=\frac{c_2}{r},$$

a linear equation, which gives

$$p=c_3\frac{1}{r^2}+\frac{c_2}{2}.$$

The coefficients c_2 and c_3 must be found from the conditions of the surface of the solid. If the pressure on the exterior cylindric surface whose radius is a_1 be denoted by h_1, and that on the interior surface whose radius is a_2 by h_2,

then $p=h_1$ when $r=a_1$

and $p=h_2$ when $r=a_2$

and the general value of p is

$$p=\frac{a_1^2h_1-a_2^2h_2}{a_1^2-a_2^2}-\frac{a_1^2a_2^2}{r^2}\frac{h_1-h_2}{a_1^2-a_2^2}\ \dots\dots\dots\dots\dots(22).$$

$$r\frac{dp}{dr}=q-p=2\frac{a_1^2a_2^2}{r^2}\frac{h_1-h_2}{a_1^2-a_2^2}\ \text{by (21)}.$$

$$q=\frac{a_1^2h_1-a_2^2h}{a_1^2-a_2^2}+\frac{a_1^2a_2^2}{r^2}\frac{h_1-h_2}{a_1^2-a_2^2}\ \dots\dots\dots\dots\dots(23).$$

$$I=b\omega\,(p-q)=-2b\omega\,\frac{a_1^2a_2^2}{r^2}\frac{h_1-h_2}{a_1^2-a_2^2}\dots\dots\dots\dots(24).$$

This last equation gives the optical effect of the pressure at any point. The law of the magnitude of this quantity is the inverse square of the radius, as in

Case I. ; but the direction of the principal axes is different, as in this case they are parallel and perpendicular to the radius. The dark bands seen by polarized light will therefore be parallel and perpendicular to the plane of polarization, instead of being inclined at an angle of 45°, as in Case I.

By substituting in Equations (18) and (20), the values of p and q given in (22) and (23), we find that when $r = a_1$,

$$\left.\begin{aligned}
\frac{\delta x}{x} &= \left(\frac{1}{9\mu}\right)\left(o + 2\,\frac{a_1^2 h_1 - a_2^2 h_2}{a_1^2 - a_2^2}\right) + \frac{2}{3m}\left(o - \frac{a_1^2 h_1 - a_2^2 h_2}{a_1^2 - a_2^2}\right) \\
&= o\left(\frac{1}{9\mu} + \frac{2}{3m}\right) + 2\left(h_1 a_1^2 - h_1 a_2^2\right)\frac{1}{a_1^2 - a_2^2}\left(\frac{1}{9\mu} - \frac{1}{3m}\right)
\end{aligned}\right\}\quad\ldots\ldots\ldots(25).$$

When $r = a_1$,

$$\left.\begin{aligned}
\frac{\delta r}{r} &= \frac{1}{9\mu}\left(o + 2\,\frac{a_1^2 h_1 - a_2^2 h_2}{a_1^2 - a_2^2}\right) + \frac{1}{3m}\left(\frac{a_1^2 h_1 + 3a_2^2 h_1 - 4a_2^2 h_2}{a_1^2 - a_2^2} - o\right) \\
&= o\left(\frac{1}{9\mu} - \frac{1}{3m}\right) + h_1\,\frac{1}{a_1^2 - a_2^2}\left(\frac{2a_1^2}{9\mu} + \frac{a_1^2 + 3a_2^2}{3m}\right) - h_2\,\frac{a_2^2}{a_1^2 - a_2^2}\left(\frac{2}{9\mu} + \frac{4}{3m}\right)
\end{aligned}\right\}\quad\ldots\ldots(26).$$

From these equations it appears that the longitudinal compression of cylindric tubes is proportional to the longitudinal pressure referred to unit of surface when the lateral pressures are constant, so that for a given pressure the compression is inversely as the sectional area of the tube.

These equations may be simplified in the following cases :—

1. When the external and internal pressures are equal, or $h_1 = h_2$.

2. When the external pressure is to the internal pressure as the square of the interior diameter is to that of the exterior diameter, or when $a_1^2 h_1 = a_2^2 h_2$.

3. When the cylinder is solid, or when $a_2 = 0$.

4. When the solid becomes an indefinitely extended plate with a cylindric hole in it, or when a_2 becomes infinite.

5. When pressure is applied only at the plane surfaces of the solid cylinder, and the cylindric surface is prevented from expanding by being inclosed in a strong case, or when $\dfrac{\delta r}{r} = 0$.

6. When pressure is applied to the cylindric surface, and the ends are retained at an invariable distance, or when $\dfrac{\delta x}{x} = 0$.

1. When $h_1 = h_2$, the equations of compression become

$$\left.\begin{aligned}
\frac{\delta x}{x} &= \frac{1}{9\mu}\left(o + 2h_1\right) + \frac{2}{3m}\left(o - h_1\right) \\
&= o\left(\frac{1}{9\mu} + \frac{2}{3m}\right) + 2h_1\left(\frac{1}{9\mu} - \frac{1}{3m}\right) \\
\frac{\delta r}{r} &= \frac{1}{9\mu}\left(o + 2h_1\right) + \frac{1}{3m}\left(h_1 - o\right) \\
&= o\left(\frac{1}{9\mu} - \frac{1}{3m}\right) + h_1\left(\frac{2}{9\mu} + \frac{1}{3m}\right)
\end{aligned}\right\} \quad\dots\dots\dots\dots (27).$$

When $h_1 = h_2 = o$, then

$$\frac{\delta x}{x} = \frac{\delta r}{r} = \frac{h_1}{3\mu}.$$

The compression of a cylindrical vessel exposed on all sides to the same hydrostatic pressure is therefore independent of m, and it may be shewn that the same is true for a vessel of any shape.

2. When $a_1^2 h_1 = a_2^2 h_2$,

$$\left.\begin{aligned}
\frac{\delta x}{x} &= o\left(\frac{1}{9\mu} + \frac{2}{3m}\right) \\
\frac{\delta r}{r} &= \frac{1}{9\mu}\left(o\right) + \frac{1}{3m}\left(3h_1 - o\right) \\
&= o\left(\frac{1}{9\mu} - \frac{1}{3m}\right) + h_1\frac{1}{m}
\end{aligned}\right\} \quad\dots\dots\dots\dots (28).$$

In this case, when $o = 0$, the compressions are independent of μ.

3. In a solid cylinder, $a_2 = 0$,

$$p = q = h_1.$$

The expressions for $\dfrac{\delta x}{x}$ and $\dfrac{\delta r}{r}$ are the same as those in the first case, when $h_1 = h_2$.

When the longitudinal pressure o vanishes,

$$\frac{\delta x}{x} = 2h_1\left(\frac{1}{9\mu} - \frac{1}{3m}\right),$$

$$\frac{\delta r}{r} = h_1\left(\frac{2}{9\mu} - \frac{1}{3m}\right).$$

When the cylinder is pressed on the plane sides only,

$$\frac{\delta x}{x} = o\left(\frac{1}{9\mu} + \frac{2}{3m}\right),$$

$$\frac{\delta r}{r} = o\left(\frac{1}{9\mu} - \frac{1}{3m}\right).$$

4. When the solid is infinite, or when a_1 is infinite,

$$\left.\begin{aligned}
p &= h_1 - \frac{1}{r^2} a_2^2 (h_1 - h_2)\\[6pt]
q &= h_1 + \frac{1}{r^2} a_2^2 (h_1 - h_2)\\[6pt]
I &= \omega (p - q) = -\frac{2\omega}{r^2} a_2^2 (h_1 - h_2)\\[6pt]
\frac{\delta x}{x} &= \frac{1}{9\mu}(o + 2h_1) + \frac{2}{3m}(o - h_1)\\[6pt]
&= o\left(\frac{1}{9\mu} + \frac{2}{3m}\right) + 2h_1\left(\frac{1}{9\mu} - \frac{1}{3m}\right)\\[6pt]
\frac{\delta r}{r} &= \frac{1}{9\mu}(o + 2h_1) + \frac{1}{3m}(h_1 - o)\\[6pt]
&= o\left(\frac{1}{9\mu} - \frac{1}{3m}\right) + h_1\left(\frac{2}{9\mu} + \frac{1}{3m}\right).
\end{aligned}\right\} \quad \dots\dots\dots (29).$$

5. When $\delta r = 0$ in a solid cylinder,

6. When

$$\left.\begin{aligned}
\frac{\delta x}{x} &= \frac{3o}{2m + 3\mu}\\[6pt]
\frac{\delta x}{x} &= 0, \quad \frac{\delta r}{r} = \frac{3h}{m + 6\mu}
\end{aligned}\right\} \quad \dots\dots\dots\dots\dots\dots(30).$$

Since the expression for the effect of a longitudinal strain is

$$\frac{\delta x}{x} = o\left(\frac{1}{9\mu} + \frac{2}{3m}\right),$$

if we make $\qquad E = \frac{9m\mu}{m + 6\mu}, \quad$ then $\quad \frac{\delta x}{x} = o\,\frac{1}{E} \quad \dots\dots\dots\dots (31).$

The quantity E may be deduced from experiment on the extension of wires or rods of the substance, and μ is given in terms of m and E by the equation,

$$\mu = \frac{Em}{9m - 6E} \dots\dots\dots\dots\dots\dots\dots\dots(32),$$

and
$$E = \frac{Pb}{s\delta x} \dots\dots\dots\dots\dots\dots\dots\dots\dots(33),$$

P being the extending force, b the length of the rod, s the sectional area, and δx the elongation, which may be determined by the deflection of a wire, as in the apparatus of S' Gravesande, or by direct measurement.

Case IV.

The only known direct method of finding the compressibility of liquids is that employed by Canton, Œrsted, Perkins, Aimé, &c.

The liquid is confined in a vessel with a narrow neck, then pressure is applied, and the descent of the liquid in the tube is observed, so that the difference between the change of volume of liquid and the change of internal capacity of the vessel may be determined.

Now, since the substance of which the vessel is formed is compressible, a change of the internal capacity is possible. If the pressure be applied only to the contained liquid, it is evident that the vessel will be distended, and the compressibility of the liquid will appear too great. The pressure, therefore, is commonly applied externally and internally at the same time, by means of a hydrostatic pressure produced by water compressed either in a strong vessel or in the depths of the sea.

As it does not necessarily follow, from the equality of the external and internal pressures, that the capacity does not change, the equilibrium of the vessel must be determined theoretically. Œrsted, therefore, obtained from Poisson his solution of the problem, and applied it to the case of a vessel of lead. To find the cubical elasticity of lead, he applied the theory of Poisson to the numerical results of Tredgold. As the compressibility of lead thus found was greater than that of water, Œrsted expected that the apparent compressibility of water in a lead vessel would be *negative*. On making the experiment the apparent compressibility was *greater* in lead than in glass. The quantity found by Tredgold from the extension of rods was that denoted by E, and the value of μ deduced from E alone by the formulæ of Poisson cannot be true, unless $\frac{\mu}{m} = \frac{5}{6}$; and as $\frac{\mu}{m}$ for lead is probably more than 3, the calculated compressibility is much too great.

A similar experiment was made by Professor Forbes, who used a vessel of caoutchouc. As in this case the apparent compressibility vanishes, it appears that the cubical compressibility of caoutchouc is equal to that of water.

Some who reject the mathematical theories as unsatisfactory, have conjectured that if the sides of the vessel be sufficiently thin, the pressure on both sides being equal, the compressibility of the vessel will not affect the result. The following calculations shew that the apparent compressibility of the liquid depends on the compressibility of the vessel, and is independent of the thickness when the pressures are equal.

A hollow sphere, whose external and internal radii are a_1 and a_2, is acted on by external and internal normal pressures h_1 and h_2, it is required to determine the equilibrium of the elastic solid.

The pressures at any point in the solid are :—

    1.   A pressure p in the direction of the radius.

    2.   A pressure q in the perpendicular plane.

These pressures depend on the distance from the centre, which is denoted by r.

The compressions at any point are $\dfrac{d\delta r}{dr}$ in the radial direction, and $\dfrac{\delta r}{r}$ in the tangent plane, the values of these compressions are :—

$$\frac{d\delta r}{dr} = \left(\frac{1}{9\mu} - \frac{1}{3m}\right)(p + 2q) + \frac{1}{m}\,p \dotfill (34).$$

$$\frac{\delta r}{r} = \left(\frac{1}{9\mu} - \frac{1}{3m}\right)(p + 2q) + \frac{1}{m}\,q \dotfill (35).$$

Multiplying the last equation by r, differentiating with respect to r, and equating the result with that of the first equation, we find

$$r\left(\frac{1}{9\mu} - \frac{1}{3m}\right)\left(\frac{dp}{dr} + 2\frac{dq}{dr}\right) + \frac{1}{m}\left(r\frac{dq}{dr} + q - p\right) = 0.$$

Since the forces which act on the particle in the direction of the radius must balance one another, or

$$2qdrd\theta + p\,(rd\theta)^2 = \left(p + \frac{dp}{dr}\,dr\right)(r+dr)^2\theta,$$

therefore
$$q - p = \frac{r}{2}\frac{dp}{dr} \quad\dots\dots\dots\dots\dots\dots\dots\dots\dots (36).$$

Substituting this value of $q-p$ in the preceding equation, and reducing,

therefore
$$\frac{dp}{dr} + 2\frac{dq}{dr} = 0.$$

Integrating,
$$p + 2q = c_1.$$

But
$$q = \frac{r}{2}\frac{dp}{dr} + p,$$

and the equation becomes

$$\frac{dp}{dr} + 3\frac{p}{r} - \frac{c_1}{r} = 0,$$

therefore
$$p = c_2\frac{1}{r^3} + \frac{c_1}{3}.$$

Since $p = h_1$ when $r = a_1$, and $p = h_2$ when $r = a_2$, the value of p at any distance is found to be

$$p = \frac{a_1^3 h_1 - a_2^3 h_2}{a_1^3 - a_2^3} - \frac{a_1^3 a_2^2}{r^3}\frac{h_1 - h_2}{a_1^3 - a_2^3} \quad\dots\dots\dots\dots\dots (37).$$

$$q = \frac{a_1^3 h_1 - a_2^3 h_2}{a_1^3 - a_2^3} + \tfrac{1}{2}\frac{a_1^3 a_2^3}{r^3}\frac{h_1 - h_2}{a_1^3 - a_2^3} \quad\dots\dots\dots\dots\dots(38).$$

$$\frac{\delta V}{V} = 3\frac{\delta r}{r} = \frac{a_1^3 h_1 - a_2^3 h_2}{a_1^3 - a_2^3}\frac{1}{\mu} + \tfrac{3}{2}\frac{a_1^3 a_2^3}{r^3}\frac{h_1 - h_2}{a_1^3 - a_2^3}\frac{1}{m}$$

When $r = a_1$,
$$\left.\begin{aligned}\frac{\delta V}{V} &= \frac{a_1^3 h_1 - a_2^3 h_2}{a_1^3 - a_2^3}\frac{1}{\mu} + \tfrac{3}{2}a_2^3\frac{h_1 - h_2}{a_1^3 - a_2^3}\frac{1}{m}\\[2mm]
&= \frac{h_1}{a_1^3 - a_2^3}\left(\frac{a_1^3}{\mu} + \frac{3a_2^3}{2m}\right) - \frac{h_2 a_2^3}{a_1^3 - a_2^3}\left(\frac{1}{\mu} - \frac{3}{2m}\right)\end{aligned}\right\} \dots\dots\dots(39).$$

When the external and internal pressures are equal

$$h_1 = h_2 = p = q, \text{ and } \frac{\delta V}{V} = \frac{h_1}{\mu} \quad\dots\dots\dots\dots\dots\dots(40),$$

the change of internal capacity depends entirely on the cubical elasticity of the vessel, and not on its thickness or linear elasticity.

When the external and internal pressures are inversely as the cubes of the radii of the surfaces on which they act,

$$
\left.
\begin{aligned}
a_1{}^3 h_1 = a_2{}^3 h_2, \quad p = \frac{a^3}{r^3} h_1, \quad q = -\tfrac{1}{2} \frac{a_1{}^3}{r^3} h_1 \\[2mm]
\frac{\delta V}{V} = -\tfrac{3}{2} \frac{a_1{}^3}{r^3} \frac{h_1}{m} \\[2mm]
\text{when } r = a_1, \quad \frac{\delta V}{V} = -\tfrac{3}{2} \frac{h_1}{m}
\end{aligned}
\right\}
\quad \dots\dots\dots\dots (41).
$$

In this case the change of capacity depends on the linear elasticity alone.

M. Regnault, in his researches on the theory of the steam engine, has given an account of the experiments which he made in order to determine with accuracy the compressibility of mercury.

He considers the mathematical formulæ very uncertain, because the theories of molecular forces from which they are deduced are probably far from the truth; and even were the equations free from error, there would be much uncertainty in the ordinary method by measuring the elongation of a rod of the substance, for it is difficult to ensure that the material of the rod is the same as that of the hollow sphere.

He has, therefore, availed himself of the results of M. Lamé for a hollow sphere in three different cases, in the first of which the pressure acts on the interior and exterior surface at the same time, while in the other two cases the pressure is applied to the exterior or interior surface alone. Equation (39) becomes in these cases,—

1.   When $h_1 = h_2$, $\dfrac{\delta V}{V} = \dfrac{h_1}{\mu}$ and the compressibility of the enclosed liquid being μ_2, and the apparent diminution of volume $\delta' V$,

$$
\frac{\delta' V}{V} = h_1 \left(\frac{1}{\mu_2} - \frac{1}{\mu} \right) \dots\dots\dots\dots\dots\dots\dots (42).
$$

2.   When $h_1 = 0$,

$$
\frac{\delta V}{V} = \frac{\delta' V}{V} = -h_2 \frac{a_2{}^3}{a_1{}^3 - a_2{}^3} \left(\frac{1}{\mu} + \frac{3}{2m} \right) \dots\dots\dots\dots (43).
$$

3. When $h_2 = 0$,

$$\frac{\delta V}{V} = \frac{h_1}{a_1{}^3 - a_2{}^3} \left(\frac{a_1{}^3}{\mu} + \tfrac{3}{2} \frac{a_2{}^3}{V} \right),$$

$$\frac{\delta V}{V} = \frac{h_1}{a_1{}^3 - a_2{}^3} \left\{ \frac{a_1{}^3}{\mu} + \tfrac{3}{2} \frac{a_2{}^3}{m} + (a_2{}^3 - a_1{}^3) \frac{1}{\mu_2} \right\}.$$

M. Lamé's equations differ from these only in assuming that $\mu = \tfrac{5}{6} m$. If this assumption be correct, then the coefficients μ, m, and μ_2, may be found from two of these equations; but since one of these equations may be derived from the other two, the *three* coefficients cannot be found when μ is supposed independent of m. In Equations (39), the quantities which may be varied at pleasure are h_1 and h_2, and the quantities which may be deduced from the apparent compressions are,

$$c_1 = \left(\frac{1}{\mu} + \frac{3}{2m} \right) \text{ and } \left(\frac{1}{\mu} - \frac{1}{\mu_2} \right) = c_2,$$

therefore some independent equation between these quantities must be found, and this cannot be done by means of the sphere alone; some other experiment must be made on the liquid, or on another portion of the substance of which the vessel is made.

The value of μ_2, the elasticity of the liquid, may be previously known.

The linear elasticity m of the vessel may be found by twisting a rod of the material of which it is made;

Or, the value of E may be found by the elongation or bending of the rod, and $\dfrac{1}{E} = \dfrac{1}{9\mu} + \dfrac{2}{3m}$.

We have here five quantities, which may be determined by experiment.

(43) 1. $c_1 = \left(\dfrac{1}{\mu} + \dfrac{3}{2m} \right)$ by external pressure ⎫
(42) 2. $c_2 = \left(\dfrac{1}{\mu} - \dfrac{1}{\mu_2} \right)$ equal pressures. ⎬ on sphere.
⎭

(31) 3. $\dfrac{1}{E} = \left(\dfrac{1}{9m} + \dfrac{2}{3m} \right)$ by elongation of a rod.

(17) 4. m by twisting the rod.

5. μ_2 the elasticity of the liquid.

When the elastic sphere is solid, the internal radius a_2 vanishes, and $h_1 = p = q$, and $\dfrac{\delta V}{V} = \dfrac{h_1}{\mu}$.

When the case becomes that of a spherical cavity in an infinite solid, the external radius a_1 becomes infinite, and

$$\left.\begin{aligned}
p &= h_1 - \frac{a_2^3}{r^3}(h_1 - h_2) \\
q &= h_1 + \tfrac{1}{2}\frac{a_2^3}{r^2}(h_1 - h_2) \\
\frac{\delta r}{r} &= h_1\frac{1}{3\mu} + \tfrac{1}{2}\frac{a_2^3}{r^3}(h_1 - h_2)\frac{1}{m} \\
\frac{\delta V}{V} &= \frac{h_1}{\mu} + \tfrac{3}{2}\frac{h_1 - h_2}{m}
\end{aligned}\right\} \dots\dots\dots\dots (44).$$

The effect of pressure on the surface of a spherical cavity on any other part of an elastic solid is therefore inversely proportional to the cube of its distance from the centre of the cavity.

When one of the surfaces of an elastic hollow sphere has its radius rendered invariable by the support of an incompressible sphere, whose radius is a_1, then

$$\frac{\delta r}{r} = 0, \quad \text{when } r = a_1,$$

therefore
$$\left.\begin{aligned}
p &= h_2\frac{3a_2^3\mu}{2a_1^3 m + 3a_2^3\mu} + h_2\frac{a_1^3 a_2^3}{r^3}\frac{2m}{2a_1^3 m + 3a_2^3\mu} \\
q &= h_2\frac{3a_2^3\mu}{2a_1^3 m + 3a_2^3\mu} - \tfrac{1}{2}h_2\frac{a_1^3 a_2^3}{r^3}\frac{m}{2a_1^3 m + 3a_2^3\mu} \\
\frac{\delta r}{r} &= h_2\frac{a_2^3}{2a_1^3 m + 3a_2^3\mu} - h_2\frac{a_1^3 a_2^3}{r^3}\frac{1}{2a_1^3 m + 3a_2^3\mu}
\end{aligned}\right\} \dots\dots\dots\dots (45).$$

When $r = a_2$, $\dfrac{\delta V}{V} = h_2\dfrac{3a_2^3 - 3a_1^3}{2a_1^3 m + 3a_2^3\mu}$

Case V.

On the equilibrium of an elastic beam of rectangular section uniformly bent.

By supposing the bent beam to be produced till it returns into itself, we may treat it as a hollow cylinder.

Let a rectangular elastic beam, whose length is $2\pi c$, be bent into a circular form, so as to be a section of a hollow cylinder, those parts of the beam which lie towards the centre of the circle will be longitudinally compressed, while the opposite parts will be extended.

The expression for the tangential compression is therefore

$$\frac{\delta r}{r} = \frac{r-c}{c}.$$

Comparing this value of $\dfrac{\delta r}{r}$ with that of Equation (20),

$$\frac{r-c}{r} = \left(\frac{1}{9\mu} - \frac{1}{3m}\right)(o+p+q) + \frac{1}{m}\,q,$$

and by (21),

$$q = p + r\,\frac{dp}{dr}.$$

By substituting for q its value, and dividing by $r\left(\dfrac{1}{9\mu} + \dfrac{2}{3m}\right)$, the equation becomes

$$\frac{dp}{dr} + \frac{2m+3\mu}{m+6\mu}\frac{p}{r} = \frac{9m\mu - (m-3\mu)\,o}{(m+6\mu)\,r} - \frac{9m\mu}{(m+6\mu)}\frac{c}{r^2},$$

a linear differential equation, which gives

$$p = C_1 r^{-\frac{2m+3\mu}{m+6\mu}} - \frac{9m\mu}{m-3\mu}\frac{c}{r} + \frac{9\mu m - (m-3\mu)\,o}{2m+3\mu} \quad\dotfill(46).$$

C_1 may be found by assuming that when $r=a_1$, $p=h_1$, and q may be found from p by equation (21).

As the expressions thus found are long and cumbrous, it is better to use the following approximations :—

$$q = -\left(\frac{9m\mu}{m+6\mu}\right)\frac{y}{c} \quad\dotfill(47).$$

$$p = \left(\frac{9m\mu}{m+6\mu}\right)\frac{1}{2c}\left(\frac{c^2-a^2}{y+c} + c - y\right) \quad\dotfill(48).$$

In these expressions a is half the depth of the beam, and y is the distance of any part of the beam from the neutral surface, which in this case is a cylindric surface, whose radius is c.

These expressions suppose c to be large compared with a, since most substances break when $\dfrac{a}{c}$ exceeds a certain small quantity.

Let b be the breadth of the beam, then the force with which the beam resists flexure $= M$.

$$M = \int byq = \frac{9m\mu}{m+6\mu}\frac{b}{c}\frac{a^3}{3} = E\frac{a^3 b}{3c} \dots\dots\dots\dots\dots\dots(49),$$

which is the ordinary expression for the stiffness of a rectangular beam.

The stiffness of a beam of any section, the form of which is expressed by an equation between x and y, the axis of x being perpendicular to the plane of flexure, or the osculating plane of the axis of the beam at any point, is expressed by

$$Mc = E \int y^2 dx \dots\dots\dots\dots\dots\dots\dots\dots(50),$$

M being the moment of the force which bends the beam, and c the radius of the circle into which it is bent.

CASE VI.

At the meeting of the British Association in 1839, Mr James Nasmyth described his method of making concave specula of silvered glass by bending.

A circular piece of silvered plate-glass was cemented to the opening of an iron vessel, from which the air was afterwards exhausted. The mirror then became concave, and the focal distance depended on the pressure of the air.

Buffon proposed to make burning-mirrors in this way, and to produce the partial vacuum by the combustion of the air in the vessel, which was to be effected by igniting sulphur in the interior of the vessel by means of a burning-glass. Although sulphur evidently would not answer for this purpose, phosphorus might; but the simplest way of removing the air is by means of the air-pump. The mirrors which were actually made by Buffon, were bent by means of a screw acting on the centre of the glass.

To find an expression for the curvature produced in a flat, circular, elastic plate, by the difference of the hydrostatic pressures which act on each side of it,—

Let t be the thickness of the plate, which must be small compared with its diameter.

Let the form of the middle surface of the plate, after the curvature is produced, be expressed by an equation between r, the distance of any point from the axis, or normal to the centre of the plate, and x the distance of the point from the plane in which the middle of the plate originally was, and let

$$ds = \sqrt{(dx)^2 + (dr)^2}.$$

Let h_1 be the pressure on one side of the plate, and h_2 that on the other.

Let p and q be the pressures in the plane of the plate at any point, p acting in the direction of a tangent to the section of the plate by a plane passing through the axis, and q acting in the direction perpendicular to that plane.

By equating the forces which act on any particle in a direction parallel to the axis, we find

$$tp\,\frac{dr}{ds}\frac{dx}{ds}+tr\,\frac{dp}{ds}\frac{dx}{ds}+trp\,\frac{d^2x}{ds^2}+r\,(h_1-h_2)\,\frac{dr}{ds}=0.$$

By making $p=0$ when $r=0$ in this equation, when integrated,

$$p=-\frac{r}{2t}\frac{ds}{dx}(h_1-h_2)\dots\dots\dots\dots\dots\dots\dots(51).$$

The forces perpendicular to the axis are

$$tp\left(\frac{dr}{ds}\right)^2+tr\,\frac{dp}{ds}\frac{dr}{ds}+trp\,\frac{d^2r}{ds^2}-(h_1-h_2)\,r\,\frac{dx}{ds}-qt=0.$$

Substituting for p its value, the equation gives

$$q=-\frac{(h_1-h_2)}{t}\,r\left(\frac{dr}{ds}\frac{dr}{dx}+\frac{dx}{ds}\right)+\frac{(h_1-h_2)}{2t}\,r^2\left(\frac{dr}{dx}\frac{ds}{dx}\frac{d^2x}{ds^2}-\frac{ds}{dx}\frac{d^2r}{ds^2}\right)\dots.(52).$$

The equations of elasticity become

$$\frac{d\delta s}{ds}=\left(\frac{1}{9\mu}-\frac{1}{3m}\right)\left(p+q+\frac{h_1+h_2}{2}\right)+\frac{p}{m},$$

$$\frac{\delta r}{r}=\left(\frac{1}{9\mu}-\frac{1}{3m}\right)\left(p+q+\frac{h_1+h_2}{2}\right)+\frac{q}{m}.$$

Differentiating $\dfrac{d\delta r}{dr}=\dfrac{d}{dr}\left(\dfrac{\delta r}{r}r\right)$, and in this case

$$\frac{d\delta r}{dr}=1-\frac{dr}{ds}+\frac{dr}{ds}\frac{d\delta s}{ds}.$$

By a comparison of these values of $\dfrac{d\delta r}{ds}$,

$$\left(1-\frac{dr}{ds}\right)\left(\frac{1}{9\mu}-\frac{1}{3m}\right)\left(p+q+\frac{h_1+h_2}{2}\right)+\frac{q}{m}+\frac{dr}{ds}\frac{p}{m}+r\left(\frac{1}{9\mu}-\frac{1}{3m}\right)\left(\frac{dp}{dr}+\frac{dq}{dr}\right)$$

$$+\frac{r}{m}\frac{dq}{dr}+\frac{dr}{ds}-1=0.$$

To obtain an expression for the curvature of the plate at the vertex, let a be the radius of curvature, then, as an approximation to the equation of the plate, let

$$x = \frac{r^2}{2a}.$$

By substituting the value of x in the values of p and q, and in the equation of elasticity, the approximate value of a is found to be

$$a = \frac{t}{h_1 - h_2} \frac{(h_1 + h_2)\left(\frac{1}{9\mu} - \frac{1}{3m}\right) - 2}{10\left(\frac{1}{9\mu} - \frac{1}{3m}\right) + \frac{9}{m}},$$

$$a = \frac{t}{h_1 - h_2}\frac{-18m\mu}{10m + 51\mu} + t\frac{h_1 + h_2}{h_1 - h_2}\frac{m - 3\mu}{10m + 51\mu} \quad \dots\dots\dots\dots(53).$$

Since the focal distance of the mirror, or $\frac{a}{2}$, depends on the difference of pressures, a telescope on Mr Nasmyth's principle would act as an aneroid barometer, the focal distance varying inversely as the pressure of the atmosphere.

Case VII.

To find the conditions of torsion of a cylinder composed of a great number of parallel wires bound together without adhering to one another.

Let x be the length of the cylinder, a its radius, r the radius at any point, $\delta\theta$ the angle of torsion, M the force producing torsion, δx the change of length, and P the longitudinal force. Each of the wires becomes a helix whose radius is r, its angular rotation $\delta\theta$, and its length along the axis $x - \delta x$.

Its length is therefore $\sqrt{(r\delta\theta)^2 + x^2\left(1 - \frac{\delta x}{x}\right)^2}$,

and the tension is $= E\left\{1 - \sqrt{\left(1 - \frac{\delta x}{x}\right)^2 + r^2\left(\frac{\delta\theta}{x}\right)^2}\right\}.$

This force, resolved parallel to the axis, is

$$\frac{1}{r}\frac{d}{d\theta}\frac{d}{dr}P = E\left\{\frac{1}{\sqrt{\left(1 - \frac{\delta x}{x}\right)^2 + r^2\left(\frac{\delta\theta}{x}\right)^2}} - 1\right\},$$

and since $\dfrac{\delta x}{x}$ and $r\dfrac{\delta\theta}{x}$ are small, we may assume

$$\frac{d}{d\theta}\frac{d}{dr}P = Er\left\{\frac{\delta x}{x} - \frac{r^2}{2}\left(\frac{\delta\theta}{x}\right)^2\right\},$$

$$P = \pi E\left\{r^2\frac{\delta x}{x} - \frac{r^4}{4}\left(\frac{\delta\theta}{x}\right)^2\right\} \quad\ldots\ldots\ldots\ldots\ldots\ldots (54),$$

The force, when resolved in the tangential direction, is approximately

$$\frac{1}{r}\frac{d}{d\theta}\frac{d}{dr}M = Er\left\{r\frac{\delta\theta}{x}\frac{\delta x}{x} - \frac{r^3}{2}\left(\frac{\delta\theta}{x}\right)^3\right\},$$

$$M = \pi E\left\{\frac{r^4}{2}\frac{\delta\theta}{x}\cdot\frac{\delta x}{x} - \frac{r^6}{6}\left(\frac{\delta\theta}{x}\right)^3\right\} \quad\ldots\ldots\ldots\ldots\ldots\ldots\ldots(55).$$

By eliminating $\dfrac{\delta x}{x}$ between (54) and (55) we have

$$M = \frac{r^2}{2}\frac{\delta\theta}{x}P - E\pi\frac{r^6}{24}\left(\frac{\delta\theta}{x}\right)^3\ldots\ldots\ldots\ldots\ldots\ldots (56).$$

When $P = 0$, M depends on the sixth power of the radius and the cube of the angle of torsion, when the cylinder is composed of separate filaments.

Since the force of torsion for a homogeneous cylinder depends on the fourth power of the radius and the first power of the angle of torsion, the torsion of a wire having a fibrous texture will depend on both these laws.

The parts of the force of torsion which depend on these two laws may be found by experiment, and thus the difference of the elasticities in the direction of the axis and in the perpendicular directions may be determined.

A calculation of the force of torsion, on this supposition, may be found in Young's *Mathematical Principles of Natural Philosophy;* and it is introduced here to account for the variations from the law of Case II., which may be observed in a twisted rod.

CASE VIII.

It is well known that grindstones and fly-wheels are often broken by the centrifugal force produced by their rapid rotation. I have therefore calculated the strains and pressure acting on an elastic cylinder revolving round its axis, and acted on by the centrifugal force alone.

The equation of the equilibrium of a particle [see Equation (21)], becomes

$$q - p = r\frac{dp}{dr} - \frac{4\pi^2 k}{gt^2}r^2 \,;$$

where q and p are the tangential and radial pressures, k is the weight in pounds of a cubic inch of the substance, g is twice the height in inches that a body falls in a second, t is the time of revolution of the cylinder in seconds.

By substituting the value of q and $\frac{dq}{dr}$ in Equations (19), (20), and neglecting o,

$$0 = \left(\frac{1}{9\mu} - \frac{1}{3m}\right)\left(3\frac{dp}{dr} - 2\frac{4\pi^2 k}{gt^2}r + r\frac{d^2p}{dr^2}\right) + \frac{1}{m}\left(3\frac{dp}{dr} - 3\frac{4\pi^2 k}{gt^2}r + r\frac{d^2p}{dr^2}\right)$$

which gives

$$\left.\begin{aligned} p &= c_1\frac{1}{r^2} + \frac{\pi^2 k}{2gt^2}\left(2 + \frac{E}{m}\right)r^2 + c_2 \\ \therefore\ q - p &= -c_1\frac{1}{r^2} + \frac{\pi^2 k}{2gt}\left(-4 + \frac{2E}{m}\right)r^2 \\ q &= -c_1\frac{1}{r^2} + \frac{\pi^2 k}{2gt^2}\left(-2 + \frac{3E}{m_2}\right)r^2 + c_2 \end{aligned}\right\} \quad\cdots\cdots\cdots\cdots\cdots\cdots (57).$$

If the radii of the surfaces of the hollow cylinder be a_1 and a_2, and the pressures acting on them h_1 and h_2, then the values of c_1 and c_2 are

$$\left.\begin{aligned} c_1 &= a_1^2 a_2^2 \frac{\pi^2 k}{2gt^2}\left(2 + \frac{E}{m}\right) - a_1^2 a_2^2\frac{h_1 - h_2}{a_1^2 - a_2^2} \\ c_2 &= \frac{a_1^2 h_1 - a_2^2 h_2}{a_1^2 - a_2^2} - (a_1^2 + a_2^2)\frac{\pi^2 k}{2gt^2}\left(2 + \frac{E}{m}\right) \end{aligned}\right\} \quad\cdots\cdots\cdots\cdots\cdots (58).$$

When $a_2 = 0$, as in the case of a solid cylinder, $c_1 = 0$, and

$$c_2 = h_1 - a_1^2\frac{\pi^2 k}{2gt^2}\left(2 + \frac{E}{m}\right),$$

$$q = h_1 + \frac{\pi^2 k}{2gt^2}\left\{2\left(r^2 + a_1^2\right) + \frac{E}{m}(3r^2 - a_1^2)\right\}\cdots\cdots\cdots\cdots (59).$$

When $h_1 = 0$, and $r = a_1$,

$$q = \frac{\pi^2 k a^2}{gt^2}\left(\frac{E}{m} - 2\right)\cdots\cdots\cdots\cdots\cdots\cdots\cdots (60).$$

When q exceeds the tenacity of the substance in pounds per square inch, the cylinder will give way; and by making q equal to the number of pounds which a square inch of the substance will support, the velocity may be found at which the bursting of the cylinder will take place.

Since $I = b\omega (q-p) = \dfrac{\pi^2 k \omega}{gt} \left(\dfrac{E}{m} - 2 \right) br^2$, a transparent revolving cylinder, when polarized light is transmitted parallel to the axis, will exhibit rings whose diameters are as the square roots of an arithmetical progression, and brushes parallel and perpendicular to the plane of polarization.

Case IX.

A hollow cylinder or tube is surrounded by a medium of a constant temperature while a liquid of a different temperature is made to flow through it. The exterior and interior surfaces are thus kept each at a constant temperature till the transference of heat through the cylinder becomes uniform.

Let v be the temperature at any point, then when this quantity has reached its limit,

$$\frac{r\, dv}{dr} = c_1,$$

$$v = c_1 \log r + c_2 \quad\dotfill (61).$$

Let the temperatures at the surfaces be θ_1 and θ_2, and the radii of the surfaces a_1 and a_2, then

$$c_1 = \frac{\theta_1 - \theta_2}{\log a_1 - \log a_2}, \quad c_2 = \frac{\log a_1 \theta_2 - \log a_2 \theta_1}{\log a_1 - \log a_2}.$$

Let the coefficient of linear dilatation of the substance be c_3, then the proportional dilatation at any point will be expressed by $c_3 v$, and the equations of elasticity (18), (19), (20), become

$$\frac{d\delta x}{dx} = \left(\frac{1}{9\mu} - \frac{1}{3m} \right)(o + p + q) + \frac{o}{m} - c_3 v,$$

$$\frac{d\delta r}{dr} = \left(\frac{1}{9\mu} - \frac{1}{3m} \right)(o + p + q) + \frac{p}{m} - c_3 v,$$

$$\frac{\delta r}{r} = \left(\frac{1}{9\mu} - \frac{1}{3m} \right)(o + p + q) + \frac{q}{m} - c_3 v.$$

The equation of equilibrium is

$$q = p + r \frac{dp}{dr} \quad\dotfill (21),$$

and since the tube is supposed to be of a considerable length

$$\frac{d\delta x}{dx} = c_4 \text{ a constant quantity.}$$

From these equations we find that

$$o = \frac{c_4 + c_3 v - \left(\frac{1}{9\mu} - \frac{1}{3m}\right)\left(2p + r\frac{dp}{dr}\right)}{\frac{1}{9\mu} + \frac{2}{3m}},$$

and hence $v = c_1 \log r + c_2$, p may be found in terms of r.

$$p = \left(\frac{2}{9\mu} + \frac{1}{3m}\right)^{-1} c_1 c_3 \log r + c_5 \frac{1}{r^2} + c_6.$$

Hence $\qquad q = \left(\frac{2}{9\mu} + \frac{1}{3m}\right)^{-1} c_1 c_3 \log r - c_5 \frac{1}{r^2} + c_6 + \left(\frac{2}{9\mu} + \frac{1}{3m}\right) c_1 c_3.$

Since $\qquad I = b\omega\,(q - p) = b\omega\left(\frac{2}{9\mu} + \frac{1}{3m}\right)^{-1} c_1 c_3 - 2b\omega c_5 \frac{1}{r^2},$

the rings seen in this case will differ from those described in Case III. only by the addition of a constant quantity.

When no pressures act on the exterior and interior surfaces of the tube $h_1 = h_2 = 0$, and

$$\left.\begin{aligned}
p &= \left(\frac{2}{9\mu} + \frac{1}{3m}\right)^{-1} c_1 c_3 \left(\log r + \frac{a_1^2 a_2^2}{r^2}\frac{\log a_1 - \log a_2}{a_1^2 - a_2^2} + \frac{a_1^2 \log a_1 - a_2^2 \log a_2}{a_1^2 - a_2^2}\right), \\
q &= \left(\frac{2}{9\mu} + \frac{1}{3m}\right)^{-1} c_1 c_3 \left(\log r - \frac{a_1^2 a_2^2}{r^2}\frac{\log a_1 - \log a_2}{a_1^2 - a_2^2} + \frac{a_1^2 \log a_1 - a_2^2 \log a_2}{a_1^2 - a_2^2} + 1\right), \\
I &= b\left(\frac{2}{9\mu} + \frac{1}{3m}\right)^{-1} c_1 c_3 \omega \left(1 - 2\frac{a_1^2 a_2^2}{r^2}\frac{\log a_1 - \log a_2}{a_1^2 - a_2^2}\right).
\end{aligned}\right\} \dots (62).$$

There will, therefore, be no action on polarized light for the ring whose radius is r when

$$r^2 = 2\frac{a_1^2 a_2^2}{a_1^2 - a_2^2} \log \frac{a_1}{a_2}.$$

CASE X.

Sir David Brewster has observed (*Edinburgh Transactions*, Vol. VIII.), that when a solid cylinder of glass is suddenly heated at the cylindric surface a polarizing force is developed, which is at any point proportional to the square of the distance from the axis of the cylinder; that is to say, that the difference of retardation of the oppositely polarized rays of light is proportional to the square of the radius r, or

$$I = bc_1 \omega r^2 = b\omega \, (q-p) = b\omega r \, \frac{dp}{dr},$$

$$\therefore \; \frac{dp}{dr} = c_1 r, \quad \therefore \; p = \frac{c_1}{2} \, r^2 + c_2.$$

Since if a be the radius of the cylinder, $p = 0$ when $r = a$,

$$p = \frac{c_1}{2} \, (r^2 - a^2).$$

Hence $q = \frac{c_1}{2} \, (3r^2 - a^2).$

By substituting these values of p and q in equations (19) and (20), and making $\dfrac{d}{dr} \dfrac{\delta r}{r} \, r = \dfrac{d\delta r}{dr}$, I find,

$$v = \frac{2c_1}{c_3} \left(\frac{1}{9\mu} + \frac{2}{3m} \right) r^2 + c_4, \quad \dotsc\dotsc\dotsc\dotsc(63).$$

c_4 being the temperature of the axis of the cylinder, and c_3 the coefficient of linear expansion for glass.

CASE XI.

Heat is passing uniformly through the sides of a spherical vessel, such as the ball of a thermometer, it is required to determine the mechanical state of the sphere. As the methods are nearly the same as in Case IX., it will be sufficient to give the results, using the same notation.

$$r^2 \frac{dv}{dr} = c_1, \quad \therefore \; v = c_2 - \frac{c_1}{r},$$

$$c_1 = a_1 a_2 \frac{\theta_1 - \theta_2}{a_1 - a_2}, \quad c_2 = \frac{\theta_1 a_1 - \theta_2 a_2}{a_1 - a_2},$$

$$p = c_4 \frac{1}{r^3} - \left(\frac{2}{9\mu} + \frac{1}{3m} \right)^{-1} c_1 c_3 \frac{1}{r} + c_5.$$

When $h_1 = h_2 = 0$ the expression for p becomes

$$p = \left(\frac{2}{9\mu} + \frac{1}{3m} \right)^{-1} c_3 \, (\theta_1 - \theta_2) \left\{ \frac{a_1^3 a_2^3}{a_1^3 - a_2^3} \frac{1}{r^3} - \frac{a_1 a_2}{a_1 - a_2} \frac{1}{r} + a_1 a_2 \frac{a_1^2 - a_2^2}{(a_1 - a_2) \, (a_1^3 - a_2^3)} \right\} \dotsc\dotsc (64).$$

From this value of p the other quantities may be found, as in Case IX., from the equations of Case IV.

Case XII.

When a long beam is bent into the form of a closed circular ring (as in Case V.), all the pressures act either parallel or perpendicular to the direction of the length of the beam, so that if the beam were divided into planks, there would be no tendency of the planks to slide on one another.

But when the beam does not form a closed circle, the planks into which it may be supposed to be divided will have a tendency to slide on one another, and the amount of sliding is determined by the linear elasticity of the substance. The deflection of the beam thus arises partly from the bending of the whole beam, and partly from the sliding of the planks; and since each of these deflections is small compared with the length of the beam, the total deflection will be the sum of the deflections due to bending and sliding.

Let
$$A = Mc = E \int xy^2 dy \dots\dots\dots\dots\dots\dots\dots\dots (65).$$

A is the stiffness of the beam as found in Case V., the equation of the transverse section being expressed in terms of x and y, y being measured from the neutral surface.

Let a horizontal beam, whose length is $2l$, and whose weight is $2w$, be supported at the extremities and loaded at the middle with a weight W.

Let the deflection at any point be expressed by $\delta_1 y$, and let this quantity be small compared with the length of the beam.

At the middle of the beam, $\delta_1 y$ is found by the usual methods to be

$$\delta_1 y = \frac{1}{A} \left(\tfrac{5}{24} l^3 w + \tfrac{1}{6} l^3 W \right) \dots\dots\dots\dots\dots\dots (66).$$

Let
$$B = \frac{m}{2} \int x dy = \frac{m}{2} \text{ (sectional area)} \dots\dots\dots\dots\dots (67).$$

B is the resistance of the beam to the sliding of the planks. The deflection of the beam arising from this cause is

$$\delta_2 y = \frac{l}{2B} (w + W) \dots\dots\dots\dots\dots\dots\dots (68).$$

This quantity is small compared with $\delta_1 y$, when the depth of the beam is small compared with its length.

The whole deflection $\Delta y = \delta_1 y + \delta_2 y$

$$\Delta y = \frac{l^3}{6A}\left(\tfrac{5}{4}w + W\right) + \frac{l}{2B}\left(w + W\right)$$

$$\Delta y = w\left(\tfrac{5}{24}\frac{l^3}{A} + \tfrac{1}{2}\frac{l}{B}\right) + W\left(\frac{l^3}{6A} + \tfrac{1}{2}\frac{l}{B}\right) \quad\ldots\ldots\ldots\ldots\ldots (69).$$

Case XIII.

When the values of the compressions at any point have been found, when two different sets of forces act on a solid separately, the compressions, when the forces act at the same time, may be found by the composition of compressions, because the small compressions are independent of one another.

It appears from Case I., that if a cylinder be twisted as there described, the compressions will be inversely proportional to the square of the distance from the centre.

If *two* cylindric surfaces, whose axes are perpendicular to the plane of an indefinite elastic plate, be equally twisted in the same direction, the resultant compression in any direction may be found by adding the compression due to each resolved in that direction.

The result of this operation may be thus stated geometrically. Let A_1 and A_2 (Fig. 1) be the centres of the twisted cylinders. Join $A_1 A_2$, and bisect $A_1 A_2$ in O. Draw OBC at right angles, and cut off OB_1 and OB_2 each equal to OA_1.

Then the difference of the retardation of oppositely polarized rays of light passing perpendicularly through any point of the plane varies directly as the product of its distances from B_1 and B_2, and inversely as the square of the product of its distances from A_1 and A_2.

The isochromatic lines are represented in the figure.

The retardation is infinite at the points A_1 and A_2; it vanishes at B_1 and B_2; and if the retardation at O be taken for unity, the isochromatic curves 2, 4, surround A_1 and A_2; that in which the retardation is unity has two loops, and passes through O; the curves $\frac{1}{2}$, $\frac{1}{4}$ are continuous, and have points of contrary flexure; the curve $\frac{1}{8}$ has multiple points at C_1 and C_2, where

$A_1C_1 = A_1A_2$, and two loops surrounding B_1 and B_2; the other curves, for which $I = \frac{1}{16}$, $\frac{1}{32}$, &c., consist each of two ovals surrounding B_1 and B_2, and an exterior portion surrounding all the former curves.

Fig. 1.

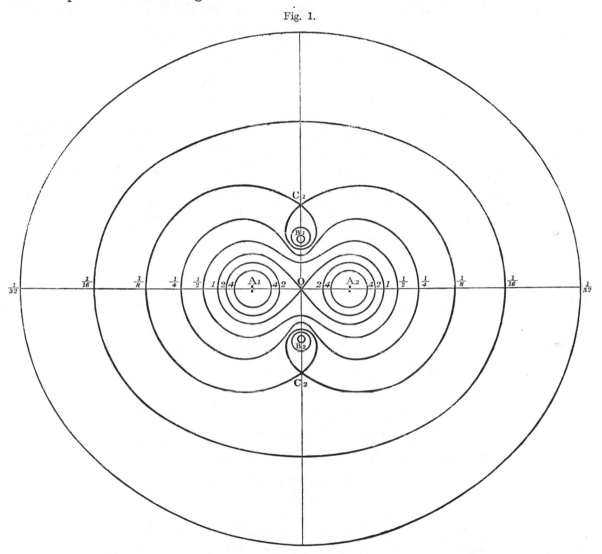

I have produced these curves in the jelly of isinglass described in Case I. They are best seen by using circularly polarized light, as the curves are then seen without interruption, and their resemblance to the calculated curves is more apparent. To avoid crowding the curves toward the centre of the figure, I have taken the values of I for the different curves, not in an arithmetical, but in a geometrical progression, ascending by powers of 2.

CASE XIV.

On the determination of the pressures which act in the interior of transparent solids, from observations of the action of the solid on polarized light.

Sir David Brewster has pointed out the method by which polarized light might be made to indicate the strains in elastic solids; and his experiments on bent glass confirm the theories of the bending of beams.

The phenomena of heated and unannealed glass are of a much more complex nature, and they cannot be predicted and explained without a knowledge of the laws of cooling and solidification, combined with those of elastic equilibrium.

In Case X. I have given an example of the inverse problem, in the case of a cylinder in which the action on light followed a simple law; and I now go on to describe the method of determining the pressures in a general case, applying it to the case of a triangle of unannealed plate-glass.

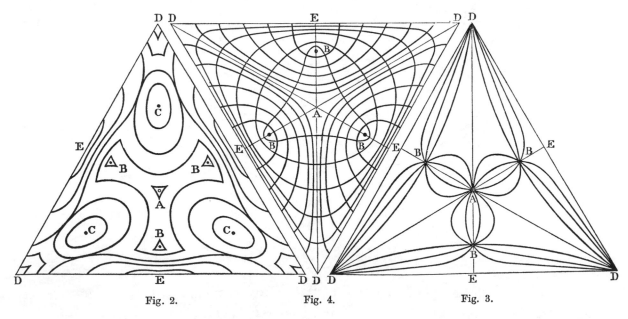

Fig. 2.   Fig. 4.   Fig. 3.

The lines of equal intensity of the action on light are seen without interruption, by using circularly polarized light. They are represented in Fig. 2, where *A, BBB, DDD* are the neutral points, or points of no action on light, and *CCC, EEE* are the points where that action is greatest; and the intensity of the action at any other point is determined by its position with respect to the isochromatic curves.

The direction of the principal axes of pressure at any point is found by transmitting plane polarized light, and analysing it in the plane perpendicular to that of polarization. The light is then restored in every part of the triangle, except in those points at which one of the principal axes is parallel to the plane of polarization. A dark band formed of all these points is seen, which shifts its position as the triangle is turned round in its own plane. Fig. 3 represents these curves for every fifteenth degree of inclination. They correspond to the lines of equal variation of the needle in a magnetic chart.

From these curves others may be found which shall indicate, by their own direction, the direction of the principal axes at any point. These curves of direction of compression and dilatation are represented in Fig. 4; the curves whose direction corresponds to that of *compression* are concave toward the centre of the triangle, and intersect at right angles the curves of dilatation.

Let the isochromatic lines in Fig. 2 be determined by the equation

$$\phi_1(x, y) = I\frac{1}{z} = \omega\,(q - p)\,\frac{1}{z}\,,$$

where I is the difference of retardation of the oppositely polarized rays, and q and p the pressures in the principal axes at any point, z being the thickness of the plate.

Let the lines of equal inclination be determined by the equation

$$\phi_2(x, y) = \tan\theta,$$

θ being the angle of inclination of the principal axes; then the differential equation of the curves of direction of compression and dilatation (Fig. 4) is

$$\phi_2(x, y) = \frac{dy}{dx}.$$

By considering any particle of the plate as a portion of a cylinder whose axis passes through the centre of curvature of the curve of compression, we find

$$q - p = r\,\frac{dp}{dr} \quad\text{...(21)}.$$

Let R denote the radius of curvature of the curve of compression at any point, and let S denote the length of the curve of dilatation at the same point,

$$\phi_3(x, y) = R, \qquad \phi_4(x, y) = S,$$

$$q - p = R\frac{dp}{dS},$$

and since $(q - p)$, R and S are known, and since at the surface, where $\phi_5(x, y) = 0$, $p = 0$, all the data are given for determining the absolute value of p by integration.

Though this is the best method of finding p and q by graphic construction, it is much better, when the equations of the curves have been found, that is, when ϕ_1 and ϕ_2 are known, to resolve the pressures in the direction of the axes.

The new quantities are p_1, p_2, and q_3; and the equations are

$$\tan\theta = \frac{q_3}{p_1 - p_2}, \quad (p - q)^2 = q_3^2 + (p_1 - p_2)^2, \quad p_1 + p_2 = p + q.$$

It is therefore possible to find the pressures from the curves of equal tint and equal inclination, in any case in which it may be required. In the meantime the curves of Figs. 2, 3, 4 shew the correctness of Sir John Herschell's ingenious explanation of the phenomena of heated and unannealed glass.

Note A.

As the mathematical laws of compressions and pressures have been very thoroughly investigated, and as they are demonstrated with great elegance in the very complete and elaborate memoir of MM. Lamé and Clapeyron, I shall state as briefly as possible their results.

Let a solid be subjected to compressions or pressures of any kind, then, if through any point in the solid lines be drawn whose lengths, measured from the given point, are proportional to the compression or pressure at the point resolved in the directions in which the lines are drawn, the extremities of such lines will be in the surface of an ellipsoid, whose centre is the given point.

The properties of the system of compressions or pressures may be deduced from those of the ellipsoid.

There are three diameters having perpendicular ordinates, which are called the *principal axes* of the ellipsoid.

Similarly, there are always three directions in the compressed particle in which there is no tangential action, or tendency of the parts to slide on one another. These directions are called the *principal axes* of *compression* or of *pressure*, and in homogeneous solids they always coincide with each other.

The compression or pressure in any other direction is equal to the sum of the products of the compressions or pressures in the principal axes multiplied into the squares of the cosines of the angles which they respectively make with that direction.

NOTE B.

The fundamental equations of this paper differ from those of Navier, Poisson, &c., only in not assuming an invariable ratio between the linear and the cubical elasticity; but since I have not attempted to deduce them from the laws of molecular action, some other reasons must be given for adopting them.

The experiments from which the laws are deduced are—

1st. Elastic solids put into motion vibrate isochronously, so that the sound does not vary with the amplitude of the vibrations.

2nd. Regnault's experiments on hollow spheres shew that both linear and cubic compressions are proportional to the pressures.

3rd. Experiments on the elongation of rods and tubes immersed in water, prove that the elongation, the decrease of diameter, and the increase of volume, are proportional to the tension.

4th. In Coulomb's balance of torsion, the angles of torsion are proportional to the twisting forces.

It would appear from these experiments, that compressions are always proportional to pressures.

Professor Stokes has expressed this by making one of his coefficients depend on the cubical elasticity, while the other is deduced from the displacement of shifting produced by a given tangential force.

M. Cauchy makes one coefficient depend on the linear compression produced by a force acting in one direction, and the other on the change of volume produced by the same force.

Both of these methods lead to a correct result; but the coefficients of Stokes seem to have more of a real signification than those of Cauchy; I have therefore adopted those of Stokes, using the symbols m and μ, and the fundamental equations (4) and (5), which define them.

Note C.

As the coefficient ω, which determines the optical effect of pressure on a substance, varies from one substance to another, and is probably a function of the linear elasticity, a determination of its value in different substances might lead to some explanation of the action of media on light.

This paper commenced by pointing out the insufficiency of all theories of elastic solids, in which the equations do not contain two independent constants deduced from experiments. One of these constants is common to liquids and solids, and is called the modulus of *cubical* elasticity. The other is peculiar to solids, and is here called the modulus of *linear* elasticity. The equations of Navier, Poisson, and Lamé and Clapeyron, contain only one coefficient; and Professor G. G. Stokes of Cambridge, seems to have formed the first theory of elastic solids which recognised the independence of cubical and linear elasticity, although M. Cauchy seems to have suggested a modification of the old theories, which made the ratio of linear to cubical elasticity the same for all substances. Professor Stokes has deduced the theory of elastic solids from that of the motion of fluids, and his equations are identical with those of this paper, which are deduced from the two following assumptions.

In an element of an elastic solid, acted on by three pressures at right angles to one another, as long as the compressions do not pass the limits of perfect elasticity—

1st. The sum of the pressures, in three rectangular axes, is proportional to the sum of the compressions in those axes.

2nd. The difference of the pressures in two axes at right angles to one another, is proportional to the difference of the compressions in those axes.

Or, in symbols:

$$1. \quad (P_1 + P_2 + P_3) = 3\mu \left(\frac{\delta x}{x} + \frac{\delta y}{y} + \frac{\delta z}{z} \right).$$

$$2. \quad \begin{cases} (P_1 - P_2) = m \left(\dfrac{\delta x}{x} - \dfrac{\delta y}{y} \right), \\[2mm] (P_2 - P_3) = m \left(\dfrac{\delta y}{y} - \dfrac{\delta z}{z} \right), \\[2mm] (P_3 - P_1) = m \left(\dfrac{\delta z}{z} - \dfrac{\delta x}{x} \right), \end{cases}$$

μ being the modulus of *cubical*, and m that of *linear* elasticity.

These equations are found to be very convenient for the solution of problems, some of which were given in the latter part of the paper.

These particular cases were—

That of an elastic hollow cylinder, the exterior surface of which was fixed, while the interior was turned through a small angle. The action of a transparent solid thus twisted on polarized light, was calculated, and the calculation confirmed by experiment.

The second case related to the torsion of cylindric rods, and a method was given by which m may be found. The quantity $E = \dfrac{9mn}{m + 6n}$ was found by elongating, or by bending the rod used to determine m, and μ is found by the equation,

$$\mu = \frac{Em}{9m - 6E}.$$

The effect of pressure on the surfaces of a hollow sphere or cylinder was calculated, and the result applied to the determination of the cubical compressibility of liquids and solids.

An expression was found for the curvature of an elastic plate exposed to pressure on one side; and the state of cylinders acted on by centrifugal force and by heat was determined.

The principle of the superposition of compressions and pressures was applied to the case of a bent beam, and a formula was given to determine E from the deflection of a beam supported at both ends and loaded at the middle.

The paper concluded with a conjecture, that as the quantity ω (which expresses the relation of the inequality of pressure in a solid to the doubly-refracting force produced) is probably a function of m, the determination of these quantities for different substances might lead to a more complete theory of double refraction, and extend our knowledge of the laws of optics.

[Extracted from the *Cambridge and Dublin Mathematical Journal*, Vol. VIII. p. 188, *February*, 1854.]

Solutions of Problems.

1. If from a point in the circumference of a vertical circle two heavy particles be successively projected along the curve, their initial velocities being equal and either in the same or in opposite directions, the subsequent motion will be such that a straight line joining the particles at any instant will touch a circle.

Note. The particles are supposed not to interfere with each other's motion.

THE direct analytical proof would involve the properties of elliptic integrals, but it may be made to depend upon the following geometrical theorems.

(1) If from a point in one of two circles a right line be drawn cutting the other, the rectangle contained by the segments so formed is double of the rectangle contained by a line drawn from the point perpendicular to the *radical axis* of the two circles, and the line joining their centres.

The radical axis is the line joining the points of intersection of the two circles. It is always a real line, whether the points of intersection of the circles be real or imaginary, and it has the geometrical property—that if from any point on the radical axis, straight lines be drawn cutting the circles, the rectangle contained by the segments formed by one of the circles is equal to the rectangle contained by the segments formed by the other.

The analytical proof of these propositions is very simple, and may be resorted to if a geometrical proof does not suggest itself as soon as the requisite figure is constructed.

If A, B be the centres of the circles, P the given point in the circle whose centre is A, a line drawn from P cuts the first circle in p, the second in Q

and q, and the radical axis in R. If PH be drawn perpendicular to the radical axis, then

$$PQ \cdot Pq = 2AB \cdot HP.$$

COR. If the line be drawn from P to *touch* the circle in T, instead of cutting it in Q and q, then the square of the tangent PT is equal to the rectangle $2AB \cdot HP$.

Similarly, if ph be drawn from p perpendicular to the radical axis

$$pT^2 = 2AB \cdot hp.$$

Hence, if a line be drawn touching one circle in T, and cutting the other in P and p, then

$$(PT)^2 : (pT)^2 :: HP : hp.$$

(2) If two straight lines touching one circle and cutting another be made to approach each other indefinitely, the small arcs intercepted by their intersections with the second circle will be ultimately proportional to their distances from the point of contact.

This result may easily be deduced from the properties of the similar triangles $P'TP$ and $p'pT$.

COR. If particles P, p be constrained to move in the circle A, while the line Pp joining them continually touches the circle B, then the velocity of P at any instant is to that of p as PT to pT; and conversely, if the velocity of P at any instant be to that of P as PT to pT, then the line Pp will continue to be a tangent to the circle B.

Now let the plane of the circles be vertical and the radical axis horizontal, and let gravity act on the particles P, p. The particles were projected from the same point with the same velocity. Let this velocity be that due to the depth of the point of projection below the radical axis. Then the square of the velocity at any other point will be proportional to the perpendicular from that point on the radical axis; or, by the corollary to (1), if P and p be at any time at the extremities of the line PTp, the square of the velocity of P will be to the square of the velocity of p as PH to ph, that is, as $(PT)^2$ to $(pT)^2$. Hence, the velocities of P and p are in the proportion of PT to pT, and therefore, by the corollary to (2), the line joining them will continue a tangent to the circle B during each instant, and will therefore remain a tangent during the motion.

10—2

The circle A, the radical axis, and one position of the line Pp, are given by the circumstances of projection of P and p. From these data it is easy to determine the circle B by a geometrical construction.

It is evident that the character of the motion will determine the position of the circle B. If the motion is oscillatory, B will intersect A. If P and p make complete revolutions in the same direction, B will lie entirely within A, but if they move in opposite directions, B will lie entirely above the radical axis.

If any number of such particles be projected from the same point at equal intervals of time with the same direction and velocity, the lines joining successive particles at any instant will be tangents to the same circle; and if the time of a complete revolution, or oscillation, contain n of these intervals, then these lines will form a polygon of n sides, and as this is true at any instant, any number of such polygons may be formed.

Hence, the following geometrical theorem is true:

"If two circles be such that n lines can be drawn touching one of them and having their successive intersections, including that of the last and first, on the circumference of the other, the construction of such a system of lines will be possible, at whatever point of the first circle we draw the first tangent."

---

2. A transparent medium is such that the path of a ray of light within it is a given circle, the index of refraction being a function of the distance from a given point in the plane of the circle.

Find the form of this function and shew that for light of the same refrangibility—

(1) The path of *every ray within the medium* is a circle.

(2) All the rays proceeding from any point in the medium will meet accurately in another point.

(3) If rays diverge from a point without the medium and enter it through a spherical surface having that point for its centre, they will be made to converge accurately to a point within the medium.

LEMMA I. Let a transparent medium be so constituted, that the refractive index is the same at the same distance from a fixed point, then the path of any ray of light within the medium will be in one plane, and the perpendicular from the fixed point on the tangent to the path of the ray at any point will vary inversely as the refractive index of the medium at that point.

We may easily prove that when a ray of light passes through a spherical surface, separating a medium whose refractive index is μ_1 from another where it is μ_2, the plane of incidence and refraction passes through the centre of the sphere, and the perpendiculars on the direction of the ray before and after refraction are in the ratio of μ_2 to μ_1. Since this is true of any number of spherical shells of different refractive powers, it is also true when the index of refraction varies continuously from one shell to another, and therefore the proposition is true.

LEMMA II. If from any fixed point in the plane of a circle, a perpendicular be drawn to the tangent at any point of the circumference, the rectangle contained by this perpendicular and the diameter of the circle is equal to the square of the line joining the point of contact with the fixed point, together with the rectangle contained by the segments of any chord through the fixed point.

Let APB be the circle, O the fixed point; then

$$OY \cdot PR = OP^2 + AO \cdot OB.$$

Produce PO to Q, and join QR, then the triangles OYP, PQR are similar; therefore

$$OY \cdot PR = OP \cdot PQ$$
$$= OP^2 + OP \cdot OQ;$$
$$\therefore OY \cdot PR = OP^2 + AO \cdot OB.$$

If we put in this expression $AO \cdot OB = a^2$,

$$PO = r, \quad OY = p, \quad PR = 2\rho,$$

it becomes
$$2p\rho = r^2 + a^2,$$
$$p = \frac{r^2 + a^2}{2\rho}.$$

To find the law of the index of refraction of the medium, so that a ray from A may describe the circle APB, μ must be made to vary inversely as p by Lemma I.

Let $AO = r_1$, and let the refractive index at $A = \mu_1$; then generally

$$\mu = \frac{C}{p} = \frac{2C\rho}{a^2 + r^2};$$

but at A

$$\mu_1 = \frac{2C\rho}{a^2 + r_1^2},$$

therefore

$$\mu = \mu_1 \frac{a^2 + r_1^2}{a^2 + r^2}.$$

The value of μ at any point is therefore independent of ρ, the radius of the given circle; so that the same law of refractive index will cause any other ray to describe another circle, for which the value of a^2 is the same. The value of OB is $\dfrac{a^2}{r}$, which is also independent of ρ; so that every ray which proceeds from A must pass through B.

Again, if we assume μ_0 as the value of μ when $r = 0$,

$$\mu_0 = \mu_1 \frac{a^2 + r_1^2}{a^2};$$

therefore

$$\mu = \mu_0 \frac{a^2}{a^2 + r^2},$$

a result independent of r_1. This shews that any point A' may be taken as the origin of the ray instead of A, and that the path of the ray will still be circular, and will pass through another point B' on the other side of O, such that $OB' = \dfrac{a^2}{OA'}$.

Next, let CP be a ray from C, a point without the medium, falling at P on a spherical surface whose centre is C.

Let O be the fixed point in the medium as before. Join PO, and produce to Q till $OQ = \dfrac{a^2}{OP}$. Through Q draw a circle touching CP in P, and cutting CO in A and B; then PBQ is the path of the ray within the medium.

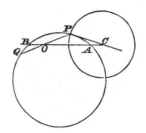

Since CP touches the circle, we have

$$CP^2 = CA \cdot CB,$$
$$= (CO - OA)(CO + OB);$$

but

$$OA = \frac{a^2}{OB};$$

therefore

$$CP^2 = CO^2 + CO\left(OB - \frac{a^2}{OB}\right) - a^2,$$

an equation whence OB may be found, B being the point in the medium through which all rays from C pass.

NOTE. The possibility of the existence of a medium of this kind possessing remarkable optical properties, was suggested by the contemplation of the structure of the crystalline lens in fish; and the method of searching for these properties was deduced by analogy from Newton's *Principia*, Lib. I. Prop. VII.

It would require a more accurate investigation into the law of the refractive index of the different coats of the lens to test its agreement with the supposed medium, which is an optical instrument theoretically perfect for homogeneous light, and might be made achromatic by proper adaptation of the dispersive power of each coat.

On the other hand, we find that the law of the index of refraction which would give a minimum of aberration for a sphere of this kind placed in water, gives results not discordant with facts, so far as they can be readily ascertained.

[From the *Transactions of the Cambridge Philosophical Society*, Vol. IX. Part IV.]

IV. *On the Transformation of Surfaces by Bending.*

EUCLID has given two definitions of a surface, which may be taken as examples of the two methods of investigating their properties.

That in the first book of the Elements is—

" A superficies is that which has only length and breadth."

The superficies differs from a line in having breadth as well as length, and the conception of a third dimension is excluded without being explicitly introduced.

In the eleventh book, where the definition of a solid is first formally given, the definition of the superficies is made to depend on that of the solid—

" That which bounds a solid is a superficies."

Here the conception of three dimensions in space is employed in forming a definition more perfect than that belonging to plane Geometry.

In our analytical treatises on geometry a surface is defined by a function of three independent variables equated to zero. The surface is therefore the boundary between the portion of space in which the value of the function is positive, and that in which it is negative; so that we may now define a surface to be the boundary of any assigned portion of space.

Surfaces are thus considered rather with reference to the figures which they limit than as having any properties belonging to themselves.

But the conception of a surface which we most readily form is that of a portion of matter, extended in length and breadth, but of which the thickness may be neglected. By excluding the thickness altogether, we arrive at Euclid's first definition, which we may state thus—

"A surface is a lamina of which the thickness is diminished so as to become evanescent."

We are thus enabled to consider a surface by itself, without reference to the portion of space of which it is a boundary. By drawing figures on the surface, and investigating their properties, we might construct a system of theorems, which would be true independently of the position of the surface in space, and which might remain the same even when the form of the solid of which it is the boundary is changed.

When the properties of a surface with respect to space are changed, while the relations of lines and figures in the surface itself are unaltered, the surface may be said to preserve its identity, so that we may consider it, after the change has taken place, as the same surface.

When a thin material lamina is made to assume a new form it is said to be *bent*. In certain cases this process of bending is called *development*, and when one surface is bent so as to coincide with another it is said to be *applied* to it.

By considering the lamina as deprived of rigidity, elasticity, and other mechanical properties, and neglecting the thickness, we arrive at a mathematical definition of this kind of transformation.

"The operation of bending is a continuous change of the form of a surface, without extension or contraction of any part of it."

The following investigations were undertaken with the hope of obtaining more definite conceptions of the nature of such transformations by the aid of those geometrical methods which appear most suitable to each particular case. The order of arrangement is that in which the different parts of the subject presented themselves at first for examination, and the methods employed form parts of the original plan, but much assistance in other matters has been derived from the works of Gauss*, Liouville†, Bertrand‡, Puiseux§, &c., references to which will be given in the course of the investigation.

---

* *Disquisitiones générales circa superficies curvas.* Presented to the Royal Society of Gottingen, 8th October, 1827. *Commentationes Recentiores*, Tom. VI.

† Liouville's *Journal*, XII.   ‡ *Ibid.* XIII.   § *Ibid.*

I.

On the Bending of Surfaces generated by the motion of a straight line in space.

If a straight line can be drawn in any surface, we may suppose that part of the surface which is on one side of the straight line to be fixed, while the other part is turned about the straight line as an axis.

In this way the surface may be bent about any number of generating lines as axes successively, till the form of every part of the surface is altered.

The mathematical conditions of this kind of bending may be obtained in the following manner.

Let the equations of the generating line be expressed so that the constants involved in them are functions of one independent variable u, by the variation of which we pass from one position of the line to another.

If in the equations of the generating line Aa, $u = u_1$, then in the equations of the line Bb we may put $u = u_2$, and from the equations of these lines we may find by the common methods the equations of the shortest line PQ between Aa and Bb, and its length, which we may call $\delta\zeta$. We may also find the angle between the directions of Aa and Bb, and let this angle be $\delta\theta$.

In the same way from the equations of Cc, in which $u = u_3$, we may deduce the equations of RS, the shortest line between Bb and Cc, its length $\delta\zeta_2$, and the angle $\delta\theta_2$ between the directions of Bb and Cc. We may also find the value of QR, the distance between the points at which PQ and RS cut Bb. Let $QR = \delta\sigma$, and let the angle between the directions of PQ and RS be $\delta\phi$.

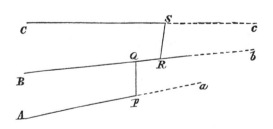

Now suppose the part of the surface between the lines Aa and Bb to be fixed, while the part between Bb and Cc is turned round Bb as an axis. The line RS will then revolve round the point R, remaining perpendicular to Bb, and Cc will still be at the same distance from Bb, and will make the same angle with it. Hence of the four quantities $\delta\zeta_2$, $\delta\theta_2$, $\delta\sigma$ and $\delta\phi$, $\delta\phi$ alone will be changed by the process of bending. $\delta\phi$, however, may be varied in a perfectly arbitrary manner, and may even be made to vanish.

For, PQ and RS being both perpendicular to Bb, RS may be turned about Bb till it is parallel to PQ, in which case $\delta\phi$ becomes $= 0$.

By repeating this process, we may make all the "shortest lines" parallel to one another, and then all the generating lines will be parallel to the same plane.

We have hitherto considered generating lines situated at finite distances from one another; but what we have proved will be equally true when their distances are indefinitely diminished. Then in the limit

$$\frac{\delta\zeta}{u_2 - u_1} \text{ becomes } \frac{d\zeta}{du},$$

$$\frac{\delta\theta}{u_2 - u_1} \quad " \quad \frac{d\theta}{du},$$

$$\frac{\delta\sigma}{u_2 - u_1} \quad " \quad \frac{d\sigma}{du},$$

$$\frac{\delta\phi}{u_2 - u_1} \quad " \quad \frac{d\phi}{du}.$$

All these quantities being functions of u, ζ, θ, σ and ϕ, are functions of u and of each other; and if the forms of these functions be known, the positions of all the generating lines may be successively determined, and the equation to the surface may be found by integrating the equations containing the values of ζ, θ, σ and ϕ.

When the surface is bent in any manner about the generating lines, ζ, θ, and σ remain unaltered, but ϕ is changed at every point.

The form of ϕ as a function of u will depend on the nature of the bending; but since this is perfectly arbitrary, ϕ may be any arbitrary function of u. In this way we may find the form of any surface produced by bending the given surface along its generating lines.

By making $\phi = 0$, we make all the generating lines parallel to the same plane. Let this plane be that of xy, and let the first generating line coincide with the axis of x, then ζ will be the height of any other generating line above the plane of xy, and θ the angle which its projection on that plane makes with the axis of x. The ultimate intersections of the projections of the generating lines on the plane of xy will form a curve, whose length, measured from the axis of x, will be σ.

11—2

Since in this case the quantities ζ, θ, and σ are represented by distinct geometrical quantities, we may simplify the consideration of all surfaces generated by straight lines by reducing them by bending to the case in which those lines are parallel to a given plane.

In the class of surfaces in which the generating lines ultimately intersect, $\frac{d\zeta}{du}=0$, and ζ constant. If these surfaces be bent so that $\phi=0$, the whole of the generating lines will lie in one plane, and their ultimate intersections will form a plane curve. The surface is thus reduced to one plane, and therefore belongs to the class usually described as "developable surfaces." The form of a developable surface may be defined by means of the three quantities θ, σ and ϕ. The generating lines form by their ultimate intersections a curve of double curvature to which they are all tangents. This curve has been called the cuspidal edge. The length of this curve is represented by σ, its absolute curvature at any point by $\frac{d\theta}{d\sigma}$, and its torsion at the same point by $\frac{d\phi}{d\sigma}$.

When the surface is developed, the cuspidal edge becomes a plane curve, and every part of the surface coincides with the plane. But it does not follow that every part of the plane is capable of being bent into the original form of the surface. This may be easily seen by considering the surface when the position of the cuspidal edge nearly coincides with the plane curve but is not confounded with it. It is evident that if from any point in space a tangent can be drawn to the cuspidal edge, a sheet of the surface passes through that point. Hence the number of sheets which pass through one point is the same as the number of tangents to the cuspidal edge which pass through that point ; and since the same is true in the limit, the number of sheets which coincide at any point of the plane is the same as the number of tangents which can be drawn from that point to the plane curve. In constructing a developable surface of paper, we must remove those parts of the sheet from which no real tangents can be drawn, and provide additional sheets where more than one tangent can be drawn.

In the case of developable surfaces we see the importance of attending to the position of the lines of bending; for though all developable surfaces may be produced from the same plane surface, their distinguishing properties depend on the form of the plane curve which determines the lines of bending.

II.

On the Bending of Surfaces of Revolution.

In the cases previously considered, the bending in one part of the surface may take place independently of that in any other part. In the case now before us the bending must be simultaneous over the whole surface, and its nature must be investigated by a different method.

The position of any point P on a surface of revolution may be determined by the distance PV from the vertex, measured along a generating line, and the angle AVO which the plane of the generating line makes with a fixed plane through the axis. Let $PV = s$ and $AVO = \theta$. Let r be the distance (Pp) of P from the axis; r will be a function of s depending on the form of the generating curve.

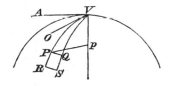

Now consider the small rectangular element of the surface at P. Its length $PR = \delta s$, and its breadth $PQ = r\delta\theta$, where r is a function of s.

If in another surface of revolution r' is some other function of s, then the length and breadth of the new element will be δs and $r'\delta\theta'$, and if

$$r' = \mu r, \quad \text{and} \quad \theta' = \frac{1}{\mu}\theta,$$

$$r'\delta\theta' = r\delta\theta,$$

and the dimensions of the two elements will be the same.

Hence the one element may be applied to the other, and the one surface may be applied to the other surface, element to element, by bending it. To effect this, the surface must be divided by cutting it along one of the generating lines, and the parts opened out, or made to overlap, according as μ is greater or less than unity.

To find the effect of this transformation on the form of the surface we must find the equation to the original form of the generating line in terms of s and r, then putting $r' = \mu r$, the equation between s and r' will give the form of the generating line after bending.

When μ is greater than 1 it may happen that for some values of s, $\dfrac{dr}{ds}$ is greater than $\dfrac{1}{\mu}$. In this case

$$\frac{dr'}{ds} = \mu \frac{dr}{ds} \text{ is greater than } 1 ;$$

a result which indicates that the curve becomes impossible for such values of s and μ.

The transformation is therefore impossible for the corresponding part of the surface. If, however, that portion of the original surface be removed, the remainder may be subjected to the required transformation.

The theory of bending when applied to the case of surfaces of revolution presents no geometrical difficulty, and little variety; but when we pass to the consideration of surfaces of a more general kind, we discover the insufficiency of the methods hitherto employed, by the vagueness of our ideas with respect to the nature of bending in such cases. In the former case the bending is of one kind only, and depends on the variation of one variable; but the surfaces we have now to consider may be bent in an infinite variety of ways, depending on the variation of three variables, of which we do not yet know the nature or interdependence.

We have therefore to discover some method sufficiently general to be applicable to every possible case, and yet so definite as to limit each particular case to one kind of bending easily understood.

The method adopted in the following investigations is deduced from the consideration of the surface as the limit of the inscribed polyhedron, when the size of the sides is indefinitely diminished, and their number indefinitely increased.

A method is then described by which such a polyhedron may be inscribed in any surface so that all the sides shall be triangles, and all the solid angles composed of six plane angles.

The problem of the bending of such a polyhedron is a question of trigonometry, and equations might be found connecting the angles of the different edges which meet in each solid angle of the polyhedron. It will be shewn that the conditions thus obtained would be equivalent to three equations between the six angles of the edges belonging to each solid angle. Hence three additional conditions would be necessary to determine the value of every such angle, and the problem would remain as indefinite as before. But if by any means we can reduce the number of edges meeting in a point to four, only one condition would be necessary to determine them all, and the problem would be reduced to the consideration of one kind of bending only.

This may be done by drawing the polyhedron in such a manner that the planes of adjacent triangles coincide two and two, and form quadrilateral facets, four of which meet in every solid angle. The bending of such a polyhedron can take place only in one way, by the increase of the angles of two of the edges which meet in a point, and the diminution of the angles of the other two.

The condition of such a polyhedron being inscribed in any surface is then found, and it is shewn that when two forms of the same surface are given, a perfectly definite rule may be given by which two corresponding polyhedrons of this kind may be inscribed, one in each surface.

Since the *kind* of bending completely defines the nature of the quadrilateral polyhedron which must be described, the lines formed by the edges of the quadrilateral may be taken as an indication of the kind of bending performed on the surface.

These lines are therefore defined as "*Lines of Bending.*"

When the lines of bending are given, the forms of the quadrilateral facets are completely determined; and if we know the angle which any two adjacent facets make with one another, we may determine the angles of the three edges which meet it at one of its extremities. From each of these we may find the angles of three other edges, and so on, so that the form of the polyhedron after bending will be completely determined when the angle of one edge is given. The bending is thus made to depend on the change of one variable only.

In this way the angle of any edge may be calculated from that of any given edge; but since this may be done in two different ways, by passing along two different sets of edges, we must have the condition that these results may be consistent with each other. This condition is satisfied by the method of inscribing the polyhedron. Another condition will be necessary that the *change* of the angle of any edge due to a small change of the given angle, produced by bending, may be the same by both calculations. This is the condition of "Instantaneous Lines of Bending." That this condition may continue to be satisfied during the whole process we must have another, which is the condition for "Permanent Lines of Bending."

The use of these lines of bending in simplifying the theory of surfaces is the only part of the present method which is new, although the investigations connected with them naturally led to the employment of other methods which had been used by those who have already treated of this subject. A statement of the principal methods and results of these mathematicians will save repetition, and will indicate the different points of view under which the subject may present itself.

The first and most complete memoir on the subject is that of M. Gauss, already referred to.

The method which he employs consists in referring every point of the surface to a corresponding point of a sphere whose radius is unity. Normals are drawn at the several points of the surface toward the same side of it, then lines drawn through the centre of the sphere in the direction of each of these normals intersect the surface of the sphere in points corresponding to those points of the original surface at which the normals were drawn.

If any line be drawn on the surface, each of its points will have a corresponding point on the sphere, so that there will be a corresponding line on the sphere.

If the line on the surface return into itself, so as to enclose a finite area of the surface, the corresponding curve on the sphere will enclose an area on the sphere, the extent of which will depend on the form of the surface.

This area on the sphere has been defined by M. Gauss as the measure of the "entire curvature" of the area on the surface. This mathematical quantity is of great use in the theory of surfaces, for it is the only quantity connected with curvature which is capable of being expressed as the sum of all its parts.

The sum of the entire curvatures of any number of areas is the entire curvature of their sum, and the entire curvature of any area depends on the form of its boundary only, and is not altered by any change in the form of the surface within the boundary line.

The curvature of the surface may even be discontinuous, so that we may speak of the entire curvature of a portion of a polyhedron, and calculate its amount.

If the dimensions of the closed curve be diminished so that it may be treated as an element of the surface, the ultimate ratio of the entire curvature to the area of the element on the surface is taken as the measure of the "specific curvature" at that point of the surface.

The terms "entire" and "specific" curvature when used in this paper are adopted from M. Gauss, although the use of the sphere and the areas on its surface formed an essential part of the original design. The use of these terms will save much explanation, and supersede several very cumbrous expressions.

M. Gauss then proceeds to find several analytical expressions for the measure of specific curvature at any point of a surface, by the consideration of three points very near each other.

The co-ordinates adopted are first rectangular, x and y, or x, y and z, being regarded as independent variables.

Then the points on the surface are referred to two systems of curves drawn on the surface, and their position is defined by the values of two independent variables p and q, such that by varying p while q remains constant, we obtain the different points of a line of the first system, while p constant and q variable defines a line of the second system.

By means of these variables, points on the surface may be referred to lines on the surface itself instead of arbitrary co-ordinates, and the measure of curvature may be found in terms of p and q when the surface is known.

In this way it is shewn that the specific curvature at any point is the reciprocal of the product of the principal radii of curvature at that point, a result of great interest.

From the condition of bending, that the length of any element of the curve must not be altered, it is shewn that the specific curvature at any point is not altered by bending.

The rest of the memoir is occupied with the consideration of particular modes of describing the two systems of lines. One case is when the lines of the first system are geodesic, or "shortest" lines having their origin in a point, and the second system is drawn so as to cut off equal lengths from the curves of the first system.

The angle which the tangent at the origin of a line of the first system makes with a fixed line is taken as one of the co-ordinates, and the distance of the point measured along that line as the other.

It is shewn that the two systems intersect at right angles, and a simple expression is found for the specific curvature at any point.

M. Liouville (*Journal*, Tom. XII.) has adopted a different mode of simplifying the problem. He has shewn that on every surface it is possible to find two systems of curves intersecting at right angles, such that the length and breadth of every element into which the surface is thus divided shall be equal, and that an infinite number of such systems may be found. By means of these curves he has found a much simpler expression for the specific curvature than that given by M. Gauss.

He has also given, in a note to his edition of Monge, a method of testing two given surfaces in order to determine whether they are applicable to one another. He first draws on both surfaces lines of equal specific curvature, and determines the distance between two corresponding consecutive lines of curvature in both surfaces.

If by assuming the origin properly these distances can be made equal for every part of the surface, the two surfaces can be applied to each other. He has developed the theorem analytically, of which this is only the geometrical interpretation.

When the lines of equal specific curvature are equidistant throughout their whole length, as in the case of surfaces of revolution, the surfaces may be applied to one another in an infinite variety of ways.

When the specific curvature at every point of the surface is positive and equal to a^2, the surface may be applied to a sphere of radius a, and when the specific curvature is negative $= -a^2$ it may be applied to the surface of revolution which cuts at right angles all the spheres of radius a, and whose centres are in a straight line.

M. Bertrand has given in the XIIIth Vol. of Liouville's *Journal* a very simple and elegant proof of the theorem of M. Gauss about the product of the radii of curvature.

He supposes one extremity of an inextensible thread to be fixed at a point in a surface, and a closed curve to be described on the surface by the other extremity, the thread being stretched all the while. It is evident that the length of such a curve cannot be altered by bending the surface. He then calculates the length of this curve, considering the length of the thread small, and finds that it depends on the product of the principal radii of curvature of the surface at the fixed point. His memoir is followed by a note of M. Diguet, who deduces the same result from a consideration of the area of the same curve; and by an independent memoir of M. Puiseux, who seems to give the same proof at somewhat greater length.

NOTE. Since this paper was written, I have seen the Rev. Professor Jellett's Memoir, *On the Properties of Inextensible Surfaces*. It is to be found in the *Transactions of the Royal Irish Academy*, Vol. XXII. *Science*, &c., and was read May 23, 1853.

Professor Jellett has obtained a system of three partial differential equations which express the conditions to which the displacements of a continuous inextensible membrane are subject. From these he has deduced the two theorems of Gauss, relating to the invariability of the product of the radii of curvature at any point, and of the " entire curvature" of a finite portion of the surface.

He has then applied his method to the consideration of cases in which the flexibility of the surface is limited by certain conditions, and he has obtained the following results :—

If the displacements of an inextensible surface be all parallel to the same plane, the surface moves as a rigid body.

Or, more generally,

If the movement of an inextensible surface, parallel to any one line, be that of a rigid body, the entire movement is that of a rigid body.

The following theorems relate to the case in which a curve traced on the surface is rendered rigid :—

If any curve be traced upon an inextensible surface whose principal radii of curvature are finite and of the same sign, and if this curve be rendered immoveable, the entire surface will become immoveable also.

In a developable surface composed of an inextensible membrane, any one of its rectilinear sections may be fixed without destroying the flexibility of the membrane.

In convexo-concave surfaces, there are two directions passing through every point of the surface, such that the curvature of a normal section taken in these directions vanishes. We may therefore conceive the entire surface to be crossed by two series of curves, such that a tangent drawn to either of them at any point shall coincide with one of these directions. These curves Professor Jellett has denominated *Curves of Flexure*, from the following properties :—

Any curve of flexure may be fixed without destroying the flexibility of the surface.

If an arc of a curve traced upon an inextensible surface be rendered fixed or rigid, the entire of the quadrilateral, formed by drawing the two curves of flexure through each extremity of the curve, becomes fixed or rigid also.

Professor Jellett has also investigated the properties of partially inextensible surfaces, and of thin material laminæ whose extensibility is small, and in a note he has demonstrated the following theorem :—

If a closed oval surface be perfectly inextensible, it is also perfectly rigid.

A demonstration of one of Professor Jellett's theorems will be found at the end of this paper.

J. C. M.

Aug. 30, 1854.

12—2

ON THE PROPERTIES OF A SURFACE CONSIDERED AS THE LIMIT OF THE INSCRIBED
POLYHEDRON.

1. *To inscribe a polyhedron in a given surface, all whose sides shall be
triangles, and all whose solid angles shall be hexahedral.*

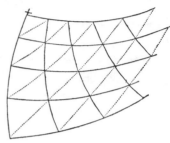

On the given surface describe a series of curves
according to any assumed law. Describe a second series
intersecting these in any manner, so as to divide the
whole surface into quadrilaterals. Lastly, describe a
third series (the dotted lines in the figure), so as to
pass through all the intersections of the first and second
series, forming the diagonals of the quadrilaterals.

The surface is now covered with a network of curvilinear triangles. The
plane triangles which have the same angular points will form a polyhedron
fulfilling the required conditions. By increasing the number of the curves in
each series, and diminishing their distance, we may make the polyhedron
approximate to the surface without limit. At the same time the polygons
formed by the edges of the polyhedron will approximate to the three systems
of intersecting curves.

2. *To find the measure of the "entire curvature" of a solid angle of the
polyhedron, and of a finite portion of its surface.*

From the centre of a sphere whose radius is unity draw perpendiculars to
the planes of the six sides forming the solid angle. These lines will meet the
surface in six points on the same side of the centre, which being joined by
arcs of great circles will form a hexagon on the surface of the sphere.

The area of this hexagon represents the entire curvature of the solid angle.

It is plain by spherical geometry that the angles of this hexagon are the
supplements of the six plane angles which form the solid angle, and that the
arcs forming the sides are the supplements of those subtended by the angles
of the six edges formed by adjacent sides.

The area of the hexagon is equal to the excess of the sum of its angles
above eight right angles, or to the defect of the sum of the six plane angles
from four right angles, which is the same thing. Since these angles are

invariable, the bending of the polyhedron cannot alter the measure of curvature of each of its solid angles.

If perpendiculars be drawn to the sides of the polyhedron which contain other solid angles, additional points on the sphere will be found, and if these be joined by arcs of great circles, a network of hexagons will be formed on the sphere, each of which corresponds to a solid angle of the polyhedron and represents its " entire curvature."

The entire curvature of any assigned portion of the polyhedron is the sum of the entire curvatures of the solid angles it contains. It is therefore represented by a polygon on the sphere, which is composed of all the hexagons corresponding to its solid angles.

If a polygon composed of the edges of the polyhedron be taken as the boundary of the assigned portion, the sum of its exterior angles will be the same as the sum of the exterior angles of the polygon on the sphere; but the area of a spherical polygon is equal to the defect of the sum of its exterior angles from four right angles, and this is the measure of entire curvature.

Therefore the entire curvature of the portion of the polyhedron enclosed by the polygon is equal to the defect of the sum of its exterior angles from four right angles.

Since the entire curvature of each solid angle is unaltered by bending, that of a finite portion of the surface must be also invariable.

3. *On the " Conic of Contact," and its use in determining the curvature of normal sections of a surface.*

Suppose the plane of one of the triangular facets of the polyhedron to be produced till it cuts the surface. The form of the curve of intersection will depend on the nature of the surface, and when the size of the triangle is indefinitely diminished, it will approximate to the form of a conic section.

For we may suppose a surface of the second order constructed so as to have a contact of the second order with the given surface at a point within the angular points of the triangle. The curve of intersection with this surface will be the conic section to which the other curve of intersection approaches. This curve will be henceforth called the " Conic of Contact," for want of a better name.

To find the radius of curvature of a normal section of the surface.

Let ARa be the conic of contact, C its centre, and CP perpendicular to its plane. rPR a normal section, and O its centre of curvature, then

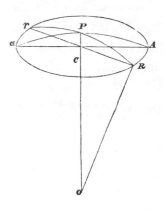

$$PO = \tfrac{1}{2}\frac{PR^2}{CP}$$

$$= \tfrac{1}{2}\frac{CR^2}{CP} \text{ in the limit, when } CR \text{ and } PR \text{ coincide,}$$

$$= \tfrac{1}{8}\frac{rR^2}{CP},$$

or calling CP the "sagitta," we have this theorem :

"The radius of curvature of a normal section is equal to the square of the corresponding diameter of the conic of contact divided by eight times the sagitta."

4. *To inscribe a polyhedron in a given surface, all whose sides shall be plane quadrilaterals, and all whose solid angles shall be tetrahedral.*

Suppose the three systems of curves drawn as described in sect. (1), then each of the quadrilaterals formed by the intersection of the first and second systems is divided into two triangles by the third system. If the planes of these two triangles coincide, they form a plane quadrilateral, and if every such pair of triangles coincide, the polyhedron will satisfy the required condition.

Let abc be one of these triangles, and acd the other, which is to be in the same plane with abc. Then if the plane of abc be produced to meet the surface in the conic of contact, the curve will pass through abc and d. Hence $abcd$ must be a quadrilateral inscribed in the conic of contact.

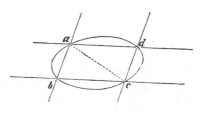

But since ab and dc belong to the same system of curves, they will be ultimately parallel when the size of the facets is diminished, and for a similar reason, ad and bc will be ultimately parallel. Hence $abcd$ will become a parallelogram, but the sides of a parallelogram inscribed in a conic are parallel to conjugate diameters.

Therefore the directions of two curves of the first and second system at their point of intersection must be parallel to two conjugate diameters of the conic of contact at that point in order that such a polyhedron may be inscribed.

Systems of curves intersecting in this manner will be referred to as "conjugate systems."

5. *On the elementary conditions of the applicability of two surfaces.*

It is evident, that if one surface is capable of being applied to another by bending, every point, line, or angle in the first has its corresponding point, line, or angle in the second.

If the transformation of the surface be effected without the extension or contraction of any part, no line drawn on the surface can experience any change in its length, and if this condition be fulfilled, there can be no extension or contraction.

Therefore the condition of bending is, that if any line whatever be drawn on the first surface, the corresponding curve on the second surface is equal to it in length. All other conditions of bending may be deduced from this.

6. *If two curves on the first surface intersect, the corresponding curves on the second surface intersect at the same angle.*

On the first surface draw any curve, so as to form a triangle with the curves already drawn, and let the sides of this triangle be indefinitely diminished, by making the new curve approach to the intersection of the former curves. Let the same thing be done on the second surface. We shall then have two corresponding triangles whose sides are equal each to each, by (5), and since their sides are indefinitely small, we may regard them as straight lines. Therefore by Euclid I. 8, the angle of the first triangle formed by the intersection of the two curves is equal to the corresponding angle of the second.

7. *At any given point of the first surface, two directions can be found, which are conjugate to each other with respect to the conic of contact at that point, and continue to be conjugate to each other when the first surface is transformed into the second.*

For let the first surface be transferred, without changing its form, to a position such that the given point coincides with the corresponding point of the second surface, and the normal to the first surface coincides with that of the second at the same point. Then let the first surface be turned about the normal as an axis till the tangent of any line through the point coincides with the tangent of the corresponding line in the second surface.

Then by (6) any pair of corresponding lines passing through the point will have a common tangent, and will therefore coincide in direction at that point.

If we now draw the conics of contact belonging to each surface we shall have two conics with the same centre, and the problem is to determine a pair of conjugate diameters of the first which coincide with a pair of conjugate diameters of the second. The analytical solution gives two directions, real, coincident, or impossible, for the diameters required.

In our investigations we can be concerned only with the case in which these directions are real.

When the conics intersect in four points, P, Q, R, S, $PQRS$ is a parallelogram inscribed in both conics, and the axes CA, CB, parallel to the sides, are conjugate in both conics.

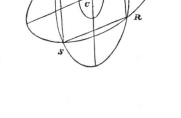

If the conics do not intersect, describe, through any point P of the second conic, a conic similar to and concentric with the first. If the conics intersect in four points, we must proceed as before; if they touch in two points, the diameter through those points and its conjugate must be taken. If they intersect in two points only, then the problem is impossible; and if they coincide altogether, the conics are similar and similarly situated, and the problem is indeterminate.

8. *Two surfaces being given as before, one pair of conjugate systems of curves may be drawn on the first surface, which shall correspond to a pair of conjugate systems on the second surface.*

By article (7) we may find at every point of the first surface two directions conjugate to one another, corresponding to two conjugate directions on the second surface. These directions indicate the directions of the two systems of curves which pass through that point.

Knowing the direction which every curve of each system must have at every point of its course, the systems of curves may be either drawn by some direct geometrical method, or constructed from their equations, which may be found by solving their differential equations.

Two systems of curves being drawn on the first surface, the corresponding systems may be drawn on the second surface. These systems being conjugate to each other, fulfil the condition of Art. (4), and may therefore be made the means of constructing a polyhedron with quadrilateral facets, by the bending of which the transformation may be effected.

These systems of curves will be referred to as the "first and second systems of Lines of Bending."

9. *General considerations applicable to Lines of Bending.*

It has been shewn that when two forms of a surface are given, one of which may be transformed into the other by bending, the nature of the lines of bending is completely determined. Supposing the problem reduced to its analytical expression, the equations of these curves would appear under the form of double solutions of differential equations of the first order and second degree, each of which would involve one arbitrary quantity, by the variation of which we should pass from one curve to another of the same system.

Hence the position of any curve of either system depends on the value assumed for the arbitrary constant; to distinguish the systems, let us call one the first system, and the other the second, and let all quantities relating to the second system be denoted by accented letters.

Let the arbitrary constants introduced by integration be u for the first system, and u' for the second.

Then the value of u will determine the position of a curve of the first system, and that of u' a curve of the second system, and therefore u and u' will suffice to determine the point of intersection of these two curves.

Hence we may conceive the position of any point on the surface to be determined by the values of u and u' for the curves of the two systems which intersect at that point.

By taking into account the equation to the surface, we may suppose x, y, and z the co-ordinates of any point, to be determined as functions of the two variables u and u'. This being done, we shall have materials for calculating everything connected with the surface, and its lines of bending. But before entering on such calculations let us examine the principal properties of these lines which we must take into account.

Suppose a series of values to be given to u and u', and the corresponding curves to be drawn on the surface.

The surface will then be covered with a system of quadrilaterals, the size of which may be diminished indefinitely by interpolating values of u and u' between those already assumed; and in the limit each quadrilateral may be regarded as a parallelogram coinciding with a facet of the inscribed polyhedron.

The *length*, the *breadth*, and the *angle* of these parallelograms will vary at different parts of the surface, and will therefore depend on the values of u and u'.

The *curvature* of a line drawn on a surface may be investigated by considering the curvature of two other lines depending on it.

The first is the projection of the line on a tangent plane to the surface at a given point in the line. The curvature of the projection at the point of contact may be called the *tangential curvature* of the line on the surface. It has also been called the *geodesic* curvature, because it is the measure of its deviation from a geodesic or shortest line on the surface.

The other projection necessary to define the curvature of a line on the surface is on a plane passing through the tangent to the curve and the normal to the surface at the point of contact. The curvature of this projection at that point may be called the *normal curvature* of the line on the surface.

It is easy to shew that this normal curvature is the same as the curvature of a normal section of the surface passing through a tangent to the curve at the same point.

10. *General considerations applicable to the inscribed polyhedron.*

When two series of lines of bending belonging to the first and second systems have been described on the surface, we may proceed, as in Art. (1), to describe a third series of curves so as to pass through all their intersections and form the diagonals of the quadrilaterals formed by the first pair of systems.

Plane triangles may then be constituted within the surface, having these points of intersection for angles, and the size of the facets of this polyhedron may be diminished indefinitely by increasing the number of curves in each series.

But by Art. (8) the first and second systems of lines of bending are conjugate to each other, and therefore by Art. (4) the polygon just constructed will have every pair of triangular facets in the same plane, and may therefore be considered as a polyhedron with plane quadrilateral facets all whose solid angles are formed by four of these facets meeting in a point.

When the number of curves in each system is increased and their distance diminished indefinitely, the plane facets of the polyhedron will ultimately coincide with the curved surface, and the polygons formed by the successive edges between the facets, will coincide with the lines of bending.

These quadrilaterals may then be considered as parallelograms, the length of which is determined by the portion of a curve of the second system intercepted between two curves of the first, while the breadth is the distance of two curves of the second system measured along a curve of the first. The expressions for these quantities will be given when we come to the calculation of our results along with the other particulars which we only specify at present.

The angle of the sides of these parallelograms will be ultimately the same as the angle of intersection of the first and second systems, which we may call ϕ; but if we suppose the dimensions of the facets to be small quantities of the first order, the angles of the four facets which meet in a point will differ from the angle of intersection of the curves at that point by small angles of the first order depending on the tangential curvature of the lines of bending. The sum of these four angles will differ from four right angles by a small angle of the second order, the circular measure of which expresses the entire curvature of the solid angle as in Art. (2).

The angle of inclination of two adjacent facets will depend on the normal curvature of the lines of bending, and will be that of the projection of two consecutive sides of the polygon of one system on a plane perpendicular to a side of the other system.

11. *Explanation of the Notation to be employed in calculation.*

Suppose each system of lines of bending to be determined by an equation containing one arbitrary parameter.

Let this parameter be u for the first system, and u' for the second.

Let two curves, one from each system, be selected as curves of reference, and let their parameters be u_0 and u'_0.

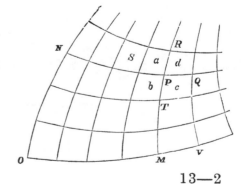

Let ON and OM in the figure represent these two curves.

Let PM be any curve of the first system whose parameter is u, and PN any curve of the second whose parameter is u', then their intersection P may be defined as the point (u, u'), and all quantities referring to the point P may be expressed as functions of u and u'.

Let PN, the length of a curve of the second system (u'), from N (u_0) to P (u), be expressed by s, and PM the length of the curve (u) from (u'_0) to (u'), by s', then s and s' will be functions of u and u'.

Let $(u+\delta u)$ be the parameter of the curve QV of the first system consecutive to PM. Then the length of PQ, the part of the curve of the second system intercepted between the curves (u) and $(u+\delta u)$, will be

$$\frac{ds}{du}\,\delta u.$$

Similarly PR may be expressed by

$$\frac{ds'}{du'}\,\delta u'.$$

These values of PQ and PR will be the ultimate values of the length and breadth of a quadrilateral facet.

The angle between these lines will be ultimately equal to ϕ, the angle of intersection of the system; but when the values of δu and $\delta u'$ are considered as finite though small, the angles a, b, c, d of the facets which form a solid angle will depend on the tangential curvature of the two systems of lines.

Let r be the tangential curvature of a curve of the first system at the given point measured in the direction in which u increases, and let r', that of the second system, be measured in the direction in which u' increases.

Then we shall have for the values of the four plane angles which meet at P,

$$a = \pi - \phi + \frac{1}{2r}\frac{ds'}{du'}\,\delta u' - \frac{1}{2r'}\frac{ds}{du}\delta u,$$

$$b = \phi \quad\;\; + \frac{1}{2r}\frac{ds'}{du'}\,\delta u' + \frac{1}{2r'}\frac{ds}{du}\,\delta u,$$

$$c = \pi - \phi - \frac{1}{2r}\frac{ds'}{du'}\,\delta u' + \frac{1}{2r'}\frac{ds}{du}\,\delta u,$$

$$d = \phi \quad\;\; - \frac{1}{2r}\frac{ds'}{du'}\,\delta u' - \frac{1}{2r'}\frac{ds}{du}\,\delta u.$$

These values are correct as far as the first order of small quantities. Those corrections which depend on the curvature of the surface are of the second order.

Let ρ be the normal curvature of a curve of the first system, and ρ' that of a curve of the second, then the inclination l of the plane facets a and b, separated by a curve of the second system, will be

$$l = \frac{1}{\rho \sin \phi} \frac{ds'}{du'} \delta u',$$

as far as the first order of small angles, and the inclination l' of b and c will be

$$l' = \frac{1}{\rho' \sin \phi} \frac{ds}{du} \delta u$$

to the same order of exactness.

12. *On the corresponding polygon on the surface of the sphere of reference.*

By the method described in Art. (2) we may find a point on the sphere corresponding to each facet of the polyhedron.

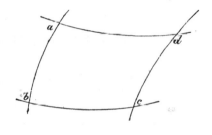

In the annexed figure, let a, b, c, d be the points on the sphere corresponding to the four facets which meet at the solid angle P. Then the area of the spherical quadrilateral a, b, c, d will be the measure of the entire curvature of the solid angle P.

This area is measured by the defect of the sum of the exterior angles from four right angles; but these exterior angles are equal to the four angles a, b, c, d, which form the solid angle P, therefore the entire curvature is measured by

$$k = 2\pi - (a + b + c + d).$$

Since a, b, c, d are invariable, it is evident, as in Art. (2), that the entire curvature at P is not altered by bending.

By the last article it appears that when the facets are small the angles b and d are approximately equal to ϕ, and a and c to $(\pi - \phi)$, and since the sides of the quadrilateral on the sphere are small, we may regard it as approximately a plane parallelogram whose angle $bad = \phi$.

The sides of this parallelogram will be l and l', the supplements of the angles of the edges of the polyhedron, and we may therefore express its area as a plane parallelogram

$$k = ll' \sin \phi.$$

By the expression for l and l' in the last article, we find

$$k = \frac{1}{\rho\rho' \sin\phi} \frac{ds}{du} \frac{ds'}{du'} \delta u\, \delta u'$$

for the entire curvature of one solid angle.

Since the whole number of solid angles is equal to the whole number of facets, we may suppose a quarter of each of the facets of which it is composed to be assigned to each solid angle. The area of these will be the same as that of one whole facet, namely,

$$\sin\phi \frac{ds}{du} \frac{ds'}{du'} \delta u\, \delta u' \, ;$$

therefore dividing the expression for k by this quantity, we find for the value of the specific curvature at P

$$p = \frac{1}{\rho\rho' \sin^2\phi} \, ;$$

which gives the specific curvature in terms of the normal curvatures of the lines of bending and their angle of intersection.

13. *Further reduction of this expression by means of the " Conic of Contact," as defined in Art.* (3).

Let a and b be the semiaxes of the conic of contact, and h the sagitta or perpendicular to its plane from the centre to the surface.

Let CP, CQ be semidiameters parallel to the lines of bending of the first and second systems, and therefore conjugate to each other.

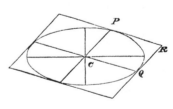

By (Art. 3), $\rho = \frac{1}{2} \dfrac{CP^2}{h}$,

and $\rho' = \frac{1}{2} \dfrac{CQ^2}{h}$;

and the expression for p in Art. (12), becomes

$$p = \frac{4h^2}{(CP \cdot CQ \sin\phi)^2} \, .$$

But $CP \cdot CQ \sin\phi$ is the area of the parallelogram $CPRQ$, which is one quarter of the circumscribed parallelogram, and therefore by a well-known theorem

$$CP \cdot CQ \sin\phi = ab,$$

and the expression for p becomes

$$p = \frac{4h^2}{a^2b^2};$$

or if the area of the circumscribing parallelogram be called A,

$$p = \frac{16h^2}{A^2}.$$

The principal radii of curvature of the surface are parallel to the axes of the conic of contact. Let R and R' denote these radii, then

$$R = \tfrac{1}{2}\frac{a^2}{h} \text{ and } R' = \tfrac{1}{2}\frac{b^2}{h};$$

and therefore substituting in the expression for p,

$$p = \frac{1}{RR'};$$

or the specific curvature is the reciprocal of the product of the principal radii of curvature.

This remarkable expression was introduced by Gauss in the memoir referred to in a former part of this paper. His method of investigation, though not so elementary, is more direct than that here given, and will shew how this result can be obtained without reference to the geometrical methods necessary to a more extended inquiry into the *modes* of bending.

14. *On the variation of normal curvature of the lines of bending as we pass from one point of the surface to another.*

We have determined the relation between the normal curvatures of the lines of bending of the two systems at their points of intersection; we have now to find the variation of normal curvature when we pass from one line of the first system to another, along a line of the second.

In analytical language we have to find the value of

$$\frac{d}{du}\left(\frac{1}{\rho}\right).$$

Referring to the figure in Art. (11), we shall see that this may be done if we can determine the difference between the angle of inclination of the facets a and b, and that of c and d: for the angle l between a and b is

$$l = \frac{1}{\rho \sin \phi}\frac{ds'}{du'}\,\delta u',$$

and therefore the difference between the angle of a and b and that of c and d is

$$\delta l = \frac{dl}{du}\, \delta u = \frac{d}{du} \left(\frac{1}{\rho \sin \phi} \frac{ds'}{du'} \right) \delta u \, \delta u' \, ;$$

whence the differential of ρ with respect to u may be found.

We must therefore find δl, and this is done by means of the quadrilateral on the sphere described in Art. (12).

15.   *To find the values of δl and $\delta l'$.*

In the annexed figure let $abcd$ represent the small quadrilateral on the surface
of the sphere. The exterior angles a, b,
c, d are equal to those of the four facets
which meet at the point P of the surface,
and the sides represent the angles which
the planes of those facets make with each
other ; so that

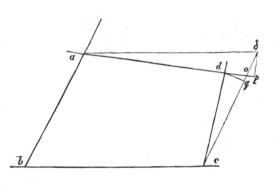

$$ab = l, \quad bc = l', \quad cd = l + \delta l, \quad da = l' + \delta l',$$

and the problem is to determine δl and $\delta l'$ in terms of the sides l and l' and
the angles a, b, c, d.

On the sides ba, bc complete the parallelogram $abcd$.

    Produce ad to p, so that $ap = a\delta$. Join δp.

     Make $cq = cd$ and join dq.

     then $\delta l = cd - ab$,

         $= cq - c\delta$,

         $= -(qo + o\delta)$.

   Now $qo = qd \tan qdo$

        $= cd \sin qcd \cot qod$,

but $cd = l$ nearly, $\sin qcd = qcd = (c + b - \pi)$ and $qod = \phi$;

    $\therefore \; qo = l\,(c + b - \pi) \cot \phi.$

Also $\quad o\delta = \dfrac{p\delta}{\sin \delta op}$

$$= a\delta\,(\delta ap)\,\frac{1}{\sin \phi}$$

$$= l'\,(a+b-\pi)\,\frac{1}{\sin \phi}.$$

Substituting the values of a, b, c, d from Art. (11),

$$\delta l = -(qo + o\delta)$$

$$= -l\,\frac{1}{r'}\,\frac{ds}{du}\,\cot \phi\,\delta u - l'\,\frac{1}{r}\,\frac{ds'}{du'}\,\frac{1}{\sin \phi}\,\delta u'.$$

Finally, substituting the values of l, l', and δl from Art. (14),

$$\frac{d}{du}\left(\frac{1}{\rho \sin \phi}\,\frac{ds'}{du'}\right)\delta u\,\delta u' = -\frac{\cot \phi}{\rho \sin \phi}\,\frac{ds'}{du'}\,\frac{1}{r'}\,\frac{ds}{du}\,\delta u\,\delta u' - \frac{1}{\rho' \sin^2 \phi}\,\frac{ds}{du}\,\frac{1}{r}\,\frac{ds'}{du'}\,\delta u\,\delta u';$$

which may be put under the more convenient form

$$\frac{d}{du}\,(\log \rho) = \frac{d}{du}\log\left(\frac{1}{\sin \phi}\,\frac{ds'}{du'}\right) + \frac{1}{r'}\,\frac{ds}{du}\,\cot \phi + \frac{\rho}{\rho'}\,\frac{1}{r}\,\frac{ds}{du}\,\frac{1}{\sin \phi};$$

and from the value of $\delta l'$ we may similarly obtain

$$\frac{d}{du'}\,(\log \rho') = \frac{d}{du'}\log\left(\frac{1}{\sin \phi}\,\frac{ds}{du}\right) + \frac{1}{r}\,\frac{ds'}{du'}\,\cot \phi + \frac{\rho'}{\rho}\,\frac{1}{r'}\,\frac{ds'}{du'}\,\frac{1}{\sin \phi}.$$

We may simplify these equations by putting p for the specific curvature of the surface, and q for the ratio $\dfrac{\rho}{\rho'}$, which is the only quantity altered by bending.

We have then

$$p = \frac{1}{\rho\rho' \sin^2 \phi}, \quad \text{and} \quad q = \frac{\rho}{\rho'},$$

whence $\quad \rho^2 = q\,\dfrac{1}{p \sin^2 \phi}, \qquad \rho'^2 = \dfrac{1}{q}\,\dfrac{1}{p \sin^2 \phi},$

and the equations become

$$\frac{d}{du}\,(\log q) = \frac{d}{du}\log\left(p\,\overline{\frac{ds'}{du'}}\Big.^{2}\right) + \frac{2}{r'}\,\frac{ds}{du}\,\cot \phi + \frac{2}{r}\,\frac{ds}{du}\,\frac{1}{\sin \phi}\,q,$$

$$\frac{d}{du'}\,(\log q) = -\frac{d}{du'}\log\left(p\,\overline{\frac{ds}{du}}\Big.^{2}\right) - \frac{2}{r}\,\frac{ds'}{du'}\,\cot \phi - \frac{2}{r'}\,\frac{ds'}{du'}\,\frac{1}{\sin \phi}\,\frac{1}{q}.$$

In this way we may reduce the problem of bending a surface to the consideration of one variable q, by means of the lines of bending.

16.   *To obtain the condition of Instantaneous lines of bending.*

We have now obtained the values of the differential coefficients of q with respect to each of the variables u, u'.

From the equation

$$\frac{d^2}{du\,du'}(\log q) = \frac{d^2}{du'\,du}(\log q),$$

we might find an equation which would give certain conditions of lines of bending. These conditions however would be equivalent to those which we have already assumed when we drew the systems of lines so as to be conjugate to each other.

To find the true conditions of bending we must suppose the form of the surface to vary continuously, so as to depend on some variable t which we may call the time.

Of the different quantities which enter into our equations, none are changed by the operation of bending except q, so that in differentiating with respect to t all the rest may be considered constant, q being the only variable.

Differentiating the equations of last article with respect to t, we obtain

$$\frac{d^2}{du\,dt}(\log q) = \frac{2}{r}\frac{ds}{du}\frac{1}{\sin\phi}\,q\,\frac{d}{dt}(\log q),$$

$$\frac{d^2}{du'\,dt}(\log q) = \frac{2}{r'}\frac{ds'}{du'}\frac{1}{\sin\phi}\frac{1}{q}\frac{d}{dt}(\log q).$$

Whence

$$\frac{d^3}{du\,du'\,dt}(\log q) =$$

$$\left\{\frac{d}{du'}\left(\frac{2}{r}\frac{ds}{du}\frac{1}{\sin\phi}\right) + \frac{2}{r}\frac{ds}{du}\frac{1}{\sin\phi}\frac{d}{du'}(\log q)\right\}q\,\frac{d}{dt}(\log q) + \frac{2}{r}\frac{ds}{du}\frac{1}{\sin\phi}\,q\,\frac{d}{du'dt}(\log q),$$

and

$$\frac{d^3}{du\,du'\,dt}(\log q) =$$

$$\left\{\frac{d}{du}\left(\frac{2}{r'}\frac{ds'}{du'}\frac{1}{\sin\phi}\right) - \frac{2}{r'}\frac{ds'}{du'}\frac{1}{\sin\phi}\frac{d}{du}\log q\right\}\frac{1}{q}\frac{d}{dt}(\log q) + \frac{2}{r'}\frac{ds'}{du'}\frac{1}{\sin\phi}\frac{1}{q}\frac{d}{du\,dt}(\log q),$$

two independent values of the same quantity, whence the required conditions may be obtained.

Substituting in these equations the values of those quantities which occur in the original equations, we obtain

$$q \frac{1}{r} \frac{ds}{du} \left\{ \frac{d}{du'} \log \left(pr \frac{ds}{du} \sin \phi \right) + \frac{2}{r} \frac{ds'}{du'} \cot \phi \right\}$$

$$= \frac{1}{q} \frac{1}{r'} \frac{ds'}{du'} \left\{ \frac{d}{du'} \log \left(pr' \frac{ds'}{du'} \sin \phi \right) + \frac{2}{r'} \frac{ds}{du} \cot \phi \right\},$$

which is the condition which must hold at every instant during the process of bending for the lines about which the bending takes place at that instant. When the bending is such that the position of the lines of bending on the surface alters at every instant, this is the only condition which is required. It is therefore called the condition of Instantaneous lines of bending.

17.   *To find the condition of Permanent lines of bending.*

Since q changes with the time, the equation of last article will not be satisfied for any finite time unless both sides are separately equal to zero.   In that case we have the two conditions

$$\left. \begin{array}{c} \dfrac{d}{du'} \log \left(pr \dfrac{ds}{du} \sin \phi \right) + \dfrac{2}{r} \dfrac{ds'}{du'} \cot \phi = 0, \\[2ex] \text{or } \dfrac{1}{r} \dfrac{ds}{du} = 0. \end{array} \right\} \dots\dots\dots\dots(1).$$

$$\left. \begin{array}{c} \dfrac{d}{du} \log \left(pr' \dfrac{ds'}{du'} \sin \phi \right) + \dfrac{2}{r'} \dfrac{ds}{du} \cot \phi = 0, \\[2ex] \text{or } \dfrac{1}{r'} \dfrac{ds'}{du'} = 0. \end{array} \right\} \dots\dots\dots\dots(2).$$

If the lines of bending satisfy these conditions, a finite amount of bending may take place without changing the position of the system on the surface. Such lines are therefore called Permanent lines of bending.

The only case in which the phenomena of bending may be exhibited by means of the polyhedron with quadrilateral facets is that in which permanent lines of bending are chosen as the boundaries of the facets.   In all other cases the bending takes place about an instantaneous system of lines which is continually in motion with respect to the surface, so that the nature of the polyhedron would need to be altered at every instant.

We are now able to determine whether any system of lines drawn on a given surface is a system of instantaneous or permanent lines of bending.

We are also able, by the method of Article (8), to deduce from two consecutive forms of a surface, the lines of bending about which the transformation must have taken place.

If our analytical methods were sufficiently powerful, we might apply our results to the determination of such systems of lines on any known surface, but the necessary calculations even in the simplest cases are so complicated, that, even if useful results were obtained, they would be out of place in a paper of this kind, which is intended to afford the means of forming distinct conceptions rather than to exhibit the results of mathematical labour.

18. *On the application of the ordinary methods of analytical geometry to the consideration of lines of bending.*

It may be interesting to those who may hesitate to accept results derived from the consideration of a polyhedron, when applied to a curved surface, to inquire whether the same results may not be obtained by some independent method.

As the following method involves only those operations which are most familiar to the analyst, it will be sufficient to give the rough outline, which may be filled up at pleasure.

The proof of the invariability of the specific curvature may be taken from any of the memoirs above referred to, and its value in terms of the equation of the surface will be found in the memoir of Gauss.

Let the equation to the surface be put under the form

$$z = f(xy),$$

then the value of the specific curvature is

$$p = \frac{\dfrac{d^2z}{dx^2}\dfrac{d^2z}{dy^2} - \overline{\dfrac{d^2z}{dx\,dy}}\Big|^2}{\sqrt{1 + \overline{\dfrac{dz}{dx}}\Big|^2 + \overline{\dfrac{dz}{dy}}\Big|^2}}.$$

The definition of conjugate systems of curves may be rendered independent of the reasoning formerly employed by the following modification.

Let a tangent plane move along any line of the first system, then if the line of ultimate intersection of this plane with itself be always a tangent to some line of the second system, the second system is said to be conjugate to the first.

It is easy to show that the first system is also conjugate to the second.

Let the system of curves be projected on the plane of xy, and at the point (x, y) let α be the angle which a projected curve of the first system makes with the axis of x, and β the angle which the projected curve of the second system which intersects it at that point makes with the same axis. Then the condition of the systems being conjugate will be found to be

$$\frac{d^2z}{dx^2} \cos \alpha \cos \beta + \frac{d^2z}{dx\,dy} \sin (\alpha + \beta) + \frac{d^2z}{dy^2} \sin \alpha \sin \beta = 0 \; ;$$

α and β being known as functions of x and y, we may determine the nature of the curves projected on the plane of xy.

Supposing the surface to touch that plane at the origin, the length and tangential curvature of the lines on the surface near the point of contact may be taken the same as those of their projections on the plane, and any change of form of the surface due to bending will not alter the form of the projected lines indefinitely near the point of contact. We may therefore consider z as the only variable altered by bending; but in order to apply our analysis with facility, we may assume

$$\frac{d^2z}{dx^2} = PQ \sin^2 \alpha + PQ^{-1} \sin^2 \beta,$$

$$\frac{d^2z}{dx\,dy} = - PQ \sin \alpha \cos \alpha - PQ^{-1} \sin \beta \cos \beta,$$

$$\frac{d^2z}{dy^2} = PQ \cos^2 \alpha + PQ^{-1} \cos^2 \beta.$$

It will be seen that these values satisfy the condition last given. Near the origin we have

$$p = \frac{d^2z}{dx^2} \frac{d^2z}{dy^2} - \overline{\frac{d^2z}{dx\,dy}}\Big|^2 = P^2 \sin^2 (\alpha - \beta),$$

and $q = Q^{-2}$.

Differentiating these values of $\dfrac{d^2z}{dx^2}$, &c., we shall obtain two values of $\dfrac{d^3z}{dx^2\,dy}$ and of $\dfrac{d^3z}{dx\,dy^2}$, which being equated will give two equations of condition.

Now if s' be measured along a curve of the first system, and R be any function of x and y, then

$$\frac{dR}{ds'} = \frac{dR}{dx}\cos a + \frac{dR}{dy}\sin a,$$

$$\text{and } \frac{dR}{du'} = \frac{dR}{ds'}\frac{ds'}{du'}.$$

We may also show that

$$\frac{da}{ds'} = \frac{1}{r},$$

and that

$$\cos a\,\frac{da}{dy} - \sin a\,\frac{da}{dx} = \frac{d}{ds}\log\left(\frac{ds'}{du'}\sin\phi\right).$$

By substituting these values in the equations thus obtained, they are reduced to the two equations given at the end of (Art. 15). This method of investigation introduces no difficulty except that of somewhat long equations, and is therefore satisfactory as supplementary to the geometrical method given at length.

As an example of the method given in page (2), we may apply it to the case of the surface whose equation is

$$\left(\frac{x}{c-z}\right) + \left(\frac{y}{c+z}\right)^2 = \left(\frac{a}{c}\right)^2.$$

This surface may be generated by the motion of a straight line whose equation is of the form

$$x = a\cos t\left(1 - \frac{z}{c}\right), \qquad y = a\sin t\left(1 + \frac{z}{c}\right),$$

t being the variable, by the change of which we pass from one position of the line to another. This line always passes through the circle

$$z = 0, \qquad x^2 + y^2 = a^2,$$

and the straight lines $\quad z = c, \qquad x = 0,$

$$\text{and } z = -c, \quad y = 0,$$

which may therefore be taken as the directors of the surface.

Taking two consecutive positions of this line, in which the values of t are t and $t + \delta t$, we may find by the ordinary methods the equation to the shortest line between them, its length, and the co-ordinates of the point in which it intersects the first line.

Calling the length $\delta\zeta$,

$$\delta\zeta = \frac{ac}{\sqrt{a^2 + c^2}}\sin 2t\, \delta t,$$

and the co-ordinates of the point of intersection are

$$x = 2a\cos^3 t, \qquad y = 2a\sin^3 t, \qquad z = -c\cos 2t.$$

The angle $\delta\theta$ between the consecutive lines is

$$\delta\theta = \frac{a}{\sqrt{a^2 + c^2}}\,\delta t.$$

The distance $\delta\sigma$ between consecutive shortest lines is

$$\delta\sigma = \frac{3a^2 + 2c^2}{\sqrt{a^2 + c^2}}\sin 2t\, \delta t,$$

and the angle $\delta\phi$ between these latter lines is

$$\delta\phi = \frac{c}{\sqrt{a^2 + c^2}}\,\delta t.$$

Hence if we suppose ζ, θ, σ, ϕ, and t to vanish together, we shall have by integration

$$\zeta = \frac{ac}{2\sqrt{a^2 + c^2}}(1 - \cos 2t),$$

$$\theta = \frac{a}{\sqrt{a^2 + c^2}}\,t,$$

$$\sigma = \frac{3a^2 + 2c^2}{2\sqrt{a^2 + c^2}}(1 - \cos 2t),$$

$$\phi = \frac{c}{\sqrt{a^2 + c^2}}\,t.$$

By bending the surface about its generating lines we alter the value of ϕ in any manner without changing ζ, θ, or σ. For instance, making $\phi = 0$, all the generating lines become parallel to the same plane. Let this plane be that of xy, then ζ is the distance of a generating line from that plane. The projections of the generating lines on the plane of xy will, by their ultimate intersections, form a curve, the length of which is measured by σ, and the angle which its tangent makes with the axis of x by θ, θ and σ being connected by the equation

$$\sigma = \frac{3a^2 + 2c^2}{2\sqrt{a^2+c^2}}\left(1 - \cos\frac{2\sqrt{a^2+c^2}}{a}\theta\right),$$

which shows the curve to be an epicycloid.

The generating lines of the surface when bent into this form are therefore tangents to a cylindrical surface on an epicycloidal base, touching that surface along a curve which is always equally inclined to the plane of the base, the tangents themselves being drawn parallel to the base.

We may now consider the bending of the surface of revolution

$$\sqrt[3]{x^2+y^2} + z^{\frac{2}{3}} = c^{\frac{2}{3}}.$$

Putting $r = \sqrt{x^2+y^2}$, then the equation of the generating line is

$$r^{\frac{2}{3}} + z^{\frac{2}{3}} = c^{\frac{2}{3}}.$$

This is the well-known hypocycloid of four cusps.

Let s be the length of the curve measured from the cusp in the axis of z, then,

$$s = \tfrac{3}{2}c^{\frac{1}{3}}r^{\frac{2}{3}},$$

wherefore,

$$r = \left(\tfrac{2}{3}\right)^{\frac{3}{2}} c^{-\frac{1}{2}} s^{\frac{3}{2}}.$$

Let θ be the angle which the plane of any generating line makes with that of xz, then s and θ determine the position of any point on the surface. The length and breadth of an element of the surface will be δs and $r\delta\theta$.

Now let the surface be bent in the manner formerly described, so that θ becomes θ', and r, r', when

$$\theta' = \mu\theta \text{ and } r' = \frac{1}{\mu}r,$$

then

$$r' = \left(\tfrac{2}{3}\right)^{\frac{3}{2}} c^{-\frac{1}{2}} \mu^{-1} s^{\frac{3}{2}}$$

$$= \left(\tfrac{2}{3}\right)^{\frac{3}{2}} c'^{-\frac{1}{2}} s^{\frac{3}{2}},$$

provided

$$c' = \mu^2 c.$$

The equation between r' and s being of the same form as that between r and s shows that the surface when bent is similar to the original surface, its dimensions being multiplied by μ^2.

This, however, is true only for one half of the surface when bent. The other half is precisely symmetrical, but belongs to a surface which is not continuous with the first.

The surface in its original form is divided by the plane of xy into two parts which meet in that plane, forming a kind of cuspidal edge of a circular form which limits the possible value of s and r.

After being bent, the surface still consists of the same two parts, but the edge in which they meet is no longer of the cuspidal form, but has a finite angle $= 2 \cos^{-1} \dfrac{1}{\mu}$, and the two sheets of the surface become parts of two different surfaces which meet but are not continuous.

NOTE.

As an example of the application of the more general theory of "lines of bending," let us consider the problem which has been already solved by Professor Jellett.

To determine the conditions under which one portion of a surface may be rendered rigid, while the remainder is flexible.

Suppose the lines of bending to be traced on the surface, and the corresponding polyhedron to be formed, as in (9) and (10), then if the angle of one of the four edges which meet at any solid angle of the polyhedron be altered by bending, those of the other three must be also altered. These edges terminate in other solid angles, the forms of which will also be changed, and therefore the effect of the alteration of one angle of the polyhedron will be communicated to every other angle within the system of lines of bending which defines the form of the polyhedron.

If any portion of the surface remains unaltered it must lie beyond the limits of the system of lines of bending. We must therefore investigate the conditions of such a system being bounded.

The boundary of any system of lines on a surface is the curve formed by the ultimate intersection of those lines, and therefore at any given point coincides in direction with the curve of the system which passes through that point. In this case there are two systems of lines of bending, which are necessarily coincident in extent, and must therefore have the same boundary. At any point of this boundary therefore the directions of the lines of bending of the first and second systems are coincident.

But, by (7), these two directions must be "conjugate" to each other, that is, must correspond to conjugate diameters of the "Conic of Contact." Now the only case in which conjugate diameters of a conic can coincide, is when the conic is an hyperbola, and both diameters coincide with one of the asymptotes; therefore the boundary of the system of lines of bending must be a curve at every point of which the conic of contact is an hyperbola, one of whose asymptotes lies in the direction of the curve. The radius of "normal curvature" must therefore by (3) be infinite at every point of the curve. This is the geometrical property of what Professor Jellett calls a "Curve of Flexure," so that we may express the result as follows:

If one portion of a surface be fixed, while the remainder is bent, the boundary of the fixed portion is a curve of flexure.

This theorem includes those given at p. (92), relative to a fixed curve on a surface, for in a surface whose curvatures are of the same sign, there can be no "curves of flexure," and in a developable surface, they are the rectilinear sections. Although the cuspidal edge, or *arête de rebroussement*, satisfies the analytical condition of a curve of flexure, yet, since its form determines that of the whole surface, it cannot remain fixed while the form of the surface is changed.

In concavo-convex surfaces, the curves of flexure must either have tangential curvature or be straight lines. Now if we put $\phi = 0$ in the equations of Art. (17), we find that the lines of bending of both systems have no tangential curvature at the point where they touch the curve of flexure. They must therefore lie entirely on the convex side of that curve, and therefore

If a curve of flexure be fixed, the surface on the concave side of the curve is not flexible.

I have not yet been able to determine whether the surface is inflexible on the convex side of the curve. It certainly is so in some cases which I have been able to work out, but I have no general proof.

When a surface has one or more rectilinear sections, the portions of the surface between them may revolve as rigid bodies round those lines as axes in any manner, but no other motion is possible. The case in which the rectilinear sections form an infinite series has been discussed in Sect. (I.).

[From the *Cambridge and Dublin Mathematical Journal*, Vol. IX.]

V. *On a particular case of the descent of a heavy body in a resisting medium.*

EVERY one must have observed that when a slip of paper falls through the air, its motion, though undecided and wavering at first, sometimes becomes regular. Its general path is not in the vertical direction, but inclined to it at an angle which remains nearly constant, and its fluttering appearance will be found to be due to a rapid rotation round a horizontal axis. The direction of deviation from the vertical depends on the direction of rotation.

If the positive directions of an axis be toward the right hand and upwards, and the positive angular direction opposite to the direction of motion of the hands of a watch, then, if the rotation is in the positive direction, the horizontal part of the mean motion will be positive.

These effects are commonly attributed to some accidental peculiarity in the form of the paper, but a few experiments with a rectangular slip of paper (about two inches long and one broad), will shew that the direction of rotation is determined, not by the irregularities of the paper, but by the initial circumstances of projection, and that the symmetry of the form of the paper greatly increases the distinctness of the phenomena. We may therefore assume that if the form of the body were accurately that of a plane rectangle, the same effects would be produced.

The following investigation is intended as a general explanation of the true cause of the phenomenon.

I suppose the resistance of the air caused by the motion of the plane to be in the direction of the normal and to vary as the square of the velocity estimated in that direction.

Now though this may be taken as a sufficiently near approximation to the magnitude of the resisting force on the plane taken as a whole, the pressure

15—2

on any given element of the surface will vary with its position so that the resultant force will not generally pass through the centre of gravity.

It is found by experiment that the position of the centre of pressure depends on the tangential part of the motion, that it lies on that side of the centre of gravity towards which the tangential motion of the plane is directed, and that its distance from that point increases as the tangential velocity increases.

I am not aware of any mathematical investigation of this effect. The explanation may be deduced from experiment.

Place a body similar in shape to the slip of paper obliquely in a current of some visible fluid. Call the edge where the fluid first meets the plane the first edge, and the edge where it leaves the plane, the second edge, then we may observe that

(1) On the anterior side of the plane the velocity of the fluid increases as it moves along the surface from the first to the second edge, and therefore by a known law in hydrodynamics, the pressure must diminish from the first to the second edge.

(2) The motion of the fluid behind the plane is very unsteady, but may be observed to consist of a series of eddies diminishing in rapidity as they pass behind the plane from the first to the second edge, and therefore relieving the posterior pressure most at the first edge.

Both these causes tend to make the total resistance greatest at the first edge, and therefore to bring the centre of pressure nearest to that edge.

Hence the moment of the resistance about the centre of gravity will always tend to turn the plane towards a position perpendicular to the direction of the current, or, in the case of the slip of paper, to the path of the body itself. It will be shewn that it is this moment that maintains the rotatory motion of the falling paper.

When the plane has a motion of rotation, the resistance will be modified on account of the unequal velocities of different parts of the surface. The magnitude of the whole resistance at any instant will not be sensibly altered if the velocity of any point due to angular motion be small compared with that due to the motion of the centre of gravity. But there will be an additional moment of the resistance round the centre of gravity, which will always act in the direction opposite to that of rotation, and will vary directly as the normal and angular velocities together.

The part of the moment due to the obliquity of the motion will remain nearly the same as before.

We are now prepared to give a general explanation of the motion of the slip of paper after it has become regular.

Let the angular position of the paper be determined by the angle between the normal to its surface and the axis of x, and let the angular motion be such that the normal, at first coinciding with the axis of x, passes towards that of y.

The motion, speaking roughly, is one of descent, that is, in the negative direction along the axis of y.

The resolved part of the resistance in the vertical direction will always act upwards, being greatest when the plane of the paper is horizontal, and vanishing when it is vertical.

When the motion has become regular, the effect of this force during a whole revolution will be equal and opposite to that of gravity during the same time.

Since the resisting force increases while the normal is in its first and third quadrants, and diminishes when it is in its second and fourth, the maxima of velocity will occur when the normal is in its first and third quadrants, and the minima when it is in the second and fourth.

The resolved part of the resistance in the horizontal direction will act in the positive direction along the axis of x in the first and third quadrants, and in the negative direction during the second and fourth; but since the resistance increases with the velocity, the whole effect during the first and third quadrants will be greater than the whole effect during the second and fourth. Hence the horizontal part of the resistance will act on the whole in the positive direction, and will therefore cause the general path of the body to incline in that direction, that is, toward the right.

That part of the moment of the resistance about the centre of gravity which depends on the angular velocity will vary in magnitude, but will always act in the negative direction. The other part, which depends on the obliquity of the plane of the paper to the direction of motion, will be positive in the first and third quadrants and negative in the second and fourth; but as its magnitude increases with the velocity, the positive effect will be greater than the negative.

When the motion has become regular, the effect of this excess in the positive direction will be equal and opposite to the negative effect due to the angular velocity during a whole revolution.

The motion will then consist of a succession of equal and similar parts performed in the same manner, each part corresponding to half a revolution of the paper.

These considerations will serve to explain the lateral motion of the paper, and the maintenance of the rotatory motion.

Similar reasoning will shew that whatever be the initial motion of the paper, it cannot remain uniform.

Any accidental oscillations will increase till their amplitude exceeds half a revolution. The motion will then become one of rotation, and will continually approximate to that which we have just considered.

It may be also shewn that this motion will be unstable unless it take place about the longer axis of the rectangle.

If this axis is inclined to the horizon, or if one end of the slip of paper be different from the other, the path will not be straight, but in the form of a helix. There will be no other essential difference between this case and that of the symmetrical arrangement.

Trinity College, April 5, 1853.

[From the *Transactions of the Royal Scottish Society of Arts*, Vol. IV. Part III.]

VI. *On the Theory of Colours in relation to Colour-Blindness.*
A letter to Dr G. Wilson.

DEAR SIR,—As you seemed to think that the results which I have obtained in the theory of colours might be of service to you, I have endeavoured to arrange them for you in a more convenient form than that in which I first obtained them. I must premise, that the first distinct statement of the theory of colour which I adopt, is to be found in *Young's Lectures on Natural Philosophy* (p. 345, Kelland's Edition); and the most philosophical enquiry into it which I have seen is that of Helmholtz, which may be found in the Annals of Philosophy for 1852.

It is well known that a ray of light, from any source, may be divided by means of a prism into a number of rays of different refrangibility, forming a series called a spectrum. The intensity of the light is different at different points of this spectrum; and the law of intensity for different refrangibilities differs according to the nature of the incident light. In Sir John F. W. Herschel's *Treatise on Light*, diagrams will be found, each of which represents completely, by means of a curve, the law of the intensity and refrangibility of a beam of solar light after passing through various coloured media.

I have mentioned this mode of defining and registering a beam of light, because it is the perfect expression of what a beam of light is in itself, considered with respect to all its properties as ascertained by the most refined instruments. When a beam of light falls on the human eye, certain sensations are produced, from which the possessor of that organ judges of the colour and intensity of the light. Now, though every one experiences these sensations, and though they are the foundation of all the phenomena of sight, yet, on account of their absolute simplicity, they are incapable of analysis, and can never become in themselves objects of thought. If we attempt to discover them, we must do so by artificial means; and our reasonings on them must be guided by some theory.

The most general form in which the existing theory can be stated is this,—

There are certain sensations, finite in number, but infinitely variable in degree, which may be excited by the different kinds of light. The compound sensation resulting from all these is the object of consciousness, is a simple act of vision.

It is easy to see that the *number* of these sensations corresponds to what may be called in mathematical language the number of independent variables, of which sensible colour is a function.

This will be readily understood by attending to the following cases:—

1. When objects are illuminated by homogeneous yellow light, the only thing which can be distinguished by the eye is difference of intensity or brightness.

If we take a horizontal line, and colour it black at one end, with increasing degrees of intensity of yellow light towards the other, then every visible object will have a brightness corresponding to some point in this line.

In this case there is nothing to prove the existence of more than one sensation in vision.

In those photographic pictures in which there is only one tint of which the different intensities correspond to the different degrees of illumination of the object, we have another illustration of an optical effect depending on one variable only.

2. Now, suppose that different kinds of light are emanating from different sources, but that each of these sources gives out perfectly homogeneous light, then there will be two things on which the nature of each ray will depend:— (1) its intensity or brightness; (2) its hue, which may be estimated by its position in the spectrum, and measured by its wave length.

If we take a rectangular plane, and illuminate it with the different kinds of homogeneous light, the intensity at any point being proportional to its horizontal distance along the plane, and its wave length being proportional to its height above the foot of the plane, then the plane will display every possible variety of homogeneous light, and will furnish an instance of an optical effect depending on two variables.

3. Now, let us take the case of nature. We find that colours differ not only in intensity and hue, but also in tint; that is, they are more or less pure. We might arrange the varieties of each colour along a line, which should begin with the homogeneous colour as seen in the spectrum, and pass through all gradations of tint, so as to become continually purer, and terminate in white.

We have, therefore, three elements in our sensation of colour, each of which may vary independently. For distinctness sake I have spoken of intensity, hue, and tint; but if any other three independent qualities had been chosen, the one set might have been expressed in terms of the other, and the results identified.

The theory which I adopt assumes the existence of three elementary sensations, by the combination of which all the actual sensations of colour are produced. It will be shewn that it is not necessary to specify any given colours as typical of these sensations. Young has called them red, green, and violet; but any other three colours might have been chosen, provided that white resulted from their combination in proper proportions.

Before going farther I would observe, that the important part of the theory is not that three elements enter into our sensation of colour, but that there are only three. Optically, there are as many elements in the composition of a ray of light as there are different kinds of light in its spectrum; and, therefore, strictly speaking, its nature depends on an infinite number of independent variables.

I now go on to the geometrical form into which the theory may be thrown. Let it be granted that the three pure sensations correspond to the colours red, green, and violet, and that we can estimate the intensity of each of these sensations numerically.

Let v, r, g be the angular points of a triangle, and conceive the three sensations as having their positions at these points. If we find the numerical measure of the red, green, and violet parts of the sensation of a given colour, and then place weights proportional to these parts at r, g, and v, and find the centre of gravity of the three weights by the ordinary process, that point will be the position of the given colour, and the numerical measure of its intensity will be the sum of the three primitive sensations.

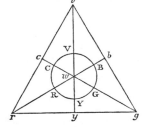

In this way, every possible colour may have its position and intensity ascertained; and it is easy to see that when two compound colours are combined, their centre of gravity is the position of the new colour.

The idea of this geometrical method of investigating colours is to be found in Newton's *Opticks* (Book I., Part 2, Prop. 6), but I am not aware that it has been ever employed in practice, except in the reduction of the experiments which I have just made. The accuracy of the method depends entirely on the truth of the theory of three sensations, and therefore its success is a testimony in favour of that theory.

Every possible colour must be included within the triangle *rgv*. White will be found at some point, *w*, within the triangle. If lines be drawn through *w* to any point, the colour at that point will vary in hue according to the angular position of the line drawn to *w*, and the purity of the tint will depend on the length of that line.

Though the homogeneous rays of the prismatic spectrum are absolutely pure in themselves, yet they do not give rise to the "pure sensations" of which we are speaking. Every ray of the spectrum gives rise to all three sensations, though in different proportions; hence the position of the colours of the spectrum is not at the boundary of the triangle, but in some curve *C R Y G B V* considerably within the triangle. The nature of this curve is not yet determined, but may form the subject of a future investigation*.

All natural colours must be within this curve, and all ordinary pigments do in fact lie very much within it. The experiments on the colours of the spectrum which I have made are not brought to the same degree of accuracy as those on coloured papers. I therefore proceed at once to describe the mode of making those experiments which I have found most simple and convenient.

The coloured paper is cut into the form of discs, each with a small hole in the centre, and divided along a radius, so as to admit of several of them being placed on the same axis, so that part of each is exposed. By slipping one disc over another, we can expose any given portion of each colour. These discs are placed on a little top or teetotum, consisting of a flat disc of tin-plate and a vertical axis of ivory. This 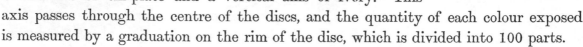 axis passes through the centre of the discs, and the quantity of each colour exposed is measured by a graduation on the rim of the disc, which is divided into 100 parts.

* [See the author's Memoir in the *Philosophical Transactions*, 1860, on the Theory of Compound Colours, and on the relations of the Colours of the Spectrum.]

By spinning the top, each colour is presented to the eye for a time pro- portional to the angle of the sector exposed, and I have found by independent experiments, that the colour produced by fast spinning is identical with that produced by causing the light of the different colours to fall on the retina at once.

By properly arranging the discs, any given colour may be imitated and afterwards registered by the graduation on the rim of the top. The principal use of the top is to obtain colour-equations. These are got by producing, by two different combinations of colours, the same mixed tint. For this purpose there is another set of discs, half the diameter of the others, which lie above them, and by which the second combination of colours is formed.

The two combinations being close together, may be accurately compared, and when they are made sensibly identical, the proportions of the different colours in each is registered, and the results equated.

These equations in the case of ordinary vision, are always between four colours, not including black.

From them, by a very simple rule, the different colours and compounds have their places assigned on the triangle of colours. The rule for finding the position is this :—Assume any three points as the positions of your three standard colours, whatever they are; then form an equation between the three standard colours, the given colour and black, by arranging these colours on the inner and outer circles so as to produce an identity when spun. Bring the given colour to the left-hand side of the equation, and the three standard colours to the right hand, leaving out black, then the position of the given colour is the centre of gravity of three masses, whose weights are as the number of degrees of each of the standard colours, taken positive or negative, as the case may be.

In this way the triangle of colours may be constructed by scale and compass from experiments on ordinary vision. I now proceed to state the results of experiments on Colour-Blind vision.

If we find two combinations of colours which appear identical to a Colour- Blind person, and mark their positions on the triangle of colours, then the straight line passing through these points will pass through all points corre- sponding to other colours, which, to such a person, appear identical with the first two.

We may in the same way find other lines passing through the series of

16—2

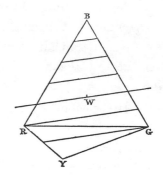

colours which appear alike to the Colour-Blind. All these lines either pass through one point or are parallel, according to the standard colours which we have assumed, and the other arbitrary assumptions we may have made. Knowing this law of Colour-Blind vision, we may predict any number of equations which will be true for eyes having this defect.

The mathematical expression of the difference between Colour-Blind and ordinary vision is, that colour to the former is a function of two independent variables, but to an ordinary eye, of three; and that the relation of the two kinds of vision is not arbitrary, but indicates the absence of a determinate sensation, depending perhaps upon some undiscovered structure or organic arrangement, which forms one-third of the apparatus by which we receive sensations of colour.

Suppose the absent structure to be that which is brought most into play when red light falls on our eyes, then to the Colour-Blind red light will be visible only so far as it affects the other two sensations, say of blue and green. It will, therefore, appear to them much less bright than to us, and will excite a sensation not distinguishable from that of a bluish-green light.

I cannot at present recover the results of all my experiments; but I recollect that the neutral colours for a Colour-Blind person may be produced by combining 6 degrees of ultramarine with 94 of vermilion, or 60 of emerald-green with 40 of ultramarine. The first of these, I suppose to represent to our eyes the kind of red which belongs to the red sensation. It excites the other two sensations, and is, therefore, visible to the Colour-Blind, but it appears very dark to them and of no definite colour. I therefore suspect that one of the three sensations in perfect vision will be found to correspond to a red of the same hue, but of much greater purity of tint. Of the nature of the other two, I can say nothing definite, except that one must correspond to a blue, and the other to a green, verging to yellow.

I hope that what I have written may help you in any way in your experiments. I have put down many things simply to indicate a way of thinking about colours which belongs to this theory of triple sensation. We are indebted to Newton for the original design; to Young for the suggestion of the means of working it out; to Prof. Forbes* for a scientific history of its application

*Phil. Mag. 1848.

to practice; to Helmholtz for a rigorous examination of the facts on which it rests; and to Prof. Grassman (in the *Phil. Mag.* for 1852), for an admirable theoretical exposition of the subject. The colours given in Hay's *Nomenclature of Colours* are illustrations of a similar theory applied to mixtures of pigments, but the results are often different from those in which the colours are combined by the eye alone. I hope soon to have results with pigments compared with those given by the prismatic spectrum, and then, perhaps, some more definite results may be obtained. Yours truly,

J. C. MAXWELL.

EDINBURGH, 4th *Jan.* 1855.

[From the *Transactions of the Royal Society of Edinburgh*, Vol. XXI. Part II.]

VII. *Experiments on Colour, as perceived by the Eye, with remarks on Colour-Blindness.* Communicated by Dr Gregory.

THE object of the following communication is to describe a method by which every variety of visible colour may be exhibited to the eye in such a form as to admit of accurate comparison; to shew how experiments so made may be registered numerically; and to deduce from these numerical results certain laws of vision.

The different tints are produced by means of a combination of discs of paper, painted with the pigments commonly used in the arts, and arranged round an axis, so that a sector of any required angular magnitude of each colour may be exposed. When this system of discs is set in rapid rotation, the sectors of the different colours become indistinguishable, and the whole appears of one uniform tint. The resultant tints of two different combinations of colours may be compared by using a second set of discs of a smaller size, and placing these over the centre of the first set, so as to leave the outer portion of the larger discs exposed. The resultant tint of the first combination will then appear in a ring round that of the second, and may be very carefully compared with it.

The form in which the experiment is most manageable is that of the common top. An axis, of which the lower extremity is conical, carries a circular plate, which serves as a support for the discs of coloured paper. The circumference of this plate is divided into 100 equal parts, for the purpose of ascertaining the proportions of the different colours which form the combination. When the discs have been properly arranged, the upper part of the axis is screwed down, so as to prevent any alteration in the proportions of the colours.

The instrument used in the first series of experiments (at Cambridge, in November, 1854) was constructed by myself, with coloured papers procured from

Mr D. R. Hay.   The experiments made in the present year were with the improved top made by Mr J. M. Bryson, Edinburgh, and coloured papers prepared by Mr T. Purdie, with the unmixed pigments used in the arts.   A number of Mr Bryson's tops, with Mr Purdie's coloured papers has been prepared, so as to afford different observers the means of testing and comparing results independently obtained.

The colour used for Mr Purdie's papers were—

Vermilion	.	.	V	Ultramarine	.	.	U	Emerald Green	.	.	EG
Carmine	.	.	C	Prussian Blue	.	.	PB	Brunswick Green	.	.	BG
Red Lead	.	.	RL	Verditer Blue	.	.	VB	Mixture of Ultramarine
Orange Orpiment	.	OO					and Chrome	.	.	UC

Vermilion . . V
Carmine . . . C
Red Lead . . RL
Orange Orpiment . OO
Orange Chrome . OC
Chrome Yellow . CY
Gamboge . . Gam
Pale Chrome . . PC

Ultramarine . . U
Prussian Blue . . PB
Verditer Blue . . VB

Emerald Green . . EG
Brunswick Green . . BG
Mixture of Ultramarine
  and Chrome . . UC

Ivory Black . . Bk
Snow White . . SW
White Paper (Pirie, Aberdeen).

The colours in the first column are reds, oranges, and yellows; those in the second, blues; and those in the third, greens.   Vermilion, ultramarine, and emerald green, seem the best colours to adopt in referring the rest to a uniform standard.   They are therefore put at the head of the list, as types of three convenient divisions of colour, red, blue, and green.

It may be asked, why some variety of yellow was not chosen in place of green, which is commonly placed among the secondary colours, while yellow ranks as a primary?   The reason for this deviation from the received system is, that the colours on the discs do not represent primary colours at all, but are simply specimens of different kinds of paint, and the choice of these was determined solely by the power of forming the requisite variety of combinations.   Now, if red, blue, and yellow, had been adopted, there would have been a difficulty in forming green by any compound of blue and yellow, while the yellow formed by vermilion and emerald green is tolerably distinct.   This will be more clearly perceived after the experiments have been discussed, by referring to the diagram.

As an example of the method of experimenting, let us endeavour to form a neutral gray by the combination of vermilion, ultramarine, and emerald green. The most perfect results are obtained by two persons acting in concert, when the operator arranges the colours and spins the top, leaving the eye of the observer free from the distracting effect of the bright colours of the papers when at rest.

After placing discs of these three colours on the circular plate of the top, and smaller discs of white and black above them, the operator must spin the top, and demand the opinion of the observer respecting the relation of the outer ring to the inner circle. He will be told that the outer circle is too red, too blue, or too green, as the case may be, and that the inner one is too light or too dark, as compared with the outer. The arrangement must then be changed, so as to render the resultant tint of the outer and inner circles more nearly alike. Sometimes the observer will see the inner circle tinted with the complementary colour of the outer one. In this case the operator must interpret the observation with respect to the outer circle, as the inner circle contains only black and white.

By a little experience the operator will learn how to put his questions, and how to interpret their answers. The observer should not look at the coloured papers, nor be told the proportions of the colours during the experiments. When these adjustments have been properly made, the resultant tints of the outer and inner circles ought to be perfectly indistinguishable, when the top has a sufficient velocity of rotation. The number of divisions occupied by the different colours must then be read off on the edge of the plate, and registered in the form of an equation. Thus, in the preceding experiment we have vermilion, ultramarine, and emerald green outside, and black and white inside. The numbers, as given by an experiment on the 6th March 1855, in daylight without sun, are—

$$\cdot 37 \ V + \cdot 27 \ U + \cdot 36 \ EG = \cdot 28 \ SW + \cdot 72 \ Bk \dots\dots\dots\dots(1).$$

The method of treating these equations will be given when we come to the theoretical view of the subject.

In this way we have formed a neutral gray by the combination of the three standard colours. We may also form neutral grays of different intensities by the combination of vermilion and ultramarine with the other greens, and thus obtain the quantities of each necessary to neutralize a given quantity of the proposed green. By substituting for each standard colour in succession one of the colours which stand under it, we may obtain equations, each of which contains two standard colours, and one of the remaining colours.

Thus, in the case of pale chrome, we have, from the same set of experiments,

$$\cdot 34 \text{ PC} + \cdot 55 \text{ U} + \cdot 12 \text{ EG} = \cdot 37 \text{ SW} + \cdot 63 \text{ Bk}\ldots\ldots\ldots\ldots(2).$$

We may also make experiments in which the resulting tint is not a neutral gray, but a decided colour. Thus we may combine ultramarine, pale chrome, and black, so as to produce a tint identical with that of a compound of vermilion and emerald-green. Experiments of this sort are more difficult, both from the inability of the observer to express the difference which he detects in two tints which have, perhaps, the same hue and intensity, but differ in purity; and also from the complementary colours which are produced in the eye after gazing too long at the colours to be compared.

The best method of arriving at a result in the case before us, is to render the *hue* of the red and green combination something like that of the yellow, to reduce the *purity* of the yellow by the admixture of blue, and to diminish its *intensity* by the addition of black. These operations must be repeated and adjusted, till the two tints are not merely varieties of the same colour, but absolutely the same. An experiment made 5th March gives—

$$\cdot 39 \text{ PC} + \cdot 21 \text{ U} + \cdot 40 \text{ Bk} = \cdot 59 \text{ V} + \cdot 41 \text{ EG}\ldots\ldots\ldots\ldots (3).$$

That these experiments are really evidence relating to the constitution of the eye, and not mere comparisons of two things which are in themselves identical, may be shewn by observing these resultant tints through coloured glasses, or by using gas-light instead of day-light. The tints which before appeared identical will now be manifestly different, and will require alteration, to reduce them to equality.

Thus, in the case of carmine, we have by day-light,

$$\cdot 44 \text{ C} + \cdot 22 \text{ U} + \cdot 34 \text{ EG} = \cdot 17 \text{ SW} + \cdot 83 \text{ Bk},$$

while by gas-light (Edinburgh)

$$\cdot 47 \text{ C} + \cdot 08 \text{ U} + \cdot 45 \text{ EG} = \cdot 25 \text{ SW} + \cdot 75 \text{ Bk},$$

which shews that the yellowing effect of the gas-light tells more on the white than on the combination of colours. If we examine the two resulting tints which appeared identical in experiment (3), observing the whirling discs through a blue glass, the combination of yellow, blue, and black, appears redder than the other, while through a yellow glass, the red and green mixture appears redder. So also a red glass makes the first side of the equation too dark, and a green glass makes it too light.

The apparent identity of the tints in these experiments is therefore not real, but a consequence of a determinate constitution of the eye, and hence arises the importance of the results, as indicating the laws of human vision.

The first result which is worthy of notice is, that the equations, as observed by different persons of ordinary vision, agree in a remarkable manner. If care be taken to secure the same kind of light in all the experiments, the equations, as determined by two independent observers, will seldom shew a difference of more than three divisions in any part of the equation containing the bright standard colours. As the duller colours are less active in changing the resultant tint, their true proportions cannot be so well ascertained. The accuracy of vision of each observer may be tested by repeating the same experiment at different times, and comparing the equations so found.

Experiments of this kind, made at Cambridge in November 1854, shew that of ten observers, the best were accurate to within $1\frac{1}{2}$ division, and agreed within 1 division of the mean of all; and the worst contradicted themselves to the extent of 6 degrees, but still were never more than 4 or 5 from the mean of all the observations.

We are thus led to conclude—

1st. That the human eye is capable of estimating the likeness of colours with a precision which in some cases is very great.

2nd. That the judgment thus formed is determined, not by the real identity of the colours, but by a cause residing in the eye of the observer.

3rd. That the eyes of different observers vary in accuracy, but agree with each other so nearly as to leave no doubt that the law of colour-vision is identical for all ordinary eyes.

Investigation of the Law of the Perception of Colour.

Before proceeding to the deduction of the elementary laws of the perception of colour from the numerical results previously obtained, it will be desirable to point out some general features of the experiments which indicate the form which these laws must assume.

Returning to experiment (1), in which a neutral gray was produced from red, blue, and green, we may observe, that, while the adjustments were incomplete, the difference of the tints could be detected only by one circle appearing more red, more green, or more blue than the other, or by being lighter or darker, that is, having an excess or defect of all the three colours together. Hence it appears that the nature of a colour may be considered as dependent on *three* things, as, for instance, redness, blueness, and greenness. This is confirmed by the fact that any tint may be imitated by mixing red, blue, and green alone, provided that tint does not exceed a certain brilliancy.

Another way of shewing that colour depends on *three* things is by considering how two tints, say two lilacs, may differ. In the first place, one may be *lighter* or *darker* than the other, that is, the tints may differ in *shade*. Secondly, one may be more *blue* or more *red* than the other, that is, they may differ in *hue*. Thirdly, one may be more or less *decided* in its colour; it may vary from purity on the one hand, to neutrality on the other. This is sometimes expressed by saying that they may differ in *tint*.

Thus, in shade, hue, and tint, we have another mode of reducing the elements of colour to three. It will be shewn that these two methods of considering colour may be deduced one from the other, and are capable of exact numerical comparison.

On a Geographical Method of Exhibiting the Relations of Colours.

The method which exhibits to the eye most clearly the results of this theory of the three elements of colour, is that which supposes each colour to be represented by a point in space, whose distances from three co-ordinate planes are proportional to the three elements of colour. But as any method by which the operations are confined to a plane is preferable to one requiring space of three dimensions, we shall only consider for the present that which has been adopted for convenience, founded on Newton's Circle of colours and Mayer and Young's Triangle.

Vermilion, ultramarine, and emerald-green, being taken (for convenience) as standard colours, are conceived to be represented by three points, taken (for convenience) at the angles of an equilateral triangle. Any colour compounded of these three is to be represented by a point found by conceiving masses proportional to the several components of the colour placed at their respective angular points, and taking the centre of gravity of the three masses. In this way, each

17—2

colour will indicate by its position the proportions of the elements of which it is composed. The total intensity of the colour is to be measured by the whole number of divisions of V, U, and EG, of which it is composed. This may be indicated by a number or coefficient appended to the name of the colour, by which the number of divisions it occupies must be multiplied to obtain its mass in calculating the results of new combinations.

This will be best explained by an example on the diagram (No. 1). We have, by experiment (1),

$$\cdot37\ V + \cdot27\ U + \cdot36\ EG = \cdot28\ SW + \cdot72\ Bk.$$

To find the position of the resultant neutral tint, we must conceive a mass of $\cdot37$ at V, of $\cdot27$ at U, and of $\cdot36$ at EG, and find the centre of gravity. This may be done by taking the line UV, and dividing it in the proportion of $\cdot37$ to $\cdot27$ at the point a, where

$$aV\ :\ aU\ ::\ \cdot27\ :\ \cdot37.$$

Then, joining a with EG, divide the joining line in W in the proportion of $\cdot36$ to $(\cdot37 + \cdot27)$, W will be the position of the neutral tint required, which is not white, but $0\cdot28$ of white, diluted with $0\cdot72$ of black, which has hardly any effect whatever, except in decreasing the amount of the other colour. The total intensity of our white paper will be represented by $\frac{1}{0\cdot28} = 3\cdot57$; so that, whenever white enters into an equation, the number of divisions must be multiplied by the coefficient $3\cdot57$ before any true results can be obtained.

We may take, as the next example, the method of representing the relation of pale chrome to the standard colours on our diagram, by making use of experiment (2), in which pale chrome, ultramarine, and emerald-green, produced a neutral gray. The resulting equation was

$$\cdot33\ PC + \cdot55\ U + \cdot12\ EG = \cdot37\ SW + \cdot63\ Bk \ \dots\dots\dots\dots\dots(2).$$

In order to obtain the total intensity of white, we must multiply the number of divisions, $\cdot37$, by the proper coefficient, which is $3\cdot57$. The result is $1\cdot32$, which therefore measures the total intensity on both sides of the equation.

Subtracting the intensity of $\cdot55\ U + \cdot12\ EG$, or $\cdot67$ from $1\cdot32$, we obtain $\cdot65$ as the *corrected* value of $\cdot33\ PC$. It will be convenient to use these corrected values of the different colours, taking care to distinguish them by small initials instead of capitals.

Equation (2) then becomes

$$\cdot65 \text{ pc} + \cdot55 \text{ U} + \cdot12 \text{ EG} = 1\cdot32 \text{ w}.$$

Hence pc must be situated at a point such that w is the centre of gravity of $\cdot65$ pc + $\cdot55$ U + $\cdot12$ EG.

To find it, we begin by determining β the centre of gravity of $\cdot55$ U + $\cdot12$ EG, then, joining βw, the point we are seeking must lie at a certain distance on the other side of w from c. This distance may be found from the proportion,

$$\cdot65 \;:\; (\cdot55 + \cdot12) \;::\; \overline{\beta\text{w}} \;:\; \overline{\text{w pc}},$$

which determines the position of pc. The proper coefficient, by which the observed values of PC must be corrected, is $\frac{65}{33}$, or $1\cdot97$.

We have thus determined the position and coefficient of a colour by a single experiment, in which it was made to produce a neutral tint along with two of the standard colours. As this may be done with every possible colour, the method is applicable wherever we can obtain a disc of the proposed colour. In this way the diagram (No. 1) has been laid down from observations made in daylight, by a good eye of the ordinary type.

It has been observed that experiments, in which the resultant tint is neutral, are more accurate than those in which the resulting tint has a decided colour, as in experiment (3), owing to the effects of accidental colours produced in the eye in the latter case. These experiments, however, may be repeated till a very good mean result has been obtained.

But since the elements of every colour have been already fixed by our previous observations and calculations, the agreement of these results with those calculated from the diagram forms a test of the correctness of our method.

By experiment (No. 3), made at the same time with (1) and (2), we have

$$\cdot39 \text{ PC} + \cdot21 \text{ U} + \cdot40 \text{ Bk} = \cdot59 \text{ V} + \cdot41 \text{ EG} \ \dots\dots\dots\dots\dots \ (3).$$

Now, joining U with pc, and V with EG, the only common point is that at which they cross, namely γ.

Measuring the parts of the line $\overline{\text{V EG}}$, we find them in the proportion of

$$\cdot58 \text{ V and } \cdot42 \text{ EG} = 1\cdot00 \ \gamma.$$

Similarly, the line $\overline{\text{U pc}}$ is divided in the proportion

$$\cdot78 \text{ pc and } \cdot22 \text{ U} = 1\cdot00 \ \gamma.$$

But ·78 pc must be divided by 1·97, to reduce it to PC, as was previously explained. The result of calculation is, therefore,

$$\cdot 39 \text{ PC} + \cdot 22 \text{ U} + \cdot 39 \text{ Bk} = \cdot 58 \text{ V} + \cdot 42 \text{ EG},$$

the black being introduced simply to fill up the circle.

This result differs very little from that of experiment (3), and it must be recollected that these are single experiments, made independently of theory, and chosen at random.

Experiments made at Cambridge, with all the combinations of five colours, shew that theory agrees with calculation always within 0·012 of the whole, and sometimes within 0·002. By the repetition of these experiments at the numerous opportunities which present themselves, the accuracy of the results may be rendered still greater. As it is, I am not aware that the judgments of the human eye with respect to colour have been supposed capable of so severe a test.

Further consideration of the Diagram of Colours.

We have seen how the composition of any tint, in terms of our three standard colours, determines its position on the diagram and its proper coefficient. In the same way, the result of mixing any other colours, situated at other points of the diagram, is to be found by taking the centre of gravity of their *reduced masses*, as was done in the last calculation (experiment 3).

We have now to turn our attention to the general aspect of the diagram.

The standard colours, V, U, and EG, occupy the angles of an equilateral triangle, and the rest are arranged in the order in which they participate in red, blue, and green, the neutral tint being at the point w within the triangle. If we now draw lines through w to the different colours ranged round it, we shall find that, if we pass from one line to another in the order in which they lie from red to green, and through blue back again to red, the order will be—

	Coefficient.		Coefficient.
Carmine	0·4	Pale Chrome	2·0
Vermilion	1·0	Mixed Green (U C) . . .	0·4
Red Lead	1·3	Brunswick Green . . .	0·2
Orange Orpiment . . .	1·0	*Emerald Green*	1·0
Orange Chrome . . .	1·6	Verditer Blue	0·8
Chrome Yellow . . .	1·5	Prussian Blue	0·1
Gamboge	1·8	*Ultramarine*	1·0

It may be easily seen that this arrangement of the colours corresponds to that of the prismatic spectrum; the only difference being that the spectrum is deficient in those fine purples which lie between ultramarine and vermilion, and which are easily produced by mixture. The experiments necessary for determining the exact relation of this list to the lines in the spectrum are not yet completed.

If we examine the colours represented by different points in one of these lines through w, we shall find the purest and most decided colours at its outer extremity, and the faint tints approaching to neutrality nearer to w.

If we also study the coefficients attached to each colour, we shall find that the brighter and more luminous colours have higher numbers for their coefficients than those which are dark.

In this way, the qualities which we have already distinguished as hue, tint, and shade, are represented on the diagram by angular position with respect to w, distance from w, and coefficient; and the relation between the two methods of reducing the elements of colour to three becomes a matter of geometry.

Theory of the Perception of Colour.

Opticians have long been divided on this point; those who trusted to popular notions and their own impressions adopting some theory of three primary colours, while those who studied the phenomena of light itself proved that no such theory could explain the constitution of the spectrum. Newton, who was the first to demonstrate the actual existence of a series of kinds of light, countless in number, yet all perfectly distinct, was also the first to propound a method of calculating the effect of the mixture of various coloured light; and this method was substantially the same as that which we have just verified. It is true, that the directions which he gives for the construction of his circle of colours are somewhat arbitrary, being probably only intended as an indication of the general nature of the method, but the method itself is mathematically reducible to the theory of three elements of the colour-sensation*.

* See Note III. For a confirmation of Newton's analysis of Light, see Helmholtz, Pogg. *Ann.* 1852; and *Phil. Mag.* 1852, Part II.

Young, who made the next great step in the establishment of the theory of light, seems also to have been the first to follow out the necessary consequences of Newton's suggestion on the mixture of colours. He saw that, since this triplicity has no foundation in the theory of light, its cause must be looked for in the constitution of the eye; and, by one of those bold assumptions which sometimes express the result of speculation better than any cautious trains of reasoning, he attributed it to the existence of three distinct modes of sensation in the retina, each of which he supposed to be produced in different degrees by the different rays. These three elementary effects, according to his view, correspond to the three sensations of red, green, and violet, and would separately convey to the sensorium the sensation of a red, a green, and a violet picture; so that by the superposition of these pictures, the actual variegated world is represented*.

In order fully to understand Young's theory, the function which he attributes to each system of nerves must be carefully borne in mind. Each nerve acts, not, as some have thought, by conveying to the mind the knowledge of the length of an undulation of light, or of its periodic time, but simply by being *more* or *less* affected by the rays which fall on it. The sensation of each elementary nerve is capable only of increase and diminution, and of no other change. We must also observe, that the nerves corresponding to the red sensation are affected chiefly by the red rays, but in some degree also by those of every other part of the spectrum; just as red glass transmits red rays freely, but also suffers those of other colours to pass in smaller quantity.

This theory of colour may be illustrated by a supposed case taken from the art of photography. Let it be required to ascertain the colours of a landscape, by means of impressions taken on a preparation equally sensitive to rays of every colour.

Let a plate of red glass be placed before the camera, and an impression taken. The positive of this will be transparent wherever the red light has been abundant in the landscape, and opaque where it has been wanting. Let it now be put in a magic lantern, along with the red glass, and a red picture will be thrown on the screen.

Let this operation be repeated with a green and a violet glass, and, by

* Young's *Lectures*, p. 345, Kelland's Edition. See also Helmholtz's statement of Young's Theory, in his Paper referred to in Note I.; and Herschel's *Light*, Art. 518.

means of three magic lanterns, let the three images be superimposed on the screen. The colour of any point on the screen will then depend on that of the corresponding point of the landscape; and, by properly adjusting the intensities of the lights, &c., a complete copy of the landscape, as far as visible colour is concerned, will be thrown on the screen. The only apparent difference will be, that the copy will be more subdued, or less pure in tint, than the original. Here, however, we have the process performed twice—first on the screen, and then on the retina.

This illustration will shew how the functions which Young attributes to the three systems of nerves may be imitated by optical apparatus. It is therefore unnecessary to search for any direct connection between the lengths of the undulations of the various rays of light and the sensations as felt by us, as the threefold partition of the properties of light may be effected by physical means. The remarkable correspondence between the results of experiments on different individuals would indicate some anatomical contrivance identical in all. As there is little hope of detecting it by dissection, we may be content at present with any subsidiary evidence which we may possess. Such evidence is furnished by those individuals who have the defect of vision which was described by Dalton, and which is a variety of that which Dr G. Wilson has lately investigated, under the name of Colour-Blindness.

Testimony of the Colour-Blind with respect to Colour.

Dr George Wilson has described a great number of cases of colour-blindness, some of which involve a general indistinctness in the appreciation of colour, while in others, the errors of judgment are plainly more numerous in those colours which approach to red and green, than among those which approach to blue and yellow. In these more definite cases of colour-blindness, the phenomena can be tolerably well accounted for by the hypothesis of an insensibility to red light; and this is, to a certain extent, confirmed by the fact, that red objects appear to these eyes decidedly more obscure than to ordinary eyes. But by experiments made with the pure spectrum, it appears that though the red appears much more obscure than other colours, it is not wholly invisible, and, what is more curious, resembles the green more than any other colour. The spectrum to them appears faintly luminous in the red;

bright yellow from orange to yellow, bright but not coloured from yellow-green to blue, and then strongly coloured in the extreme blue and violet, after which it seems to approach the neutral obscure tint of the red. It is not easy to see why an insensibility to red *rays* should deprive the green rays, which have no *optical* connection with them, of their distinctive appearance. The phenomena seem rather to lead to the conclusion that it is the red *sensation* which is wanting, that is, that supposed system of nerves which is affected in various degrees by all light, but chiefly by red. We have fortunately the means of testing this hypothesis by numerical results.

Of the subjects of my experiments at Cambridge, four were decided cases of colour-blindness. Of these two, namely, Mr R. and Mr S., were not sufficiently critical in their observations to afford any results consistent within 10 divisions of the colour-top. The remaining two, Mr N. and Mr X., were as consistent in their observations as any persons of ordinary vision can be, while the results shewed all the more clearly how completely their sensations must differ from ours.

The method of experimenting was the same as that adopted with ordinary eyes, except that in these cases the operator can hardly influence the result by yielding to his own impressions, as he has no perception whatever of the similarity of the two tints as seen by the observer. The questions which he must ask are two, Which circle appears most blue or yellow? Which appears lightest and which darkest? By means of the answers to these questions he must adjust the resulting tints to equality in these respects as it appears to the observer, and then ascertain that these tints now present no difference of colour whatever to his eyes. The equations thus obtained do not require five colours including black, but four only. For instance, the mean of several observations gives—

$$\cdot 19\,G + \cdot 05\,B + \cdot 76\,Bk = 100\,R \dots\dots\dots\dots(4).$$

[In these experiments R, B, G, Y, stand for red, blue, green, and yellow papers prepared by Mr D. R. Hay. I am not certain that they are identical with his standard colours, but I believe so. Their relation to vermilion, ultramarine, and emerald-green is given in diagram (1). Their relations to each other are very accurately given in diagram (2).]

It appears, then, that the dark blue-green of the left side of the equation is equivalent to the full red of the right side.

Hence, if we divide the line BG in the proportion 19 to 5 at the point β, and join Rβ, the tint at β will differ from that at R (to the colour-blind) only in being more brilliant in the proportion of 100 to 24, and all intermediate tints on the line Rβ will appear to them of the same hue, but of intermediate intensities.

Now, if we take a point D, so that RD is to Rβ in the proportion of 24 to $100 - 24$, or 76, the tint of D, if producible, should be invisible to the colour-blind. D, therefore, represents the pure sensation which is unknown to the colour-blind, and the addition of this sensation to any others cannot alter it in their estimation. It is for them equivalent to black.

Hence, if we draw lines through D in different directions, the colours belonging to any line ought to differ only in intensity as seen by them, so that one of them may be reduced to the other by the addition of black only. If we draw DW and produce it, all colours on the upper side of DW will be varieties of blue, and those on the under side varieties of yellow, so that the line DW is a boundary line between their two kinds of colour, blue and yellow being the names by which they call them.

The accuracy of this theory will be evident from the comparison of the experiments which I had an opportunity of making on Mr N. and Mr X. with each other, and with measurements taken from the diagram No. 2, which was constructed from the observations of ordinary eyes only, the point D alone being ascertained from a series of observations by Mr N.

Taking the point γ, between R and B, it appears, by measurement of the lines Rγ and Bγ, that γ corresponds to

$$\cdot 07 \, B + \cdot 93 \, R.$$

By measurement of Wγ and Dγ, and correction by means of the coefficient of W, and calling D black in the colour-blind language, γ corresponds to

$$\cdot 105 \, W + \cdot 895 \, Bk.$$

Therefore

By measurement	$\cdot 93 \, R + \cdot 07 \, B = \cdot 105 \, W + \cdot 895 \, Bk$	⎫
By observation N. & X. together	$\cdot 94 \, R + \cdot 06 \, B = \cdot 10 \;\; W + \cdot 90 \;\; Bk$	⎬ ......(5).
By X. alone	$\cdot 93 \, R + \cdot 07 \, B = \cdot 10 \;\; W + \cdot 90 \;\; Bk$	⎭

The agreement here is as near as can be expected.

18—2

By a similar calculation with respect to the point δ, between B and G,

By measurement .... ·43 B + ·57 G = ·335 W + ·665 Bk ⎫
Observed by N. and X. ...... ·41 B + ·59 G = ·34 W + ·66 Bk ⎬ ......(6).
By X. alone .... ·42 B + ·58 G = ·32 W + ·68 Bk ⎭

We may also observe, that the line GD crosses RY. At the point of inter-section we have—

By calculation .... ·87 R + ·13 Y = ·34 G + ·66 Bk ⎫
Observed by N. and X. .... ·86 R + ·14 Y = ·40 G + ·60 Bk ⎪
„   „   X. .... ·84 R + ·16 Y = ·31 G + ·69 Bk ⎬ ......(7).
„   „   X. .... ·90 R + ·10 Y = ·27 G + ·73 Bk ⎭

Here observations are at variance, owing to the decided colours produced affecting the state of the retina, but the mean agrees well with calculation.

Drawing the line BY, we find that it cuts lines through D drawn to every colour. Hence all colours appear to the colour-blind as if composed of blue and yellow. By measurement on the diagram, we find for red

Measured ... ... ·138 Y + ·123 B + ·749 Bk = 100 R ⎫
Observed by N.... ·15 Y + ·11 B + ·74 Bk = 100 R ⎬ ....(8).
...   X.... ·13 Y + ·11 B + ·76 Bk = 100 R ⎭

For green we have in the same way—

Measured ... ... ·705 Y + ·295 B = ·95 G + ·05 Bk ⎫
Observed by N.... ·70 Y + ·30 B = ·86 G + ·14 Bk ⎬ ....(9).
...   X.... ·70 Y + ·30 B = ·83 G + ·17 Bk ⎭

For white—

Measured ... ... ·407 Y + ·593 B = ·326 W + ·674 Bk
Observed by N.... ·40 Y + ·60 B = ·33 W + ·67 Bk
...   X.... ·44 Y + ·56 B = ·33 W + ·67 Bk

The accuracy of these results shews that, whether the hypothesis of the want of one element out of three necessary to perfect vision be actually true or not, it affords a most trustworthy foundation on which to build a theory of colour-blindness, as it expresses completely the observed facts of the case. They also furnish us with a datum for our theory of perfect vision, namely, the point D, which points out the exact nature of the colour-sensation, which must be added to the colour-blind eye to render it perfect. I am not aware of any method of determining by a legitimate process the nature of the other two sensations, although Young's reasons for adopting something like green and violet appear to me worthy of attention.

The only remaining subject to which I would call the attention of the Society is the effect of coloured glasses on the colour-blind. Although they cannot distinguish reds and greens from varieties of gray, the transparency of red and green glasses for those kinds of light is very different. Hence, after finding a case such as that in equation (4), in which a red and a green appear identical, on looking through a red glass they see the red clearly and the green obscurely, while through a green glass the red appears dark and the green light.

By furnishing Mr X. with a red and a green glass, which he could distinguish only by their shape, I enabled him to make judgments in previously doubtful cases of colour with perfect certainty. I have since had a pair of spectacles constructed with one eye-glass red and the other green. These Mr X. intends to use for a length of time, and he hopes to acquire the habit of discriminating red from green tints by their different effects on his two eyes. Though he can never acquire our sensation of red, he may then discern for himself what things are red, and the mental process may become so familiar to him as to act unconsciously like a new sense.

In one experiment, after looking at a bright light, with a red glass over one eye and a green over the other, the two tints in experiment (4) appeared to him altered, so that the outer circle was lighter according to one eye, and the inner according to the other. As far as I could ascertain, it appeared as if the eye which had used the red glass saw the red circle brightest. This result, which seems at variance with what might be expected, I have had no opportunity of verifying.

This paper is already longer than was originally intended. For further information I would refer the reader to Newton's *Opticks*, Book I. Part II., to Young's *Lectures on Natural Philosophy*, page 345, to Mr D. R. Hay's works on Colours, and to Professor Forbes on the "Classification of Colours" (*Phil. Mag.*, March, 1849).

The most remarkable paper on the subject is that of M. Helmholtz, in the *Philosophical Magazine* for 1852, in which he discusses the different theories of primary colours, and describes his method of mixing the colours of the spectrum. An examination of the results of M. Helmholtz with reference to the theory of three elements of colour, by Professor Grassmann, is translated in the *Phil. Mag.*, April, 1854.

References to authors on colour-blindness are given in Dr G. Wilson's papers on that subject. A valuable Letter of Sir J. F. W. Herschel to Dalton on his peculiarity of vision, is to be found in the *Life of Dalton* by Dr Henry.

I had intended to describe some experiments on the propriety of the method of mixing colours by rotation, which might serve as an extension of Mr Swan's experiments on instantaneous impressions on the eye. These, together with the explanation of some phenomena which seem to be at variance with the theory of vision here adopted, must be deferred for the present. On some future occasion, I hope to be able to connect these simple experiments on the colours of pigments with others in which the pure hues of the spectrum are used. I have already constructed a model of apparatus for this purpose, and the results obtained are sufficiently remarkable to encourage perseverance.

NOTE I.

On different Methods of Exhibiting the Mixtures of Colours.

(1) *Mechanical Mixture of Coloured Powders.*

By grinding coloured powders together, the differently-coloured particles may be so intermingled that the eye cannot distinguish the colours of the separate powders, but receives the impression of a uniform tint, depending on the nature and proportions of the pigments used. In this way, Newton mixed the powders of orpiment, purple, bise, and *viride æris*, so as to form a gray, which, in sunlight, resembled white paper in the shade. (Newton's *Opticks*, Book I. Part II., Exp. xv.) This method of mixture, besides being adopted by all painters, has been employed by optical writers as a means of obtaining numerical results. The specimens of such mixtures given by D. R. Hay in his works on Colour, and the experiments of Professor J. D. Forbes on the same subject, shew the importance of the method as a means of classifying colours. There are two objections, however, to this method of exhibiting colours to the eye. When two powders of unequal fineness are mixed, the particles of the finer powder cover over those of the coarser, so as to produce more than their due effect in influencing the resultant tint. For instance, a small quantity of lamp-black, mixed with a large quantity of chalk, will produce a mixture which is nearly black. Although the powders generally used are not so different in this respect as lamp-black and chalk, the results of mixing given weights of any coloured powders must be greatly modified by the mode in which these powders have been prepared.

Again, the light which reaches the eye from the surface of the mixed powders consists partly of light which has fallen on one of the substances mixed without being modified by the other, and partly of light which, by repeated reflection or transmission, has been acted on by both substances. The colour of these rays will not be a mixture of those of the substances, but will be the result of the absorption due to both substances successively. Thus, a mixture of yellow and blue produces a neutral tint tending towards red, but the remainder of white light, after passing through both, is green; and this green is generally sufficiently powerful to overpower the reddish gray due to the separate colours of the substances mixed. This curious result has been ably investigated by Professor Helmholtz of Königsberg, in his *Memoir on the Theory of Compound Colours*, a translation of which may be found in the *Annals of Philosophy* for 1852, Part 2.

(2) *Mixture of differently-coloured Beams of Light by Superposition on an Opaque Screen.*

When we can obtain light of sufficient intensity, this method produces the most beautiful results. The best series of experiments of this kind are to be found in Newton's *Opticks*, Book I. Part II. The different arrangements for mixing the rays of the spectrum on a screen, as described by Newton, form a very complete system of combinations of lenses and prisms, by which almost every possible modification of coloured light may be produced. The principal objections to the use of this method are—(1) The difficulty of obtaining a constant supply of uniformly intense light; (2) The uncertainty of the effect of the position of the screen with respect to the incident beams and the eye of the observer; (3) The possible change in the colour of the incident light due to the *fluorescence* of the substance of the screen. Professor Stokes has found that many substances, when illuminated by homogeneous light of one refrangibility, become themselves luminous, so as to emit light of lower refrangibility. This phenomenon must be carefully attended to when screens are used to exhibit light.

(3) *Union of Coloured Beams by a Prism so as to form one Beam.*

The mode of viewing the beam of light directly, without first throwing it on a screen, was not much used by the older experimenters, but it possesses the advantage of saving much light, and admits of examining the rays before they have been stopped in any way. In Newton's 11th proposition of the 2nd Book, an experiment is described, in which a beam is analysed by a prism, concentrated by a lens, and recombined by another prism, so as to form a beam of white light similar to the incident beam. By stopping the coloured rays at the lens, any proposed combination may be made to pass into the emergent beam, where it may be received directly by the eye, or on a screen, at pleasure.

The experiments of Helmholtz on the colours of the spectrum were made with the ordinary apparatus for directly viewing the pure spectrum, two oblique slits crossing one another being employed to admit the light instead of one vertical slit. Two pure spectra were then seen crossing each other, and so exhibiting at once a large number of combinations. The proportions of these combinations were altered by varying the inclination of the slits to the plane of refraction, and in this way a number of very remarkable results were obtained,— for which see his Memoir, before referred to.

In experiments of the same kind made by myself in August 1852 (independently of M. Helmholtz), I used a combination of three moveable vertical slits to admit the light, instead of two cross slits, and observed the compound ray through a slit made in a screen on which the pure spectrum is formed. In this way a considerable field of view was filled with the mixed light, and might be compared with another part of the field illuminated by light proceeding from a second system of slits, placed below the first set. The general character of the results agreed with those of M. Helmholtz. The chief difficulties seemed to arise from the defects of the optical apparatus of my own eye, which rendered apparent the compound nature of the light, by analysing it as a prism or an ordinary lens would do, whenever the lights mixed differed much in refrangibility.

(4) *Union of two beams by means of a transparent surface, which reflects the first and transmits the second.*

The simplest experiment of this kind is described by M. Helmholtz. He places two coloured wafers on a table, and then, taking a piece of transparent glass, he places it between them, so that the reflected image of one apparently coincides with the other as seen through the glass. The colours are thus mixed, and, by varying the angle of reflection, the relative intensities of the reflected and transmitted beams may be varied at pleasure.

In an instrument constructed by myself for photometrical purposes two reflecting plates were used. They were placed in a square tube, so as to polarize the incident light, which entered through holes in the sides of the tubes, and was reflected in the direction of the axis. In this way two beams oppositely polarized were mixed, either of which could be coloured in any way by coloured glasses placed over the holes in the tube. By means of a Nicol's prism placed at the end of the tube, the relative intensities of the two colours as they entered the eye could be altered at pleasure.

(5) *Union of two coloured beams by means of a doubly-refracting Prism.*

I am not aware that this method has been tried, although the opposite polarization of the emergent rays is favourable to the variation of the experiment.

(6) *Successive presentation of the different Colours to the Retina.*

It has long been known, that light does not produce its full effect on the eye at once, and that the effect, when produced, remains visible for some time after the light has ceased to act. In the case of the rotating disc, the various colours become indistinguishable, and the disc appears of a uniform tint, which is in some sense the resultant of the colours so blended. This method of combining colours has been used since the time of Newton, to exhibit the results of theory. The experiments of Professor J. D. Forbes, which I witnessed in 1849, first encouraged me to think that the laws of this kind of mixture might be discovered by special experiments. After repeating the well-known experiment in which a series of colours representing those of the spectrum are combined to form gray, Professor Forbes endeavoured to form a neutral tint, by the combination of three colours only. For this purpose, he combined the three so-called primary colours, red, blue, and yellow, but the resulting tint could not be rendered neutral by any combination of these colours; and the reason was found to be, that blue and yellow do not make green, but a pinkish tint, when neither prevails in the combination. It was plain, that no addition of *red* to this, could produce a neutral tint.

This result of mixing blue and yellow was, I believe, not previously known. It directly contradicted the received theory of colours, and seemed to be at variance with the fact, that the same blue and yellow paint, when ground together, do make green. Several experiments were proposed by Professor Forbes, in order to eliminate the effect of motion, but he was not then able to undertake them. One of these consisted in viewing alternate stripes of blue and yellow, with a telescope out of focus. I have tried this, and find the resultant tint pink as before*. I also found that the beams of light coloured by transmission through blue and yellow glasses appeared pink, when mixed on a screen, while a beam of light, after passing through both glasses, appeared green. By the help of the theory of absorption, given by Herschel†, I made out the complete explanation of this phenomenon. Those of pigments were, I think, first explained by Helmholtz in the manner above referred to‡.

It may still be asked, whether the effect of successive presentation to the eye is identical with that of simultaneous presentation, for if there is any action of the one kind of light on the other, it can take place only in the case of simultaneous presentation. An experiment tending to settle this point is recorded by Newton (Book I. Part II., Exp. 10). He used a comb with large teeth to intercept various rays of the spectrum. When it was moved slowly, the various colours could be perceived, but when the speed was increased the result was perfect whiteness. For another form of this experiment, see Newton's *Sixth Letter to Oldenburg* (Horsley's Edition, Vol. IV., page 335).

In order more fully to satisfy myself on this subject, I took a disc in which were cut a number of slits, so as to divide it into spokes. In a plane, nearly passing through the axis of this disc, I placed a blue glass, so that one

---

* See however *Encyc. Metropolitana*, Art. "Light," section 502.   † *Ib.* sect. 516.

‡ I have lately seen a passage in Moigno's *Cosmos*, stating that M. Plateau, in 1819, had obtained gray by whirling together gamboge and Prussian blue. *Correspondance Math. et Phys.* de M. Quetelet, Vol. v., p. 221.

half of the disc might be seen by transmitted light—blue, and the other by reflected light—white. In the course of the reflected light I placed a yellow glass, and in this way I had two nearly coincident images of the slits, one yellow and one blue. By turning the disc slowly, I observed that in some parts the yellow slits and the blue slits appeared to pass over the field alternately, while in others they appeared superimposed, so as to produce alternately their mixture, which was pale pink, and complete darkness. As long as the disc moved slowly I could perceive this, but when the speed became great, the whole field appeared uniformly coloured pink, so that those parts in which the colours were seen successively were indistinguishable from those in which they were presented together to the eye.

Another form in which the experiment may be tried requires only the colour-top above described. The disc should be covered with alternate sectors of any two colours, say red and green, disposed alternately in four quadrants. By placing a piece of glass above the top, in the plane of the axis, we make the image of one half seen by reflection coincide with that of the other seen by transmission. It will then be seen that, in the diameters of the top which are parallel and perpendicular to the plane of reflection, the transmitted green coincides with the reflected green, and the transmitted red with the reflected red, so that the result is always either pure red or pure green. But in the diameters intermediate to these, the transmitted red coincides with the reflected green, and *vice versa*, so that the pure colours are never seen, but only their mixtures. As long as the top is spun slowly, these parts of the disc will appear more steady in colour than those in which the greatest alternations take place; but when the speed is sufficiently increased, the disc appears perfectly uniform in colour. From these experiments it appears, that the apparent mixture of colours is not due to a mechanical superposition of vibrations, or to any mutual action of the mixed rays, but to some cause residing in the constitution of the apparatus of vision.

(7) *Presentation of the Colours to be mixed one to each Eye.*

This method is said not to succeed with some people; but I have always found that the mixture of *colours* was perfect, although it was difficult to conceive the *objects* seen by the two eyes as identical. In using the spectacles,

19—2

of which one eye is green and the other red, I have found, when looking at an arrangement of green and red papers, that some looked metallic and others transparent. This arises from the very different relations of brightness of the two colours as seen by each eye through the spectacles, which suggests the false conclusion, that these differences are the result of reflection from a polished surface, or of light transmitted through a clear one.

NOTE II.

Results of Experiments with Mr Hay's *Papers at Cambridge, November,* 1854.

The mean of ten observations made by six observers gave

$$\cdot449 \text{ R} + \cdot299 \text{ G} + \cdot252 \text{ B} = \cdot224 \text{ W} + \cdot776 \text{ Bk} \dots\dots\dots\dots(1).$$
$$\cdot696 \text{ R} + \cdot304 \text{ G} = \cdot181 \text{ B} + \cdot327 \text{ Y} + \cdot492 \text{ Bk} \dots\dots\dots\dots(2).$$

These two equations served to determine the positions of white and yellow in diagram No. 2. The coefficient of W is 4·447, and that of yellow 2·506.

From these data we may deduce three other equations, either by calculation, or by measurement on the diagram (No. 2).

Eliminating green from the equations, we find

$$\cdot565 \text{ B} + \cdot435 \text{ Y} = \cdot307 \text{ R} + \cdot304 \text{ W} + \cdot389 \text{ Bk}\dots\dots\dots\dots(3).$$

The mean of three observations by three different observers gives

$$\cdot573 \text{ B} + \cdot477 \text{ Y} = \cdot313 \text{ R} + \cdot297 \text{ W} + \cdot390 \text{ Bk}.$$

Errors of calculation $\quad -\cdot008 \text{ B} + \cdot008 \text{ Y} - \cdot006 \text{ R} + \cdot007 \text{ W} - \cdot001 \text{ Bk}.$

The point on the diagram to which this equation corresponds is the intersection of the lines BY and RW, and the resultant tint is a pinkish-gray.

Eliminating red from the equations, we obtain

Calculation	$\cdot533 \text{ B} + \cdot150 \text{ G} + \cdot317 \text{ Y} = \cdot337 \text{ W} + \cdot663 \text{ Bk}$	
By 10 observations	$\cdot537 \text{ B} + \cdot146 \text{ G} + \cdot317 \text{ Y} = \cdot337 \text{ W} + \cdot663 \text{ Bk}$	......(4).
Errors	$-\cdot004 \quad +\cdot004 \quad\;\; - \quad\quad\;\; - \quad\quad\;\; -$	
Eliminating blue	$\cdot660 \text{ R} + \cdot340 \text{ G} = \cdot218 \text{ Y} + \cdot108 \text{ W} + \cdot682 \text{ Bk}$	
By 5 observations	$\cdot672 \text{ R} + \cdot328 \text{ G} = \cdot224 \text{ Y} + \cdot094 \text{ W} + \cdot672 \text{ Bk}$	......(5).
Errors	$-\cdot012 \quad +\cdot012 \quad -\cdot006 \quad +\cdot014 \quad +\cdot008$	

Note III.

On the Theory of Compound Colours.

Newton's theorem on the mixture of colours is to be found in his *Opticks,* Book I., Part II., Prop. VI.

In a mixture of primary colours, the quantity and quality of each being given, to know the colour of the compound.

He divides the circumference of a circle into parts proportional to the seven musical intervals, in accordance with his opinion of the divisions of the spectrum. He then conceives the colours of the spectrum arranged round the circle, and at the centre of gravity of each of the seven arcs he places a little circle, the area of which represents the number of rays of the corresponding colour which enter into the given mixture. He takes the centre of gravity of all these circles to represent the colour formed by the mixture. The *hue* is determined by drawing a line through the centre of the circle and this point to the circumference. The position of this line points out the colour of the spectrum which the mixture most resembles, and the distance of the resultant tint from the centre determines the fulness of its colour.

Newton, by this construction (for which he gives no reasons), plainly shews that he considered it possible to find a place within his circle for every possible colour, and that the entire nature of any compound colour may be known from its place in the circle. It will be seen that the same colour may be compounded from the colours of the spectrum in an infinite variety of ways. The apparent identity of all these mixtures, which are optically different, as may be shewn by the prism, implies some law of vision not explicitly stated by Newton. This law, if Newton's method be true, must be that which I have endeavoured to establish, namely, the threefold nature of sensible colour.

With respect to Newton's construction, we now know that the proportions of the colours of the spectrum vary with the nature of the refracting medium. The only *absolute* index of the kind of light is the *time* of its vibration. The *length* of its vibration depends on the medium in which it is; and if any proportions are to be sought among the wave-lengths of the colours, they must be determined for those tissues of the eye in which their physical effects are supposed to terminate. It may be remarked, that the apparent colour of the spectrum changes most rapidly at three points, which lie respectively in the yellow, between blue and green, and between violet and blue. The wave-lengths of the corresponding rays *in water* are in the proportions of three geometric means between 1 and 2 very nearly. This result, however, is not to be considered established, unless confirmed by better observations than mine.

The only safe method of completing Newton's construction is by an examination of the colours of the spectrum and their mixtures, and subsequent calculation by the method used in the experiments with coloured papers. In this way I hope to determine the relative positions in the colour-diagram of every ray of the spectrum, and its relative intensity in the solar light. The spectrum will then form a curve not necessarily circular or even re-entrant, and its peculiarities so ascertained may form the foundation of a more complete theory of the colour-sensation.

On the relation of the pure rays of the Spectrum to the three assumed Elementary Sensations.

If we place the three elementary colour-sensations (which we may call, after Young, red, green, and violet) at the angles of a triangle, all colours which the eye can possibly perceive (whether by the action of light, or by pressure, disease, or imagination) must be somewhere within this triangle, those which lie farthest from the centre being the fullest and purest colours. Hence the colours which lie at the middle of the sides are the purest of their kind which the eye can see, although not so pure as the elementary sensations.

It is natural to suppose that the pure red, green, and violet rays of the spectrum produce the sensations which bear their names in the highest purity. But from this supposition it would follow that the yellow, composed of the red and green of the spectrum, would be the most intense yellow possible, while it is the result of experiment, that the yellow of the spectrum itself is much more full in colour. Hence the sensations produced by the pure red and green rays of the spectrum are not the pure sensations of our theory. Newton has remarked, that no two colours of the spectrum produce, when mixed, a colour equal in fulness to the intermediate colour. The colours of the spectrum are all more intense than any compound ones. Purple is the only colour which must be produced by combination. The experiments of Helmholtz lead to the same conclusion; and hence it would appear that we can find no part of the spectrum which produces a pure sensation.

An additional, though less satisfactory evidence of this, is supplied by the observation of the colours of the spectrum when excessively bright. They then appear to lose their peculiar colour, and to merge into pure whiteness. This is probably due to the want of capacity of the organ to take in so strong an impression; one sensation becomes first saturated, and the other two speedily follow it, the final effect being simple brightness.

From these facts I would conclude, that every ray of the spectrum is capable of producing all three pure sensations, though in different degrees. The curve, therefore, which we have supposed to represent the spectrum will be quite within the triangle of colour. All natural or artificial colours, being compounded of the colours of the spectrum, must lie within this curve, and, therefore, the colours corresponding to those parts of the triangle beyond this curve must be for ever unknown to us. The determination of the exact nature of the pure sensations, or of their relation to ordinary colours, is therefore impossible, unless we can prevent them from interfering with each other as they do. It may be possible to experience sensations more pure than those directly produced by the spectrum, by first exhausting the sensibility to one colour by protracted gazing, and then suddenly turning to its opposite. But if, as I suspect, colour-blindness be due to the absence of one of these sensations, then the point D in diagram (2), which indicates their absent sensation, indicates also our pure sensation, which we may call red, but which we can never experience, because all kinds of light excite the other sensations.

Newton has stated one objection to his theory, as follows:—"*Also, if only two of the primary colours, which in the circle are opposite to one another, be mixed in an equal proportion, the point Z*" (the resultant tint) "*shall fall upon the centre O*" (neutral tint); "*and yet the colour compounded of these two shall not be perfectly white, but some faint anonymous colour. For I could never yet, by mixing only two primary colours, produce a perfect white.*" This is confirmed by the experiments of Helmholtz; who, however, has succeeded better with some pairs of colours than with others.

In my experiments on the spectrum, I came to the same result; but it appeared to me that the very peculiar appearance of the neutral tints produced was owing to some optical effect taking place in the transparent part of the eye on the mixture of two rays of very different refrangibility. Most eyes are by no means achromatic, so that the images of objects illuminated with mixed light of this kind appear divided into two different colours; and even when there is no distinct object, the mixtures become in some degree analysed, so as to present a very strange, and certainly "anonymous" appearance.

Additional Note on the more recent experiments of M. Helmholtz*.

In his former memoir on the Theory of Compound Colours†, M. Helmholtz arrived at the conclusion that only one pair of homogeneous colours, orange-yellow and indigo-blue, were strictly complementary. This result was shewn by Professor Grassmann‡ to be at variance with Newton's theory of compound colours; and although the reasoning was founded on intuitive rather than experimental truths, it pointed out the tests by which Newton's theory must be verified or overthrown. In applying these tests, M. Helmholtz made use of an apparatus similar to that described by M. Foucault§, by which a screen of white paper is illuminated by the mixed light. The field of mixed colour is much larger than in M. Helmholtz's former experiments, and the facility of forming combinations is much increased. In this memoir the mathematical theory of Newton's circle, and of the curve formed by the spectrum, with its possible transformations, is completely stated, and the form of this curve is in some degree indicated, as far as the determination of the colours which lie on opposite sides of white, and of those which lie opposite the part of the curve which is wanting. The colours between red and yellow-green are complementary to colours between blue-green and violet, and those between yellow-green and blue-green have no homogeneous complementaries, but must be neutralized by various hues of purple, i.e., mixtures of red and violet. The names of the complementary colours, with their wave-lengths in air, as deduced from Fraunhofer's measurements, are given in the following table :—

---

* Poggendorff's *Annalen*, Bd. XCIV.   (I am indebted for the perusal of this Memoir to Professor Stokes.)

† *Ib.* Bd. LXXXVII. *Annals of Philosophy*, 1852, Part II.

‡ *Ib.* Bd. LXXXIX. *Ann. Phil.*, 1854, April.

§ *Ib.* Bd. LXXXVIII. Moigno, *Cosmos*, 1853, Tom. II., p. 232.

Colour	Wave-length	Complementary Colour	Wave-length	Ratio of wave-lengths
Red	2425	Green-blue . .	1818	1·334
Orange . . .	2244	Blue	1809	1·240
Gold-yellow .	2162	Blue	1793	1·206
Gold yellow .	2120	Blue	1781	1·190
Yellow . . .	2095	Indigo-blue .	1716	1·221
Yellow . . .	2085	Indigo-blue .	1706	1·222
Green-yellow .	2082	Violet . . .	1600 –	1·301

(The wave-lengths are expressed in millionths of a Paris inch.)

(In order to reduce these wave-lengths to their actual lengths in the eye, each must be divided by the index of refraction for that kind of light in the medium in which the physical effect of the vibrations is supposed to take place.)

Although these experiments are not in themselves sufficient to give the complete theory of the curve of homogeneous colours, they determine the most important element of that theory in a way which seems very accurate, and I cannot doubt that when a philosopher who has so fully pointed out the importance of general theories in physics turns his attention to the theory of sensation, he will at least establish the principle that the laws of sensation can be successfully investigated only after the corresponding physical laws have been ascertained, and that the connection of these two kinds of laws can be apprehended only when the distinction between them is fully recognised.

NOTE IV.

Description of the Figures. Plate I.

No. 1. is the colour-diagram already referred to, representing, on Newton's principle, the relations of different coloured papers to the three standard colours—vermilion, emerald-green, and ultramarine. The initials denoting the colours are explained in the list at page 276, and the numbers belonging to them are their coefficients of intensity, the use of which has been explained. The initials H.R., H.B., and H.G., represent the red, blue and green papers of Mr HAY, and serve to connect this diagram with No. (2), which takes these colours for its standards.

No. 2. represents the relations of Mr HAY's red, blue, green, white, and yellow papers, as determined by a large number of experiments at Cambridge.—(See Note II.). The use of the point D, in calculating the results of colour-blindness, is explained in the Paper.

Fig. 3. represents a disc of the larger size, with its slit.

Fig. 4. shows the mode of combining two discs of the smaller size.

Fig. 5. shows the combination of discs, as placed on the top, in the first experiment described in the Paper.

Fig. 6. represents the method of spinning the top, when speed is required.

The last four figures are half the actual size.

Colour-tops of the kind used in these experiments, with paper discs of the colours whose relations are represented in No. 1, are to be had of Mr J. M. BRYSON, Optician, Edinburgh.

FIG. 6

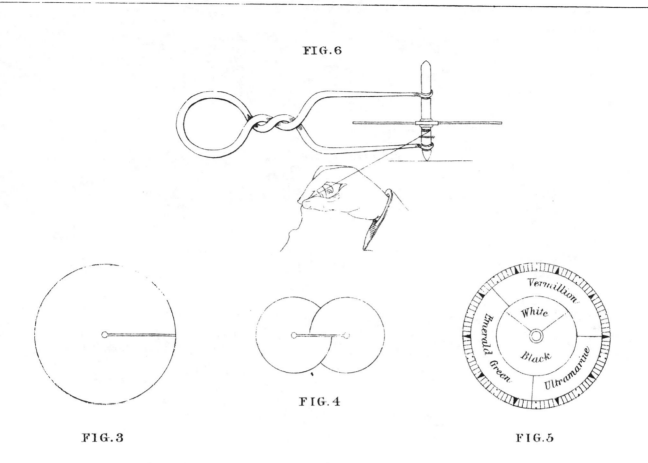

FIG. 3

FIG. 4

FIG. 5

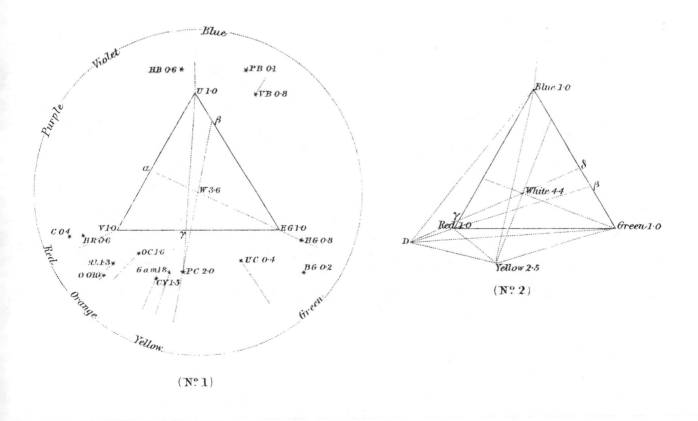

(Nº 1)

(Nº 2)

Cambridge University Press

[From the *Transactions of the Cambridge Philosophical Society*, Vol. x. Part I.]

VIII. *On Faraday's Lines of Force.*

[Read *Dec.* 10, 1855, and *Feb.* 11, 1856.]

THE present state of electrical science seems peculiarly unfavourable to speculation. The laws of the distribution of electricity on the surface of conductors
have been analytically deduced from experiment; some parts of the mathematical
theory of magnetism are established, while in other parts the experimental data
are wanting; the theory of the conduction of galvanism and that of the mutual
attraction of conductors have been reduced to mathematical formulæ, but have
not fallen into relation with the other parts of the science. No electrical theory
can now be put forth, unless it shews the connexion not only between electricity
at rest and current electricity, but between the attractions and inductive effects
of electricity in both states. Such a theory must accurately satisfy those laws,
the mathematical form of which is known, and must afford the means of calculating the effects in the limiting cases where the known formulæ are inapplicable.
In order therefore to appreciate the requirements of the science, the student
must make himself familiar with a considerable body of most intricate mathematics, the mere retention of which in the memory materially interferes with
further progress. The first process therefore in the effectual study of the science,
must be one of simplification and reduction of the results of previous investigation to a form in which the mind can grasp them. The results of this simplification may take the form of a purely mathematical formula or of a physical
hypothesis. In the first case we entirely lose sight of the phenomena to be
explained; and though we may trace out the consequences of given laws, we
can never obtain more extended views of the connexions of the subject. If,
on the other hand, we adopt a physical hypothesis, we see the phenomena only
through a medium, and are liable to that blindness to facts and rashness in

20—2

assumption which a partial explanation encourages. We must therefore discover some method of investigation which allows the mind at every step to lay hold of a clear physical conception, without being committed to any theory founded on the physical science from which that conception is borrowed, so that it is neither drawn aside from the subject in pursuit of analytical subtleties, nor carried beyond the truth by a favourite hypothesis.

In order to obtain physical ideas without adopting a physical theory we must make ourselves familiar with the existence of physical analogies. By a physical analogy I mean that partial similarity between the laws of one science and those of another which makes each of them illustrate the other. Thus all the mathematical sciences are founded on relations between physical laws and laws of numbers, so that the aim of exact science is to reduce the problems of nature to the determination of quantities by operations with numbers. Passing from the most universal of all analogies to a very partial one, we find the same resemblance in mathematical form between two different phenomena giving rise to a physical theory of light.

The changes of direction which light undergoes in passing from one medium to another, are identical with the deviations of the path of a particle in moving through a narrow space in which intense forces act. This analogy, which extends only to the direction, and not to the velocity of motion, was long believed to be the true explanation of the refraction of light; and we still find it useful in the solution of certain problems, in which we employ it without danger, as an artificial method. The other analogy, between light and the vibrations of an elastic medium, extends much farther, but, though its importance and fruitfulness cannot be over-estimated, we must recollect that it is founded only on a resemblance *in form* between the laws of light and those of vibrations. By stripping it of its physical dress and reducing it to a theory of "transverse alternations," we might obtain a system of truth strictly founded on observation, but probably deficient both in the vividness of its conceptions and the fertility of its method. I have said thus much on the disputed questions of Optics, as a preparation for the discussion of the almost universally admitted theory of attraction at a distance.

We have all acquired the mathematical conception of these attractions. We can reason about them and determine their appropriate forms or formulæ. These formulæ have a distinct mathematical significance, and their results are found to be in accordance with natural phenomena. There is no formula in applied mathematics more consistent with nature than the formula of attractions, and no theory better established in the minds of men than that of the action of bodies on one another at a distance. The laws of the conduction of heat in uniform media appear at first sight among the most different in their physical relations from those relating to attractions. The quantities which enter into them are *temperature, flow of heat, conductivity.* The word *force* is foreign to the subject. Yet we find that the mathematical laws of the uniform motion of heat in homogeneous media are identical in form with those of attractions varying inversely as the square of the distance. We have only to substitute *source of heat* for *centre of attraction, flow of heat* for *accelerating effect of attraction* at any point, and *temperature* for *potential,* and the solution of a problem in attractions is transformed into that of a problem in heat.

This analogy between the formulæ of heat and attraction was, I believe, first pointed out by Professor William Thomson in the *Camb. Math. Journal,* Vol. III.

Now the conduction of heat is supposed to proceed by an action between contiguous parts of a medium, while the force of attraction is a relation between distant bodies, and yet, if we knew nothing more than is expressed in the mathematical formulæ, there would be nothing to distinguish between the one set of phenomena and the other.

It is true, that if we introduce other considerations and observe additional facts, the two subjects will assume very different aspects, but the mathematical resemblance of some of their laws will remain, and may still be made useful in exciting appropriate mathematical ideas.

It is by the use of analogies of this kind that I have attempted to bring before the mind, in a convenient and manageable form, those mathematical ideas which are necessary to the study of the phenomena of electricity. The methods are generally those suggested by the processes of reasoning which are found in the researches of Faraday[*], and which, though they have been interpreted mathematically by Prof. Thomson and others, are very generally supposed to be of an indefinite and unmathematical character, when compared with those employed by the professed mathematicians. By the method which I adopt, I hope to render it evident that I am not attempting to establish any physical theory of a science in which I have hardly made a single experiment, and that the limit of my design is to shew how, by a strict application of the ideas and

---

[*] See especially Series XXXVIII. of the *Experimental Researches,* and *Phil. Mag.* 1852.

methods of Faraday, the connexion of the very different orders of phenomena which he has discovered may be clearly placed before the mathematical mind. I shall therefore avoid as much as I can the introduction of anything which does not serve as a direct illustration of Faraday's methods, or of the mathematical deductions which may be made from them. In treating the simpler parts of the subject I shall use Faraday's mathematical methods as well as his ideas. When the complexity of the subject requires it, I shall use analytical notation, still confining myself to the development of ideas originated by the same philosopher.

I have in the first place to explain and illustrate the idea of "lines of force."

When a body is electrified in any manner, a small body charged with positive electricity, and placed in any given position, will experience a force urging it in a certain direction. If the small body be now negatively electrified, it will be urged by an equal force in a direction exactly opposite.

The same relations hold between a magnetic body and the north or south poles of a small magnet. If the north pole is urged in one direction, the south pole is urged in the opposite direction.

In this way we might find a line passing through any point of space, such that it represents the direction of the force acting on a positively electrified particle, or on an elementary north pole, and the reverse direction of the force on a negatively electrified particle or an elementary south pole. Since at every point of space such a direction may be found, if we commence at any point and draw a line so that, as we go along it, its direction at any point shall always coincide with that of the resultant force at that point, this curve will indicate the direction of that force for every point through which it passes, and might be called on that account a *line of force*. We might in the same way draw other lines of force, till we had filled all space with curves indicating by their direction that of the force at any assigned point.

We should thus obtain a geometrical model of the physical phenomena, which would tell us the *direction* of the force, but we should still require some method of indicating the *intensity* of the force at any point. If we consider these curves not as mere lines, but as fine tubes of variable section carrying an incompressible fluid, then, since the velocity of the fluid is inversely as the section of the tube, we may make the velocity vary according to any given law, by regulating the section of the tube, and in this way we might represent the intensity of the force as well as its direction by the motion of the fluid in these tubes. This method of representing the intensity of a force by the velocity of an imaginary fluid in a tube is applicable to any conceivable system of forces, but it is capable of great simplification in the case in which the forces are such as can be explained by the hypothesis of attractions varying inversely as the square of the distance, such as those observed in electrical and magnetic phenomena. In the case of a perfectly arbitrary system of forces, there will generally be interstices between the tubes; but in the case of electric and magnetic forces it is possible to arrange the tubes so as to leave no interstices. The tubes will then be mere surfaces, directing the motion of a fluid filling up the whole space. It has been usual to commence the investigation of the laws of these forces by at once assuming that the phenomena are due to attractive or repulsive forces acting between certain points. We may however obtain a different view of the subject, and one more suited to our more difficult inquiries, by adopting for the definition of the forces of which we treat, that they may be represented in magnitude and direction by the uniform motion of an incompressible fluid.

I propose, then, first to describe a method by which the motion of such a fluid can be clearly conceived; secondly to trace the consequences of assuming certain conditions of motion, and to point out the application of the method to some of the less complicated phenomena of electricity, magnetism, and galvanism; and lastly to shew how by an extension of these methods, and the introduction of another idea due to Faraday, the laws of the attractions and inductive actions of magnets and currents may be clearly conceived, without making any assumptions as to the physical nature of electricity, or adding anything to that which has been already proved by experiment.

By referring everything to the purely geometrical idea of the motion of an imaginary fluid, I hope to attain generality and precision, and to avoid the dangers arising from a premature theory professing to explain the cause of the phenomena. If the results of mere speculation which I have collected are found to be of any use to experimental philosophers, in arranging and interpreting their results, they will have served their purpose, and a mature theory, in which physical facts will be physically explained, will be formed by those who by interrogating Nature herself can obtain the only true solution of the questions which the mathematical theory suggests.

I. *Theory of the Motion of an incompressible Fluid.*

(1) The substance here treated of must not be assumed to possess any of the properties of ordinary fluids except those of freedom of motion and resistance to compression. It is not even a hypothetical fluid which is introduced to explain actual phenomena. It is merely a collection of imaginary properties which may be employed for establishing certain theorems in pure mathematics in a way more intelligible to many minds and more applicable to physical problems than that in which algebraic symbols alone are used. The use of the word "Fluid" will not lead us into error, if we remember that it denotes a purely imaginary substance with the following property:

The portion of fluid which at any instant occupied a given volume, will at any succeeding instant occupy an equal volume.

This law expresses the incompressibility of the fluid, and furnishes us with a convenient measure of its quantity, namely its volume. The unit of quantity of the fluid will therefore be the unit of volume.

(2) The direction of motion of the fluid will in general be different at different points of the space which it occupies, but since the direction is determinate for every such point, we may conceive a line to begin at any point and to be continued so that every element of the line indicates by its direction the direction of motion at that point of space. Lines drawn in such a manner that their direction always indicates the direction of fluid motion are called *lines of fluid motion.*

If the motion of the fluid be what is called *steady motion*, that is, if the direction and velocity of the motion at any fixed point be independent of the time, these curves will represent the paths of individual particles of the fluid, but if the motion be variable this will not generally be the case. The cases of motion which will come under our notice will be those of steady motion.

(3) If upon any surface which cuts the lines of fluid motion we draw a closed curve, and if from every point of this curve we draw a line of motion, these lines of motion will generate a tubular surface which we may call a *tube of fluid motion.* Since this surface is generated by lines in the direction of fluid

motion no part of the fluid can flow across it, so that this imaginary surface is as impermeable to the fluid as a real tube.

(4) The quantity of fluid which in unit of time crosses any fixed section of the tube is the same at whatever part of the tube the section be taken. For the fluid is incompressible, and no part runs through the sides of the tube, therefore the quantity which escapes from the second section is equal to that which enters through the first.

If the tube be such that unit of volume passes through any section in unit of time it is called a *unit tube of fluid motion*.

(5) In what follows, various units will be referred to, and a finite number of lines or surfaces will be drawn, representing in terms of those units the motion of the fluid. Now in order to define the motion in every part of the fluid, an infinite number of lines would have to be drawn at indefinitely small intervals; but since the description of such a system of lines would involve continual reference to the theory of limits, it has been thought better to suppose the lines drawn at intervals depending on the assumed unit, and afterwards to assume the unit as small as we please by taking a small submultiple of the standard unit.

(6) To define the motion of the whole fluid by means of a system of unit tubes.

Take any fixed surface which cuts all the lines of fluid motion, and draw upon it any system of curves not intersecting one another. On the same surface draw a second system of curves intersecting the first system, and so arranged that the quantity of fluid which crosses the surface within each of the quadrilaterals formed by the intersection of the two systems of curves shall be unity in unit of time. From every point in a curve of the first system let a line of fluid motion be drawn. These lines will form a surface through which no fluid passes. Similar impermeable surfaces may be drawn for all the curves of the first system. The curves of the second system will give rise to a second system of impermeable surfaces, which, by their intersection with the first system, will form quadrilateral tubes, which will be tubes of fluid motion. Since each quadrilateral of the cutting surface transmits unity of fluid in unity of time, every tube in the system will transmit unity of fluid through any of its sections in unit of time. The motion of the fluid at every part of the space it occupies is determined by this system of unit tubes; for the direction of motion is that of the tube through the point in question, and the velocity is the reciprocal of the area of the section of the unit tube at that point.

(7) We have now obtained a geometrical construction which completely defines the motion of the fluid by dividing the space it occupies into a system of unit tubes. We have next to shew how by means of these tubes we may ascertain various points relating to the motion of the fluid.

A unit tube may either return into itself, or may begin and end at different points, and these may be either in the boundary of the space in which we investigate the motion, or within that space. In the first case there is a continual circulation of fluid in the tube, in the second the fluid enters at one end and flows out at the other. If the extremities of the tube are in the bounding surface, the fluid may be supposed to be continually supplied from without from an unknown source, and to flow out at the other into an unknown reservoir; but if the origin of the tube or its termination be within the space under consideration, then we must conceive the fluid to be supplied by a *source* within that space, capable of creating and emitting unity of fluid in unity of time, and to be afterwards swallowed up by a *sink* capable of receiving and destroying the same amount continually.

There is nothing self-contradictory in the conception of these sources where the fluid is created, and sinks where it is annihilated. The properties of the fluid are at our disposal, we have made it incompressible, and now we suppose it produced from nothing at certain points and reduced to nothing at others. The places of production will be called *sources*, and their numerical value will be the number of units of fluid which they produce in unit of time. The places of reduction will, for want of a better name, be called *sinks*, and will be estimated by the number of units of fluid absorbed in unit of time. Both places will sometimes be called sources, a source being understood to be a sink when its sign is negative.

(8) It is evident that the amount of fluid which passes any fixed surface is measured by the number of unit tubes which cut it, and the direction in which the fluid passes is determined by that of its motion in the tubes. If the surface be a closed one, then any tube whose terminations lie on the same side of the surface must cross the surface as many times in the one direction as in the other, and therefore must carry as much fluid out of the surface as

it carries in. A tube which begins within the surface and ends without it will carry out unity of fluid; and one which enters the surface and terminates within it will carry in the same quantity. In order therefore to estimate the amount of fluid which flows out of the closed surface, we must subtract the number of tubes which end within the surface from the number of tubes which begin there. If the result is negative the fluid will on the whole flow inwards.

If we call the beginning of a unit tube a unit source, and its termination a unit sink, then the quantity of fluid produced within the surface is estimated by the number of unit sources minus the number of unit sinks, and this must flow out of the surface on account of the incompressibility of the fluid.

In speaking of these unit tubes, sources and sinks, we must remember what was stated in (5) as to the magnitude of the unit, and how by diminishing their size and increasing their number we may distribute them according to any law however complicated.

(9) If we know the direction and velocity of the fluid at any point in two different cases, and if we conceive a third case in which the direction and velocity of the fluid at any point is the resultant of the velocities in the two former cases at corresponding points, then the amount of fluid which passes a given fixed surface in the third case will be the algebraic sum of the quantities which pass the same surface in the two former cases. For the rate at which the fluid crosses any surface is the resolved part of the velocity normal to the surface, and the resolved part of the resultant is equal to the sum of the resolved parts of the components.

Hence the number of unit tubes which cross the surface outwards in the third case must be the algebraical sum of the numbers which cross it in the two former cases, and the number of sources within any closed surface will be the sum of the numbers in the two former cases. Since the closed surface may be taken as small as we please, it is evident that the distribution of sources and sinks in the third case arises from the simple superposition of the distributions in the two former cases.

II. *Theory of the uniform motion of an imponderable incompressible fluid through a resisting medium.*

(10) The fluid is here supposed to have no inertia, and its motion is opposed by the action of a force which we may conceive to be due to the resistance of a

medium through which the fluid is supposed to flow. This resistance depends on the nature of the medium, and will in general depend on the direction in which the fluid moves, as well as on its velocity. For the present we may restrict ourselves to the case of a uniform medium, whose resistance is the same in all directions. The law which we assume is as follows.

Any portion of the fluid moving through the resisting medium is directly opposed by a retarding force proportional to its velocity.

If the velocity be represented by v, then the resistance will be a force equal to kv acting on unit of volume of the fluid in a direction contrary to that of motion. In order, therefore, that the velocity may be kept up, there must be a greater pressure behind any portion of the fluid than there is in front of it, so that the difference of pressures may neutralise the effect of the resistance. Conceive a cubical unit of fluid (which we may make as small as we please, by (5)), and let it move in a direction perpendicular to two of its faces. Then the resistance will be kv, and therefore the difference of pressures on the first and second faces is kv, so that the pressure diminishes in the direction of motion at the rate of kv for every unit of length measured along the line of motion; so that if we measure a length equal to h units, the difference of pressure at its extremities will be kvh.

(11) Since the pressure is supposed to vary continuously in the fluid, all the points at which the pressure is equal to a given pressure p will lie on a certain surface which we may call the *surface (p) of equal pressure*. If a series of these surfaces be constructed in the fluid corresponding to the pressures 0, 1, 2, 3 &c., then the number of the surface will indicate the pressure belonging to it, and the surface may be referred to as the surface 0, 1, 2 or 3. The unit of pressure is that pressure which is produced by unit of force acting on unit of surface. In order therefore to diminish the unit of pressure as in (5) we must diminish the unit of force in the same proportion.

(12) It is easy to see that these surfaces of equal pressure must be perpendicular to the lines of fluid motion; for if the fluid were to move in any other direction, there would be a resistance to its motion which could not be balanced by any difference of pressures. (We must remember that the fluid here considered has no inertia or mass, and that its properties are those only which are formally assigned to it, so that the resistances and pressures are the only things

to be considered.) There are therefore two sets of surfaces which by their intersection form the system of unit tubes, and the system of surfaces of equal pressure cuts both the others at right angles. Let h be the distance between two consecutive surfaces of equal pressure measured along a line of motion, then since the difference of pressures $= 1$,

$$kvh = 1,$$

which determines the relation of v to h, so that one can be found when the other is known. Let s be the sectional area of a unit tube measured on a surface of equal pressure, then since by the definition of a unit tube

$$vs = 1,$$

we find by the last equation

$$s = kh.$$

(13) The surfaces of equal pressure cut the unit tubes into portions whose length is h and section s. These elementary portions of unit tubes will be called *unit cells*. In each of them unity of volume of fluid passes from a pressure p to a pressure $(p-1)$ in unit of time, and therefore overcomes unity of resistance in that time. The work spent in overcoming resistance is therefore unity in every cell in every unit of time.

(14) If the surfaces of equal pressure are known, the direction and magnitude of the velocity of the fluid at any point may be found, after which the complete system of unit tubes may be constructed, and the beginnings and endings of these tubes ascertained and marked out as the sources whence the fluid is derived, and the sinks where it disappears. In order to prove the converse of this, that if the distribution of sources be given, the pressure at every point may be found, we must lay down certain preliminary propositions.

(15) If we know the pressures at every point in the fluid in two different cases, and if we take a third case in which the pressure at any point is the sum of the pressures at corresponding points in the two former cases, then the velocity at any point in the third case is the resultant of the velocities in the other two, and the distribution of sources is that due to the simple superposition of the sources in the two former cases.

For the velocity in any direction is proportional to the rate of decrease of the pressure in that direction; so that if two systems of pressures be added together, since the rate of decrease of pressure along any line will be the sum of the combined rates, the velocity in the new system resolved in the same direction will be the sum of the resolved parts in the two original systems. The velocity in the new system will therefore be the resultant of the velocities at corresponding points in the two former systems.

It follows from this, by (9), that the quantity of fluid which crosses any fixed surface is, in the new system, the sum of the corresponding quantities in the old ones, and that the sources of the two original systems are simply combined to form the third.

It is evident that in the system in which the pressure is the difference of pressure in the two given systems the distribution of sources will be got by changing the sign of all the sources in the second system and adding them to those in the first.

(16) If the pressure at every point of a closed surface be the same and equal to p, and if there be no sources or sinks within the surface, then there will be no motion of the fluid within the surface, and the pressure within it will be uniform and equal to p.

For if there be motion of the fluid within the surface there will be tubes of fluid motion, and these tubes must either return into themselves or be terminated either within the surface or at its boundary. Now since the fluid always flows from places of greater pressure to places of less pressure, it cannot flow in a re-entering curve; since there are no sources or sinks within the surface, the tubes cannot begin or end except on the surface; and since the pressure at all points of the surface is the same, there can be no motion in tubes having both extremities on the surface. Hence there is no motion within the surface, and therefore no difference of pressure which would cause motion, and since the pressure at the bounding surface is p, the pressure at any point within it is also p.

(17) If the pressure at every point of a given closed surface be known, and the distribution of sources within the surface be also known, then only one distribution of pressures can exist within the surface.

For if two different distributions of pressures satisfying these conditions could be found, a third distribution could be formed in which the pressure at any point should be the difference of the pressures in the two former distributions. In this case, since the pressures at the surface and the sources within it are the same in both distributions, the pressure at the surface in the third distribution would be zero, and all the sources within the surface would vanish, by (15).

Then by (16) the pressure at every point in the third distribution must be zero; but this is the difference of the pressures in the two former cases, and therefore these cases are the same, and there is only one distribution of pressure possible.

(18) Let us next determine the pressure at any point of an infinite body of fluid in the centre of which a unit source is placed, the pressure at an infinite distance from the source being supposed to be zero.

The fluid will flow out from the centre symmetrically, and since unity of volume flows out of every spherical surface surrounding the point in unit of time, the velocity at a distance r from the source will be

$$v = \frac{1}{4\pi r^2}.$$

The rate of decrease of pressure is therefore kv or $\frac{k}{4\pi r^2}$, and since the pressure $= 0$ when r is infinite, the actual pressure at any point will be

$$p = \frac{k}{4\pi r}.$$

The pressure is therefore inversely proportional to the distance from the source.

It is evident that the pressure due to a unit sink will be negative and equal to $-\frac{k}{4\pi r}$.

If we have a source formed by the coalition of S unit sources, then the resulting pressure will be $p = \frac{kS}{4\pi r}$, so that the pressure at a given distance varies as the resistance and number of sources conjointly.

(19) If a number of sources and sinks coexist in the fluid, then in order to determine the resultant pressure we have only to add the pressures which each source or sink produces. For by (15) this will be a solution of the problem, and by (17) it will be the only one. By this method we can determine the pressures due to any distribution of sources, as by the method

of (14) we can determine the distribution of sources to which a given distribution of pressures is due.

(20) We have next to shew that if we conceive any imaginary surface as fixed in space and intersecting the lines of motion of the fluid, we may substitute for the fluid on one side of this surface a distribution of sources upon the surface itself without altering in any way the motion of the fluid on the other side of the surface.

For if we describe the system of unit tubes which defines the motion of the fluid, and wherever a tube enters through the surface place a unit source, and wherever a tube goes out through the surface place a unit sink, and at the same time render the surface impermeable to the fluid, the motion of the fluid in the tubes will go on as before.

(21) If the system of pressures and the distribution of sources which produce them be known in a medium whose resistance is measured by k, then in order to produce the same system of pressures in a medium whose resistance is unity, the rate of production at each source must be multiplied by k. For the pressure at any point due to a given source varies as the rate of production and the resistance conjointly; therefore if the pressure be constant, the rate of production must vary inversely as the resistance.

(22) *On the conditions to be fulfilled at a surface which separates two media whose coefficients of resistance are k and k'.*

These are found from the consideration, that the quantity of fluid which flows out of the one medium at any point flows into the other, and that the pressure varies continuously from one medium to the other. The velocity normal to the surface is the same in both media, and therefore the rate of diminution of pressure is proportional to the resistance. The direction of the tubes of motion and the surfaces of equal pressure will be altered after passing through the surface, and the law of this refraction will be, that it takes place in the plane passing through the direction of incidence and the normal to the surface, and that the tangent of the angle of incidence is to the tangent of the angle of refraction as k' is to k.

(23) Let the space within a given closed surface be filled with a medium different from that exterior to it, and let the pressures at any point of this compound system due to a given distribution of sources within and without

the surface be given; it is required to determine a distribution of sources which would produce the same system of pressures in a medium whose coefficient of resistance is unity.

Construct the tubes of fluid motion, and wherever a unit tube enters either medium place a unit source, and wherever it leaves it place a unit sink. Then if we make the surface impermeable all will go on as before.

Let the resistance of the exterior medium be measured by k, and that of the interior by k'. Then if we multiply the rate of production of all the sources in the exterior medium (including those in the surface), by k, and make the coefficient of resistance unity, the pressures will remain as before, and the same will be true of the interior medium if we multiply all the sources in it by k', including those in the surface, and make its resistance unity.

Since the pressures on both sides of the surface are now equal, we may suppose it permeable if we please.

We have now the original system of pressures produced in a uniform medium by a combination of three systems of sources. The first of these is the given external system multiplied by k, the second is the given internal system multiplied by k', and the third is the system of sources and sinks on the surface itself. In the original case every source in the external medium had an equal sink in the internal medium on the other side of the surface, but now the source is multiplied by k and the sink by k', so that the result is for every external unit source on the surface, a source $=(k-k')$. By means of these three systems of sources the original system of pressures may be produced in a medium for which $k=1$.

(24) Let there be no resistance in the medium within the closed surface, that is, let $k'=0$, then the pressure within the closed surface is uniform and equal to p, and the pressure at the surface itself is also p. If by assuming any distribution of pairs of sources and sinks within the surface in addition to the given external and internal sources, and by supposing the medium the same within and without the surface, we can render the pressure at the surface uniform, the pressures so found for the external medium, together with the uniform pressure p in the internal medium, will be the true and only distribution of pressures which is possible.

For if two such distributions could be found by taking different imaginary distributions of pairs of sources and sinks within the medium, then by taking the difference of the two for a third distribution, we should have the pressure of the bounding surface constant in the new system and as many sources as sinks within it, and therefore whatever fluid flows in at any point of the surface, an equal quantity must flow out at some other point.

In the external medium all the sources destroy one another, and we have an infinite medium without sources surrounding the internal medium. The pressure at infinity is zero, that at the surface is constant. If the pressure at the surface is positive, the motion of the fluid must be outwards from every point of the surface; if it be negative, it must flow inwards towards the surface. But it has been shewn that neither of these cases is possible, because if any fluid enters the surface an equal quantity must escape, and therefore the pressure at the surface is zero in the third system.

The pressure at all points in the boundary of the internal medium in the third case is therefore zero, and there are no sources, and therefore the pressure is everywhere zero, by (16).

The pressure in the bounding surface of the internal medium is also zero, and there is no resistance, therefore it is zero throughout; but the pressure in the third case is the difference of pressures in the two given cases, therefore these are equal, and there is only one distribution of pressure which is possible, namely, that due to the imaginary distribution of sources and sinks.

(25) When the resistance is infinite in the internal medium, there can be no passage of fluid through it or into it. The bounding surface may therefore be considered as impermeable to the fluid, and the tubes of fluid motion will run along it without cutting it.

If by assuming any arbitrary distribution of sources within the surface in addition to the given sources in the outer medium, and by calculating the resulting pressures and velocities as in the case of a uniform medium, we can fulfil the condition of there being no velocity across the surface, the system of pressures in the outer medium will be the true one. For since no fluid passes through the surface, the tubes in the interior are independent of those outside, and may be taken away without altering the external motion.

(26) If the extent of the internal medium be small, and if the difference of resistance in the two media be also small, then the position of the unit tubes will not be much altered from what it would be if the external medium filled the whole space.

On this supposition we can easily calculate the kind of alteration which the introduction of the internal medium will produce; for wherever a unit tube enters the surface we must conceive a source producing fluid at a rate $\dfrac{k'-k}{k}$, and wherever a tube leaves it we must place a sink annihilating fluid at the rate $\dfrac{k'-k}{k}$, then calculating pressures on the supposition that the resistance in both media is k, the same as in the external medium, we shall obtain the true distribution of pressures very approximately, and we may get a better result by repeating the process on the system of pressures thus obtained.

(27) If instead of an abrupt change from one coefficient of resistance to another we take a case in which the resistance varies continuously from point to point, we may treat the medium as if it were composed of thin shells each of which has uniform resistance. By properly assuming a distribution of sources over the surfaces of separation of the shells, we may treat the case as if the resistance were equal to unity throughout, as in (23). The sources will then be distributed continuously throughout the whole medium, and will be positive whenever the motion is from places of less to places of greater resistance, and negative when in the contrary direction.

(28) Hitherto we have supposed the resistance at a given point of the medium to be the same in whatever direction the motion of the fluid takes place; but we may conceive a case in which the resistance is different in different directions. In such cases the lines of motion will not in general be perpendicular to the surfaces of equal pressure. If a, b, c be the components of the velocity at any point, and α, β, γ the components of the resistance at the same point, these quantities will be connected by the following system of linear equations, which may be called "*equations of conduction,*" and will be referred to by that name.

$$a = P_1 \alpha + Q_3 \beta + R_2 \gamma,$$
$$b = P_2 \beta + Q_1 \gamma + R_3 \alpha,$$
$$c = P_3 \gamma + Q_2 \alpha + R_1 \beta.$$

In these equations there are nine independent coefficients of conductivity. In order to simplify the equations, let us put

$$Q_1 + R_1 = 2S_1, \quad Q_1 - R_1 = 2lT,$$
$$\ldots\ldots\ldots \&c. \quad \ldots\ldots\ldots \&c.$$

22—2

where
$$4T^2 = (Q_1 - R_1)^2 + (Q_2 - R_2)^2 + (Q_3 - R_3)^2,$$
and l, m, n are direction-cosines of a certain fixed line in space.

The equations then become
$$a = P_1 a + S_3 \beta + S_2 \gamma + (n\beta - m\gamma)\,T,$$
$$b = P_2 \beta + S_1 \gamma + S_3 a + (l\gamma - na)\,T,$$
$$c = P_3 \gamma + S_2 a + S_1 \beta + (ma - l\beta)\,T.$$

By the ordinary transformation of co-ordinates we may get rid of the coefficients marked S. The equations then become
$$a = P_1' a + (n'\beta - m'\gamma)\,T,$$
$$b = P_2' \beta + (l'\gamma - n'a)\,T,$$
$$c = P_3' \gamma + (m'a - l'\beta)\,T,$$
where l', m', n' are the direction-cosines of the fixed line with reference to the new axes. If we make
$$a = \frac{dp}{dx}, \quad \beta = \frac{dp}{dy}, \quad \text{and} \quad \gamma = \frac{dp}{dz},$$
the equation of continuity
$$\frac{da}{dx} + \frac{db}{dy} + \frac{dc}{dz} = 0,$$
becomes
$$P_1' \frac{d^2 p}{dx^2} + P_2' \frac{d^2 p}{dy^2} + P_3' \frac{d^2 p}{dz^2} = 0,$$
and if we make
$$x = \sqrt{P_1'}\,\xi, \quad y = \sqrt{P_2'}\,\eta, \quad z = \sqrt{P_3'}\,\zeta,$$
then
$$\frac{d^2 p}{d\xi^2} + \frac{d^2 p}{d\eta^2} + \frac{d^2 p}{d\zeta^2} = 0,$$
the ordinary equation of conduction.

It appears therefore that the distribution of pressures is not altered by the existence of the coefficient T. Professor Thomson has shewn how to conceive a substance in which this coefficient determines a property having reference to an axis, which unlike the axes of P_1, P_2, P_3 is *dipolar*.

For further information on the equations of conduction, see Professor Stokes *On the Conduction of Heat in Crystals* (*Cambridge and Dublin Math. Journ.*), and Professor Thomson *On the Dynamical Theory of Heat*, Part v. (*Transactions of Royal Society of Edinburgh*, Vol. xxi. Part i.).

It is evident that all that has been proved in (14), (15), (16), (17), with respect to the superposition of different distributions of pressure, and there being only one distribution of pressures corresponding to a given distribution of sources, will be true also in the case in which the resistance varies from point to point, and the resistance at the same point is different in different directions. For if we examine the proof we shall find it applicable to such cases as well as to that of a uniform medium.

(29) We now are prepared to prove certain general propositions which are true in the most general case of a medium whose resistance is different in different directions and varies from point to point.

We may by the method of (28), when the distribution of pressures is known, construct the surfaces of equal pressure, the tubes of fluid motion, and the sources and sinks. It is evident that since in each cell into which a unit tube is divided by the surfaces of equal pressure unity of fluid passes from pressure p to pressure $(p-1)$ in unit of time, unity of work is done by the fluid in each cell in overcoming resistance.

The number of cells in each unit tube is determined by the number of surfaces of equal pressure through which it passes. If the pressure at the beginning of the tube be p and at the end p', then the number of cells in it will be $p-p'$. Now if the tube had extended from the source to a place where the pressure is zero, the number of cells would have been p, and if the tube had come from the sink to zero, the number would have been p', and the true number is the difference of these.

Therefore if we find the pressure at a source S from which S tubes proceed to be p, Sp is the number of cells due to the source S; but if S' of the tubes terminate in a sink at a pressure p', then we must cut off $S'p'$ cells from the number previously obtained. Now if we denote the source of S tubes by S, the sink of S' tubes may be written $-S'$, sinks always being reckoned negative, and the general expression for the number of cells in the system will be $\Sigma\,(Sp)$.

(30) The same conclusion may be arrived at by observing that unity of work is done on each cell. Now in each source S, S units of fluid are expelled against a pressure p, so that the work done by the fluid in overcoming resistance is Sp. At each sink in which S' tubes terminate, S' units of fluid sink into nothing under pressure p'; the work done upon the fluid by

the pressure is therefore $S'p'$. The whole work done by the fluid may therefore be expressed by

$$W = \Sigma Sp - \Sigma S'p',$$

or more concisely, considering sinks as negative sources,

$$W = \Sigma(Sp).$$

(31) Let S represent the rate of production of a source in any medium, and let p be the pressure at any given point due to that source. Then if we superpose on this another equal source, every pressure will be doubled, and thus by successive superposition we find that a source nS would produce a pressure np, or more generally the pressure at any point due to a given source varies as the rate of production of the source. This may be expressed by the equation

$$p = RS,$$

where R is a coefficient depending on the nature of the medium and on the positions of the source and the given point. In a uniform medium whose resistance is measured by k,

$$p = \frac{kS}{4\pi r}, \quad \therefore R = \frac{k}{4\pi r},$$

R may be called the coefficient of resistance of the medium between the source and the given point. By combining any number of sources we have generally

$$p = \Sigma(RS).$$

(32) In a uniform medium the pressure due to a source S

$$p = \frac{k}{4\pi} \frac{S}{r}.$$

At another source S' at a distance r we shall have

$$S'p = \frac{k}{4\pi} \frac{SS'}{r} = Sp',$$

if p' be the pressure at S due to S'. If therefore there be two systems of sources $\Sigma(S)$ and $\Sigma(S')$, and if the pressures due to the first be p and to the second p', then

$$\Sigma(S'p) = \Sigma(Sp').$$

For every term $S'p$ has a term Sp' equal to it.

(33) Suppose that in a uniform medium the motion of the fluid is everywhere parallel to one plane, then the surfaces of equal pressure will be perpendicular to this plane. If we take two parallel planes at a distance equal to k from each other, we can divide the space between these planes into unit tubes by means of cylindric surfaces perpendicular to the planes, and these together with the surfaces of equal pressure will divide the space into cells of which the length is equal to the breadth. For if h be the distance between consecutive surfaces of equal pressure and s the section of the unit tube, we have by (13) $s = kh$.

But s is the product of the breadth and depth; but the depth is k, therefore the breadth is h and equal to the length.

If two systems of plane curves cut each other at right angles so as to divide the plane into little areas of which the length and breadth are equal, then by taking another plane at distance k from the first and erecting cylindric surfaces on the plane curves as bases, a system of cells will be formed which will satisfy the conditions whether we suppose the fluid to run along the first set of cutting lines or the second*.

Application of the Idea of Lines of Force.

I have now to shew how the idea of lines of fluid motion as described above may be modified so as to be applicable to the sciences of statical electricity, permanent magnetism, magnetism of induction, and uniform galvanic currents, reserving the laws of electro-magnetism for special consideration.

I shall assume that the phenomena of statical electricity have been already explained by the mutual action of two opposite kinds of matter. If we consider one of these as positive electricity and the other as negative, then any two particles of electricity repel one another with a force which is measured by the product of the masses of the particles divided by the square of their distance.

Now we found in (18) that the velocity of our imaginary fluid due to a source S at a distance r varies inversely as r^2. Let us see what will be the effect of substituting such a source for every particle of positive electricity. The velocity due to each source would be proportional to the attraction due to the corresponding particle, and the resultant velocity due to all the sources would

* See *Cambridge and Dublin Mathematical Journal*, Vol. III. p. 286.

be proportional to the resultant attraction of all the particles. Now we may find the resultant pressure at any point by adding the pressures due to the given sources, and therefore we may find the resultant velocity in a given direction from the rate of decrease of pressure in that direction, and this will be proportional to the resultant attraction of the particles resolved in that direction.

Since the resultant attraction in the electrical problem is proportional to the decrease of pressure in the imaginary problem, and since we may select any values for the constants in the imaginary problem, we may assume that the resultant attraction in any direction is numerically equal to the decrease of pressure in that direction, or

$$X = -\frac{dp}{dx}.$$

By this assumption we find that if V be the potential,

$$dV = Xdx + Ydy + Zdz = -dp,$$

or since at an infinite distance $V = 0$ and $p = 0$, $V = -p$.

In the electrical problem we have

$$V = -\Sigma \left(\frac{dm}{r}\right).$$

In the fluid $p = \Sigma \left(\frac{k}{4\pi} \frac{S}{r}\right);$

$$\therefore S = \frac{4\pi}{k} dm.$$

If k be supposed very great, the amount of fluid produced by each source in order to keep up the pressures will be very small.

The potential of any system of electricity on itself will be

$$\Sigma(pdm) = \frac{k}{4\pi}, \quad \Sigma(pS) = \frac{k}{4\pi} W.$$

If $\Sigma(dm)$, $\Sigma(dm')$ be two systems of electrical particles and p, p' the potentials due to them respectively, then by (32)

$$\Sigma(pdm') = \frac{k}{4\pi} \Sigma(pS') = \frac{k}{4\pi} \Sigma(p'S) = \Sigma(p'dm),$$

or the potential of the first system on the second is equal to that of the second system on the first.

So that in the ordinary electrical problems the analogy in fluid motion is of this kind:

$$V = -p,$$

$$X = -\frac{dp}{dx} = ku,$$

$$dm = \frac{k}{4\pi} S,$$

whole potential of a system $= -\Sigma Vdm = \frac{k}{4\pi} W$, where W is the work done by the fluid in overcoming resistance.

The lines of forces are the unit tubes of fluid motion, and they may be estimated numerically by those tubes.

Theory of Dielectrics.

The electrical induction exercised on a body at a distance depends not only on the distribution of electricity in the inductric, and the form and position of the inducteous body, but on the nature of the interposed medium, or dielectric. Faraday* expresses this by the conception of one substance having a *greater inductive capacity*, or conducting the lines of inductive action more freely than another. If we suppose that in our analogy of a fluid in a resisting medium the resistance is different in different media, then by making the resistance less we obtain the analogue to a dielectric which more easily conducts Faraday's lines.

It is evident from (23) that in this case there will always be an apparent distribution of electricity on the surface of the dielectric, there being negative electricity where the lines enter and positive electricity where they emerge. In the case of the fluid there are no real sources on the surface, but we use them merely for purposes of calculation. In the dielectric there may be no real charge of electricity, but only an apparent electric action due to the surface.

If the dielectric had been of less conductivity than the surrounding medium, we should have had precisely opposite effects, namely, positive electricity where lines enter, and negative where they emerge.

* Series XI.

If the conduction of the dielectric is perfect or nearly so for the small quantities of electricity with which we have to do, then we have the case of (24). The dielectric is then considered as a conductor, its surface is a surface of equal potential, and the resultant attraction near the surface itself is perpendicular to it.

Theory of Permanent Magnets.

A magnet is conceived to be made up of elementary magnetized particles, each of which has its own north and south poles, the action of which upon other north and south poles is governed by laws mathematically identical with those of electricity. Hence the same application of the idea of lines of force can be made to this subject, and the same analogy of fluid motion can be employed to illustrate it.

But it may be useful to examine the way in which the polarity of the elements of a magnet may be represented by the unit cells in fluid motion. In each unit cell unity of fluid enters by one face and flows out by the opposite face, so that the first face becomes a unit sink and the second a unit source with respect to the rest of the fluid. It may therefore be compared to an elementary magnet, having an equal quantity of north and south magnetic matter distributed over two of its faces. If we now consider the cell as forming part of a system, the fluid flowing out of one cell will flow into the next, and so on, so that the source will be transferred from the end of the cell to the end of the unit tube. If all the unit tubes begin and end on the bounding surface, the sources and sinks will be distributed entirely on that surface, and in the case of a magnet which has what has been called a solenoidal or tubular distribution of magnetism, all the imaginary magnetic matter will be on the surface*.

Theory of Paramagnetic and Diamagnetic Induction.

Faraday† has shewn that the effects of paramagnetic and diamagnetic bodies in the magnetic field may be explained by supposing paramagnetic bodies to

---

* See Professor Thomson *On the Mathematical Theory of Magnetism*, Chapters III. and v. *Phil. Trans.* 1851.

† *Experimental Researches* (3292).

conduct the lines of force better, and diamagnetic bodies worse, than the surrounding medium. By referring to (23) and (26), and supposing sources to represent north magnetic matter, and sinks south magnetic matter, then if a paramagnetic body be in the neighbourhood of a north pole, the lines of force on entering it will produce south magnetic matter, and on leaving it they will produce an equal amount of north magnetic matter. Since the quantities of magnetic matter on the whole are equal, but the southern matter is nearest to the north pole, the result will be attraction. If on the other hand the body be diamagnetic, or a worse conductor of lines of force than the surrounding medium, there will be an imaginary distribution of northern magnetic matter where the lines pass into the worse conductor, and of southern where they pass out, so that on the whole there will be repulsion.

We may obtain a more general law from the consideration that the potential of the whole system is proportional to the amount of work done by the fluid in overcoming resistance. The introduction of a second medium increases or diminishes the work done according as the resistance is greater or less than that of the first medium. The amount of this increase or diminution will vary as the square of the velocity of the fluid.

Now, by the theory of potentials, the moving force in any direction is measured by the rate of decrease of the potential of the system in passing along that direction, therefore when k', the resistance within the second medium, is greater than k, the resistance in the surrounding medium, there is a force tending from places where the resultant force v is greater to where it is less, so that a diamagnetic body moves from greater to less values of the resultant force*.

In paramagnetic bodies k' is less than k, so that the force is now from points of less to points of greater resultant magnetic force. Since these results depend only on the relative values of k and k', it is evident that by changing the surrounding medium, the behaviour of a body may be changed from paramagnetic to diamagnetic at pleasure.

It is evident that we should obtain the same mathematical results if we had supposed that the magnetic force had a power of exciting a polarity in bodies which is in the *same* direction as the lines in paramagnetic bodies, and

---

* *Experimental Researches* (2797), (2798). See Thomson, *Cambridge and Dublin Mathematical Journal*, May, 1847.

in the *reverse* direction in diamagnetic bodies*. In fact we have not as yet come to any facts which would lead us to choose any one out of these three theories, that of lines of force, that of imaginary magnetic matter, and that of induced polarity. As the theory of lines of force admits of the most precise, and at the same time least theoretic statement, we shall allow it to stand for the present.

Theory of Magnecrystallic Induction.

The theory of Faraday† with respect to the behaviour of crystals in the magnetic field may be thus stated. In certain crystals and other substances the lines of magnetic force are conducted with different facility in different directions. The body when suspended in a uniform magnetic field will turn or tend to turn into such a position that the lines of force shall pass through it with least resistance. It is not difficult by means of the principles in (28) to express the laws of this kind of action, and even to reduce them in certain cases to numerical formulæ. The principles of induced polarity and of imaginary magnetic matter are here of little use; but the theory of lines of force is capable of the most perfect adaptation to this class of phenomena.

Theory of the Conduction of Current Electricity.

It is in the calculation of the laws of constant electric currents that the theory of fluid motion which we have laid down admits of the most direct application. In addition to the researches of Ohm on this subject, we have those of M. Kirchhoff, *Ann. de Chim.* XLI. 496, and of M. Quincke, XLVII. 203, on the Conduction of Electric Currents in Plates. According to the received opinions we have here a current of fluid moving uniformly in conducting circuits, which oppose a resistance to the current which has to be overcome by the application of an electro-motive force at some part of the circuit. On account of this resistance to the motion of the fluid the pressure must be different at different points in the circuit. This pressure, which is commonly called electrical tension,

---

* *Exp. Res.* (2429), (3320). See Weber, Poggendorff, LXXXVII. p. 145. Prof. Tyndall, *Phil. Trans.* 1856, p. 237.

† *Exp. Res.* (2836), &c.

is found to be physically identical with the *potential* in statical electricity, and thus we have the means of connecting the two sets of phenomena. If we knew what amount of electricity, measured statically, passes along that current which we assume as our unit of current, then the connexion of electricity of tension with current electricity would be completed*. This has as yet been done only approximately, but we know enough to be certain that the conducting powers of different substances differ only in degree, and that the difference between glass and metal is, that the resistance is a great but finite quantity in glass, and a small but finite quantity in metal. Thus the analogy between statical electricity and fluid motion turns out more perfect than we might have supposed, for there the induction goes on by conduction just as in current electricity, but the quantity conducted is insensible owing to the great resistance of the dielectrics†.

On Electro-motive Forces.

When a uniform current exists in a closed circuit it is evident that some other forces must act on the fluid besides the pressures. For if the current were due to difference of pressures, then it would flow from the point of greatest pressure in both directions to the point of least pressure, whereas in reality it circulates in one direction constantly. We must therefore admit the existence of certain forces capable of keeping up a constant current in a closed circuit. Of these the most remarkable is that which is produced by chemical action. A cell of a voltaic battery, or rather the surface of separation of the fluid of the cell and the zinc, is the seat of an electro-motive force which can maintain a current in opposition to the resistance of the circuit. If we adopt the usual convention in speaking of electric currents, the positive current is from the fluid through the platinum, the conducting circuit, and the zinc, back to the fluid again. If the electro-motive force act only in the surface of separation of the fluid and zinc, then the tension of electricity in the fluid must exceed that in the zinc by a quantity depending on the nature and length of the circuit and on the strength of the current in the conductor. In order to keep up this difference of pressure there must be an electro-motive force whose intensity is measured by that difference of pressure. If F be the electro-motive force, I the quantity of the current or the number of electrical

* See *Exp. Res.* (371). † *Exp. Res.* Vol. III. p. 513.

units delivered in unit of time, and K a quantity depending on the length and resistance of the conducting circuit, then

$$F = IK = p - p',$$

where p is the electric tension in the fluid and p' in the zinc.

If the circuit be broken at any point, then since there is no current the tension of the part which remains attached to the platinum will be p, and that of the other will be p', $p - p'$ or F affords a measure of the intensity of the current. This distinction of quantity and intensity is very useful [*], but must be distinctly understood to mean nothing more than this :—The quantity of a current is the amount of electricity which it transmits in unit of time, and is measured by I the number of unit currents which it contains. The intensity of a current is its power of overcoming resistance, and is measured by F or IK, where K is the resistance of the whole circuit.

The same idea of quantity and intensity may be applied to the case of magnetism[†]. The quantity of magnetization in any section of a magnetic body is measured by the number of lines of magnetic force which pass through it. The intensity of magnetization in the section depends on the resisting power of the section, as well as on the number of lines which pass through it. If k be the resisting power of the material, and S the area of the section, and I the number of lines of force which pass through it, then the whole intensity throughout the section

$$= F = I\frac{k}{S}.$$

When magnetization is produced by the influence of other magnets only, we may put p for the magnetic tension at any point, then for the whole magnetic solenoid

$$F = I\int\frac{k}{S}dx = IK = p - p'.$$

When a solenoidal magnetized circuit returns into itself, the magnetization does not depend on difference of tensions only, but on some magnetizing force of which the intensity is F.

If i be the quantity of the magnetization at any point, or the number of lines of force passing through unit of area in the section of the solenoid, then

---

* *Exp. Res.* Vol. III. p. 519.   † *Exp. Res.* (2870), (3293).

the total quantity of magnetization in the circuit is the number of lines which pass through any section, $I = \Sigma i\, dy\, dz$, where $dy\, dz$ is the element of the section, and the summation is performed over the whole section.

The intensity of magnetization at any point, or the force required to keep up the magnetization, is measured by $ki = f$, and the total intensity of magnetization in the circuit is measured by the sum of the local intensities all round the circuit,

$$F = \Sigma \left(f\, dx \right),$$

where dx is the element of length in the circuit, and the summation is extended round the entire circuit.

In the same circuit we have always $F = IK$, where K is the total resistance of the circuit, and depends on its form and the matter of which it is composed.

On the Action of closed Currents at a Distance.

The mathematical laws of the attractions and repulsions of conductors have been most ably investigated by Ampère, and his results have stood the test of subsequent experiments.

From the single assumption, that the action of an element of one current upon an element of another current is an attractive or repulsive force acting in the direction of the line joining the two elements, he has determined by the simplest experiments the mathematical form of the law of attraction, and has put this law into several most elegant and useful forms. We must recollect however that no experiments have been made on these elements of currents except under the form of closed currents either in rigid conductors or in fluids, and that the laws of closed currents can only be deduced from such experiments. Hence if Ampère's formulæ applied to closed currents give true results, their truth is not proved for *elements* of currents unless we assume that the action between two such elements must be along the line which joins them. Although this assumption is most warrantable and philosophical in the present state of science, it will be more conducive to freedom of investigation if we endeavour to do without it, and to assume the laws of closed currents as the ultimate datum of experiment.

Ampère has shewn that when currents are combined according to the law of the parallelogram of forces, the force due to the resultant current is the resultant of the forces due to the component currents, and that equal and opposite currents generate equal and opposite forces, and when combined neutralize each other.

He has also shewn that a closed circuit of any form has no tendency to turn a moveable circular conductor about a fixed axis through the centre of the circle perpendicular to its plane, and that therefore the forces in the case of a closed circuit render $Xdx + Ydy + Zdz$ a complete differential.

Finally, he has shewn that if there be two systems of circuits similar and similarly situated, the quantity of electrical current in corresponding conductors being the same, the resultant forces are equal, whatever be the absolute dimensions of the systems, which proves that the forces are, *cæteris paribus*, inversely as the square of the distance.

From these results it follows that the mutual action of two closed currents whose areas are very small is the same as that of two elementary magnetic bars magnetized perpendicularly to the plane of the currents.

The direction of magnetization of the equivalent magnet may be predicted by remembering that a current travelling round the earth from east to west as the sun appears to do, would be equivalent to that magnetization which the earth actually possesses, and therefore in the reverse direction to that of a magnetic needle when pointing freely.

If a number of closed unit currents in contact exist on a surface, then at all points in which two currents are in contact there will be two equal and opposite currents which will produce no effect, but all round the boundary of the surface occupied by the currents there will be a residual current not neutralized by any other; and therefore the result will be the same as that of a single unit current round the boundary of all the currents.

From this it appears that the external attractions of a shell uniformly magnetized perpendicular to its surface are the same as those due to a current round its edge, for each of the elementary currents in the former case has the same effect as an element of the magnetic shell.

If we examine the lines of magnetic force produced by a closed current, we shall find that they form closed curves passing round the current and *embracing* it, and that the total intensity of the magnetizing force all along the closed line of force depends on the quantity of the electric current only.

The number of unit lines* of magnetic force due to a closed current depends on the form as well as the quantity of the current, but the number of unit cells† in each complete line of force is measured simply by the number of unit currents which embrace it. The unit cells in this case are portions of space in which unit of magnetic quantity is produced by unity of magnetizing force. The length of a cell is therefore inversely as the intensity of the magnetizing force, and its section inversely as the quantity of magnetic induction at that point.

The whole number of cells due to a given current is therefore proportional to the strength of the current multiplied by the number of lines of force which pass through it. If by any change of the form of the conductors the number of cells can be increased, there will be a force tending to produce that change, so that there is always a force urging a conductor transverse to the lines of magnetic force, so as to cause more lines of force to pass through the closed circuit of which the conductor forms a part.

The number of cells due to two given currents is got by multiplying the number of lines of inductive magnetic action which pass through each by the quantity of the currents respectively. Now by (9) the number of lines which pass through the first current is the sum of its own lines and those of the second current which would pass through the first if the second current alone were in action. Hence the whole number of cells will be increased by any motion which causes more lines of force to pass through either circuit, and therefore the resultant force will tend to produce such a motion, and the work done by this force during the motion will be measured by the number of new cells produced. All the actions of closed conductors on each other may be deduced from this principle.

On Electric Currents produced by Induction.

Faraday has shewn‡ that when a conductor moves transversely to the lines of magnetic force, an electro-motive force arises in the conductor, tending to produce a current in it. If the conductor is closed, there is a continuous current, if open, tension is the result. If a closed conductor move transversely to the lines of magnetic induction, then, if the number of lines which pass

---

* *Exp. Res.* (3122). See Art. (6) of this paper. † Art. (13).
‡ *Exp. Res.* (3077), &c.

ON FARADAY'S LINES OF FORCE.

186

through it does not change during the motion, the electro-motive forces in the circuit will be in equilibrium, and there will be no current. Hence the electro-motive forces depend on the number of lines which are cut by the conductor during the motion. If the motion be such that a greater number of lines pass through the circuit formed by the conductor after than before the motion, then the electro-motive force will be measured by the increase of the number of lines, and will generate a current the reverse of that which would have produced the additional lines. When the number of lines of inductive magnetic action through the circuit is increased, the induced current will tend to diminish the number of lines, and when the number is diminished the induced current will tend to increase them.

That this is the true expression for the law of induced currents is shewn from the fact that, in whatever way the number of lines of magnetic induction passing through the circuit be increased, the electro-motive effect is the same, whether the increase take place by the motion of the conductor itself, or of other conductors, or of magnets, or by the change of intensity of other currents, or by the magnetization or demagnetization of neighbouring magnetic bodies, or lastly by the change of intensity of the current itself.

In all these cases the electro-motive force depends on the *change* in the number of lines of inductive magnetic action which pass through the circuit*.

* The electro-magnetic forces, which tend to produce motion of the material conductor, must be carefully distinguished from the electro-motive forces, which tend to produce electric currents.

Let an electric current be passed through a mass of metal of any form. The distribution of the currents within the metal will be determined by the laws of conduction. Now let a constant electric current be passed through another conductor near the first. If the two currents are in the same direction the two conductors will be attracted towards each other, and would come nearer if not held in their positions. But though the material conductors are attracted, the currents (which are free to choose any course within the metal) will not alter their original distribution, or incline towards each other. For, since no change takes place in the system, there will be no electro-motive forces to modify the original distribution of currents.

In this case we have electro-magnetic forces acting on the material conductor, without any electro-motive forces tending to modify the current which it carries.

Let us take as another example the case of a linear conductor, not forming a closed circuit, and let it be made to traverse the lines of magnetic force, either by its own motion, or by changes in the magnetic field. An electro-motive force will act in the direction of the conductor, and, as it cannot produce a current, because there is no circuit, it will produce electric tension at the extremities. There will be no electro-magnetic attraction on the material conductor, for this attraction depends on the existence of the current within it, and this is prevented by the circuit not being closed.

Here then we have the opposite case of an electro-motive force acting on the electricity in the conductor, but no attraction on its material particles.

It is natural to suppose that a force of this kind, which depends on a change in the number of lines, is due to a change of state which is measured by the number of these lines. A closed conductor in a magnetic field may be supposed to be in a certain state arising from the magnetic action. As long as this state remains unchanged no effect takes place, but, when the state changes, electro-motive forces arise, depending as to their intensity and direction on this change of state. I cannot do better here than quote a passage from the first series of Faraday's *Experimental Researches*, Art. (60).

"While the wire is subject to either volta-electric or magno-electric induction it appears to be in a peculiar state, for it resists the formation of an electrical current in it; whereas, if in its common condition, such a current would be produced; and when left uninfluenced it has the power of originating a current, a power which the wire does not possess under ordinary circumstances. This electrical condition of matter has not hitherto been recognised, but it probably exerts a very important influence in many if not most of the phenomena produced by currents of electricity. For reasons which will immediately appear (7) I have, after advising with several learned friends, ventured to designate it as the *electro-tonic* state." Finding that all the phenomena could be otherwise explained without reference to the electro-tonic state, Faraday in his second series rejected it as not necessary; but in his recent researches* he seems still to think that there may be some physical truth in his conjecture about this new state of bodies.

The conjecture of a philosopher so familiar with nature may sometimes be more pregnant with truth than the best established experimental law discovered by empirical inquirers, and though not bound to admit it as a physical truth, we may accept it as a new idea by which our mathematical conceptions may be rendered clearer.

In this outline of Faraday's electrical theories, as they appear from a mathematical point of view, I can do no more than simply state the mathematical methods by which I believe that electrical phenomena can be best comprehended and reduced to calculation, and my aim has been to present the mathematical ideas to the mind in an embodied form, as systems of lines or surfaces, and not as mere symbols, which neither convey the same ideas, nor readily adapt themselves to the phenomena to be explained. The idea of the electro-tonic state, however, has not yet presented itself to my mind in such a

* (3172) (3269).

24—2

form that its nature and properties may be clearly explained without reference to mere symbols, and therefore I propose in the following investigation to use symbols freely, and to take for granted the ordinary mathematical operations. By a careful study of the laws of elastic solids and of the motions of viscous fluids, I hope to discover a method of forming a mechanical conception of this electro-tonic state adapted to general reasoning*.

PART II.

On Faraday's "Electro-tonic State."

When a conductor moves in the neighbourhood of a current of electricity, or of a magnet, or when a current or magnet near the conductor is moved, or altered in intensity, then a force acts on the conductor and produces electric tension, or a continuous current, according as the circuit is open or closed. This current is produced only by *changes* of the electric or magnetic phenomena surrounding the conductor, and as long as these are constant there is no observed effect on the conductor. Still the conductor is in different states when near a current or magnet, and when away from its influence, since the removal or destruction of the current or magnet occasions a current, which would not have existed if the magnet or current had not been previously in action.

Considerations of this kind led Professor Faraday to connect with his discovery of the induction of electric currents the conception of a state into which all bodies are thrown by the presence of magnets and currents. This state does not manifest itself by any known phenomena as long as it is undisturbed, but any change in this state is indicated by a current or tendency towards a current. To this state he gave the name of the "Electro-tonic State," and although he afterwards succeeded in explaining the phenomena which suggested it by means of less hypothetical conceptions, he has on several occasions hinted at the probability that some phenomena might be discovered which would render the electro-tonic state an object of legitimate induction. These speculations, into which Faraday had been led by the study of laws which he has well established, and which he abandoned only for want of experi-

---

* See Prof. W. Thomson *On a Mechanical Representation of Electric, Magnetic and Galvanic Forces. Camb. and Dub. Math. Jour.* Jan. 1847.

mental data for the direct proof of the unknown state, have not, I think, been made the subject of mathematical investigation. Perhaps it may be thought that the quantitative determinations of the various phenomena are not sufficiently rigorous to be made the basis of a mathematical theory; Faraday, however, has not contented himself with simply stating the numerical results of his experiments and leaving the law to be discovered by calculation. Where he has perceived a law he has at once stated it, in terms as unambiguous as those of pure mathematics; and if the mathematician, receiving this as a physical truth, deduces from it other laws capable of being tested by experiment, he has merely assisted the physicist in arranging his own ideas, which is confessedly a necessary step in scientific induction.

In the following investigation, therefore, the laws established by Faraday will be assumed as true, and it will be shewn that by following out his speculations other and more general laws can be deduced from them. If it should then appear that these laws, originally devised to include one set of phenomena, may be generalized so as to extend to phenomena of a different class, these mathematical connexions may suggest to physicists the means of establishing physical connexions; and thus mere speculation may be turned to account in experimental science.

On Quantity and Intensity as Properties of Electric Currents.

It is found that certain effects of an electric current are equal at whatever part of the circuit they are estimated. The quantities of water or of any other electrolyte decomposed at two different sections of the same circuit, are always found to be equal or equivalent, however different the material and form of the circuit may be at the two sections. The magnetic effect of a conducting wire is also found to be independent of the form or material of the wire in the same circuit. There is therefore an electrical effect which is equal at every section of the circuit. If we conceive of the conductor as the channel along which a fluid is constrained to move, then the quantity of fluid transmitted by each section will be the same, and we may define the *quantity* of an electric current to be the quantity of electricity which passes across a complete section of the current in unit of time. We may for the present measure quantity of electricity by the quantity of water which it would decompose in unit of time.

In order to express mathematically the electrical currents in any conductor, we must have a definition, not only of the entire flow across a complete section, but also of the flow at a given point in a given direction.

DEF. The quantity of a current at a given point and in a given direction is measured, when uniform, by the quantity of electricity which flows across unit of area taken at that point perpendicular to the given direction, and when variable by the quantity which would flow across this area, supposing the flow uniformly the same as at the given point.

In the following investigation, the quantity of electric current at the point (xyz) estimated in the directions of the axes x, y, z respectively will be denoted by a_2, b_2, c_2.

The quantity of electricity which flows in unit of time through the elementary area dS

$$= dS\,(la_2 + mb_2 + nc_2),$$

where l, m, n are the direction-cosines of the normal to dS.

This flow of electricity at any point of a conductor is due to the electromotive forces which act at that point. These may be either external or internal.

External electro-motive forces arise either from the relative motion of currents and magnets, or from changes in their intensity, or from other causes acting at a distance.

Internal electro-motive forces arise principally from difference of electric tension at points of the conductor in the immediate neighbourhood of the point in question. The other causes are variations of chemical composition or of temperature in contiguous parts of the conductor.

Let p_2 represent the electric tension at any point, and X_2, Y_2, Z_2 the sums of the parts of all the electro-motive forces arising from other causes resolved parallel to the co-ordinate axes, then if a_2, β_2, γ_2 be the effective electro-motive forces

$$\left.\begin{aligned} a_2 &= X_2 - \frac{dp_2}{dx} \\[2ex] \beta_2 &= Y_2 - \frac{dp_2}{dy} \\[2ex] \gamma_2 &= Z_2 - \frac{dp_2}{dz} \end{aligned}\right\} \quad\dots\dots\dots\dots\dots\dots\dots\dots\dots\text{(A)}.$$

Now the quantity of the current depends on the electro-motive force and on the resistance of the medium. If the resistance of the medium be uniform in all directions and equal to k_2,

$$a_2 = k_2 a_2, \qquad \beta_2 = k_2 b_2, \qquad \gamma_2 = k_2 c_2 \dotfill \text{(B)},$$

but if the resistance be different in different directions, the law will be more complicated.

These quantities a_2, β_2, γ_2 may be considered as representing the intensity of the electric action in the directions of x, y, z.

The intensity measured along an element $d\sigma$ of a curve is given by

$$\epsilon = la + m\beta + n\gamma,$$

where l, m, n are the direction-cosines of the tangent.

The integral $\int \epsilon d\sigma$ taken with respect to a given portion of a curve line, represents the total intensity along that line. If the curve is a closed one, it represents the total intensity of the electro-motive force in the closed curve.

Substituting the values of a, β, γ from equations (A)

$$\int \epsilon d\sigma = \int (X dx + Y dy + Z dz) - p + C.$$

If therefore $(X dx + Y dy + Z dz)$ is a complete differential, the value of $\int \epsilon d\sigma$ for a closed curve will vanish, and in all closed curves

$$\int \epsilon d\sigma = \int (X dx + Y dy + Z dz),$$

the integration being effected along the curve, so that in a closed curve the total intensity of the effective electro-motive force is equal to the total intensity of the impressed electro-motive force.

The total *quantity* of conduction through any surface is expressed by

$$\int e dS,$$

where

$$e = la + mb + nc,$$

l, m, n being the direction-cosines of the normal,

$$\therefore \int e dS = \iint a\, dy dz + \iint b\, dz dx + \iint c\, dx dy,$$

the integrations being effected over the given surface. When the surface is a closed one, then we may find by integration by parts

$$\int e dS = \iiint \left(\frac{da}{dx} + \frac{db}{dy} + \frac{dc}{dz} \right) dx dy dz.$$

If we make

$$\frac{da}{dx} + \frac{db}{dy} + \frac{dc}{dz} = 4\pi\rho \dots\dots\dots\dots\dots\dots\dots\dots\text{(C)},$$

$$\int e\, dS = 4\pi \iiint \rho\, dx\, dy\, dz,$$

where the integration on the right side of the equation is effected over every part of space within the surface. In a large class of phenomena, including all cases of uniform currents, the quantity ρ disappears.

Magnetic Quantity and Intensity.

From his study of the lines of magnetic force, Faraday has been led to the conclusion that in the tubular surface* formed by a system of such lines, the quantity of magnetic induction across any section of the tube is constant, and that the alteration of the character of these lines in passing from one substance to another, is to be explained by a difference of *inductive capacity* in the two substances, which is analogous to conductive power in the theory of electric currents.

In the following investigation we shall have occasion to treat of magnetic quantity and intensity in connection with electric. In such cases the magnetic symbols will be distinguished by the suffix 1, and the electric by the suffix 2. The equations connecting a, b, c, k, α, β, γ, p, and ρ, are the same in form as those which we have just given. a, b, c are the symbols of magnetic induction with respect to quantity; k denotes the resistance to magnetic induction, and may be different in different directions; α, β, γ, are the effective magnetizing forces, connected with a, b, c, by equations (B); p is the magnetic tension or potential which will be afterwards explained; ρ denotes the density of *real magnetic matter* and is connected with a, b, c by equations (C). As all the details of magnetic calculations will be more intelligible after the exposition of the connexion of magnetism with electricity, it will be sufficient here to say that all the definitions of total quantity, with respect to a surface, the total intensity to a curve, apply to the case of magnetism as well as to that of electricity.

* *Exp. Res.* 3271, definition of "Sphondyloid."

Electro-magnetism.

Ampère has proved the following laws of the attractions and repulsions of electric currents :

I. Equal and opposite currents generate equal and opposite forces.

II. A crooked current is equivalent to a straight one, provided the two currents nearly coincide throughout their whole length.

III. Equal currents traversing similar and similarly situated closed curves act with equal forces, whatever be the linear dimensions of the circuits.

IV. A closed current exerts no force tending to turn a circular conductor about its centre.

It is to be observed, that the currents with which Ampère worked were constant and therefore re-entering. All his results are therefore deduced from experiments on closed currents, and his expressions for the mutual action of the elements of a current involve the assumption that this action is exerted in the direction of the line joining those elements. This assumption is no doubt warranted by the universal consent of men of science in treating of attractive forces considered as due to the mutual action of particles; but at present we are proceeding on a different principle, and searching for the explanation of the phenomena, not in the currents alone, but also in the surrounding medium.

The first and second laws shew that currents are to be combined like velocities or forces.

The third law is the expression of a property of all attractions which may be conceived of as depending on the inverse square of the distance from a fixed system of points; and the fourth shews that the electro-magnetic forces may always be reduced to the attractions and repulsions of imaginary matter properly distributed.

In fact, the action of a very small electric circuit on a point in its neighbourhood is identical with that of a small magnetic element on a point outside it. If we divide any given portion of a surface into elementary areas, and cause equal currents to flow in the same direction round all these little areas, the effect on a point not in the surface will be the same as that of a shell coinciding with the surface, and uniformly magnetized normal to its surface. But by the first law all the currents forming the little circuits will destroy one another, and leave a single current running round the bounding line. So that the magnetic effect of a uniformly magnetized shell is equivalent to that of an electric current round the edge of the shell. If the direction of the current coincide with that of the apparent motion of the sun, then the direction of magnetization of the imaginary shell will be the same as that of the real magnetization of the earth*.

The total intensity of magnetizing force in a closed curve passing through and embracing the closed current is constant, and may therefore be made a measure of the quantity of the current. As this intensity is independent of the form of the closed curve and depends only on the quantity of the current which passes through it, we may consider the elementary case of the current which flows through the elementary area $dydz$.

Let the axis of x point towards the west, z towards the south, and y upwards. Let x, y, z be the coordinates of a point in the middle of the area $dydz$, then the total intensity measured round the four sides of the element is

$$+\left(\beta_1 + \frac{d\beta_1}{dz}\frac{dz}{2}\right)dy,$$

$$-\left(\gamma_1 + \frac{d\gamma_1}{dy}\frac{dy}{2}\right)dz,$$

$$-\left(\beta_1 - \frac{d\beta_1}{dz}\frac{dz}{2}\right)dy,$$

$$+\left(\gamma_1 - \frac{d\gamma_1}{dy}\frac{dy}{2}\right)dz,$$

$$\text{Total intensity} = \left(\frac{d\beta_1}{dz} - \frac{d\gamma_1}{dy}\right)dy\,dz.$$

The quantity of electricity conducted through the elementary area $dydz$ is a_2dydz, and therefore if we define the measure of an electric current to be the total intensity of magnetizing force in a closed curve embracing it, we shall have

$$a_2 = \frac{d\beta_1}{dz} - \frac{d\gamma_1}{dy},$$

$$b_2 = \frac{d\gamma_1}{dx} - \frac{da_1}{dz},$$

$$c_2 = \frac{da_1}{dy} - \frac{d\beta_1}{dx}.$$

* See *Experimental Researches* (3265) for the relations between the electrical and magnetic circuit, considered as *mutually embracing* curves.

These equations enable us to deduce the distribution of the currents of electricity whenever we know the values of α, β, γ, the magnetic intensities. If α, β, γ be exact differentials of a function of x, y, z with respect to x, y and z respectively, then the values of a_2, b_2, c_2 disappear; and we know that the magnetism is not produced by electric currents in that part of the field which we are investigating. It is due either to the presence of permanent magnetism within the field, or to magnetizing forces due to external causes.

We may observe that the above equations give by differentiation

$$\frac{da_2}{dx} + \frac{db_2}{dy} + \frac{dc_2}{dz} = 0,$$

which is the equation of continuity for closed currents. Our investigations are therefore for the present limited to closed currents; and we know little of the magnetic effects of any currents which are not closed.

Before entering on the calculation of these electric and magnetic states it may be advantageous to state certain general theorems, the truth of which may be established analytically.

THEOREM I.

The equation

$$\frac{d^2V}{dx^2} + \frac{d^2V}{dy^2} + \frac{d^2V}{dz^2} + 4\pi\rho = 0,$$

(where V and ρ are functions of x, y, z never infinite, and vanishing for all points at an infinite distance), can be satisfied by one, and only one, value of V. See Art. (17) above.

THEOREM II.

The value of V which will satisfy the above conditions is found by integrating the expression

$$\iiint \frac{\rho\,dx\,dy\,dz}{\left((x-x')^2 + (y-y')^2 + (z-z')^2\right)^{\frac{1}{2}}},$$

where the limits of x, y, z are such as to include every point of space where ρ is finite.

The proofs of these theorems may be found in any work on attractions or electricity, and in particular in Green's *Essay on the Application of Mathematics to Electricity*. See Arts. 18, 19 of this paper. See also Gauss, *on Attractions*, translated in Taylor's *Scientific Memoirs*.

Theorem III.

Let U and V be two functions of x, y, z, then

$$\iiint U \left(\frac{d^2V}{dx^2} + \frac{d^2V}{dy^2} + \frac{d^2V}{dz^2}\right) dxdydz = -\iiint \left(\frac{dU}{dx}\frac{dV}{dx} + \frac{dU}{dy}\frac{dV}{dy} + \frac{dU}{dz}\frac{dV}{dz}\right) dxdydz$$

$$= \iiint \left(\frac{d^2U}{dx^2} + \frac{d^2U}{dy^2} + \frac{d^2U}{dz^2}\right) Vdx\,dy\,dz;$$

where the integrations are supposed to extend over all the space in which U and V have values differing from 0.—(Green, p. 10.)

This theorem shews that if there be two attracting systems the actions between them are equal and opposite. And by making $U = V$ we find that the potential of a system on itself is proportional to the integral of the square of the resultant attraction through all space; a result deducible from Art. (30), since the volume of each cell is inversely as the square of the velocity (Arts. 12, 13), and therefore the number of cells in a given space is directly as the square of the velocity.

Theorem IV.

Let a, β, γ, ρ be quantities finite through a certain space and vanishing in the space beyond, and let k be given for all parts of space as a continuous or discontinuous function of x, y, z, then the equation in p

$$\frac{d}{dx}\frac{1}{k}\left(a - \frac{dp}{dx}\right) + \frac{d}{dy}\frac{1}{k}\left(\beta - \frac{dp}{dy}\right) + \frac{d}{dz}\frac{1}{k}\left(\gamma - \frac{dp}{dz}\right) + 4\pi\rho = 0,$$

has one, and only one solution, in which p is always finite and vanishes at an infinite distance.

The proof of this theorem, by Prof. W. Thomson, may be found in the *Cambridge and Dublin Mathematical Journal*, Jan. 1848.

If a, β, γ be the electro-motive forces, p the electric tension, and k the coefficient of resistance, then the above equation is identical with the equation of continuity

$$\frac{da_2}{dx} + \frac{db_2}{dy} + \frac{dc_2}{dz} + 4\pi\rho = 0 \; ;$$

and the theorem shews that when the electro-motive forces and the rate of production of electricity at every part of space are given, the value of the electric tension is determinate.

Since the mathematical laws of magnetism are identical with those of electricity, as far as we now consider them, we may regard a, β, γ as magnetizing forces, p as *magnetic tension*, and ρ as *real magnetic density*, k being the coefficient of resistance to magnetic induction.

The proof of this theorem rests on the determination of the minimum value of

$$Q = \iiint \left\{ \frac{1}{k}\left(a - \frac{dp}{dx} - k\frac{dV}{dx}\right)^2 + \frac{1}{k}\left(\beta - \frac{dp}{dy} - k\frac{dV}{dy}\right)^2 + \frac{1}{k}\left(\gamma - \frac{dp}{dz} - k\frac{dV}{dz}\right)^2 \right\} dx\,dy\,dz \; ;$$

where V is got from the equation

$$\frac{d^2V}{dx^2} + \frac{d^2V}{dy^2} + \frac{d^2V}{dz^2} + 4\pi\rho = 0,$$

and p has to be determined.

The meaning of this integral in electrical language may be thus brought out. If the presence of the media in which k has various values did not affect the distribution of forces, then the "quantity" resolved in x would be simply $\frac{dV}{dx}$ and the intensity $k\frac{dV}{dx}$. But the actual quantity and intensity are $\frac{1}{k}\left(a - \frac{dp}{dx}\right)$ and $a - \frac{dp}{dx}$, and the parts due to the distribution of media alone are therefore

$$\frac{1}{k}\left(a - \frac{dp}{dx}\right) - \frac{dV}{dx} \text{ and } a - \frac{dp}{dx} - k\frac{dV}{dx} \; .$$

Now the product of these represents the work done on account of this distribution of media, the distribution of sources being determined, and taking in the terms in y and z we get the expression Q for the total work done by that part of the whole effect at any point which is due to the distribution of conducting media, and not directly to the presence of the sources.

This quantity Q is rendered a minimum by one and only one value of p, namely, that which satisfies the original equation.

THEOREM V.

If a, b, c be three functions of x, y, z satisfying the equation

$$\frac{da}{dx} + \frac{db}{dy} + \frac{dc}{dz} = 0,$$

it is always possible to find three functions α, β, γ which shall satisfy the equations

$$\frac{d\beta}{dz} - \frac{d\gamma}{dy} = a,$$

$$\frac{d\gamma}{dx} - \frac{d\alpha}{dz} = b,$$

$$\frac{d\alpha}{dy} - \frac{d\beta}{dx} = c.$$

Let $A = \int c\,dy$, where the integration is to be performed upon c considered as a function of y, treating x and z as constants. Let $B = \int a\,dz$, $C = \int b\,dx$, $A' = \int b\,dz$, $B' = \int c\,dx$, $C' = \int a\,dy$, integrated in the same way.

Then

$$\alpha = A - A' + \frac{d\psi}{dx},$$

$$\beta = B - B' + \frac{d\psi}{dy},$$

$$\gamma = C - C' + \frac{d\psi}{dz}$$

will satisfy the given equations; for

$$\frac{d\beta}{dz} - \frac{d\gamma}{dy} = \int \frac{da}{dy}\,dz - \int \frac{dc}{dz}\,dx - \int \frac{db}{dy}\,dx + \int \frac{da}{dy}\,dy,$$

and

$$0 = \int \frac{da}{dx}\,dx + \int \frac{db}{dy}\,dx + \int \frac{dc}{dz}\,dx;$$

$$\therefore \frac{d\beta}{dz} - \frac{d\gamma}{dy} = \int \frac{da}{dx}\,dx + \int \frac{da}{dy}\,dy + \int \frac{da}{dz}\,dz,$$

$$= a.$$

In the same way it may be shewn that the values of a, β, γ satisfy the other given equations. The function ψ may be considered at present as perfectly indeterminate.

The method here given is taken from Prof. W. Thomson's memoir on Magnetism (*Phil. Trans.* 1851, p. 283).

As we cannot perform the required integrations when a, b, c are discontinuous functions of x, y, z, the following method, which is perfectly general though more complicated, may indicate more clearly the truth of the proposition.

Let A, B, C be determined from the equations

$$\frac{d^2A}{dx^2} + \frac{d^2A}{dy^2} + \frac{d^2A}{dz^2} + a = 0,$$

$$\frac{d^2B}{dx^2} + \frac{d^2B}{dy^2} + \frac{d^2B}{dz^2} + b = 0,$$

$$\frac{d^2C}{dx^2} + \frac{d^2C}{dy^2} + \frac{d^2C}{dz^2} + c = 0,$$

by the methods of Theorems I. and II., so that A, B, C are never infinite, and vanish when x, y, or z is infinite.

Also let

$$a = \frac{dB}{dz} - \frac{dC}{dy} + \frac{d\psi}{dx},$$

$$\beta = \frac{dC}{dx} - \frac{dA}{dz} + \frac{d\psi}{dy},$$

$$\gamma = \frac{dA}{dy} - \frac{dB}{dx} + \frac{d\psi}{dz},$$

then

$$\frac{d\beta}{dz} - \frac{d\gamma}{dy} = \frac{d}{dx}\left(\frac{dA}{dx} + \frac{dB}{dy} + \frac{dC}{dz}\right) - \left(\frac{d^2A}{dx^2} + \frac{d^2A}{dy^2} + \frac{d^2A}{dz^2}\right)$$

$$= \frac{d}{dx}\left(\frac{dA}{dx} + \frac{dB}{dy} + \frac{dC}{dz}\right) + a.$$

If we find similar equations in y and z, and differentiate the first by x, the second by y, and the third by z, remembering the equation between a, b, c, we shall have

$$\left(\frac{d^2}{dx^2} + \frac{d^2}{dy^2} + \frac{d^2}{dz^2}\right)\left(\frac{dA}{dx} + \frac{dB}{dy} + \frac{dC}{dz}\right) = 0;$$

and since A, B, C are always finite and vanish at an infinite distance, the only solution of this equation is

$$\frac{dA}{dx} + \frac{dB}{dy} + \frac{dC}{dz} = 0,$$

and we have finally

$$\frac{d\beta}{dz} - \frac{d\gamma}{dy} = a,$$

with two similar equations, shewing that a, β, γ have been rightly determined.

The function ψ is to be determined from the condition

$$\frac{da}{dx} + \frac{d\beta}{dy} + \frac{d\gamma}{dz} = \left(\frac{d^2}{dx^2} + \frac{d^2}{dy^2} + \frac{d^2}{dz^2} \right) \psi :$$

if the left-hand side of this equation be always zero, ψ must be zero also.

THEOREM VI.

Let a, b, c be any three functions of x, y, z, it is possible to find three functions a, β, γ and a fourth V, so that

$$\frac{da}{dx} + \frac{d\beta}{dy} + \frac{d\gamma}{dz} = 0,$$

and

$$a = \frac{d\beta}{dz} - \frac{d\gamma}{dy} + \frac{dV}{dx},$$

$$b = \frac{d\gamma}{dx} - \frac{da}{dz} + \frac{dV}{dy},$$

$$c = \frac{da}{dy} - \frac{d\beta}{dx} + \frac{dV}{dz}.$$

Let

$$\frac{da}{dx} + \frac{db}{dy} + \frac{dc}{dz} = -4\pi\rho,$$

and let V be found from the equation

$$\frac{d^2V}{dx^2} + \frac{d^2V}{dy^2} + \frac{d^2V}{dz^2} = -4\pi\rho,$$

then

$$a' = a - \frac{dV}{dx},$$

$$b' = b - \frac{dV}{dy},$$

$$c' = c - \frac{dV}{dz},$$

satisfy the condition

$$\frac{da'}{dx} + \frac{db'}{dy} + \frac{dc'}{dz} = 0 ;$$

and therefore we can find three functions A, B, C, and from these a, β, γ, so as to satisfy the given equations.

THEOREM VII.

The integral throughout infinity

$$Q = \iiint (a_1 a_1 + b_1 \beta_1 + c_1 \gamma_1) \, dx \, dy \, dz,$$

where $a_1 b_1 c_1$, $a_1 \beta_1 \gamma_1$ are any functions whatsoever, is capable of transformation into

$$Q = + \iiint \{4\pi p \rho_1 - (a_0 a_2 + \beta_0 b_2 + \gamma_0 c_2)\} \, dx \, dy \, dz,$$

in which the quantities are found from the equations

$$\frac{da_1}{dx} + \frac{db_1}{dy} + \frac{dc_1}{dz} + 4\pi \rho_1 = 0,$$

$$\frac{da_1}{dx} + \frac{d\beta_1}{dy} + \frac{d\gamma_1}{dz} + 4\pi \rho_1' = 0 ;$$

$a_0 \beta_0 \gamma_0 V$ are determined from $a_1 b_1 c_1$ by the last theorem, so that

$$a_1 = \frac{d\beta_0}{dz} - \frac{d\gamma_0}{dy} + \frac{dV}{dx} ;$$

$a_2 b_2 c_2$ are found from $a_1 \beta_1 \gamma_1$ by the equations

$$a_2 = \frac{d\beta_1}{dz} - \frac{d\gamma_1}{dy} \text{ \&c.,}$$

and p is found from the equation

$$\frac{d^2 p}{dx^2} + \frac{d^2 p}{dy^2} + \frac{d^2 p}{dz^2} + 4\pi \rho_1' = 0.$$

For, if we put a_1 in the form

$$\frac{d\beta_0}{dz} - \frac{d\gamma_0}{dy} + \frac{dV}{dx},$$

and treat b_1 and c_1 similarly, then we have by integration by parts through infinity, remembering that all the functions vanish at the limits,

$$Q = -\iiint \left\{ V\left(\frac{da_1}{dx} + \frac{d\beta_1}{dy} + \frac{d\gamma_1}{dz}\right) + a_0\left(\frac{d\beta_1}{dz} - \frac{d\gamma_1}{dy}\right) + \beta_0\left(\frac{d\gamma_1}{dx} - \frac{da_1}{dz}\right) \right.$$
$$\left. + \gamma_0\left(\frac{da_1}{dy} - \frac{d\beta_1}{dx}\right) \right\} dx\,dy\,dz,$$

or $Q = +\iiint \{(4\pi V\rho') - (a_0 a_2 + \beta_0 b_2 + \gamma_0 c_2)\}\, dx\,dy\,dz,$

and by Theorem III.

$$\iiint V\rho'\, dx\,dy\,dz = \iiint p\rho\, dx\,dy\,dz,$$

so that finally

$$Q = \iiint \{4\pi p\rho - (a_0 a_2 + \beta_0 b_2 + \gamma_0 c_2)\}\, dx\,dy\,dz.$$

If $a_1 b_1 c_1$ represent the components of magnetic quantity, and $a_1\beta_1\gamma_1$ those of magnetic intensity, then ρ will represent the *real magnetic density*, and p the magnetic potential or tension. $a_2 b_2 c_2$ will be the components of quantity of electric currents, and $a_0\beta_0\gamma_0$ will be three functions deduced from $a_1 b_1 c_1$, which will be found to be the mathematical expression for Faraday's Electrotonic state.

Let us now consider the bearing of these analytical theorems on the theory of magnetism. Whenever we deal with quantities relating to magnetism, we shall distinguish them by the suffix $(_1)$. Thus $a_1 b_1 c_1$ are the components resolved in the directions of x, y, z of the quantity of magnetic induction acting through a given point, and $a_1\beta_1\gamma_1$ are the resolved intensities of magnetization at the same point, or, what is the same thing, the components of the force which would be exerted on a unit south pole of a magnet placed at that point without disturbing the distribution of magnetism.

The electric currents are found from the magnetic intensities by the equations

$$a_2 = \frac{d\beta_1}{dz} - \frac{d\gamma_1}{dy} \text{ &c.}$$

When there are no electric currents, then

$$a_1 dx + \beta_1 dy + \gamma_1 dz = dp_1,$$

a perfect differential of a function of x, y, z. On the principle of analogy we may call p_1 the magnetic tension.

The forces which act on a mass m of south magnetism at any point are

$$-m \frac{dp_1}{dx}, \quad -m \frac{dp_1}{dy}, \text{ and } -m \frac{dp_1}{dz},$$

in the direction of the axes, and therefore the whole work done during any displacement of a magnetic system is equal to the decrement of the integral

$$Q = \iiint \rho_1 p_1 dx dy dz$$

throughout the system.

Let us now call Q the *total potential of the system on itself*. The increase or decrease of Q will measure the work lost or gained by any displacement of any part of the system, and will therefore enable us to determine the forces acting on that part of the system.

By Theorem III. Q may be put under the form

$$Q = + \frac{1}{4\pi} \iiint (a_1 \alpha_1 + b_1 \beta_1 + c_1 \gamma_1)\, dx dy dz,$$

in which $\alpha_1 \beta_1 \gamma_1$ are the differential coefficients of p_1 with respect to x, y, z respectively.

If we now assume that this expression for Q is true whatever be the values of α_1, β_1, γ_1, we pass from the consideration of the magnetism of permanent magnets to that of the magnetic effects of electric currents, and we have then by Theorem VII.

$$Q = \iiint \left\{ p_1 \rho_1 - \frac{1}{4\pi} (\alpha_0 a_2 + \beta_0 b_2 + \gamma_0 c_2) \right\} dx dy dz.$$

So that in the case of electric currents, the components of the currents have to be multiplied by the functions α_0, β_0, γ_0 respectively, and the summations of all such products throughout the system gives us the part of Q due to those currents.

We have now obtained in the functions α_0, β_0, γ_0 the means of avoiding the consideration of the quantity of magnetic induction which *passes through* the circuit. Instead of this artificial method we have the natural one of considering the current with reference to quantities existing in the same space with the current itself. To these I give the name of *Electro-tonic functions*, or *components of the Electro-tonic intensity*.

Let us now consider the conditions of the conduction of the electric currents within the medium during changes in the electro-tonic state. The method which we shall adopt is an application of that given by Helmholtz in his memoir on the Conservation of Force*.

Let there be some external source of electric currents which would generate in the conducting mass currents whose quantity is measured by a_2, b_2, c_2 and their intensity by a_2, β_2, γ_2.

Then the amount of work due to this cause in the time dt is

$$dt \iiint (a_2 a_2 + b_2 \beta_2 + c_2 \gamma_2)\, dx\, dy\, dz$$

in the form of resistance overcome, and

$$\frac{dt}{4\pi} \frac{d}{dt} \iiint (a_2 a_0 + b_2 \beta_0 + c_2 \gamma_0)\, dx\, dy\, dz$$

in the form of work done mechanically by the electro-magnetic action of these currents. If there be no external cause producing currents, then the quantity representing the whole work done by the external cause must vanish, and we have

$$dt \iiint (a_2 a_2 + b_2 \beta_2 + c_2 \gamma_2)\, dx\, dy\, dz + \frac{dt}{4\pi} \frac{d}{dt} \iiint (a_2 a_0 + b_2 \beta_0 + c_2 \gamma_0)\, dx\, dy\, dz,$$

where the integrals are taken through any arbitrary space. We must therefore have

$$a_2 a_2 + b_2 \beta_2 + c_2 \gamma_2 = \frac{1}{4\pi} \frac{d}{dt} (a_2 a_0 + b_2 \beta_0 + c_2 \gamma_0)$$

for every point of space; and it must be remembered that the variation of Q is supposed due to variations of a_0, β_0, γ_0, and not of a_2, b_2, c_2. We must therefore treat a_2, b_2, c_2 as constants, and the equation becomes

$$a_2 \left(a_2 + \frac{1}{4\pi} \frac{da_0}{dt} \right) + b_2 \left(\beta_2 + \frac{1}{4\pi} \frac{d\beta_0}{dt} \right) + c_2 \left(\gamma_2 + \frac{1}{4\pi} \frac{d\gamma_0}{dt} \right) = 0.$$

In order that this equation may be independent of the values of a_2, b_2, c_2, each of these coefficients must $= 0$; and therefore we have the following expressions for the electro-motive forces due to the action of magnets and currents at a distance in terms of the electro-tonic functions,

$$a_2 = -\frac{1}{4\pi} \frac{da_0}{dt}, \qquad \beta_2 = -\frac{1}{4\pi} \frac{d\beta_0}{dt}, \qquad \gamma_2 = -\frac{1}{4\pi} \frac{d\gamma_0}{dt}.$$

* Translated in Taylor's *New Scientific Memoirs*, Part II.

It appears from experiment that the expression $\dfrac{da_0}{dt}$ refers to the change of electro-tonic state of a *given particle of the conductor*, whether due to change in the electro-tonic functions themselves or to the motion of the particle.

If a_0 be expressed as a function of x, y, z and t, and if x, y, z be the co-ordinates of a moving particle, then the electro-motive force measured in the direction of x is

$$a_2 = -\frac{1}{4\pi}\left(\frac{da_0}{dx}\frac{dx}{dt} + \frac{da_0}{dy}\frac{dy}{dt} + \frac{da_0}{dz}\frac{dz}{dt} + \frac{da_0}{dt}\right).$$

The expressions for the electro-motive forces in y and z are similar. The distribution of currents due to these forces depends on the form and arrangement of the conducting media and on the resultant electric tension at any point.

The discussion of these functions would involve us in mathematical formulæ, of which this paper is already too full. It is only on account of their physical importance as the mathematical expression of one of Faraday's conjectures that I have been induced to exhibit them at all in their present form. By a more patient consideration of their relations, and with the help of those who are engaged in physical inquiries both in this subject and in others not obviously connected with it, I hope to exhibit the theory of the electro-tonic state in a form in which all its relations may be distinctly conceived without reference to analytical calculations.

Summary of the Theory of the Electro-tonic State.

We may conceive of the electro-tonic state at any point of space as a quantity determinate in magnitude and direction, and we may represent the electro-tonic condition of a portion of space by any mechanical system which has at every point some quantity, which may be a velocity, a displacement, or a force, whose direction and magnitude correspond to those of the supposed electro-tonic state. This representation involves no physical theory, it is only a kind of artificial notation. In analytical investigations we make use of the three components of the electro-tonic state, and call them electro-tonic functions. We take the resolved part of the electro-tonic intensity at every point of a closed curve, and find by integration what we may call the *entire electro-tonic intensity round the curve.*

PROP. I. *If on any surface a closed curve be drawn, and if the surface within it be divided into small areas, then the entire intensity round the closed curve is equal to the sum of the intensities round each of the small areas, all estimated in the same direction.*

For, in going round the small areas, every boundary line between two of them is passed along twice in opposite directions, and the intensity gained in the one case is lost in the other. Every effect of passing along the interior divisions is therefore neutralized, and the whole effect is that due to the exterior closed curve.

LAW I. *The entire electro-tonic intensity round the boundary of an element of surface measures the quantity of magnetic induction which passes through that surface, or, in other words, the number of lines of magnetic force which pass through that surface.*

By PROP. I. it appears that what is true of elementary surfaces is true also of surfaces of finite magnitude, and therefore any two surfaces which are bounded by the same closed curve will have the same quantity of magnetic induction through them.

LAW II. *The magnetic intensity at any point is connected with the quantity of magnetic induction by a set of linear equations, called the equations of conduction*.*

LAW III. *The entire magnetic intensity round the boundary of any surface measures the quantity of electric current which passes through that surface.*

LAW IV. *The quantity and intensity of electric currents are connected by a system of equations of conduction.*

By these four laws the magnetic and electric quantity and intensity may be deduced from the values of the electro-tonic functions. I have not discussed the values of the units, as that will be better done with reference to actual experiments. We come next to the attraction of conductors of currents, and to the induction of currents within conductors.

* See Art. (28).

LAW V. *The total electro-magnetic potential of a closed current is measured by the product of the quantity of the current multiplied by the entire electro-tonic intensity estimated in the same direction round the circuit.*

Any displacement of the conductors which would cause an increase in the potential will be assisted by a force measured by the rate of increase of the potential, so that the mechanical work done during the displacement will be measured by the increase of potential.

Although in certain cases a displacement in direction or alteration of intensity *of the current* might increase the potential, such an alteration would not itself produce work, and there will be no tendency towards this displacement, for alterations in the current are due to electro-motive force, not to electro-magnetic attractions, which can only act on the conductor.

LAW VI. *The electro-motive force on any element of a conductor is measured by the instantaneous rate of change of the electro-tonic intensity on that element, whether in magnitude or direction.*

The electro-motive force in a closed conductor is measured by the rate of change of the entire electro-tonic intensity round the circuit referred to unit of time. It is independent of the nature of the conductor, though the current produced varies inversely as the resistance; and it is the same in whatever way the change of electro-tonic intensity has been produced, whether by motion of the conductor or by alterations in the external circumstances.

In these six laws I have endeavoured to express the idea which I believe to be the mathematical foundation of the modes of thought indicated in the *Experimental Researches.* I do not think that it contains even the shadow of a true physical theory; in fact, its chief merit as a temporary instrument of research is that it does not, even in appearance, *account for* anything.

There exists however a professedly physical theory of electro-dynamics, which is so elegant, so mathematical, and so entirely different from anything in this paper, that I must state its axioms, at the risk of repeating what ought to be well known. It is contained in M. W. Weber's *Electro-dynamic Measurements*, and may be found in the Transactions of the Leibnitz Society, and of the Royal Society of Sciences of Saxony*. The assumptions are,

* When this was written, I was not aware that part of M. Weber's Memoir is translated in Taylor's *Scientific Memoirs*, Vol. v. Art. xiv. The value of his researches, both experimental and theoretical, renders the study of his theory necessary to every electrician.

(1) That two particles of electricity when in motion do not repel each other with the same force as when at rest, but that the force is altered by a quantity depending on the relative motion of the two particles, so that the expression for the repulsion at distance r is

$$\frac{ee'}{r^2}\left(1 + a\left|\overline{\frac{dr}{dt}}\right|^2 + br\frac{d^2r}{dt^2}\right).$$

(2) That when electricity is moving in a conductor, the velocity of the positive fluid *relatively to the matter of the conductor* is equal and opposite to that of the negative fluid.

(3) The total action of one conducting element on another is the resultant of the mutual actions of the masses of electricity of both kinds which are in each.

(4) The electro-motive force at any point is the difference of the forces acting on the positive and negative fluids.

From these axioms are deducible Ampère's laws of the attraction of conductors, and those of Neumann and others, for the induction of currents. Here then is a really physical theory, satisfying the required conditions better perhaps than any yet invented, and put forth by a philosopher whose experimental researches form an ample foundation for his mathematical investigations. What is the use then of imagining an electro-tonic state of which we have no distinctly physical conception, instead of a formula of attraction which we can readily understand? I would answer, that it is a good thing to have two ways of looking at a subject, and to admit that there *are* two ways of looking at it. Besides, I do not think that we have any right at present to understand the action of electricity, and I hold that the chief merit of a temporary theory is, that it shall guide experiment, without impeding the progress of the true theory when it appears. There are also objections to making any ultimate forces in nature depend on the velocity of the bodies between which they act. If the forces in nature are to be reduced to forces acting between particles, the principle of the Conservation of Force requires that these forces should be in the line joining the particles and functions of the distance only. The experiments of M. Weber on the reverse polarity of diamagnetics, which have been recently repeated by Professor Tyndall, establish a fact which is equally a consequence of M. Weber's theory of electricity and of the theory of lines of force.

With respect to the history of the present theory, I may state that the recognition of certain mathematical functions as expressing the "electro-tonic state" of Faraday, and the use of them in determining electro-dynamic potentials and electro-motive forces is, as far as I am aware, original; but the distinct conception of the possibility of the mathematical expressions arose in my mind from the perusal of Prof. W. Thomson's papers "On a Mechanical Representation of Electric, Magnetic and Galvanic Forces," *Cambridge and Dublin Mathematical Journal*, January, 1847, and his "Mathematical Theory of Magnetism," *Philosophical Transactions*, Part I. 1851, Art. 78, &c. As an instance of the help which may be derived from other physical investigations, I may state that after I had investigated the Theorems of this paper Professor Stokes pointed out to me the use which he had made of similar expressions in his "Dynamical Theory of Diffraction," Section 1, *Cambridge Transactions*, Vol. IX. Part 1. Whether the theory of these functions, considered with reference to electricity, may lead to new mathematical ideas to be employed in physical research, remains to be seen. I propose in the rest of this paper to discuss a few electrical and magnetic problems with reference to spheres. These are intended merely as concrete examples of the methods of which the theory has been given; I reserve the detailed investigation of cases chosen with special reference to experiment till I have the means of testing their results.

EXAMPLES.

I. *Theory of Electrical Images.*

The method of Electrical Images, due to Prof. W. Thomson*, by which the theory of spherical conductors has been reduced to great geometrical simplicity, becomes even more simple when we see its connexion with the methods of this paper. We have seen that the pressure at any point in a uniform medium, due to a spherical shell (radius $= a$) giving out fluid at the rate of $4\pi P a^2$ units in unit of time, is $kP\dfrac{a^2}{r}$ outside the shell, and kPa inside it, where r is the distance of the point from the centre of the shell.

* See a series of papers "On the Mathematical Theory of Electricity," in the *Cambridge and Dublin Math. Jour.*, beginning March, 1848.

If there be two shells, one giving out fluid at a rate $4\pi P a^2$, and the other absorbing at the rate of $4\pi P' a'^2$, then the expression for the pressure will be, outside the shells,

$$p = 4\pi P \frac{a^2}{r} - 4\pi P' \frac{a'^2}{r'},$$

where r and r' are the distances from the centres of the two shells. Equating this expression to zero we have, as the surface of no pressure, that for which

$$\frac{r'}{r} = \frac{P'a'^2}{Pa^2}.$$

Now the surface, for which the distances to two fixed points have a given ratio, is a sphere of which the centre O is in the line joining the centres of the shells CC' produced, so that

$$C'O = CC' \frac{\overline{P'a'^2}\vert^2}{\overline{Pa^2}\vert^2 - \overline{P'a'^2}\vert^2},$$

and its radius

$$= CC' \frac{Pa^2 \cdot P'a'^2}{\overline{Pa^2}\vert^2 - \overline{P'a'^2}\vert^2},$$

If at the centre of this sphere we place another source of the fluid, then the pressure due to this source must be added to that due to the other two; and since this additional pressure depends only on the distance from the centre, it will be constant at the surface of the sphere, where the pressure due to the two other sources is zero.

We have now the means of arranging a system of sources within a given sphere, so that when combined with a given system of sources outside the sphere, they shall produce a given constant pressure at the surface of the sphere.

Let a be the radius of the sphere, and p the given pressure, and let the given sources be at distances b_1, b_2, &c. from the centre, and let their rates of production be $4\pi P_1$, $4\pi P_2$, &c.

Then if at distances $\dfrac{a^2}{b_1}$, $\dfrac{a^2}{b_2}$, &c. (measured in the same direction as b_1, b_2, &c. from the centre) we place negative sources whose rates are

$$-4\pi P_1 \frac{a}{b_1}, \quad -4\pi P_2 \frac{a}{b_2}, \text{ &c.,}$$

the pressure at the surface $r=a$ will be reduced to zero. Now placing a source $4\pi\dfrac{pa}{k}$ at the centre, the pressure at the surface will be uniform and equal to p.

The whole amount of fluid emitted by the surface $r=a$ may be found by adding the rates of production of the sources within it. The result is

$$4\pi a\left\{\frac{p}{k}-\frac{P_1}{b_1}-\frac{P_2}{b_2}-\&\text{c.}\right\}.$$

To apply this result to the case of a conducting sphere, let us suppose the external sources $4\pi P_1$, $4\pi P_2$ to be small electrified bodies, containing e_1, e_2 of positive electricity. Let us also suppose that the whole charge of the conducting sphere is $=E$ previous to the action of the external points. Then all that is required for the complete solution of the problem is, that the surface of the sphere shall be a surface of equal potential, and that the total charge of the surface shall be E.

If by any distribution of imaginary sources within the spherical surface we can effect this, the value of the corresponding potential outside the sphere is the true and only one. The potential inside the sphere must really be constant and equal to that at the surface.

We must therefore find the *images* of the external electrified points, that is, for every point at distance b from the centre we must find a point on the same radius at a distance $\dfrac{a^2}{b_1}$, and at that point we must place a quantity $=-e\dfrac{a}{b_1}$ of imaginary electricity.

At the centre we must put a quantity E' such that

$$E'=E+e_1\frac{a}{b_1}+e_2\frac{a}{b_2}+\&\text{c.};$$

then if R be the distance from the centre, r_1, r_2, &c. the distances from the electrified points, and r'_1, r'_2, &c. the distances from their images at any point outside the sphere, the potential at that point will be

$$p=\frac{E'}{R}+e_1\left(\frac{1}{r_1}-\frac{a}{b_1}\frac{1}{r'_1}\right)+e_2\left(\frac{1}{r_2}-\frac{a}{b_2}\frac{1}{r'_2}\right)+\&\text{c.}$$

$$=\frac{E}{R}+\frac{e_1}{b_1}\left(\frac{a}{R}+\frac{b_1}{r_1}-\frac{a}{r'_1}\right)+\frac{e_2}{b_2}\left(\frac{a}{R}+\frac{b_2}{r_2}-\frac{a}{r'_2}\right)+\&\text{c.}$$

This is the value of the potential outside the sphere. At the surface we have

$$R = a \text{ and } \frac{b_1}{r_1} = \frac{a}{r'_1}, \qquad \frac{b_2}{r_2} = \frac{a}{r'_2}, \quad \&c.,$$

so that at the surface

$$p = \frac{E}{a} + \frac{e_1}{b_1} + \frac{e_2}{b_2} + \&c.,$$

and this must also be the value of p for any point within the sphere.

For the application of the principle of electrical images the reader is referred to Prof. Thomson's papers in the *Cambridge and Dublin Mathematical Journal*. The only case which we shall consider is that in which $\frac{e_1}{b_1^2} = I$, and b_1 is infinitely distant along the axis of x, and $E = 0$.

The value p outside the sphere becomes then

$$p = Ix\left(-\frac{a^3}{r^3}\right),$$

and inside $p = 0$.

II. *On the effect of a paramagnetic or diamagnetic sphere in a uniform field of magnetic force*.*

The expression for the potential of a small magnet placed at the origin of co-ordinates in the direction of the axis of x is

$$l\frac{d}{dx}\left(\frac{m}{r}\right) = -lm\frac{x}{r^3}.$$

The effect of the sphere in disturbing the lines of force may be supposed as a first hypothesis to be similar to that of a small magnet at the origin, whose strength is to be determined. (We shall find this to be accurately true.)

---

* See Prof. Thomson, on the Theory of Magnetic Induction, *Phil. Mag.* March, 1851. The *inductive capacity* of the sphere, according to that paper, is the ratio of the *quantity* of magnetic induction (not the *intensity*) within the sphere to that without. It is therefore equal to $\frac{1}{I}B\frac{k'}{k} = \frac{3k'}{2k+k'}$, according to our notation.

Let the value of the potential undisturbed by the presence of the sphere be

$$p = Ix.$$

Let the sphere produce an additional potential, which for external points is

$$p' = A \frac{a^3}{r^3} x,$$

and let the potential within the sphere be

$$p_1 = Bx.$$

Let k' be the coefficient of resistance outside, and k inside the sphere, then the conditions to be fulfilled are, that the interior and exterior potentials should coincide at the surface, and that the induction through the surface should be the same whether deduced from the external or the internal potential. Putting $x = r \cos \theta$, we have for the external potential

$$P = \left(Ir + A \frac{a^3}{r^2} \right) \cos \theta,$$

and for the internal

$$p_1 = Br \cos \theta,$$

and these must be identical when $r = a$, or

$$I + A = B.$$

The induction through the surface in the external medium is

$$\frac{1}{k'} \frac{dp}{dr}_{r=a} = \frac{1}{k'} (I - 2A) \cos \theta,$$

and that through the interior surface is

$$\frac{1}{k} \frac{dp_1}{dr}_{r=a} = \frac{1}{k} B \cos \theta;$$

and $\therefore \frac{1}{k'} (I - 2A) = \frac{1}{k} B.$

These equations give

$$A = \frac{k - k'}{2k + k'} I, \qquad B = \frac{3k}{2k + k'} I.$$

The effect outside the sphere is equal to that of a little magnet whose length is l and moment ml, provided

$$ml = \frac{k - k'}{2k + k'} a^3 I.$$

Suppose this uniform field to be that due to terrestrial magnetism, then, if k is less than k' as in paramagnetic bodies, the marked end of the equivalent magnet will be turned to the north. If k is greater than k' as in diamagnetic bodies, the unmarked end of the equivalent magnet would be turned to the north.

III.   *Magnetic field of variable Intensity.*

Now suppose the intensity in the undisturbed magnetic field to vary in magnitude and direction from one point to another, and that its components in x, y, z are represented by a, β, γ, then, if as a first approximation we regard the intensity within the sphere as sensibly equal to that at the centre, the change of potential outside the sphere arising from the presence of the sphere, disturbing the lines of force, will be the same as that due to three small magnets at the centre, with their axes parallel to x, y, and z, and their moments equal to

$$\frac{k-k'}{2k+k'}\,a^3 a, \qquad \frac{k-k'}{2k+k'}\,a^3\beta, \qquad \frac{k-k'}{2k+k'}\,a^3\gamma.$$

The actual distribution of potential within and without the sphere may be conceived as the result of a distribution of imaginary magnetic matter on the surface of the sphere; but since the external effect of this superficial magnetism is exactly the same as that of the three small magnets at the centre, the mechanical effect of external attractions will be the same as if the three magnets really existed.

Now let three small magnets whose lengths are l_1, l_2, l_3, and strengths m_1, m_2, m_3, exist at the point x, y, z with their axes parallel to the axes of x, y, z; then resolving the forces on the three magnets in the direction of X, we have

$$-X = m_1 \left\{ \begin{array}{l} a + \dfrac{da}{dx}\dfrac{l_1}{2} \\[2mm] -a + \dfrac{da}{dx}\dfrac{l_1}{2} \end{array} \right\} + m_2 \left\{ \begin{array}{l} a + \dfrac{da}{dy}\dfrac{l_2}{2} \\[2mm] -a + \dfrac{da}{dy}\dfrac{l_2}{2} \end{array} \right\} + m_3 \left\{ \begin{array}{l} a + \dfrac{da}{dz}\dfrac{l_3}{2} \\[2mm] -a + \dfrac{da}{dz}\dfrac{l_2}{2} \end{array} \right\}$$

$$= m_1 l_1 \frac{da}{dx} + m_2 l_2 \frac{da}{dy} + m_3 l_3 \frac{da}{dz}.$$

Substituting the values of the moments of the imaginary magnets

$$-X = \frac{k-k'}{2k+k'}\, a^3 \left(a\frac{d\alpha}{dx} + \beta\frac{d\beta}{dx} + \gamma\frac{d\gamma}{dx} \right) = \frac{k-k'}{2k+k'}\, \frac{a^3}{2}\, \frac{d}{dx}(\alpha^2 + \beta^2 + \gamma^2).$$

The force impelling the sphere in the direction of x is therefore dependent on the variation of the square of the intensity or $(\alpha^2 + \beta^2 + \gamma^2)$, as we move along the direction of x, and the same is true for y and z, so that the law is, that the force acting on diamagnetic spheres is from places of greater to places of less intensity of magnetic force, and that in similar distributions of magnetic force it varies as the mass of the sphere and the square of the intensity.

It is easy by means of Laplace's Coefficients to extend the approximation to the value of the potential as far as we please, and to calculate the attraction. For instance, if a north or south magnetic pole whose strength is M, be placed at a distance b from a diamagnetic sphere, radius a, the repulsion will be

$$R = M^2\,(k-k')\, \frac{a^3}{b^5} \left(\frac{2 \cdot 1}{2k+k'} + \frac{3 \cdot 2}{3k+2k'}\, \frac{a^2}{b^2} + \frac{4 \cdot 3}{4k+3k'}\, \frac{a^4}{b^4} + \&c. \right).$$

When $\dfrac{a}{b}$ is small, the first term gives a sufficient approximation. The repulsion is then as the square of the strength of the pole, and the mass of the sphere directly and the fifth power of the distance inversely, considering the pole as a point.

IV. *Two Spheres in uniform field.*

Let two spheres of radius a be connected together so that their centres are kept at a distance b, and let them be suspended in a uniform magnetic field, then, although each sphere by itself would have been in equilibrium at any part of the field, the disturbance of the field will produce forces tending to make the balls set in a particular direction.

Let the centre of one of the spheres be taken as origin, then the undisturbed potential is

$$p = Ir \cos\theta,$$

and the potential due to the sphere is

$$p' = I\,\frac{k-k'}{2k+k'}\, \frac{a^3}{r^2}\cos\theta.$$

The whole potential is therefore equal to

$$I\left(r + \frac{k-k'}{2k+k'}\frac{a^3}{r^2}\right)\cos\theta = p,$$

$$\frac{dp}{dr} = I\left(1 - 2\frac{k-k'}{2k+k'}\frac{a^3}{r^3}\right)\cos\theta,$$

$$\frac{1}{r}\frac{dp}{d\theta} = -I\left(1 + \frac{k-k'}{2k+k'}\frac{a^3}{r^3}\right)\sin\theta, \qquad \frac{dp}{d\phi} = 0,$$

$$\therefore \quad i^2 = \overline{\frac{dp}{dr}}\Big|^2 + \frac{1}{r^2}\overline{\frac{dp}{d\theta}}\Big|^2 + \frac{1}{r^2\sin^2\theta}\overline{\frac{dp}{d\phi}}\Big|^2$$

$$= I^2\left\{1 + \frac{k-k'}{2k+k'}\frac{a^3}{r^3}(1 - 3\cos^2\theta) + \overline{\frac{k-k'}{2k+k'}}\Big|^2\frac{a^6}{r^6}(1 + 3\cos^2\theta)\right\}.$$

This is the value of the square of the intensity at any point. The moment of the couple tending to turn the combination of balls in the direction of the original force

$$L = \tfrac{1}{2}\frac{d}{d\theta}i^2\left(\frac{k-k'}{2k+k'}a^3\right) \text{ when } r = b,$$

$$L = \tfrac{3}{2}I^2\overline{\frac{k-k'}{2k+k'}}\Big|^2\frac{a^6}{b^3}\left(1 - \frac{k-k'}{2k+k'}\frac{a^3}{b^3}\right)\sin 2\theta.$$

This expression, which must be positive, since b is greater than a, gives the moment of a force tending to turn the line joining the centres of the spheres towards the original lines of force.

Whether the spheres are magnetic or diamagnetic they tend to set in the *axial* direction, and that without distinction of north and south. If, however, one sphere be magnetic and the other diamagnetic, the line of centres will set equatoreally. The magnitude of the force depends on the square of $(k-k')$, and is therefore quite insensible except in iron[*].

V. *Two Spheres between the poles of a Magnet.*

Let us next take the case of the same balls placed not in a uniform field but between a north and a south pole, $\pm M$, distant $2c$ from each other in the direction of x.

---

[*] See Prof. Thomson in *Phil. Mag.* March, 1851.

The expression for the potential, the middle of the line joining the poles being the origin, is

$$p = M \left(\frac{1}{\sqrt{c^2 + r^2 - 2cr \cos \theta}} - \frac{1}{\sqrt{c^2 + r^2 + 2cr \cos \theta}} \right).$$

From this we find as the value of I^2,

$$I^2 = \frac{4M^2}{c^4} \left(1 - 3\frac{r^2}{c^2} + 9\frac{r^2}{c^2} \cos^2 \theta \right);$$

$$\therefore I\frac{dI}{d\theta} = -18\frac{M^2}{c^6} r^2 \sin 2\theta,$$

and the moment to turn a pair of spheres (radius a, distance $2b$) in the direction in which θ is increased is

$$-36 \frac{k-k'}{2k+k'} \frac{M^2 a^3 b^2}{c^6} \sin 2\theta.$$

This force, which tends to turn the line of centres equatoreally for diamagnetic and axially for magnetic spheres, varies directly as the square of the strength of the magnet, the cube of the radius of the spheres and the square of the distance of their centres, and inversely as the sixth power of the distance of the poles of the magnet, considered as points. As long as these poles are near each other this action of the poles will be much stronger than the mutual action of the spheres, so that as a general rule we may say that elongated bodies set axially or equatoreally between the poles of a magnet according as they are magnetic or diamagnetic. If, instead of being placed between two poles very near to each other, they had been placed in a uniform field such as that of terrestrial magnetism or that produced by a spherical electro-magnet (see Ex. VIII.), an elongated body would set axially whether magnetic or diamagnetic.

In all these cases the phenomena depend on $k-k'$, so that the sphere conducts itself magnetically or diamagnetically according as it is more or less magnetic, or less or more diamagnetic than the medium in which it is placed.

VI. *On the Magnetic Phenomena of a Sphere cut from a substance whose coefficient of resistance is different in different directions.*

Let the axes of magnetic resistance be parallel throughout the sphere, and let them be taken for the axes of x, y, z. Let k_1, k_2, k_3, be the coefficients of resistance in these three directions, and let k' be that of the external medium, and a the radius of the sphere. Let I be the undisturbed magnetic intensity of the field into which the sphere is introduced, and let its direction-cosines be l, m, n.

Let us now take the case of a homogeneous sphere whose coefficient is k_1 placed in a uniform magnetic field whose intensity is lI in the direction of x. The resultant potential outside the sphere would be

$$p' = lI \left(1 + \frac{k_1 - k'}{2k_1 + k'} \frac{a^3}{r^3}\right) x,$$

and for internal points

$$p_1 = lI \frac{3k_1}{2k_1 + k'} x.$$

So that in the interior of the sphere the magnetization is entirely in the direction of x. It is therefore quite independent of the coefficients of resistance in the directions of x and y, which may be changed from k_1 into k_2 and k_3 without disturbing this distribution of magnetism. We may therefore treat the sphere as homogeneous for each of the three components of I, but we must use a different coefficient for each. We find for external points

$$p' = I \left\{ lx + my + nz + \left(\frac{k_1 - k'}{2k_1 + k'} lx + \frac{k_2 - k'}{2k_2 + k'} my + \frac{k_3 - k'}{2k_3 + k'} nz\right) \frac{a^3}{r^3}\right\},$$

and for internal points

$$p_1 = I \left(\frac{3k_1}{2k_1 + k'} lx + \frac{3k_2}{2k_2 + k'} my + \frac{3k_3}{2k_3 + k'} nz\right).$$

The external effect is the same as that which would have been produced if the small magnet whose moments are

$$\frac{k_1 - k'}{2k_1 + k'} lIa^3, \qquad \frac{k_2 - k'}{2k_2 + k'} mIa^3, \qquad \frac{k_3 - k'}{2k_3 + k'} nIa^3,$$

had been placed at the origin with their directions coinciding with the axes of x, y, z. The effect of the original force I in turning the sphere about the axis of x may be found by taking the moments of the components of that force on these equivalent magnets. The moment of the force in the direction of y acting on the third magnet is

$$\frac{k_3 - k'}{2k_3 + k'} mnI^2a^3,$$

and that of the force in z on the second magnet is

$$-\frac{k_2 - k'}{2k_2 + k'} mnI^2a^3.$$

The whole couple about the axis of x is therefore

$$\frac{3k'(k_3 - k_2)}{(2k_3 + k')(2k_2 + k')} mnI^2a^3,$$

tending to turn the sphere round from the axis of y towards that of z. Suppose the sphere to be suspended so that the axis of x is vertical, and let I be horizontal, then if θ be the angle which the axis of y makes with the direction of I, $m = \cos\theta$, $n = -\sin\theta$, and the expression for the moment becomes

$$\frac{3}{2} \frac{k'(k_2 - k_3)}{(2k_2 + k')(2k_3 + k')} I^2a^3 \sin 2\theta,$$

tending to increase θ. The axis of least resistance therefore sets axially, but with either end indifferently towards the north.

Since in all bodies, except iron, the values of k are nearly the same as in a vacuum, the coefficient of this quantity can be but little altered by changing the value of k' to k, the value in space. The expression then becomes

$$\frac{1}{6} \frac{k_2 - k_3}{k} I^2a^3 \sin 2\theta,$$

independent of the external medium*.

VII. *Permanent magnetism in a spherical shell.*

The case of a homogeneous shell of a diamagnetic or paramagnetic substance presents no difficulty. The intensity within the shell is less than what it would have been if the shell were away, whether the substance of the shell be diamagnetic or paramagnetic. When the resistance of the shell is infinite, and when it vanishes, the intensity within the shell is zero.

In the case of no resistance the entire effect of the shell on any point, internal or external, may be represented by supposing a superficial stratum of

---

* Taking the more general case of magnetic induction referred to in Art. (28), we find, in the expression for the moment of the magnetic forces, a constant term depending on T, besides those terms which depend on sines and cosines of θ. The result is, that in every complete revolution in the negative direction round the axis of T, a certain positive amount of work is gained; but, since no inexhaustible source of work can exist in nature, we must admit that $T = 0$ in all substances, with respect to magnetic induction. This argument does not hold in the case of electric conduction, or in the case of a body through which heat or electricity is passing, for such states are maintained by the continual expenditure of work. See Prof. Thomson, *Phil. Mag.* March, 1851, p. 186.

magnetic matter spread over the outer surface, the density being given by the equation

$$\rho = 3I \cos \theta.$$

Suppose the shell now to be converted into a permanent magnet, so that the distribution of imaginary magnetic matter is invariable, then the external potential due to the shell will be

$$p' = -I \frac{a^3}{r^2} \cos \theta,$$

and the internal potential $\qquad p_1 = -Ir \cos \theta.$

Now let us investigate the effect of filling up the shell with some substance of which the resistance is k, the resistance in the external medium being k'. The thickness of the magnetized shell may be neglected. Let the magnetic moment of the permanent magnetism be Ia^3, and that of the imaginary superficial distribution due to the medium $k = Aa^3$. Then the potentials are

$$\text{external } p' = (I+A) \frac{a^3}{r^2} \cos \theta, \qquad \text{internal } p_1 = (I+A) r \cos \theta.$$

The distribution of real magnetism is the same before and after the introduction of the medium k, so that

$$\frac{1}{k'} I + \frac{2}{k'} I = \frac{1}{k} (I+A) + \frac{2}{k} (I+A),$$

$$\text{or } A = \frac{k-k'}{2k+k'} I.$$

The external effect of the magnetized shell is increased or diminished according as k is greater or less than k'. It is therefore increased by filling up the shell with diamagnetic matter, and diminished by filling it with paramagnetic matter, such as iron.

VIII. *Electro-magnetic spherical shell.*

Let us take as an example of the magnetic effects of electric currents, an electro-magnet in the form of a thin spherical shell. Let its radius be a, and its thickness t, and let its external effect be that of a magnet whose moment is Ia^3. Both within and without the shell the magnetic effect may be represented by a potential, but within the substance of the shell, where there are electric currents, the magnetic effects cannot be represented by a potential. Let p', p_1 be the external and internal potentials,

$$p' = I \frac{a^3}{r^2} \cos \theta, \qquad p_1 = Ar \cos \theta,$$

and since there is no permanent magnetism, $\dfrac{dp'}{dr} = \dfrac{dp_1}{dr}$, when $r = a$,

$$A = -2I.$$

If we draw any closed curve cutting the shell at the equator, and at some other point for which θ is known, then the total magnetic intensity round this curve will be $3Ia \cos \theta$, and as this is a measure of the total electric current which flows through it, the quantity of the current at any point may be found by differentiation. The quantity which flows through the element $td\theta$ is $-3Ia \sin \theta d\theta$, so that the quantity of the current referred to unit of area of section is

$$-3I \frac{a}{t} \sin \theta.$$

If the shell be composed of a wire coiled round the sphere so that the number of coils to the inch varies as the sine of θ, then the external effect will be nearly the same as if the shell had been made of a uniform conducting substance, and the currents had been distributed according to the law we have just given.

If a wire conducting a current of strength I_2 be wound round a sphere of radius a so that the distance between successive coils measured along the axis of x is $\dfrac{2a}{n}$, then there will be n coils altogether, and the value of I_1 for the resulting electro-magnet will be

$$I_1 = \frac{n}{6a} I_2.$$

The potentials, external and internal, will be

$$p' = I_2 \frac{n}{6} \frac{a^2}{r^2} \cos \theta, \qquad p_1 = -2I_2 \frac{n}{6} \frac{r}{a} \cos \theta.$$

The interior of the shell is therefore a uniform magnetic field.

IX. *Effect of the core of the electro-magnet.*

Now let us suppose a sphere of diamagnetic or paramagnetic matter introduced into the electro-magnetic coil. The result may be obtained as in the last case, and the potentials become

$$p' = I_2 \frac{n}{6} \frac{3k'}{2k+k'} \frac{a^2}{r^2} \cos\theta, \quad p_1 = -2I_2 \frac{n}{6} \frac{3k}{2k+k'} \frac{r}{a} \cos\theta.$$

The external effect is greater or less than before, according as k' is greater or less than k, that is, according as the interior of the sphere is magnetic or diamagnetic with respect to the external medium, and the internal effect is altered in the opposite direction, being greatest for a diamagnetic medium.

This investigation explains the effect of introducing an iron core into an electro-magnet. If the value of k for the core were to vanish altogether, the effect of the electro-magnet would be three times that which it has without the core. As k has always a finite value, the effect of the core is less than this.

In the interior of the electro-magnet we have a uniform field of magnetic force, the intensity of which may be increased by surrounding the coil with a shell of iron. If $k' = 0$, and the shell infinitely thick, the effect on internal points would be tripled.

The effect of the core is greater in the case of a cylindric magnet, and greatest of all when the core is a ring of soft iron.

X. *Electro-tonic functions in spherical electro-magnet.*

Let us now find the electro-tonic functions due to this electro-magnet.

They will be of the form

$$\alpha_0 = 0, \qquad \beta_0 = \omega z, \qquad \gamma_0 = -\omega y,$$

where ω is some function of r. Where there are no electric currents, we must have a_2, b_2, c_2 each $= 0$, and this implies

$$\frac{d}{dr}\left(3\omega + r\frac{d\omega}{dr}\right) = 0,$$

the solution of which is

$$\omega = C_1 + \frac{C_2}{r^3}.$$

Within the shell ω cannot become infinite; therefore $\omega = C_1$ is the solution, and outside a must vanish at an infinite distance, so that

$$\omega = \frac{C_2}{r^3}$$

is the solution outside. The magnetic quantity within the shell is found by last article to be

$$-2I_2 \frac{n}{6a} \frac{3}{2k+k'} = a_1 = \frac{d\beta_0}{dz} - \frac{d\gamma_0}{dy} = 2C_1;$$

therefore within the sphere

$$\omega_0 = -\frac{I_2 n}{2a} \frac{1}{3k+k'}.$$

Outside the sphere we must determine ω so as to coincide at the surface with the internal value. The external value is therefore

$$\omega = -\frac{I_2 n}{2a} \frac{1}{3k+k'} \frac{a^3}{r^3},$$

where the shell containing the currents is made up of n coils of wire, conducting a current of total quantity I_2.

Let another wire be coiled round the shell according to the same law, and let the total number of coils be n'; then the total electro-tonic intensity EI_2 round the second coil is found by integrating

$$EI_2 = \int_0^{2\pi} \omega a \sin \theta ds,$$

along the whole length of the wire. The equation of the wire is

$$\cos \theta = \frac{\phi}{n'\pi},$$

where n' is a large number; and therefore

$$ds = a \sin \theta d\phi,$$

$$= -an'\pi \sin^2 \theta d\theta,$$

$$\therefore EI_2 = \frac{4\pi}{3} \omega a^2 n' = -\frac{2\pi}{3} ann'I \frac{1}{3k+k'}.$$

E may be called the electro-tonic coefficient for the particular wire.

XI. *Spherical electro-magnetic Coil-Machine.*

We have now obtained the electro-tonic function which defines the action of the one coil on the other. The action of each coil on itself is found by putting n^2 or n'^2 for nn'. Let the first coil be connected with an apparatus producing a variable electro-motive force F. Let us find the effects on both wires, supposing their total resistances to be R and R', and the quantity of the currents I and I'.

Let N stand for $\dfrac{2\pi}{3}\dfrac{a}{(3k+k')}$, then the electro-motive force of the first wire on the second is

$$-Nnn'\frac{dI}{dt}.$$

That of the second on itself is

$$-Nn'^2\frac{dI'}{dt}.$$

The equation of the current in the second wire is therefore

$$-Nnn'\frac{dI}{dt}-Nn'^2\frac{dI'}{dt}=R'I' \quad\ldots\ldots\ldots\ldots\ldots\ldots \text{(1)}.$$

The equation of the current in the first wire is

$$-Nn^2\frac{dI}{dt}-Nnn'\frac{dI'}{dt}+F=RI \ldots\ldots\ldots\ldots\ldots\ldots \text{(2)}.$$

Eliminating the differential coefficients, we get

$$\frac{R}{n}I-\frac{R'}{n'}I'=\frac{F}{n},$$

and $\quad N\left(\dfrac{n^2}{R}+\dfrac{n'^2}{R'}\right)\dfrac{dI}{dt}+I=\dfrac{F}{R}+N\dfrac{n'^2}{R'}\dfrac{dF}{dt} \ldots\ldots\ldots\ldots\ldots \text{(3)},$

from which to find I and I'. For this purpose we require to know the value of F in terms of t.

Let us first take the case in which F is constant and I and I' initially $=0$. This is the case of an electro-magnetic coil-machine at the moment when the connexion is made with the galvanic trough.

Putting $\frac{1}{2}\tau$ for $N\left(\dfrac{n^2}{R}+\dfrac{n'^2}{R'}\right)$ we find

$$I=\frac{F}{R}\left(1-\epsilon^{-\frac{2t}{\tau}}\right),$$

$$I'=-F\frac{n'}{R'n}\epsilon^{-\frac{2t}{\tau}}.$$

The primary current increases very rapidly from 0 to $\dfrac{F}{R}$, and the secondary commences at $-\dfrac{F}{R'}\dfrac{n'}{n}$ and speedily vanishes, owing to the value of τ being generally very small.

The whole work done by either current in heating the wire or in any other kind of action is found from the expression

$$\int_0^\infty I^2 R\, dt.$$

The total quantity of current is

$$\int_0^\infty I\, dt.$$

For the secondary current we find

$$\int_0^\infty I'^2 R'\, dt =\frac{F^2 n'^2}{R'n^2}\frac{\tau}{4}, \qquad \int_0^\infty I'\, dt =\frac{Fn'}{R'n}\frac{\tau}{2}.$$

The work done and the quantity of the current are therefore the same as if a current of quantity $I' =\dfrac{Fn'}{2R'n}$ had passed through the wire for a time τ, where

$$\tau = 2N\left(\frac{n^2}{R}+\frac{n'^2}{R'}\right).$$

This method of considering a variable current of short duration is due to Weber, whose experimental methods render the determination of the equivalent current a matter of great precision.

Now let the electro-motive force F suddenly cease while the current in the primary wire is I_0 and in the secondary $= 0$. Then we shall have for the subsequent time

$$I=I_0\epsilon^{-\frac{2t}{\tau}}, \qquad I' =\frac{I_0}{R'}\frac{Rn'}{n}\epsilon^{-\frac{2t}{\tau}}.$$

The equivalent currents are $\frac{1}{2}I_0$ and $\frac{1}{2}I_0 \dfrac{R}{R'} \dfrac{n'}{n}$, and their duration is τ.

When the communication with the source of the current is cut off, there will be a change of R. This will produce a change in the value of τ, so that if R be suddenly increased, the strength of the secondary current will be increased, and its duration diminished. This is the case in the ordinary coil-machines. The quantity N depends on the form of the machine, and may be determined by experiment for a machine of any shape.

XII. *Spherical shell revolving in magnetic field.*

Let us next take the case of a revolving shell of conducting matter under the influence of a uniform field of magnetic force. The phenomena are explained by Faraday in his *Experimental Researches*, Series II., and references are there given to previous experiments.

Let the axis of z be the axis of revolution, and let the angular velocity be ω. Let the magnetism of the field be represented in quantity by I, inclined at an angle θ to the direction of z, in the plane of zx.

Let R be the radius of the spherical shell, and T the thickness. Let the quantities a_0, β_0, γ_0, be the electro-tonic functions at any point of space; a_1, b_1, c_1, a_1, β_1, γ_1 symbols of magnetic quantity and intensity; a_2, b_2, c_2, a_2, β_2, γ_2 of electric quantity and intensity. Let p_2 be the electric tension at any point,

$$\left.\begin{aligned}
a_2 &= \frac{dp_2}{dx} + ka_2 \\[1mm]
\beta_2 &= \frac{dp_2}{dy} + kb_2 \\[1mm]
\gamma_2 &= \frac{dp_2}{dz} + kc_2
\end{aligned}\right\} \dotfill (1),$$

$$\frac{da_2}{dx} + \frac{db_2}{dy} + \frac{dc_2}{dz} = 0 \dotfill (2);$$

$$\therefore \frac{da_2}{dx} + \frac{d\beta_2}{dy} + \frac{d\gamma_2}{dz} = \nabla^2 p.$$

The expressions for a_0, β_0, γ_0 due to the magnetism of the field are

$$a_0 = A_0 + \frac{I}{2} y \cos \theta,$$

$$\beta_0 = B_0 + \frac{I}{2} (z \sin \theta - x \cos \theta),$$

$$\gamma_0 = C_0 - \frac{I}{2} y \sin \theta,$$

A_0, B_0, C_0 being constants; and the velocities of the particles of the revolving sphere are

$$\frac{dx}{dt} = -\omega y, \quad \frac{dy}{dt} = \omega x, \quad \frac{dz}{dt} = 0.$$

We have therefore for the electro-motive forces

$$a_2 = -\frac{1}{4\pi} \frac{da_0}{dt} = -\frac{1}{4\pi} \frac{I}{2} \cos \theta \omega x,$$

$$\beta_2 = -\frac{1}{4\pi} \frac{d\beta_0}{dt} = -\frac{1}{4\pi} \frac{I}{2} \cos \theta \omega y,$$

$$\gamma_2 = -\frac{1}{4\pi} \frac{d\gamma_0}{dt} = \frac{1}{4\pi} \frac{I}{2} \sin \theta \omega x.$$

Returning to equations (1), we get

$$k\left(\frac{db_2}{dz} - \frac{dc_2}{dy}\right) = \frac{d\beta_2}{dz} - \frac{d\gamma_2}{dy} = 0,$$

$$k\left(\frac{dc_2}{dx} - \frac{da_2}{dz}\right) = \frac{d\gamma_2}{dx} - \frac{da_2}{dz} = \frac{1}{4\pi} \frac{I}{2} \sin \theta \omega,$$

$$k\left(\frac{da_2}{dy} - \frac{db_2}{dx}\right) = \frac{da_2}{dy} - \frac{d\beta_2}{dx} = 0.$$

From which with equation (2) we find

$$a_2 = -\frac{1}{k} \frac{1}{4\pi} \frac{I}{4} \sin \theta \omega z,$$

$$b_2 = 0,$$

$$c_2 = \frac{1}{k} \frac{1}{4\pi} \frac{I}{4} \sin \theta \omega x,$$

$$p_2 = -\frac{1}{16\pi} I\omega \{(x^2 + y^2) \cos \theta - xz \sin \theta\}.$$

These expressions would determine completely the motion of electricity in a revolving sphere if we neglect the action of these currents on themselves. They express a system of circular currents about the axis of y, the quantity of current at any point being proportional to the distance from that axis. The external magnetic effect will be that of a small magnet whose moment is $\frac{TR^3}{48\pi k} \omega I \sin \theta$, with its direction along the axis of y, so that the magnetism of the field would tend to turn it back to the axis of x*.

The existence of these currents will of course alter the distribution of the electro-tonic functions, and so they will react on themselves. Let the final result of this action be a system of currents about an axis in the plane of xy inclined to the axis of x at an angle ϕ and producing an external effect equal to that of a magnet whose moment is $I'R^3$.

The magnetic inductive components within the shell are

$$I_1 \sin \theta - 2I' \cos \phi \quad \text{in } x,$$
$$-2I' \sin \phi \quad \text{in } y,$$
$$I_1 \cos \theta \qquad\qquad \text{in } z.$$

Each of these would produce its own system of currents when the sphere is in motion, and these would give rise to new distributions of magnetism, which, when the velocity is uniform, must be the same as the original distribution,

$$(I_1 \sin \theta - 2I' \cos \phi) \text{ in } x \text{ produces } 2\frac{T}{48\pi k} \omega (I_1 \sin \theta - 2I' \cos \phi) \text{ in } y,$$

$$(-2I' \sin \phi) \text{ in } y \text{ produces } 2\frac{T}{48\pi k} \omega (2I' \sin \phi) \text{ in } x;$$

$I_1 \cos \theta$ in z produces no currents.

We must therefore have the following equations, since the state of the shell is the same at every instant,

$$I_1 \sin \theta - 2I' \cos \phi = I_1 \sin \theta + \frac{T}{24\pi k} \omega 2I' \sin \phi$$

$$-2I' \sin \phi = \frac{T}{24\pi k} \omega (I_1 \sin \theta - 2I' \cos \phi),$$

---

* The expression for p_2 indicates a variable electric tension in the shell, so that currents might be collected by wires touching it at the equator and poles.

whence
$$\cot \phi = -\frac{TR^3}{24\pi k}\omega, \quad I' = \tfrac{1}{2}\,\frac{\dfrac{T}{4\pi k}}{\sqrt{1+\left(\dfrac{T}{4\pi k}\omega\right)^2}}\,I_1 \sin\theta.$$

To understand the meaning of these expressions let us take a particular case.

Let the axis of the revolving shell be vertical, and let the revolution be from north to west. Let I be the total intensity of the terrestrial magnetism, and let the dip be θ, then $I\cos\theta$ is the horizontal component in the direction of magnetic north.

The result of the rotation is to produce currents in the shell about an axis inclined at a small angle $=\tan^{-1}\dfrac{T}{24\pi k}\omega$ to the south of magnetic west, and the external effect of these currents is the same as that of a magnet whose moment is

$$\tfrac{1}{2}\,\frac{T\omega}{\sqrt{24\pi k\rceil^2 + T^2\omega^2}}\,R^3 I \cos\theta.$$

The moment of the couple due to terrestrial magnetism tending to stop the rotation is

$$\frac{24\pi k}{2}\,\frac{T\omega}{24\pi k\rceil^2 + T^2\omega^2}\,R^3 I^2 \cos^2\theta,$$

and the loss of work due to this in unit of time is

$$\frac{24\pi k}{2}\,\frac{T\omega^2}{24\pi k\rceil^2 + T^2\omega^2}\,R^3 I^2 \cos^2\theta.$$

This loss of work is made up by an evolution of heat in the substance of the shell, as is proved by a recent experiment of M. Foucault (see *Comptes Rendus*, XLI. p. 450).

[From the *Transactions of the Royal Scottish Society of Arts*, Vol. IV. Part IV.]

IX. *Description of a New Form of the Platometer, an Instrument for measuring the Areas of Plane Figures drawn on Paper*.*

1. THE measurement of the area of a plane figure on a map or plan is an operation so frequently occurring in practice, that any method by which it may be easily and quickly performed is deserving of attention. A very able exposition of the principle of such instruments will be found in the article on Planimeters in the Reports of the Juries of the Great Exhibition, 1851.

2. In considering the principle of instruments of this kind, it will be most convenient to suppose the area of the figure measured by an imaginary straight line, which, by moving parallel to itself, and at the same time altering in length to suit the form of the area, accurately sweeps it out.

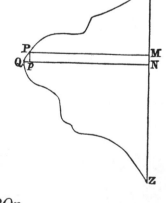

Let AZ be a fixed vertical line, $APQZ$ the boundary of the area, and let a variable horizontal line move parallel to itself from A to Z, so as to have its extremities, P, M, in the curve and in the fixed straight line. Now, suppose the horizontal line (which we shall call the generating line) to move from the position PM to QN, MN being some small quantity, say one inch for distinctness. During this movement, the generating line will have swept out the narrow strip of the surface, $PMNQ$, which exceeds the portion $PMNp$ by the small triangle PQp.

But since MN, the breadth of the strip, is one inch, the strip will contain as many square inches as PM is inches long; so that, when the generating

* Read to the Society, 22nd Jan. 1855.

line descends one inch, it sweeps out a number of square inches equal to the number of linear inches in its length.

Therefore, if we have a machine with an index of any kind, which, while the generating line moves one inch downwards, moves forward as many degrees as the generating line is inches long, and if the generating line be alternately moved an inch and altered in length, the index will mark the number of square inches swept over during the whole operation. By the ordinary method of limits, it may be shown that, if these changes be made continuous instead of sudden, the index will still measure the area of the curve traced by the extremity of the generating line.

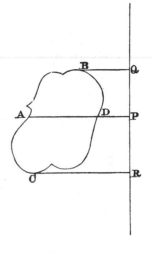

3. When the area is bounded by a closed curve, as *ABDC*, then to determine the area we must carry the tracing point from some point *A* of the curve, completely round the circumference to *A* again. Then, while the tracing point moves from *A* to *C*, the index will go forward and measure the number of square inches in *ACRP*, and, while it moves from *C* to *D*, the index will measure backwards the square inches in *CRPD*, so that it will now indicate the square inches in *ACD*. Similarly, during the other part of the motion from *D* to *B*, and from *B* to *D*, the part *DBA* will be measured; so that when the tracing point returns to *D*, the instrument will have measured the area *ACDB*. It is evident that the whole area will appear positive or negative according as the tracing point is carried round in the direction *ACDB* or *ABDC*.

4. We have next to consider the various methods of communicating the required motion to the index. The first is by means of two discs, the first having a flat horizontal rough surface, turning on a vertical axis, *OQ*, and the second vertical, with its circumference resting on the flat surface of the first at *P*, so as to be driven round by the motion of the first disc. The velocity of the second disc will depend on *OP*, the distance of the point of contact from the centre of the first disc; so that if *OP* be made always equal to the generating line, the conditions of the instrument will be fulfilled.

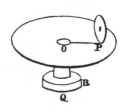

This is accomplished by causing the index-disc to slip along the radius of the horizontal disc; so that in working the instrument, the motion of the index-disc is compounded of a rolling motion due to the rotation of the first disc, and a slipping motion due to the variation of the generating line.

5. In the instrument presented by Mr Sang to the Society, the first disc is replaced by a cone, and the action of the instrument corresponds to a mathematical valuation of the area by the use of oblique co-ordinates. As he has himself explained it very completely, it will be enough here to say, that the index-wheel has still a motion of slipping as well as of rolling.

6. Now, suppose a wheel rolling on a surface, and pressing on it with a weight of a pound; then suppose the coefficient of friction to be $\frac{1}{8}$, it will require a force of 2 oz. at least to produce slipping at all, so that even if the resistance of the axis, &c., amounted to 1 oz., the rolling would be perfect. But if the wheel were forcibly pulled sideways, so as to slide along in the direction of the axis, then, if the friction of the axis, &c., opposed no resistance to the turning of the wheel, the rotation would still be that due to the forward motion; but if there were any resistance, however small, it would produce its effect in diminishing the amount of rotation.

The case is that of a mass resting on a rough surface, which requires a great force to produce the slightest motion; but when some other force acts on it and keeps it in motion, the very smallest force is sufficient to alter that motion in direction.

7. This effect of the combination of slipping and rolling has not escaped the observation of Mr Sang, who has both measured its amount, and shown how to eliminate its effect. In the improved instrument as constructed by him, I believe that the greatest error introduced in this way does not equal the ordinary errors of measurement by the old process of triangulation. This accuracy, however, is a proof of the excellence of the workmanship, and the smoothness of the action of the instrument; for if any considerable resistance had to be overcome, it would display itself in the results.

8. Having seen and admired these instruments at the Great Exhibition in 1851, and being convinced that the combination of slipping and rolling was a drawback on the perfection of the instrument, I began to search for some arrangement by which the motion should be that of perfect rolling in every

motion of which the instrument is capable. The forms of the rolling parts which I considered were—

    1. Two equal spheres.

    2. Two spheres, the diameters being as 1 to 2.

    3. A cone and cylinder, axes at right angles.

Of these, the first combination only suited my purpose. I devised several modes of mounting the spheres so as to make the principle available. That which I adopted is borrowed, as to many details, from the instruments already constructed, so that the originality of the device may be reduced to this principle— The abolition of slipping by the use of two equal spheres.

9. The instrument (Fig. 1) is mounted on a frame, which rolls on the two connected wheels, *MM*, and is thus constrained to travel up and down the paper, moving parallel to itself.

CH is a horizontal axis, passing through two supports attached to the frame, and carrying the wheel *K* and the hemisphere *LAP*. The wheel *K* rolls on the plane on which the instrument travels, and communicates its motion to the hemisphere, which therefore revolves about the axis *AH* with a velocity proportional to that with which the instrument moves backwards or forwards.

FCO is a framework (better seen in the other figures) capable of revolving about a vertical axis, *Cc*, being joined at *C* and *c* to the frame of the instrument. The parts *CF* and *CO* are at right angles to each other and horizontal. The part *CO* carries with it a ring, *SOS*, which turns about a vertical axis *Oo*. This ring supports the index-sphere *Bb* by the extremities of its axis *Ss*, just as the meridian circle carries a terrestrial globe. By this arrangement, it will be seen that the axis of the sphere is kept always horizontal, while its centre moves so as to be always at a constant distance from that of the hemisphere. This distance must be adjusted so that the spheres may always remain in contact, and the pressure at the point of contact may be regulated by means of springs or compresses at *O* and *o* acting in the direction *OC, oc*. In this way the rotation of the hemisphere is made to drive the index-sphere.

10. Now, let us consider the working of the instrument. Suppose the arm *CE* placed so as to coincide with *CD*, then *O*, the centre of the index-sphere will be in the prolongation of the axis *HA*. Suppose also that, when in this position, the equator *bB* of the index-sphere is in contact with the pole *A* of the hemisphere. Now, let the arch be turned into the position *CE* as in the

figure, then the rest of the framework will be turned through an equal angle, and the index-sphere will roll on the hemisphere till it come into the position represented in the figure. Then, if there be no slipping, the arc $AP = BP$, and the angle $ACP = BOP$.

Next, let the instrument be moved backwards or forwards, so as to turn the wheel Kk and the hemisphere Ll, then the index-sphere will be turned about its axis Ss by the action of the hemisphere, but the ratio of their velocities will depend on their relative positions. If we draw PQ, PR, perpendiculars from the point of contact on the two axes, then the angular motion of the index-sphere will be to that of the hemisphere, as PQ is to PR; that is, as PQ is to QC, by the equal triangles POQ, PQC; that is, as ED is to DC, by the similar triangles CQP, CDE.

Therefore the ratio of the angular velocities is as ED to DC, but since DC is constant, this ratio varies as ED. We have now only to contrive some way of making ED act as the generating line, and the machine is complete (see art. 2).

11. The arm CF is moved in the following manner:—Tt is a rectangular metal beam, fixed to the frame of the instrument, and parallel to the axis AH. eEe is a little carriage which rolls along it, having two rollers on one side and one on the other, which is pressed against the beam by a spring. This carriage carries a vertical pin, E, turning in its socket, and having a collar above, through which the arm CF works smoothly. The tracing point G is attached to the carriage by a jointed frame eGe, which is so arranged that the point may not bear too heavily on the paper.

12. When the machine is in action, the tracing point is placed on a point in the boundary of the figure, and made to move round it always in one direction till it arrives at the same point again. The up-and-down motion of the tracing point moves the whole instrument over the paper, turns the wheel K, the hemisphere Ll, and the index-sphere Bb; while the lateral motion of the tracing point moves the carriage E on the beam Tt, and so works the arm CF and the framework CO; and so changes the relative velocities of the two spheres, as has been explained.

13. In this way the instrument works by a perfect rolling motion, in whatever direction the tracing point is moved; but since the accuracy of the result depends on the equality of the arcs AP and BP, and since the smallest error

of adjustment would, in the course of time, produce a considerable deviation from this equality, some contrivance is necessary to secure it. For this purpose a wheel is fixed on the same axis with the ring SOs, and another of the same size is fixed to the frame of the instrument, with its centre coinciding with the vertical axis through C. These wheels are connected by two pieces of watchspring, which are arranged so as to apply closely to the edges of the wheels. The first is firmly attached to the nearer side of the fixed wheel, and to the farther side of the moveable wheel, and the second to the farther side of the fixed wheel, and the nearer side of the moveable wheel, crossing beneath the first steel band. In this way the spheres are maintained in their proper relative position; but since no instrument can be perfect, the wheels, by preventing derangement, must cause some slight slipping, depending on the errors of workmanship. This, however, does not ruin the pretensions of the instrument, for it may be shown that the error introduced by slipping depends on the distance through which the lateral slipping takes place; and since in this case it must be very small compared with its necessarily large amount in the other instruments, the error introduced by it must be diminished in the same proportion.

14. I have shewn how the rotation of the index-sphere is proportional to the area of the figure traced by the tracing point. This rotation must be measured by means of a graduated circle attached to the sphere, and read off by means of a vernier. The result, as measured in degrees, may be interpreted in the following manner:—

Suppose the instrument to be placed with the arm CF coinciding with CD, the equator Bb of the index-sphere touching the pole A of the hemisphere, and the index of the vernier at zero: then let these four operations be performed:—

(1) Let the tracing point be moved to the right till $DE = DC$, and therefore DCE, ACP, and $POB = 45°$.

(2) Let the instrument be rolled upwards till the wheel K has made a complete revolution, carrying the hemisphere with it; then, on account of the equality of the angles SOP, PCA, the index-sphere will also make a complete revolution.

(3) Let the arm CF be brought back again till F coincides with D.

(4) Let the instrument be rolled back again through a complete revolution of the wheel K. The index-sphere will not rotate, because the point of contact is at the pole of the hemisphere.

30—2

The tracing point has now traversed the boundary of a rectangle, whose length is the circumference of the wheel K, and its breadth is equal to CD; and during this operation, the index-sphere has made a complete revolution. 360° on the sphere, therefore, correspond to an area equal to the rectangle contained by the circumference of the wheel and the distance CD. The size of the wheel K being known, different values may be given to CD, so as to make the instrument measure according to any required scale. This may be done, either by shifting the position of the beam Tt, or by having several sockets in the carriage E for the pin which directs the arm to work in.

15. If I have been too prolix in describing the action of an instrument which has never been constructed, it is because I have myself derived great satisfaction from following out the mechanical consequences of the mathematical theorem on which the truth of this method depends. Among the other forms of apparatus by which the action of the two spheres may be rendered available, is one which might be found practicable in cases to which that here given would not apply. In this instrument (Fig. 4) the areas are swept out by a radius-vector of variable length, turning round a fixed point in the plane. The area is thus swept out with a velocity varying as the angular velocity of the radius-vector and the square of its length conjointly, and the construction of the machine is adapted to the case as follows :—

The hemisphere is *fixed* on the top of a vertical pillar, about which the rest of the instrument turns. The index-sphere is supported as before by a ring and framework. This framework turns about the vertical pillar along with the tracing point, but has also a motion in a vertical plane, which is communicated to it by a curved slide connected with the tracing point, and which, by means of a prolonged arm, moves the framework as the tracing point is moved to and from the pillar.

The form of the curved slide is such, that the tangent of the angle of inclination of the line joining the centres of the spheres with the vertical is proportional to the square of the distance of the tracing point from the vertical axis of the instrument. The curve which fulfils this condition is an hyperbola, one of whose asymptotes is vertical, and passes through the tracing point, and the other horizontal through the centre of the hemisphere.

The other parts of this instrument are identical with those belonging to that already described.

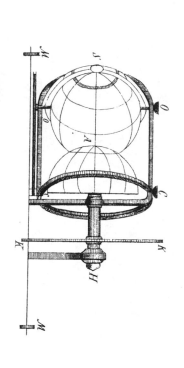

Fig 2 Front Elevation

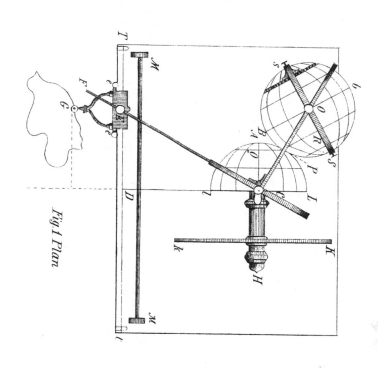

Fig.1 Plan

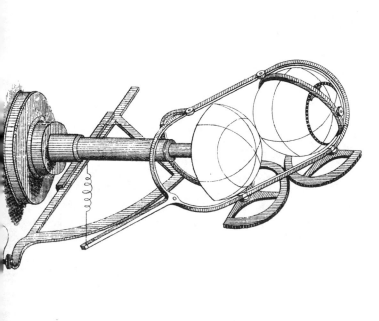

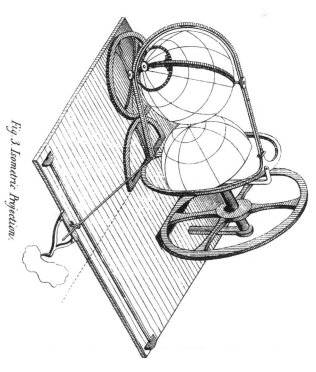

Fig.3 Isometric Projection.

Cambridge University Press

When the tracing point is made to traverse the boundary of a plane figure, there is a continued rotation of the radius-vector combined with a change of length. The rotation causes the index-sphere to roll on the fixed hemisphere, while the length of the radius-vector determines the *rate* of its motion about its axis, so that its whole motion measures the area swept out by the radius-vector during the motion of the tracing point.

The areas measured by this instrument may either lie on one side of the pillar, or they may extend all round it. In either case the action of the instrument is the same as in the ordinary case. In this form of the instrument we have the advantages of a fixed stand, and a simple motion of the tracing point; but there seem to be difficulties in the way of supporting the spheres and arranging the slide; and even then the instrument would require a tall pillar, in order to take in a large area.

16. It will be observed that I have said little or nothing about the practical details of these instruments. Many useful hints will be found in the large work on Platometers, by Professor T. Gonnellu, who has given us an account of the difficulties, as well as the results, of the construction of his most elaborate instrument. He has also given some very interesting investigations into the errors produced by various irregularities of construction, although, as far as I am aware, he has not even suspected the error which the sliding of the index-wheel over the disc must necessarily introduce. With respect to this, and other points relating to the working of the instrument, the memoir of Mr Sang, in the *Transactions* of this Society, is the most complete that I have met with. It may, however, be as well to state, that at the time when I devised the improvements here suggested, I had not seen that paper, though I had seen the instrument standing at rest in the Crystal Palace.

EDINBURGH, 30th *January*, 1855.

NOTE.—Since the design of the above instrument was submitted to the Society of Arts, I have met with a description of an instrument combining simplicity of construction with the power of adaptation to designs of any size, and at the same time more portable than any other instrument of the kind. Although it does not act by perfect rolling, and therefore belongs to a different class of instruments from that described in this paper, I think that its simplicity, and the beauty of the principle on which it acts, render it worth the attention of engineers and mechanists, whether practical or theoretical. A full account of this instrument is to be found in Moigno's "Cosmos," 5th year, Vol. VIII., Part VIII., p. 213, published 20th February 1856. *Description et Théorie du planimètre polaire, inventé par J. Amsler, de Schaffouse en Suisse.*

CAMBRIDGE, 30th *April*, 1856.

[From the *Cambridge Philosophical Society Proceedings*, Vol. I. pp. 173—175.]

X. *On the Elementary Theory of Optical Instruments.*

THE object of this communication was to shew how the magnitude and position of the image of any object seen through an optical instrument could be ascertained without knowing the construction of the instrument, by means of data derived from two experiments on the instrument. Optical questions are generally treated of with respect to the pencils of rays which pass through the instrument. A pencil is a collection of rays which have passed through one point, and may again do so, by some optical contrivance. Now if we suppose all the points of a plane luminous, each will give out a pencil of rays, and that collection of pencils which passes through the instrument may be treated as a *beam* of light. In a pencil only one ray passes through any point of space, unless that point be the focus. In a beam an infinite number of rays, corresponding each to some point in the luminous plane, passes through any point; and we may, if we choose, treat this collection of rays as a pencil proceeding from that point. Hence the same beam of light may be decomposed into pencils in an infinite variety of ways; and yet, since we regard it as the same collection of rays, we may study its properties as a beam independently of the particular way in which we conceive it analysed into pencils.

Now in any instrument the incident and emergent beams are composed of the same light, and therefore every ray in the incident beam has a corresponding ray in the emergent beam. We do not know their path within the instrument, but before incidence and after emergence they are straight lines, and therefore any two points serve to determine the direction of each.

Let us suppose the instrument such that it forms an accurate image of a plane object in a given position. Then every ray which passes through a given point of the object before incidence passes through the corresponding point of the image after emergence, and this determines one point of the emergent ray. If at any other distance from the instrument a plane object has an accurate image, then there will be two other corresponding points given in the incident and emergent rays. Hence if we know the points in which an incident ray meets the planes of the two objects, we may find the incident ray by joining the points of the two images corresponding to them.

It was then shewn, that if the image of a plane object be distinct, flat, and similar to the object for two different distances of the object, the image of any other plane object perpendicular to the axis will be distinct, flat and similar to the object.

When the object is at an infinite distance, the plane of its image is the *principal focal plane*, and the point where it cuts the axis is the *principal focus*. The line joining any point in the object to the corresponding point of the image cuts the axis at a fixed point called the *focal centre*. The distance of the principal focus from the focal centre is called the *principal focal length*, or simply the *focal length*.

There are two principal foci, etc., formed by incident parallel rays passing in opposite directions through the instrument. If we suppose light always to pass in the same direction through the instrument, then the focus of incident rays when the emergent rays are parallel is the *first* principal focus, and the focus of emergent rays when the incident rays are parallel is the *second* principal focus.

Corresponding to these we have first and second focal centres and focal lengths.

Now let Q_1 be the focus of incident rays, P_1 the foot of the perpendicular from Q_1 on the axis, Q_2 the focus of emergent rays, P_2 the foot of the corresponding perpendicular, $F_1 F_2$ the first and second principal foci, $A_1 A_2$ the first and second focal centres, then

$$\frac{P_1 F_1}{A_1 F_1} = \frac{P_1 Q_1}{P_2 Q_2} = \frac{F_2 P_2}{F_2 A_2},$$

lines being positive when measured in the direction of the light. Therefore the position and magnitude of the image of any object is found by a simple proportion.

In one important class of instruments there are no principal foci or focal centres. A telescope in which parallel rays emerge parallel is an instance. In such instruments, if m be the angular magnifying power, the linear dimensions of the image are $\dfrac{1}{m}$ of the object, and the distance of the image of the object from the image of the object-glass is $\dfrac{1}{m^2}$ of the distance of the object from the object-glass. Rules were then laid down for the composition of instruments, and suggestions for the adaptation of this method to second approximations, and the method itself was considered with reference to the labours of Cotes, Smith, Euler, Lagrange, and Gauss on the same subject.

[From the *Report of the British Association,* 1856.]

XI. *On a Method of Drawing the Theoretical Forms of Faraday's Lines of Force without Calculation.*

THE method applies more particularly to those cases in which the lines are entirely parallel to one plane, such as the lines of electric currents in a thin plate, or those round a system of parallel electric currents. In such cases, if we know the forms of the lines of force in any two cases, we may combine them by simple addition of the functions on which the equations of the lines depend. Thus the system of lines in a uniform magnetic field is a series of parallel straight lines at equal intervals, and that for an infinite straight electric current perpendicular to the paper is a series of concentric circles whose radii are in geometric progression. Having drawn these two sets of lines on two separate sheets of paper, and laid a third piece above, draw a third set of lines through the intersections of the first and second sets. This will be the system of lines in a uniform field disturbed by an electric current. The most interesting cases are those of uniform fields disturbed by a small magnet. If we draw a circle of any diameter with the magnet for centre, and join those points in which the circle cuts the lines of force, the straight lines so drawn will be parallel and equidistant; and it is easily shown that they represent the actual lines of force in a paramagnetic, diamagnetic, or crystallized body, according to the nature of the original lines, the size of the circle, &c. No one can study Faraday's researches without wishing to see the forms of the lines of force. This method, therefore, by which they may be easily drawn, is recommended to the notice of electrical students.

[From the *Report of the British Association*, 1856.]

XII.　*On the Unequal Sensibility of the Foramen Centrale to Light of different Colours.*

WHEN observing the spectrum formed by looking at a long vertical slit through a simple prism, I noticed an elongated dark spot running up and down in the blue, and following the motion of the eye as it moved *up and down* the spectrum, but refusing to pass out of the blue into the other colours. It was plain that the spot belonged both to the eye and to the blue part of the spectrum. The result to which I have come is, that the appearance is due to the yellow spot on the retina, commonly called the *Foramen Centrale* of Soemmering. The most convenient method of observing the spot is by presenting to the eye in not too rapid succession, blue and yellow glasses, or, still better, allowing blue and yellow papers to revolve slowly before the eye. In this way the spot is seen in the blue. It fades rapidly, but is renewed every time the yellow comes in to relieve the effect of the blue. By using a Nicol's prism along with this apparatus, the brushes of Haidinger are well seen in connexion with the spot, and the fact of the brushes being the spot analysed by polarized light becomes evident. If we look steadily at an object behind a series of bright bars which move in front of it, we shall see a curious bending of the bars as they come up to the ·place of the yellow spot. The part which comes over the spot seems to start in advance of the rest of the bar, and this would seem to indicate a greater rapidity of sensation at the yellow spot than in the surrounding retina. But I find the experiment difficult, and I hope for better results from more accurate observers.

[From the *Report of the British Association*, 1856.]

XIII. *On the Theory of Compound Colours with reference to Mixtures of Blue and Yellow Light.*

WHEN we mix together blue and yellow paint, we obtain green paint. This fact is well known to all who have handled colours; and it is universally admitted that blue and yellow make green. Red, yellow, and blue, being the primary colours among painters, green is regarded as a secondary colour, arising from the mixture of blue and yellow. Newton, however, found that the green of the spectrum was not the same thing as the mixture of two colours of the spectrum, for such a mixture could be separated by the prism, while the green of the spectrum resisted further decomposition. But still it was believed that yellow and blue would make a green, though not that of the spectrum. As far as I am aware, the first experiment on the subject is that of M. Plateau, who, before 1819, made a disc with alternate sectors of prussian blue and gamboge, and observed that, when spinning, the resultant tint was not green, but a neutral gray, inclining sometimes to yellow or blue, but never to green. Prof. J. D. Forbes of Edinburgh made similar experiments in 1849, with the same result. Prof. Helmholtz of Königsberg, to whom we owe the most complete investigation on visible colour, has given the true explanation of this phænomenon. The result of mixing two coloured powders is not by any means the same as mixing the beams of light which flow from each separately. In the latter case we receive all the light which comes either from the one powder or the other. In the former, much of the light coming from one powder falls on particles of the other, and we receive only that portion which has escaped absorption by one or other. Thus the light coming from a mixture of blue and yellow powder, consists partly of light coming directly from blue particles or yellow particles, and partly of light acted on by both blue and yellow particles. This latter light is green, since the blue stops the red, yellow, and orange, and the yellow stops

31—2

the blue and violet. I have made experiments on the mixture of blue and yellow *light*—by rapid rotation, by combined reflexion and transmission, by viewing them out of focus, in stripes, at a great distance, by throwing the colours of the spectrum on a screen, and by receiving them into the eye directly; and I have arranged a portable apparatus by which any one may see the result of this or any other mixture of the colours of the spectrum. In all these cases blue and yellow do *not* make green. I have also made experiments on the mixture of coloured powders. Those which I used principally were "mineral blue" (from copper) and "chrome-yellow." Other blue and yellow pigments gave curious results, but it was more difficult to make the mixtures, and the greens were less uniform in tint. The mixtures of these colours were made by weight, and were painted on discs of paper, which were afterwards treated in the manner described in my paper "On Colour as perceived by the Eye," in the *Transactions of the Royal Society of Edinburgh*, Vol. XXI. Part 2. The visible effect of the colour is estimated in terms of the standard-coloured papers:—vermilion (V), ultramarine (U), and emerald-green (E). The accuracy of the results, and their significance, can be best understood by referring to the paper before mentioned. I shall denote mineral blue by B, and chrome-yellow by Y; and $B_3 Y_5$ means a mixture of three parts blue and five parts yellow.

Given Colour.			Standard Colours.			Coefficient of brightness.
			V.	U.	E.	
B_8		, 100 =	2	36	7	 45
B_7	Y_1 , 100 =		1	18	17	 37
B_6	Y_2 , 100 =		4	11	34	 49
B_5	Y_3 , 100 =		9	5	40	 54
B_4	Y_4 , 100 =		15	1	40	 56
B_3	Y_5 , 100 =		22	− 2	44	 64
B_2	Y_6 , 100 =		35	−10	51	 76
B_1	Y_7 , 100 =		64	−19	64	 109
	Y_8 , 100 =		180	− 27	124	 277

The columns V, U, E give the proportions of the standard colours which are equivalent to 100 of the given colour; and the sum of V, U, E gives a coefficient, which gives a general idea of the brightness. It will be seen that the first admixture of yellow *diminishes* the brightness of the blue. The negative values of U indicate that a mixture of V, U, and E cannot be made equivalent to the given colour. The experiments from which these results were taken had the negative values transferred to the other side of the equation. They were all made by means of the colour-top, and were verified by repetition at different times. It may be necessary to remark, in conclusion, with reference to the mode of registering visible colours in terms of three arbitrary standard colours, that it proceeds upon that theory of three primary elements in the sensation of colour, which treats the investigation of the laws of visible colour as a branch of human physiology, incapable of being deduced from the laws of light itself, as set forth in physical optics. It takes advantage of the methods of optics to study vision itself; and its appeal is not to physical principles, but to our consciousness of our own sensations.

[From the *Report of the British Association*, 1856.]

XIV. *On an Instrument to illustrate Poinsôt's Theory of Rotation.*

IN studying the rotation of a solid body according to Poinsôt's method, we have to consider the successive positions of the instantaneous axis of rotation with reference both to directions fixed in space and axes assumed in the moving body. The paths traced out by the pole of this axis on the *invariable plane* and on the *central ellipsoid* form interesting subjects of mathematical investigation. But when we attempt to follow with our eye the motion of a rotating body, we find it difficult to determine through what point of the *body* the instantaneous axis passes at any time,—and to determine its path must be still more difficult. I have endeavoured to render visible the path of the instantaneous axis, and to vary the circumstances of motion, by means of a top of the same kind as that used by Mr Elliot, to illustrate precession*. The body of the instrument is a hollow cone of wood, rising from a ring, 7 inches in diameter and 1 inch thick. An iron axis, 8 inches long, screws into the vertex of the cone. The lower extremity has a point of hard steel, which rests in an agate cup, and forms the support of the instrument. An iron nut, three ounces in weight, is made to screw on the axis, and to be fixed at any point; and in the wooden ring are screwed four bolts, of three ounces, working horizontally, and four bolts, of one ounce, working vertically. On the upper part of the axis is placed a disc of card, on which are drawn four concentric rings. Each ring is divided into four quadrants, which are coloured red, yellow, green, and blue. The spaces between the rings are white. When the top is in motion, it is easy to see in which quadrant the instantaneous axis is at any moment and the distance between it and the axis of the instrument; and we observe,—1st. That the instantaneous axis travels in a closed curve, and returns to its original position in the body. 2ndly.

---

* *Transactions of the Royal Scottish Society of Arts*, 1855.

That by working the vertical bolts, we can make the axis of the instrument the centre of this closed curve. It will then be one of the principal axes of inertia. 3rdly. That, by working the nut on the axis, we can make the order of colours either red, yellow, green, blue, or the reverse. When the order of colours is in the *same* direction as the rotation, it indicates that the axis of the instrument is that of *greatest* moment of inertia. 4thly. That if we screw the two pairs of opposite horizontal bolts to different distances from the axis, the path of the instantaneous pole will no longer be equidistant from the axis, but will describe an ellipse, whose longer axis is in the direction of the *mean axis* of the instrument. 5thly. That if we now make one of the two horizontal axes less and the other greater than the vertical axis, the instantaneous pole will separate from the axis of the instrument, and the axis will incline more and more till the spinning can no longer go on, on account of the obliquity. It is easy to see that, by attending to the laws of motion, we may produce any of the above effects at pleasure, and illustrate many different propositions by means of the same instrument.

[From the *Transactions of the Royal Society of Edinburgh*, Vol. XXI. Part IV.]

XV. *On a Dynamical Top, for exhibiting the phenomena of the motion of a system of invariable form about a fixed point, with some suggestions as to the Earth's motion.*

(Read 20th April, 1857.)

To those who study the progress of exact science, the common spinning-top is a symbol of the labours and the perplexities of men who had successfully threaded the mazes of the planetary motions. The mathematicians of the last age, searching through nature for problems worthy of their analysis, found in this toy of their youth, ample occupation for their highest mathematical powers.

No illustration of astronomical precession can be devised more perfect than that presented by a properly balanced top, but yet the motion of rotation has intricacies far exceeding those of the theory of precession.

Accordingly, we find Euler and D'Alembert devoting their talent and their patience to the establishment of the laws of the rotation of solid bodies. Lagrange has incorporated his own analysis of the problem with his general treatment of mechanics, and since his time M. Poinsôt has brought the subject under the power of a more searching analysis than that of the calculus, in which ideas take the place of symbols, and intelligible propositions supersede equations.

In the practical department of the subject, we must notice the rotatory machine of Bohnenberger, and the nautical top of Troughton. In the first of these instruments we have the model of the Gyroscope, by which Foucault has been able to render visible the effects of the earth's rotation. The beautiful experiments by which Mr J. Elliot has made the ideas of precession so familiar to us are performed with a top, similar in some respects to Troughton's, though not borrowed from his.

The top which I have the honour to spin before the Society, differs from that of Mr Elliot in having more adjustments, and in being designed to exhibit far more complicated phenomena.

The arrangement of these adjustments, so as to produce the desired effects, depends on the mathematical theory of rotation. The method of exhibiting the motion of the axis of rotation, by means of a coloured disc, is essential to the success of these adjustments. This optical contrivance for rendering visible the nature of the rapid motion of the top, and the practical methods of applying the theory of rotation to such an instrument as the one before us, are the grounds on which I bring my instrument and experiments before the Society as my own.

I propose, therefore, in the first place, to give a brief outline of such parts of the theory of rotation as are necessary for the explanation of the phenomena of the top.

I shall then describe the instrument with its adjustments, and the effect of each, the mode of observing of the coloured disc when the top is in motion, and the use of the top in illustrating the mathematical theory, with the method of making the different experiments.

Lastly, I shall attempt to explain the nature of a possible variation in the earth's axis due to its figure. This variation, if it exists, must cause a periodic inequality in the latitude of every place on the earth's surface, going through its period in about eleven months. The amount of variation must be very small, but its character gives it importance, and the necessary observations are already made, and only require reduction.

On the Theory of Rotation.

The theory of the rotation of a rigid system is strictly deduced from the elementary laws of motion, but the complexity of the motion of the particles of a body freely rotating renders the subject so intricate, that it has never been thoroughly understood by any but the most expert mathematicians. Many who have mastered the lunar theory have come to erroneous conclusions on this subject; and even Newton has chosen to deduce the disturbance of the earth's axis from his theory of the motion of the nodes of a free orbit, rather than attack the problem of the rotation of a solid body.

The method by which M. Poinsôt has rendered the theory more manageable, is by the liberal introduction of "appropriate ideas," chiefly of a geometrical character, most of which had been rendered familiar to mathematicians by the writings of Monge, but which then first became illustrations of this branch of dynamics. If any further progress is to be made in simplifying and arranging the theory, it must be by the method which Poinsôt has repeatedly pointed out as the only one which can lead to a true knowledge of the subject,—that of proceeding from one distinct idea to another, instead of trusting to symbols and equations.

An important contribution to our stock of appropriate ideas and methods has lately been made by Mr R. B. Hayward, in a paper, "On a Direct Method of estimating Velocities, Accelerations, and all similar quantities, with respect to axes, moveable in any manner in Space." (*Trans. Cambridge Phil. Soc.* Vol. x. Part i.)

* In this communication I intend to confine myself to that part of the subject which the top is intended to illustrate, namely, the alteration of the position of the axis in a body rotating freely about its centre of gravity. I shall, therefore, deduce the theory as briefly as possible, from two considerations only,—the permanence of the original *angular momentum* in direction and magnitude, and the permanence of the original *vis viva*.

* The mathematical difficulties of the theory of rotation arise chiefly from the want of geometrical illustrations and sensible images, by which we might fix the results of analysis in our minds.

It is easy to understand the motion of a body revolving about a fixed axle. Every point in the body describes a circle about the axis, and returns to its original position after each complete revolution. But if the axle itself be in motion, the paths of the different points of the body will no longer be circular or re-entrant. Even the velocity of rotation about the axis requires a careful definition, and the proposition that, in all motion about a fixed point, there is always one line of particles forming an instantaneous axis, is usually given in the form of a very repulsive mass of calculation. Most of these difficulties may be got rid of by devoting a little attention to the mechanics and geometry of the problem before entering on the discussion of the equations.

Mr Hayward, in his paper already referred to, has made great use of the mechanical conception of Angular Momentum.

---

* 7th May, 1857. The paragraphs marked thus have been rewritten since the paper was read.

DEFINITION.—*The Angular Momentum of a particle about an axis is measured by the product of the mass of the particle, its velocity resolved in the normal plane, and the perpendicular from the axis on the direction of motion.*

* The angular momentum of any system about an axis is the algebraical sum of the angular momenta of its parts.

As the *rate of change* of the *linear momentum* of a particle measures the *moving force* which acts on it, so the *rate of change* of *angular momentum* measures the *moment* of that force about an axis.

All actions between the parts of a system, being pairs of equal and opposite forces, produce equal and opposite changes in the angular momentum of those parts. Hence the whole angular momentum of the system is not affected by these actions and re-actions.

* When a system of invariable form revolves about an axis, the angular velocity of every part is the same, and the angular momentum about the axis is the product of the *angular velocity* and the *moment of inertia* about that axis.

* It is only in particular cases, however, that the *whole* angular momentum can be estimated in this way. In general, the axis of angular momentum differs from the axis of rotation, so that there will be a residual angular momentum about an axis perpendicular to that of rotation, unless that axis has one of three positions, called the principal axes of the body.

By referring everything to these three axes, the theory is greatly simplified. The moment of inertia about one of these axes is greater than that about any other axis through the same point, and that about one of the others is a minimum. These two are at right angles, and the third axis is perpendicular to their plane, and is called the mean axis.

* Let A, B, C be the moments of inertia about the principal axes through the centre of gravity, taken in order of magnitude, and let ω_1 ω_2 ω_3 be the angular velocities about them, then the angular momenta will be $A\omega_1$, $B\omega_2$ and $C\omega_3$.

Angular momenta may be compounded like forces or velocities, by the law of the "parallelogram," and since these three are at right angles to each other, their resultant is

$$\sqrt{A^2\omega_1^2 + B^2\omega_2^2 + C^2\omega_3^2} = H \dots\dots\dots\dots\dots\dots (1),$$

and this must be constant, both in magnitude and direction in space, since no external forces act on the body.

32—2

We shall call this axis of angular momentum the *invariable axis*. It is perpendicular to what has been called the invariable plane. Poinsôt calls it the axis of the couple of impulsion. The *direction-cosines* of this axis in the body are,

$$l = \frac{A\omega_1}{H}, \qquad m = \frac{B\omega_2}{H}, \qquad n = \frac{C\omega_3}{H}.$$

Since l, m and n vary during the motion, we need some additional condition to determine the relation between them. We find this in the property of the *vis viva* of a system of invariable form in which there is no friction. The *vis viva* of such a system must be constant. We express this in the equation

$$A\omega_1^2 + B\omega_2^2 + C\omega_3^2 = V \dots\dots\dots\dots\dots\dots\dots\dots\dots(2).$$

Substituting the values of ω_1, ω_2, ω_3 in terms of l, m, n,

$$\frac{l^2}{A} + \frac{m^2}{B} + \frac{n^2}{C} = \frac{V}{H^2}.$$

Let $\dfrac{1}{A} = a^2, \qquad \dfrac{1}{B} = b^2, \qquad \dfrac{1}{C} = c^2, \qquad \dfrac{V}{H^2} = e^2,$

and this equation becomes

$$a^2 l^2 + b^2 m^2 + c^2 n^2 = e^2 \dots\dots\dots\dots\dots\dots\dots\dots\dots(3),$$

and the equation to the cone, described by the invariable axis within the body, is

$$(a^2 - e^2)\, x^2 + (b^2 - e^2)\, y^2 + (c^2 - e^2)\, z^2 = 0 \dots\dots\dots\dots\dots(4).$$

The intersections of this cone with planes perpendicular to the principal axes are found by putting x, y, or z, constant in this equation. By giving e various values, all the different paths of the pole of the invariable axis, corresponding to different initial circumstances, may be traced.

*In the figures, I have supposed $a^2 = 100$, $b^2 = 107$, and $c^2 = 110$. The first figure represents a section of the various cones by a plane perpendicular to the axis of x, which is that of greatest moment of inertia. These sections are ellipses having their major axis parallel to the axis of b. The value of e^2 corresponding to each of these curves is indicated by figures beside the curve. The ellipticity increases with the size of the ellipse, so that the section corresponding to $e^2 = 107$ would be two parallel straight lines (beyond the bounds of the figure), after which the sections would be hyperbolas.

*The second figure represents the sections made by a plane, perpendicular to the *mean* axis. They are all hyperbolas, except when $e^2 = 107$, when the section is two intersecting straight lines.

The third figure shows the sections perpendicular to the axis of least moment of inertia. From $e^2 = 110$ to $e^2 = 107$ the sections are ellipses, $e^2 = 107$ gives two parallel straight lines, and beyond these the curves are hyperbolas.

*The fourth and fifth figures show the sections of the series of cones made by a cube and a sphere respectively. The use of these figures is to exhibit the connexion between the different curves described about the three principal axes by the invariable axis during the motion of the body.

*We have next to compare the velocity of the invariable axis with respect to the body, with that of the body itself round one of the principal axes. Since the invariable axis is fixed in space, its motion relative to the body must be equal and opposite to that of the portion of the body through which it passes. Now the angular velocity of a portion of the body whose direction-cosines are l, m, n, about the axis of x is

$$\frac{\omega_1}{1 - l^2} - \frac{l}{1 - l^2} (l\omega_1 + m\omega_2 + n\omega_3).$$

Substituting the values of ω_1, ω_2, ω_3, in terms of l, m, n, and taking account of equation (3), this expression becomes

$$H \frac{(a^2 - e^2)}{1 - l^2} l.$$

Changing the sign and putting $l = \frac{\omega_1}{a^2 H}$ we have the angular velocity of the invariable axis about that of x

$$= \frac{\omega_1}{1 - l^2} \frac{e^2 - a^2}{a^2},$$

always positive about the axis of greatest moment, negative about that of least moment, and positive or negative about the mean axis according to the value of e^2. The direction of the motion in every case is represented by the arrows in the figures. The arrows on the outside of each figure indicate the direction of rotation of the body.

*If we attend to the curve described by the pole of the invariable axis on the sphere in fig. 5, we shall see that the areas described by that point, if projected on the plane of yz, are swept out at the rate

$$\omega_1 \frac{e^2 - a^2}{a^2}.$$

Now the semi-axes of the projection of the spherical ellipse described by the pole are

$$\sqrt{\frac{e^2 - a^2}{b^2 - a^2}} \quad \text{and} \quad \sqrt{\frac{e^2 - a^2}{c^2 - a^2}}.$$

Dividing the area of this ellipse by the area described during one revolution of the body, we find the number of revolutions of the body during the description of the ellipse—

$$= \frac{a^2}{\sqrt{b^2 - a^2} \sqrt{c^2 - a^2}}.$$

The projections of the spherical ellipses upon the plane of yz are all similar ellipses, and described in the same number of revolutions; and in each ellipse so projected, the area described in any time is proportional to the number of revolutions of the body about the axis of x, so that if we measure time by revolutions of the body, the motion of the projection of the pole of the invariable axis is identical with that of a body acted on by an attractive central force varying directly as the distance. In the case of the hyperbolas in the plane of the greatest and least axis, this force must be supposed repulsive. The dots in the figures 1, 2, 3, are intended to indicate roughly the progress made by the invariable axis during each revolution of the body about the axis of x, y and z respectively. It must be remembered that the rotation about these axes varies with their inclination to the invariable axis, so that the angular velocity diminishes as the inclination increases, and therefore the areas in the ellipses above mentioned are not described with uniform velocity in absolute time, but are less rapidly swept out at the extremities of the major axis than at those of the minor.

*When two of the axes have equal moments of inertia, or $b = c$, then the angular velocity ω_1 is constant, and the path of the invariable axis is circular, the number of revolutions of the body during one circuit of the invariable axis, being

$$\frac{a^2}{b^2 - a^2}.$$

The motion is in the same direction as that of rotation, or in the opposite direction, according as the axis of x is that of greatest or of least moment of inertia.

*Both in this case, and in that in which the three axes are unequal, the motion of the invariable axis in the body may be rendered very slow by diminishing the difference of the moments of inertia. The angular velocity of the axis of x about the invariable axis in space is

$$\omega_1 \frac{e^2 - a^2 l^2}{a^2 (1 - l^2)},$$

which is greater or less than ω_1, as e^2 is greater or less than a^2, and, when these quantities are nearly equal, is very nearly the same as ω_1 itself. This quantity indicates the rate of revolution of the axle of the top about its mean position, and is very easily observed.

*The _instantaneous axis_ is not so easily observed. It revolves round the invariable axis in the same time with the axis of x, at a distance which is very small in the case when a, b, c, are nearly equal. From its rapid angular motion in space, and its near coincidence with the invariable axis, there is no advantage in studying its motion in the top.

*By making the moments of inertia very unequal, and in definite proportion to each other, and by drawing a few strong lines as diameters of the disc, the combination of motions will produce an appearance of epicycloids, which are the result of the continued intersection of the successive positions of these lines, and the cusps of the epicycloids lie in the curve in which the instantaneous axis travels. Some of the figures produced in this way are very pleasing.

In order to illustrate the theory of rotation experimentally, we must have a body balanced on its centre of gravity, and capable of having its principal axes and moments of inertia altered in form and position within certain limits. We must be able to make the axle of the instrument the greatest, least, or mean principal axis, or to make it not a principal axis at all, and we must be able to _see_ the position of the invariable axis of rotation at any time. There must be three adjustments to regulate the position of the centre of gravity, three for the magnitudes of the moments of inertia, and three for the directions of the principal axes, nine independent adjustments, which may be distributed as we please among the screws of the instrument.

The form of the body of the instrument which I have found most suitable is that of a bell (p. 262, fig. 6). C is a hollow cone of brass, R is a heavy ring cast in the same piece. Six screws, with heavy heads, x, y, z, x', y', z', work horizontally in the ring, and three similar screws, l, m, n, work vertically through the ring at equal intervals. AS is the axle of the instrument, SS is a brass screw working in the upper part of the cone C, and capable of being firmly clamped by means of the nut c. B is a cylindrical brass bob, which may be screwed up or down the axis, and fixed in its place by the nut b.

The lower extremity of the axle is a fine steel point, finished without emery, and afterwards hardened. It runs in a little agate cup set in the top of the pillar P. If any emery had been embedded in the steel, the cup would soon be worn out. The upper end of the axle has also a steel point by which it may be kept steady while spinning.

When the instrument is in use, a coloured disc is attached to the upper end of the axle.

It will be seen that there are eleven adjustments, nine screws in the brass ring, the axle screwing in the cone, and the bob screwing on the axle. The advantage of the last two adjustments is, that by them large alterations can be made, which are not possible by means of the small screws.

The first thing to be done with the instrument is, to make the steel point at the end of the axle coincide with the centre of gravity of the whole. This is done roughly by screwing the axle to the right place nearly, and then balancing the instrument on its point, and screwing the bob and the horizontal screws till the instrument will remain balanced in any position in which it is placed.

When this adjustment is carefully made, the rotation of the top has no tendency to shake the steel point in the agate cup, however irregular the motion may appear to be.

The next thing to be done, is to make one of the principal axes of the central ellipsoid coincide with the axle of the top.

To effect this, we must begin by spinning the top gently about its axle, steadying the upper part with the finger at first. If the axle is already a principal axis the top will continue to revolve about its axle when the finger is removed. If it is not, we observe that the top begins to spin about some other axis, and the axle moves away from the centre of motion and then back to it again, and so on, alternately widening its circles and contracting them.

It is impossible to observe this motion successfully, without the aid of the coloured disc placed near the upper end of the axis. This disc is divided into sectors, and strongly coloured, so that each sector may be recognised by its colour when in rapid motion. If the axis about which the top is really revolving, falls within this disc, its position may be ascertained by the colour of the spot at the centre of motion. If the central spot appears red, we know that the invariable axis at that instant passes through the red part of the disc.

In this way we can trace the motion of the invariable axis in the revolving body, and we find that the path which it describes upon the disc may be a circle, an ellipse, an hyperbola, or a straight line, according to the arrangement of the instrument.

In the case in which the invariable axis coincides at first with the axle of the top, and returns to it after separating from it for a time, its true path is a circle or an ellipse having the axle in its *circumference*. The true principal axis is at the *centre* of the closed curve. It must be made to coincide with the axle by adjusting the vertical screws *l, m, n*.

Suppose that the colour of the centre of motion, when farthest from the axle, indicated that the axis of rotation passed through the sector *L*, then the principal axis must also lie in that sector at half the distance from the axle.

If this principal axis be that of *greatest* moment of inertia, we must *raise* the screw *l* in order to bring it nearer the axle *A*. If it be the axis of least moment we must *lower* the screw *l*. In this way we may make the principal axis coincide with the axle. Let us suppose that the principal axis is that of greatest moment of inertia, and that we have made it coincide with the axle of the instrument. Let us also suppose that the moments of inertia about the other axes are equal, and very little less than that about the axle. Let the top be spun about the axle and then receive a disturbance which causes it to spin about some other axis. The instantaneous axis will not remain at rest either in space or in the body. In space it will describe a right cone, completing a revolution in somewhat less than the time of revolution of the top. In the body it will describe another cone of larger angle in a period which is longer as the difference of axes of the body is smaller. The invariable axis will be fixed in space, and describe a cone in the body.

The relation of the different motions may be understood from the following illustration. Take a hoop and make it revolve about a stick which remains at rest and touches the inside of the hoop. The section of the stick represents the path of the instantaneous axis in space, the hoop that of the same axis in the body, and the axis of the stick the invariable axis. The point of contact represents the pole of the instantaneous axis itself, travelling many times round the stick before it gets once round the hoop. It is easy to see that the direction in which the instantaneous axis travels round the hoop, is in this case the same as that in which the hoop moves round the stick, so that if the top be spinning in the direction L, M, N, the colours will appear in the same order.

By screwing the bob B up the axle, the difference of the axes of inertia may be diminished, and the time of a complete revolution of the invariable axis in the body increased. By observing the number of revolutions of the top in a complete cycle of colours of the invariable axis, we may determine the ratio of the moments of inertia.

By screwing the bob up farther, we may make the axle the principal axis of *least* moment of inertia.

The motion of the instantaneous axis will then be that of the point of contact of the stick with the *outside* of the hoop rolling on it. The order of colours will be N, M, L, if the top be spinning in the direction L, M, N, and the more the bob is screwed up, the more rapidly will the colours change, till it ceases to be possible to make the observations correctly.

In calculating the dimensions of the parts of the instrument, it is necessary to provide for the exhibition of the instrument with its axle either the greatest or the least axis of inertia. The dimensions and weights of the parts of the top which I have found most suitable, are given in a note at the end of this paper.

Now let us make the axes of inertia in the plane of the ring unequal. We may do this by screwing the balance screws x and x^1 farther from the axle without altering the centre of gravity.

Let us suppose the bob B screwed up so as to make the axle the axis of least inertia. Then the mean axis is parallel to xx^1, and the greatest is at right angles to xx^1 in the horizontal plane. The path of the invariable axis on the disc is no longer a circle but an ellipse, concentric with the disc, and having its major axis parallel to the mean axis xx^1.

The smaller the difference between the moment of inertia about the axle and about the mean axis, the more eccentric the ellipse will be; and if, by screwing the bob down, the axle be made the mean axis, the path of the invariable axis will be no longer a closed curve, but an hyperbola, so that it will depart altogether from the neighbourhood of the axle. When the top is in this condition it must be spun gently, for it is very difficult to manage it when its motion gets more and more eccentric.

When the bob is screwed still farther down, the axle becomes the axis of greatest inertia, and xx^1 the least. The major axis of the ellipse described by the invariable axis will now be perpendicular to xx^1, and the farther the bob is screwed down, the eccentricity of the ellipse will diminish, and the velocity with which it is described will increase.

I have now described all the phenomena presented by a body revolving freely on its centre of gravity. If we wish to trace the motion of the invariable axis by means of the coloured sectors, we must make its motion very slow compared with that of the top. It is necessary, therefore, to make the moments of inertia about the principal axes very nearly equal, and in this case a very small change in the position of any part of the top will greatly derange the *position* of the principal axis. So that when the top is well adjusted, a single turn of one of the screws of the ring is sufficient to make the axle no longer a principal axis, and to set the true axis at a considerable inclination to the axle of the top.

All the adjustments must therefore be most carefully arranged, or we may have the whole apparatus deranged by some eccentricity of spinning. The method of making the principal axis coincide with the axle must be studied and practised, or the first attempt at spinning rapidly may end in the destruction of the top, if not of the table on which it is spun.

On the Earth's Motion.

We must remember that these motions of a body about its centre of gravity, are *not* illustrations of the theory of the precession of the Equinoxes. Precession can be illustrated by the apparatus, but we must arrange it so that the force of gravity acts the part of the attraction of the sun and moon in producing a force tending to alter the axis of rotation. This is easily done by bringing the centre of gravity of the whole a little below the point on which it spins. The theory of such motions is far more easily comprehended than that which we have been investigating.

But the earth is a body whose principal axes are unequal, and from the phenomena of precession we can determine the ratio of the polar and equatorial axes of the "central ellipsoid;" and supposing the earth to have been set in motion about any axis except the principal axis, or to have had its original

260 ON A DYNAMICAL TOP.

axis disturbed in any way, its subsequent motion would be that of the top when the bob is a little below the critical position.

The axis of angular momentum would have an invariable position in space, and would travel with respect to the earth round the axis of figure with a velocity $= \omega \, \dfrac{C-A}{A}$ where ω is the sidereal angular velocity of the earth. The apparent pole of the earth would travel (with respect to the earth) from west to east round the true pole, completing its circuit in $\dfrac{A}{C-A}$ sidereal days, which appears to be about 325·6 solar days.

The instantaneous axis would revolve about this axis in space in about a day, and would always be in a plane with the true axis of the earth and the axis of angular momentum. The effect of such a motion on the apparent position of a star would be, that its zenith distance would be increased and diminished during a period of 325·6 days. This alteration of zenith distance is the same above and below the pole, so that the polar distance of the star is unaltered. In fact the method of finding the pole of the heavens by observations of stars, gives the pole of the *invariable axis*, which is altered only by external forces, such as those of the sun and moon.

There is therefore no change in the apparent polar distance of stars due to this cause. It is the latitude which varies. The magnitude of this variation cannot be determined by theory. The periodic time of the variation may be found approximately from the known dynamical properties of the earth. The epoch of maximum latitude cannot be found except by observation, but it must be later in proportion to the east longitude of the observatory.

In order to determine the existence of such a variation of latitude, I have examined the observations of *Polaris* with the Greenwich Transit Circle in the years 1851-2-3-4. The observations of the upper transit during each month were collected, and the mean of each month found. The same was done for the lower transits. The difference of zenith distance of upper and lower transit is twice the polar distance of Polaris, and half the sum gives the co-latitude of Greenwich.

In this way I found the apparent co-latitude of Greenwich for each month of the four years specified.

There appeared a very slight indication of a maximum belonging to the set of months,

March, 51. Feb. 52. Dec. 52. Nov. 53. Sept. 54.

This result, however, is to be regarded as very doubtful, as there did not appear to be evidence for any variation exceeding half a second of space, and more observations would be required to establish the existence of so small a variation at all.

I therefore conclude that the earth has been for a long time revolving about an axis very near to the axis of figure, if not coinciding with it. The cause of this near coincidence is either the original softness of the earth, or the present fluidity of its interior. The axes of the earth are so nearly equal, that a considerable elevation of a tract of country might produce a deviation of the principal axis within the limits of observation, and the only cause which would restore the uniform motion, would be the action of a fluid which would gradually diminish the oscillations of latitude. The permanence of latitude essentially depends on the inequality of the earth's axes, for if they had been all equal, any alteration of the crust of the earth would have produced new principal axes, and the axis of rotation would travel about those axes, altering the latitudes of all places, and yet not in the least altering the position of the axis of rotation among the stars.

Perhaps by a more extensive search and analysis of the observations of different observatories, the nature of the periodic variation of latitude, if it exist, may be determined. I am not aware of any calculations having been made to prove its non-existence, although, on dynamical grounds, we have every reason to look for some very small variation having the periodic time of 325·6 days nearly, a period which is clearly distinguished from any other astronomical cycle, and therefore easily recognised.

NOTE.

Dimensions and Weights of the parts of the Dynamical Top.

I. Body of the top—
   Mean diameter of ring, 4 inches.
   Section of ring, ½ inch square.
   The conical portion rises from the upper and inner edge of the ring, a
     height of 1½ inches from the base.
   The whole body of the top weighs . . . . 1 lb. **7** oz.
   Each of the nine adjusting screws has its screw 1 inch long, and the
     screw and head together weigh 1 ounce. The whole weigh . . **9** „

II. Axle, &c.—

   Length of axle 5 inches, of which ½ inch at the bottom is occupied by
     the steel point, 3½ inches are brass with a good screw turned on it,
     and the remaining inch is of steel, with a sharp point at the top.
     The whole weighs . . . . 1½ „
   The bob B has a diameter of 1·4 inches, and a thickness of ·4. It weighs 2¾ „
   The nuts b and c, for clamping the bob and the body of the top on the
     axle, each weigh ½ oz. . . . . 1 „

                                   Weight of whole top   2 lb. 5¼ oz.

The best arrangement, for general observations, is to have the disc of card divided into four quadrants, coloured with vermilion, chrome yellow, emerald green, and ultramarine. These are bright colours, and, if the vermilion is good, they combine into a grayish tint when the revolution is about the axle, and burst into brilliant colours when the axis is disturbed. It is useful to have some concentric circles, drawn with ink, over the colours, and about 12 radii drawn in strong pencil lines. It is easy to distinguish the ink from the pencil lines, as they cross the invariable axis, by their want of lustre. In this way, the path of the invariable axis may be identified with great accuracy, and compared with theory.

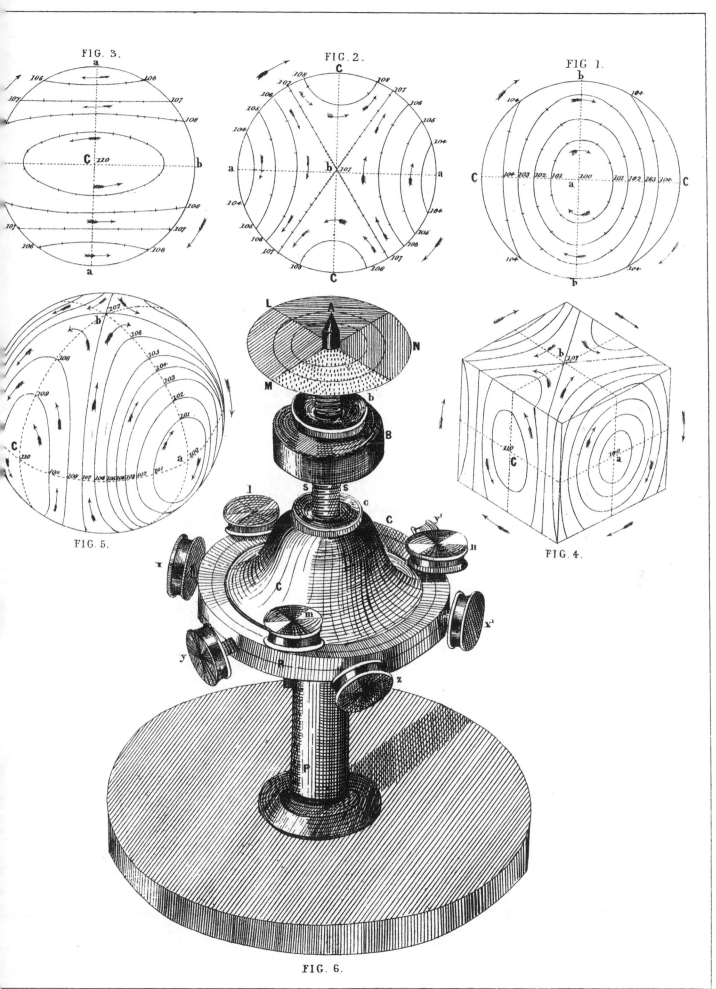

FIG. 3.

FIG. 2.

FIG 1.

FIG. 5.

FIG. 4.

FIG. 6.

[From the *Philosophical Magazine*, Vol. XIV.]

XVI. *Account of Experiments on the Perception of Colour.*

To the Editors of the Philosophical Magazine and Journal.

GENTLEMEN,

THE experiments which I intend to describe were undertaken in order to render more perfect the quantitative proof of the theory of three primary colours. According to that theory, every sensation of colour in a perfect human eye is distinguished by three, and only three, elementary qualities, so that in mathematical language the quality of a colour may be expressed as a function of three independent variables. There is very little evidence at present for deciding the precise tints of the true primaries. I have ascertained that a certain red is the sensation wanting in colour-blind eyes, but the mathematical theory relates to the number, not to the nature of the primaries. If, with Sir David Brewster, we assume red, blue, and yellow to be the primary colours, this amounts to saying that every conceivable tint may be produced by adding together so much red, so much yellow, and so much blue. This is perhaps the best method of forming a provisional notion of the theory. It is evident that if any colour could be found which could not be accurately defined as so much of each of the three primaries, the theory would fall to the ground. Besides this, the truth of the theory requires that every mathematical consequence of assuming every colour to be the result of mixture of three primaries should also be true.

I have made experiments on upwards of 100 different artificial colours, consisting of the pigments used in the arts, and their mechanical mixtures. These experiments were made primarily to trace the effects of mechanical mixture on various coloured powders; but they also afford evidence of the truth of the theory, that all these various colours can be referred to three primaries. The following experiments relate to the combinations of six well-defined colours only, and I shall describe them the more minutely, as I hope to induce those who have good eyes to subject them to the same trial of skill in distinguishing tints.

The method of performing the experiments is described in the Transactions of the Royal Society of Edinburgh, Vol. XXI. Part 2. The colour-top or teetotum which I used may be had of Mr J. M. Bryson, Edinburgh, or it may be easily extemporized. Any rotatory apparatus which will keep a disc revolving steadily and rapidly in a good light, without noise or disturbance, and can be easily stopped and shifted, will do as well as the contrivance of the spinning-top.

The essential part of the experiment consists in placing several discs of coloured paper of the same size, and slit along a radius, over one another, so that a portion of each is seen, the rest being covered by the other discs. By sliding the discs over each other the proportion of each colour may be varied, and by means of divisions on a circle on which the discs lie, the proportion of each colour may be read off. My circle was divided into 100 parts.

On the top of this set of discs is placed a smaller set of concentric discs, so that when the whole is in motion round the centre, the colour resulting from the mixture of colours of the small discs is seen in the middle of that arising from the larger discs. It is the object of the experimenter to shift the colours till the outer and inner tints appear exactly the same, and then to read off the proportions.

It is easy to deduce from the theory of three primary colours what must be the number of discs exposed at one time, and how much of each colour must appear.

Every colour placed on either circle consists of a certain proportion of each of the primaries, and in order that the outer and inner circles may have precisely the same resultant colour in every respect, there must be the same amount of each of the primary colours in the outer and inner circles. Thus we have as many conditions to fulfil as there are primary colours; and besides these we have two more, because the whole number of divisions in either the outer or the inner circle is 100, so that if there are three primary colours there will be five conditions to fulfil, and this will require five discs to be disposable, and these must be arranged so that three are matched against two, or four against one.

If we take six different colours, we may leave out any one of the six, and so form six different combinations of five colours. It is plain that these six combinations must be equivalent to two equations only, if the theory of three primaries be true.

The method which I have found most convenient for registering the result of an experiment, after an identity of tint has been obtained in the inner and outer circles, is the following :—

Write down the names or symbols of the coloured discs each at the top of a column, and underneath write the number of degrees of that colour observed, calling it + when the colour is in the outer circle, and — when it is in the inner circle; then equate the whole to zero. In this way the account of each colour is kept in a separate column, and the equations obtained are easily combined and reduced, without danger of confounding the colours of which the quantities have been measured. The following experiments were made between the 3rd and 11th of September, 1856, about noon of each day, in a room fronting the north, without curtains or any bright-coloured object near the window. The same combination was never made twice in one day, and no thought was bestowed upon the experiments except at the time of observation. Of course the graduation was never consulted, nor former experiments referred to, till each combination of colours had been fixed by the eye alone; and no reduction was attempted till all the experiments were concluded.

The coloured discs were cut from paper painted of the following colours :— Vermilion, Ultramarine, Emerald-green, Snow-white, Ivory-black, and Pale Chrome-yellow. They are denoted by the letters V, U, G, W, B, Y respectively. These colours were chosen, because each is well distinguished from the rest, so that a small change of its intensity in any combination can be observed. Two discs of each colour were prepared, so that in each combination the colours might occasionally be transposed from the outer circle to the inner.

The first equation was formed by leaving out vermilion. The remaining colours are Ultramarine-blue, Emerald-green, White, Black, and Yellow. We might suppose, that by mixing the blue and yellow in proper proportions, we should get a green of the same hue as the emerald-green, but not so intense, so that in order to match it we should have to mix the green with white to dilute it, and with black to make it darker. But it is not in this way that we have to arrange the colours, for our blue and yellow produce a pinkish tint, and never a green, so that we must add *green* to the combination of blue and yellow, to produce a neutral tint, identical with a mixture of white and black.

Blue, green, and yellow must therefore be combined on the large discs, and stand on one side of the equation, and black and white, on the small discs, must stand on the other side. In order to facilitate calculations, the colours are always put down in the same order; but those belonging to the small discs are marked negative. Thus, instead of writing

$$54U + 14G + 32Y = 32W + 68B,$$

we write $$+ 54U + 14G - 32W - 68B + 32Y = 0.$$

The sum of all the positive terms of such an equation is 100, being the whole number of divisions in the circle. The sum of the negative terms is also 100.

The second equation consists of all the colours except blue; and in this way we obtain six different combinations of five colours.

Each of these combinations was formed by the unassisted judgment of my eye, on six different occasions, so that there are thirty-six independent observations of equations between five colours.

Table I. gives the actual observations, with their dates.

Table II. gives the result of summing together each group of six equations.

Each equation in Table II. has the sums of its positive and negative co-efficients each equal to 600.

Having obtained a number of observations of each combination of colours, we have next to test the consistency of these results, since theoretically two equations are sufficient to determine all the relations among six colours. We must therefore, in the first place, determine the comparative accuracy of the different sets of observations. Table III. gives the averages of the errors of each of the six groups of observations. It appears that the combination IV. is the least accurately observed, and that VI. is the best.

Table IV. gives the averages of the errors in the observation of each colour in the whole series of experiments. This Table was computed in order to detect any tendency to colour-blindness in my own eyes, which might be less accurate in discriminating red and green, than in detecting variations of other colours. It appears, however, that my observations of red and green were more accurate than those of blue or yellow. White is the most easily observed, from the brilliancy of the colour, and black is liable to the greatest mistakes.   I would recommend this method of examining a series of experiments as a means of detecting partial colour-blindness, by the different accuracy in observing different colours.   The next operation is to combine all the equations according to their values.   Each was first multiplied by a coefficient proportional to its accuracy, and to the coefficient of white in that equation.   The result of adding all the equations so found is given in equation (W).

Equation (Y) is the result of similar operations with reference to the yellow on each equation.

We have now two equations from which to deduce six new equations, by eliminating each of the six colours in succession.   We must first combine the equations, so as to get rid of one of the colours, and then we must divide by the sum of the positive or negative coefficients, so as to reduce the equations to the form of the observed equations.   The results of these operations are given in Table V., along with the means of each group of six observations.   It will be seen that the differences between the results of calculation from two equations and the six independent observed equations are very small.   The errors in red and green are here again somewhat less than in blue and yellow, so that there is certainly no tendency to mistake red and green more than other colours. The average difference between the observed mean value of a colour and the calculated value is ·77 of a degree.   The average error of an observation in any group from the mean of that group was ·92.   No observation was attempted to be registered nearer than one degree of the top, or $\frac{1}{100}$ of a circle; so that this set of observations agrees with the theory of three primary colours quite as far as the observations can warrant us in our calculations; and I think that the human eye has seldom been subjected to so severe a test of its power of distinguishing colours.   My eyes are by no means so accurate in this respect as many eyes I have examined, but a little practice produces great improvement even in inaccurate observers.

I have laid down, according to Newton's method, the relative positions of the five positive colours with which I worked.   It will be seen that W lies within the triangle V U G, and Y outside that triangle.

The first combination, Equation I., consisted of blue, yellow, and green, taken in such proportions that their centre of gravity falls at W.

34—2

In Equation II. a mixture of red and green, represented in the diagram by the point 2, is seen to be equivalent to a mixture of white and yellow, also represented by 2, which is a pale yellow tint.

Equation III. is between a mixture of blue and yellow and another of white and red. The resulting tint is at the intersection of YU and WV; that is, at the point 3, which represents a pale pink grey.

Equation IV. is between VG and UY, that is, at 4, a dirty yellow.

Equation V. is between a mixture of white, red, and green, and a mixture of blue and yellow at the point 5, a pale dirty yellow.

Equation VI. has W. for its resulting tint.

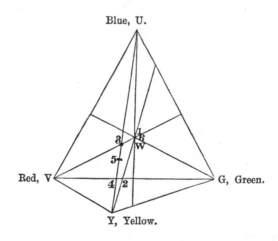

Of all the resulting tints, that of Equation IV. is the furthest from white; and we find that the observations of this equation are affected with the greatest errors. Hence the importance of reducing the resultant tint to as nearly a neutral colour as possible.

It is hardly necessary for me to observe, that the whole of the numerical results which I have given apply only to the coloured papers which I used, and to them only when illuminated by daylight from the north at mid-day in September, latitude 55°. In the evening, or in winter, or by candlelight, the results are very different. I believe, however, that the results would differ far less if observed by different persons, than if observed under different lights; for the apparatus of vision is wonderfully similar in different eyes, and even in colour-blind eyes the system of perception is not different, but defective.

TABLE I.—The observations arranged in groups.

Equation I.	V=0.	+U.	+G.	−W.	−B.	+Y.	Equation IV.	−V.	+U.	−G.	W=0.	+B.	+Y.
1856, Sept. 3.	0	54	12	34	66	34	1856, Sept. 3.	62	15	38	0	53	32
4.	0	58	14	31	69	28	4.	63	17	37	0	46	37
5.	0	55	12	32	68	33	5.	64	16	36	0	50	34
6.	0	54	14	32	68	32	6.	62	19	38	0	46	35
8.	0	54	14	32	68	32	8.	62	19	38	0	47	34
9.	0	53	15	32	68	32	9.	63	17	37	0	49	34

Equation II.	−V.	U=0.	−G.	+W.	+B.	+Y.	Equation V.	+V.	−U.	+G.	+W.	B=0.	−Y.
Sept. 3.	59	0	41	9	71	20	Sept. 3.	56	47	28	16	0	53
4.	61	0	39	9	68	23	4.	57	50	25	18	0	50
5.	61	0	39	9	67	24	5.	56	49	24	20	0	51
6.	59	0	41	10	66	24	6.	55	47	27	18	0	53
8.	60	0	40	9	69	22	8.	54	49	26	20	0	51
9.	61	0	39	9	68	23	11.	56	50	27	17	0	50

Equation III.	+V.	−U.	G=0.	+W.	+B.	−Y.	Equation VI.	+V.	+U.	+G.	−W.	−B.	Y=0.
Sept. 3.	20	56	0	28	52	44	Sept. 3.	38	27	35	24	76	0
4.	23	58	0	30	47	42	4.	39	27	34	24	76	0
5.	24	56	0	29	47	44	5.	40	26	34	24	76	0
6.	20	56	0	31	49	44	6.	38	28	34	24	76	0
8.	21	57	0	29	50	43	8.	39	28	33	24	76	0
9.	21	58	0	29	50	42	11.	39	27	34	23	77	0

TABLE II.—The sums of the observed equations.

	V.	U.	G.	W.	B.	Y.
Equation I.	0	+328	+81	−193	−407	+191
... II.	−361	0	−239	+55	+409	+136
... III.	+129	−341	0	+176	+295	−259
... IV.	−376	+103	−224	0	+291	+206
... V.	+334	−292	+157	+109	0	−308
... VI.	+233	+163	+204	−143	−457	0

TABLE III.—The averages of the errors of the several equations from the means expressed in $\frac{1}{100}$ parts of a circle.

Equations.	I.	II.	III.	IV.	V.	VI.
Errors.	·94	·85	1·05	1·17	1·08	·40

TABLE IV.—The averages of the errors of the several colours from the means in $\frac{1}{100}$ parts of a circle.

Colours.	V.	U.	G.	W.	B.	Y.
Errors.	·83	·99	·80	·61	1·15	1·09

Average error on the whole ·92.

The equations from which the reduced results were obtained were calculated as follow :—

Equation for (W) = (II) + 2 (III) + (V) − 2 (I) − 4 (VI).
Equation for (Y) = 2 (I) + 2 (II) − 3 (III) + 2 (IV) − 3 (V).

These operations being performed, gave

	V.	U.	G.	W.	B.	Y.
(W)	$+ 701$	$+ 2282$	$+ 1060$	$- 1474$	$- 3641$	$+ 1072 = 0.$
(Y)	$+ 2863$	$- 2761$	$+ 1235$	$+ 1131$	$+ 299$	$- 2767 = 0.$

From these were obtained the following results by elimination:—

TABLE V.

Equation		V.	U.	G.	W.	B.	Y.
I.	From (W) and (Y)	0	$- 54.1$	$- 13.9$	$+ 32.0$	$+ 68.0$	$- 32.0$
	From observation	0	$- 54.7$	$- 13.5$	$+ 32.1$	$+ 67.9$	$- 31.8$
II.	From (W) and (Y)	$- 59.6$	0	$- 40.4$	$+ 10.4$	$+ 66.0$	$+ 23.6$
	From observation	$- 60.2$	0	$- 39.8$	$+ 9.2$	$+ 68.2$	$+ 22.6$
III.	From (W) and (Y)	$- 21.7$	$+ 57.4$	0	$- 30.2$	$- 48.1$	$+ 42.6$
	From observation	$- 21.5$	$+ 56.8$	0	$- 29.3$	$- 49.2$	$+ 43.2$
IV.	From (W) and (Y)	$- 62.4$	$+ 18.6$	$- 37.6$	0	$+ 45.7$	$+ 35.7$
	From observation	$- 62.7$	$+ 17.2$	$- 37.3$	0	$+ 48.5$	$+ 34.3$
V.	From (W) and (Y)	$+ 55.6$	$- 49.0$	$+ 25.2$	$+ 19.2$	0	$- 51.0$
	From observation	$+ 55.7$	$- 48.7$	$+ 26.1$	$+ 18.2$	0	$- 51.3$
VI.	From (W) and (Y)	$- 39.7$	$- 26.6$	$- 33.7$	$+ 22.7$	$+ 77.3$	0
	From observation	$- 38.8$	$- 27.2$	$- 34.0$	$+ 28.3$	$+ 76.2$	0

JAMES CLERK MAXWELL.

GLENLAIR, *June* 13, 1857.

[From *The Quarterly Journal of Pure and Applied Mathematics*, Vol. II.]

XVII. *On the General Laws of Optical Instruments.*

THE optical effects of compound instruments have been generally deduced from those of the elementary parts of which they are composed. The formulæ given in most works on Optics for calculating the effect of each spherical surface are simple enough, but, when we attempt to carry on our calculations from one of these surfaces to the next, we arrive at fractional expressions so complicated as to make the subsequent steps very troublesome.

Euler (Acad. R. de Berlin, 1757, 1761. Acad. R. de Paris, 1765) has attacked these expressions, but his investigations are not easy reading. Lagrange (Acad. Berlin, 1778, 1803) has reduced the case to the theory of continued fractions and so obtained general laws.

Gauss (*Dioptrische Untersuchungen*, Göttingen, 1841) has treated the subject with that combination of analytical skill with practical ability which he displays elsewhere, and has made use of the properties of principal foci and principal planes. An account of these researches is given by Prof. Miller in the third volume of Taylor's *Scientific Memoirs*. It is also given entire in French by M. Bravais in *Liouville's Journal* for 1856, with additions by the translator.

The method of Gauss has been followed by Prof. Listing in his *Treatise on the Dioptrics of the Eye* (in Wagner's *Handworterbuch der Physiologie*) from whom I copy these references, and by Prof. Helmholtz in his *Treatise on Physiological Optics* (in Karsten's *Cyclopadie*).

The earliest general investigations are those of Cotes, given in Smith's *Optics*, II. 76 (1738). The method there is geometrical, and perfectly general, but proceeding from the elementary cases to the more complex by the method of mathematical induction. Some of his modes of expression, as for instance his measure of "apparent distance," have never come into use, although his results may easily be expressed more intelligibly; and indeed the whole fabric of

Geometrical Optics, as conceived by Cotes and laboured by Smith, has fallen into neglect, except among the writers before named. Smith tells us that it was with reference to these optical theorems that Newton said "If Mr Cotes had lived we might have known something."

The investigations which I now offer are intended to show how simple and how general the theory of instruments may be rendered, by considering the optical effects of the entire instrument, without examining the mechanism by which those effects are obtained. I have thus established a theory of "perfect instruments," geometrically complete in itself, although I have also shown, that no instrument depending on refraction and reflexion, (except the plane mirror) can be optically perfect. The first part of this theory was communicated to the Philosophical Society of Cambridge, 28th April, 1856, and an abstract will be found in the *Philosophical Magazine*, November, 1856. Propositions VIII. and IX. are now added. I am not aware that the last has been proved before.

In the following propositions I propose to establish certain rules for determining, from simple data, the path of a ray of light after passing through any optical instrument, the position of the conjugate focus of a luminous point, and the magnitude of the image of a given object. The method which I shall use does not require a knowledge of the internal construction of the instrument and derives all its data from two simple experiments.

There are certain defects incident to optical instruments from which, in the elementary theory, we suppose them to be free. A perfect instrument must fulfil three conditions:

I. Every ray of the pencil, proceeding from a single point of the object, must, after passing through the instrument, converge to, or diverge from, a single point of the image. The corresponding defect, when the emergent rays have not a common focus, has been appropriately called (by Dr Whewell) *Astigmatism*.

II. If the object is a plane surface, perpendicular to the axis of the instrument, the image of any point of it must also lie in a plane perpendicular to the axis. When the points of the image lie in a curved surface, it is said to have the defect of *curvature*.

III. The image of an object on this plane must be similar to the object, whether its linear dimensions be altered or not; when the image is not similar to the object, it is said to be *distorted*.

An image free from these three defects is said to be *perfect*.

In Fig. 1, p. 285, let $A_1a_1\alpha_1$ represent a plane object perpendicular to the axis of an instrument represented by I., then if the instrument is perfect, as regards an object at that distance, an image $A_2a_2\alpha_2$ will be formed by the emergent rays, which will have the following properties:

I. Every ray, which passes through a point a_1 of the object, will pass through the corresponding point a_2 of the image.

II. Every point of the image will lie in a plane perpendicular to the axis.

III. The figure $A_2a_2\alpha_2$ will be similar and similarly situated to the figure $A_1a_1\alpha_1$.

Now let us assume that the instrument is also perfect as regards an object in the plane $B_1b_1\beta_1$ perpendicular to the axis through B_1, and that the image of such an object is in the plane $B_2b_2\beta_2$ and similar to the object, and we shall be able to prove the following proposition:

PROP. I. If an instrument give a perfect image of a plane object at two different distances from the instrument, *all* incident rays having a common focus will have a common focus after emergence.

Let P_1 be the focus of incident rays. Let $P_1a_1b_1$ be any incident ray. Then, since every ray which passes through a_1 passes through a_2, its image after emergence, and since every ray which passes through b_1 passes through b_2, the direction of the ray $P_1a_1b_1$ after emergence must be a_2b_2.

Similarly, since a_2 and β_2 are the images of a_1 and β_1, if $P_1a_1\beta_1$ be any other ray, its direction after emergence will be $a_2\beta_2$.

Join a_1a_1, $b_1\beta_1$, a_2a_2, $b_2\beta_2$; then, since the parallel planes $A_1a_1\alpha_1$ and $B_1b_1\beta_1$ are cut by the plane of the two rays through P_1, the intersections a_1a_1 and $b_1\beta_1$ are parallel.

Also, their images, being similarly situated, are parallel to them, therefore a_2a_2 is parallel to $b_2\beta_2$, and the lines a_2b_2 and $a_2\beta_2$ are in the same plane, and therefore either meet in a point P_2 or are parallel.

Now take a third ray through P_1, not in the plane of the two former. After emergence it must either cut both, or be parallel to them. If it cuts both it must pass through the point P_2, and then every other ray must pass through P_2, for no line can intersect three lines, not in one plane, without passing through their point of intersection. If not, then all the emergent rays are parallel, which is a particular case of a perfect pencil. So that for every position of the focus of incident rays, the emergent pencil is free from *astigmatism*.

PROP. II. In an instrument, perfect at two different distances, the image of any plane object perpendicular to the axis will be free from the defects of curvature and distortion.

Through the point P_1 of the object draw any line P_1Q_1 in the plane of the object, and through P_1Q_1 draw a plane cutting the planes A_1, B_1 in the lines a_1a_1, $b_1\beta_1$. These lines will be parallel to P_1Q_1 and to each other, wherefore also their images, a_2a_2, $b_2\beta_2$, will be parallel to P_1Q_1 and to each other, and therefore in one plane.

Now suppose another plane drawn through P_1Q_1 cutting the planes A_1 and B_1 in two other lines parallel to P_1Q_1. These will have parallel images in the planes A_2 and B_2, and the intersection of the planes passing through the two pairs of images will define the line P_2Q_2 which will be parallel to them, and therefore to P_1Q_1, and will be the *image* of P_1Q_1. Therefore P_2Q_2, the image of P_1Q_1 is parallel to it, and therefore in a plane perpendicular to the axis. Now if all corresponding lines in any two figures be parallel, however the lines be drawn, the figures are similar, and similarly situated.

From these two propositions it follows that an instrument giving a perfect image at two different distances will give a perfect image at all distances. We have now only to determine the simplest method of finding the position and magnitude of the image, remembering that wherever two rays of a pencil intersect, all other rays of the pencil must meet, and that all parts of a plane object have their images in the same plane, and equally magnified or diminished.

PROP. III. A ray is incident on a perfect instrument parallel to the axis, to find its direction after emergence.

Let a_1b_1 (fig. 2) be the incident ray, A_1a_1 one of the planes at which an object has been ascertained to have a perfect image. A_2a_2 that image, similar to A_1a_1 but in magnitude such that $A_2a_2 = xA_1a_1$.

Similarly let B_2b_2 be the image of B_1b_1, and let $B_2b_2 = yB_1b_1$. Also let $A_1B_1 = c_1$ and $A_2B_2 = c_2$.

Then since a_2 and b_2 are the images of a_1 and b_1, the line $F_2a_2b_2$ will be the direction of the ray after emergence, cutting the axis in F_2, (unless $x = y$, when a_2b_2 becomes parallel to the axis). The point F_2 may be found, by remembering that $A_1a_1 = B_1b_1$, $A_2a_2 = xA_1a_1$, $B_2b_2 = yB_1b_1$. We find—

$$A_2F_2 = c_2 \frac{x}{y - x}.$$

Let g_2 be the point at which the emergent ray is at the same distance from the axis as the incident ray, draw g_2G_2 perpendicular to the axis, then we have

$$F_2G_2 = \frac{c_2}{y - x}.$$

Similarly, if $a_1\beta_1F_1$ be a ray, which, after emergence, becomes parallel to the axis; and g_1G_1 a line perpendicular to the axis, equal to the distance of the parallel emergent ray, then

$$A_1F_1 = c_1 \frac{y}{x - y}, \qquad F_1G_1 = \frac{c_1xy}{x - y}.$$

Definitions.

I. The point F_1, the focus of incident rays when the emergent rays are parallel to the axis, is called the *first principal focus* of the instrument.

II. The plane G_1g_1 at which incident rays through F_1 are at the same distance from the axis as they are after emergence, is called the *first principal plane* of the instrument. F_1G_1 is called the *first focal length*.

III. The point F_2, the focus of emergent rays when the incident rays are parallel, is called the *second principal focus*.

IV. The plane G_2g_2, at which the emergent rays are at the same distance from the axis, as before incidence, is called the *second principal plane*, and F_2G_2 is called the *second focal length*.

When $x = y$, the ray is parallel to the axis, both at incidence and emergence, and there are no such points as F and G. The instrument is then called a *telescope*. $x (= y)$ is called the *linear magnifying power* and is denoted by l, and the ratio $\frac{c_2}{c_1}$ is denoted by n, and may be called the *elongation*.

In the more general case, in which x and y are different, the principal foci and principal planes afford the readiest means of finding the position of images.

35—2

PROP. IV. Given the principal foci and principal planes of an instrument, to find the relations of the foci of the incident and emergent pencils.

Let F_1, F_2 (fig. 3) be the principal foci, G_1, G_2 the principal planes, Q_1 the focus of incident light, Q_1P_1 perpendicular to the axis.

Through Q_1 draw the ray $Q_1g_1F_1$. Since this ray passes through F_1 it emerges parallel to the axis, and at a distance from it equal to G_1g_1. Its direction after emergence is therefore Q_2g_2 where $G_2g_2 = G_1g_1$. Through Q_1 draw $Q_1\gamma_1$ parallel to the axis. The corresponding emergent ray will pass through F_2, and will cut the second principal plane at a distance $G_2\gamma_2 = G_1\gamma_1$, so that $F_2\gamma_2$ is the direction of this ray after emergence.

Since both rays pass through the focus of the emergent pencil, Q_2, the point of intersection, is that focus. Draw Q_2P_2 perpendicular to the axis. Then $P_1Q_1 = G_1\gamma_1 = G_2\gamma_2$, and $G_1g_1 = G_2g_2 = P_2Q_2$. By similar triangles $F_1P_1Q_1$ and $F_1G_1g_1$

$$P_1F_1 : F_1G_1 :: P_1Q_1 : (G_1g_1 =)\, P_2Q_2.$$

And by similar triangles $F_2P_2Q_2$ and $F_2G_2\gamma_2$

$$P_1Q_1 (= G_2\gamma_2) : P_2Q_2 :: G_2F_2 : F_2P_2.$$

We may put these relations into the concise form

$$\frac{P_1F_1}{F_1G_1} = \frac{P_1Q_1}{P_2Q_2} = \frac{G_2F_2}{F_2P_2},$$

and the values of F_2P_2 and P_2Q_2 are

$$F_2P_2 = \frac{F_1G_1 \cdot G_2F_2}{P_1F_1} \text{ and } P_2Q_2 = \frac{F_1G_1}{P_1F_1}\,P_1Q_1.$$

These expressions give the distance of the image from F_2 measured along the axis, and also the perpendicular distance from the axis, so that they serve to determine completely the position of the image of any point, when the principal foci and principal planes are known.

PROP. V. To find the focus of emergent rays, when the instrument is a *telescope*.

Let Q_1 (fig. 4) be the focus of incident rays, and let $Q_1a_1b_1$ be a ray parallel to the axis; then, since the instrument is telescopic, the emergent ray $Q_2a_2b_2$ will be parallel to the axis, and $Q_2P_2 = l \cdot Q_1P_1$.

Let $Q_1 a_1 B_1$ be a ray through B_1, the emergent ray will be $Q_2 a_2 B_2$, and $A_2 a_2 = l \cdot A_1 a_1$.

Now
$$\frac{P_2 B_2}{A_2 B_2} = \frac{P_2 Q_2}{A_2 a_2} = \frac{l \cdot P_1 Q_1}{l \cdot A_1 a_1} = \frac{P_1 Q_1}{A_1 a_1} = \frac{P_1 B_1}{A_1 B_1},$$

so that
$$\frac{P_2 B_2}{P_1 B_1} = \frac{A_2 B_2}{A_1 B_1} = n, \text{ a constant ratio.}$$

Cor. If a point C be taken on the axis of the instrument so that
$$CB_2 = \frac{A_2 B_2}{A_1 B_1 - A_2 B_2} B_1 B_2 = \frac{n}{1-n} B_1 B_2,$$

then
$$CP_2 = n \cdot CP_1.$$

Def. The point C is called the *centre* of the telescope.

It appears, therefore, that the image of an object in a telescope has its dimensions perpendicular to the axis equal to l times the corresponding dimensions of the object, and the distance of any part from the plane through C equal to n times the distance of the corresponding part of the object. Of course all longitudinal distances among objects must be multiplied by n to obtain those of their images, and the tangent of the angular magnitude of an object as seen from a given point in the axis must be multiplied by $\frac{l}{n}$ to obtain that of the image of the object as seen from the image of the given point. The quantity $\frac{l}{n}$ is therefore called the *angular magnifying power*, and is denoted by m.

PROP. VI. To find the principal foci and principal planes of a combination of two instruments having a common axis.

Let I, I' (fig. 5) be the two instruments, $G_1 F_1 F_2 G_2$ the principal foci and planes of the first, $G_1' F_1' F_2' G_2'$ those of the second, $\Gamma_1 \phi_1 \phi_2 \Gamma_2$ those of the combination. Let the ray $g_1 g_2 g_1' g_2'$ pass through both instruments, and let it be parallel to the axis before entering the first instrument. It will therefore pass through F_2 the second principal focus of the first instrument, and through g_2 so that $G_2 g_2 = G_1 g_1$.

On emergence from the second instrument it will pass through ϕ_2 the focus conjugate to F_2, and through g_2' in the second principal plane, so that

$G_2'g_2' = G_1'g_1'$. ϕ_2 is by definition the second principal focus of the combination of instruments, and if $\Gamma_2\gamma_2$ be the second principal plane, then $\Gamma_2\gamma_2 = G_1g_1$.

We have now to find the positions of ϕ_2 and Γ_2.

By Prop. IV., we have

$$F_2'\phi_2 = \frac{F_1'G_1' \cdot G_2'F_2'}{F_2F_1'}.$$

Or, the distance of the principal focus of the combination, from that of the second instrument, is equal to the product of the focal lengths of the second instrument, divided by the distance of the second principal focus of the first instrument from the first of the second.   From this we get

$$G_2'F_2' - F_2'\phi_2 = \frac{G_2'F_2' (F_2F_1' - F_1'G_1')}{F_2F_1'},$$

or

$$G_2'\phi_2 = \frac{G_2'F_2' \cdot G_1'F_2}{F_2F_1'}.$$

Now, by the pairs of similar triangles $\phi G_2'g_2'$, $\phi\Gamma_2\gamma_2$ and $F_2G_1'g_1'$, $F_2G_2g_2$,

$$\frac{\Gamma_2\phi_2}{G_2'\phi_2} = \frac{\Gamma_2\gamma_2}{G_2'g_2} = \frac{G_2g_2}{G_1'g_1'} = \frac{F_2G_2}{G_1'F_2}.$$

Multiplying the two sides of the former equation respectively by the first and last of these equal quantities, we get

$$\Gamma_2\phi_2 = \frac{G_2F_2 \cdot G_2'F_2'}{F_2F_1'}.$$

Or, the second focal distance of a combination is the product of the second focal lengths of its two components, divided by the distance of their consecutive principal foci.

If we call the focal distances of the first instrument f_1 and f_2, those of the second f_1' and f_2', and those of the combination $\overline{f_1}$, $\overline{f_2}$, and put $F_2F_1' = d$, then the positions of the principal foci are found from the values

$$\phi_1F_1 = \frac{f_1f_2}{d}, \qquad F_2'\phi_2 = \frac{f_1'f_2'}{d},$$

and the focal lengths of the combination from

$$\overline{f_1} = \frac{f_1f_1'}{d}, \qquad \overline{f_2} = \frac{f_2f_2'}{d}.$$

When $d = 0$, all these values become infinite, and the compound instrument becomes a telescope.

PROP. VII. To find the linear magnifying power, the elongation, and the centre of the instrument, when the combination becomes a telescope.

Here (fig. 6) the second principal focus of the first instrument coincides at F with the first of the second. (In the figure, the focal distances of both instruments are taken in the opposite direction from that formerly assumed. They are therefore to be regarded as *negative*.)

In the first place, F_2' is conjugate to F_1, for a pencil whose focus before incidence is F_1 will be parallel to the axis between the instruments, and will converge to F_2' after emergence.

Also if $G_1 g_1$ be an object in the first principal plane, $G_2 g_2$ will be its first image, equal to itself, and if Hh be its final image

$$F_2'H = \frac{FG_1' \cdot G_2'F_2'}{G_2F} = \frac{f_1'f_2'}{f_2},$$

$$Hh = \frac{FG_1'}{G_2F} G_2 g_2 = -\frac{f_1'}{f_2} G_1 g_1.$$

Now the linear magnifying power is $\dfrac{Hh}{G_1 g_1}$, and the elongation is $\dfrac{F_2'H}{F_1 G_1}$, because F_2' and H are the images of F_1 and G_1 respectively; therefore

$$l = -\frac{f_1'}{f_2}, \quad \text{and} \quad n = \frac{f_1'f_2'}{f_1 f_2}.$$

The angular magnifying power $= m = \dfrac{l}{n} = -\dfrac{f_1}{f_2'}$.

The centre of the telescope is at the point C, such that

$$F_2'C = \frac{n}{1-n} F_1 F_2'.$$

When n becomes 1 the telescope has no centre. The effect of the instrument is then simply to alter the position of an object by a certain distance measured along the axis, as in the case of refraction through a plate of glass bounded by parallel planes. In certain cases this constant distance itself disappears, as in the case of a combination of three convex lenses of which the focal lengths are

4, 1, 4 and the distances 4 and 4. This combination simply inverts every object without altering its magnitude or distance along the axis.

The preceding theory of perfect instruments is quite independent of the mode in which the course of the rays is changed within the instrument, as we are supposed to know only that the path of every ray is straight before it enters, and after it emerges from the instrument. We have now to consider, how far these results can be applied to actual instruments, in which the course of the rays is changed by reflexion or refraction. We know that such instruments may be made so as to fulfil approximately the conditions of a perfect instrument, but that absolute perfection has not yet been obtained. Let us inquire whether any additional general law of optical instruments can be deduced from the laws of reflexion and refraction, and whether the imperfection of instruments is necessary or removeable.

The following theorem is a necessary consequence of the known laws of reflexion and refraction, whatever theory we adopt.

If we multiply the length of the parts of a ray which are in different media by the indices of refraction of those media, and call the sum of these products the *reduced path* of the ray, then :

I. The extremities of all rays from a given origin, which have the same reduced path, lie in a surface normal to those rays.

II. When a pencil of rays is brought to a focus, the reduced path from the origin to the focus is the same for every ray of the pencil.

In the undulatory theory, the "reduced path" of a ray is the distance through which light would travel in space, during the time which the ray takes to traverse the various media, and the surface of equal "reduced paths" is the wave-surface. In *extraordinary* refraction the wave-surface is not always normal to the ray, but the other parts of the proposition are true in this and all other cases.

From this general theorem in optics we may deduce the following propositions, true for all instruments depending on refraction and reflexion.

Prop. VIII. In any optical instrument depending on refraction or reflexion, if a_1a_1, $b_1\beta_1$ (fig. 7) be two objects and a_2a_2, $b_2\beta_2$ their images, A_1B_1 the distance of the objects, A_2B_2 that of the images, μ_1 the index of refraction of the medium in which the objects are, μ_2 that of the medium in which the images are, then

$$\mu_1 \frac{a_1 a_1 \times b_1 \beta_1}{A_1 B_1} = \mu_2 \frac{a_2 a_2 \times b_2 \beta_2}{A_2 B_2},$$

approximately, when the objects are small.

Since a_2 is the image of a_1, the reduced path of the ray $a_1 b_1 a_2$ will be equal to that of $a_1 \beta_1 a_2$, and the reduced paths of the rays $a_1 \beta_1 a_2$ and $a_1 b_1 a_2$ will be equal.

Also because $b_1 \beta_1$ and $b_2 \beta_2$ are conjugate foci, the reduced paths of the rays $b_1 a_2 b_2$ and $b_1 a_2 b_2$, and of $\beta_1 a_2 \beta_2$ and $\beta_1 a_2 \beta_2$ will be equal. So that the reduced paths

$$a_1 b_1 + b_1 a_2 = a_1 \beta_1 + \beta_1 a_2$$
$$a_1 \beta_1 + \beta_1 a_2 = a_1 b_1 + b_1 a_2$$
$$b_1 a_2 + a_2 b_2 = b_1 a_2 + a_2 b_2$$
$$\beta_1 a_2 + a_2 \beta_2 = \beta_1 a_2 + a_2 \beta_2$$

$$\therefore \ a_1 b_1 + a_1 \beta_1 + a_2 b_2 + a_2 \beta_2 = a_1 \beta_1 + a_1 b_1 + a_2 b_2 + a_2 \beta_2,$$

these being still the *reduced paths* of the rays, that is, the length of each ray multiplied by the index of refraction of the medium.

If the figure is symmetrical about the axis, we may write the equation

$$\mu_1 \left(a_1 \beta_1 - a_1 b_1 \right) = \mu_2 \left(a_2 \beta_2 - a_2 b_2 \right),$$

where $a_1 \beta_1$, &c. are now the *actual lengths* of the rays so named.

Now
$$\overline{a_1 \beta_1}^2 = \overline{A_1 B_1}^2 + \tfrac{1}{4} \left(a_1 a_1 + b_1 \beta_1 \right)^2,$$

$$\overline{a_1 b_1}^2 = \overline{A_1 B_1}^2 + \tfrac{1}{4} \left(a_1 a_1 - b_1 \beta_1 \right)^2,$$

so that
$$\overline{a_1 \beta_1}^2 - \overline{a_1 b_1}^2 = a_1 a_1 \times b_1 \beta_1,$$

and
$$\mu_1 \left(a_1 \beta_1 - a_1 b_1 \right) = \mu_1 \frac{a_1 a_1 \times b_1 \beta_1}{a_1 \beta_1 + a_1 b_1}.$$

Similarly
$$\mu_2 \left(a_2 \beta_2 - a_2 b_2 \right) = \mu_2 \frac{a_2 a_2 \times b_2 \beta_2}{a_2 \beta_2 + a_2 b_2}.$$

So that the equation
$$\mu_1 \frac{a_1 a_1 \times b_1 \beta_1}{a_1 \beta_1 + a_1 b_1} = \mu_2 \frac{a_2 a_2 \times b_2 \beta_2}{a_2 \beta_2 + a_2 b_2},$$

is true accurately, and since when the objects are small, the denominators are nearly $2A_1B_1$ and $2A_2B_2$, the proposition is proved approximately true.

Using the expressions of Prop. III., this equation becomes

$$\mu_1 \frac{1}{c_1} = \mu_2 \frac{xy}{c_2}.$$

Now by Prop. III., when x and y are different, the focal lengths f_1 and f_2 are

$$f_1 = c_1 \frac{xy}{x-y}, \qquad f_2 = c_2 \frac{1}{y-x};$$

therefore
$$\frac{f_1}{f_2} = \frac{c_1 xy}{c_2} = \frac{\mu_1}{\mu_2} \text{ by the present theorem.}$$

So that in any instrument, not a telescope, the focal lengths are directly as the indices of refraction of the media to which they belong. If, as in most cases, these media are the same, then the two focal distances are *equal*.

When $x = y$, the instrument becomes a telescope, and we have, by Prop. V., $l = x$, and $n = \dfrac{c_2}{c_1}$; and therefore by this theorem

$$\frac{\mu_1}{\mu_2} = \frac{l^2}{n}.$$

We may find l experimentally by measuring the actual diameter of the image of a known near object, such as the aperture of the object glass. If O be the diameter of the aperture and o that of the circle of light at the eye-hole (which is its image), then

$$l = \frac{o}{O}.$$

From this we find the elongation and the angular magnifying power

$$n = \frac{\mu_2}{\mu_1} l^2, \text{ and } m = \frac{\mu_1}{\mu_2} \frac{1}{l}.$$

When $\mu_1 = \mu_2$, as in ordinary cases, $m = \dfrac{1}{l} = \dfrac{O}{o}$, which is Gauss' rule for determining the magnifying power of a telescope.

PROP. IX. It is impossible, by means of any combination of reflexions and refractions, to produce a *perfect* image of an object at two different distances, unless the instrument be a telescope, and

$$l = n = \frac{\mu_1}{\mu_2}, \qquad m = 1.$$

It appears from the investigation of Prop. VIII. that the results there obtained, if true when the objects are very small, will be incorrect when the objects are large, unless

$$a_1\beta_1 + a_1b_1 : a_2\beta_2 + a_2b_2 :: A_1B_1 : A_2B_2,$$

and it is easy to prove that this cannot be, unless all the lines in the one figure are proportional to the corresponding lines in the other.

In this way we might show that we cannot in general have an astigmatic, plane, undistorted image of a plane object. But we can prove that we cannot get perfectly focussed images of an object in two positions, even at the expense of curvature and distortion.

We shall first prove that if two objects have perfect images, the reduced path of the ray joining any given points of the two objects is equal to that of the ray joining the corresponding points of the images.

Let a_2 (fig. 8) be the perfect image of a_1 and β_2 of β_1. Let

$$A_1a_1 = a_1, \quad B_1\beta_1 = b_1, \quad A_2a_2 = a_2, \quad B_2\beta_2 = b_2, \quad A_1B_1 = c_1, \quad A_2B_2 = c_2.$$

Draw a_1D_1 parallel to the axis to meet the plane B_1, and a_2D_2 to the plane of B_2.

Since everything is symmetrical about the axis of the instrument we shall have the angles $D_1B_1\beta_1 = D_2B_2\beta_2 = \theta$, then in either figure, omitting the suffixes,

$$\overline{a\beta^2} = \overline{aD^2} + \overline{D\beta^2}$$

$$= c^2 + a^2 + b^2 - 2ab\cos\theta.$$

It has been shown in Prop. VIII. that the difference of the reduced paths of the rays a_1b_1, $a_1\beta_1$ in the object must be equal to the difference of the reduced paths of a_2b_2, $a_2\beta_2$ in the image. Therefore, since we may assume any value for θ

$$\mu_1\surd(a_1^2 + b_1^2 + c_1^2 - 2a_1b_1\cos\theta) - \mu_2\surd(a_2^2 + b_2^2 + c_2^2 - 2a_2b_2\cos\theta)$$

36—2

is constant for all values of θ. This can be only when

$$\mu_1 \sqrt{(a_1^2 + b_1^2 + c_1^2)} = \mu_2 \sqrt{(a_2^2 + b_2^2 + c_2^2)},$$

and
$$\mu_1 \sqrt{(a_1 b_1)} = \mu_2 \sqrt{(a_2 b_2)},$$

which shows that the constant must vanish, and that the lengths of lines joining corresponding points of the objects and of the images must be inversely as the indices of refraction before incidence and after emergence.

Next let *ABC*, *DEF* (fig. 9) represent three points in the one object and three points in the other object, the figure being drawn to a scale so that all the lines in the figure are the actual lines multiplied by μ_1. The lines of the figure represent the reduced paths of the rays between the corresponding points of the objects.

Now it may be shown that the form of this figure cannot be altered without altering the length of one or more of the nine lines joining the points *ABC* to *DEF*. Therefore since the reduced paths of the rays in the image are equal to those in the object, the figure must represent the image on a scale of μ_2 to 1, and therefore the instrument must magnify every part of the object alike and elongate the distances parallel to the axis in the same proportion. It is therefore a telescope, and $m = 1$.

If $\mu_1 = \mu_2$, the image is exactly equal to the object, which is the case in reflexion in a plane mirror, which we know to be a perfect instrument for all distances.

The only case in which by refraction at a single surface we can get a perfect image of more than one point of the object, is when the refracting surface is a sphere, radius r, index μ, and when the two objects are spherical surfaces, concentric with the sphere, their radii being $\dfrac{r}{\mu}$, and r ; and the two images also concentric spheres, radii μr, and r.

In this latter case the image is perfect, only at these particular distances and not generally.

I am not aware of any other case in which a perfect image of an object can be formed, the rays being straight before they enter, and after they emerge from the instrument. The only case in which perfect astigmatism for all pencils has hitherto been proved to exist, was suggested to me by the consideration

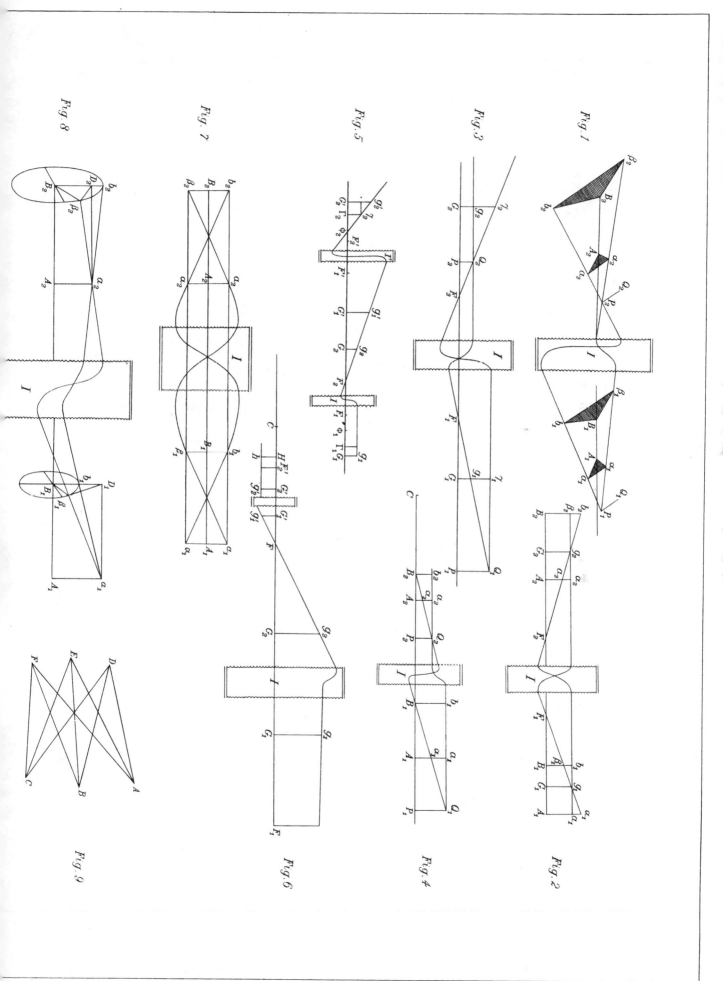

of the structure of the crystalline lens in fish, and was published in one of the problem-papers of the *Cambridge and Dublin Mathematical Journal*. My own method of treating that problem is to be found in that *Journal*, for February, 1854. The case is that of a medium whose index of refraction varies with the distance from a centre, so that if μ_0 be its value at the centre, a a given line, and r the distance of any point where the index is μ, then

$$\mu = \mu_0 \frac{a^2}{a^2 + r^2}.$$

The path of every ray within this medium is a circle in a plane passing through the centre of the medium.

Every ray from a point in the medium, distant b from the centre, will converge to a point on the opposite side of the centre and distant from it $\dfrac{a^2}{b}$.

It will be observed that both the object and the image are included in the variable medium, otherwise the images would not be perfect. This case therefore forms no exception to the result of Prop. IX., in which the object and image are supposed to be outside the instrument.

Aberdeen, 12th Jan., 1858.

[From the *Proceedings of the Royal Society of Edinburgh*, Vol. IV.]

XVIII.   *On Theories of the Constitution of Saturn's Rings.*

THE planet Saturn is surrounded by several concentric flattened rings, which appear to be quite free from any connection with each other, or with the planet, except that due to gravitation.

The exterior diameter of the whole system of rings is estimated at about 176,000 miles, the breadth from outer to inner edge of the entire system, 36,000 miles, and the thickness not more than 100 miles.

It is evident that a system of this kind, so broad and so thin, must depend for its stability upon the dynamical equilibrium between the motions of each part of the system, and the attractions which act on it, and that the cohesion of the parts of so large a body can have no effect whatever on its motions, though it were. made of the most rigid material known on earth. It is therefore necessary, in order to satisfy the demands of physical astronomy, to explain how a material system, presenting the appearance of Saturn's Rings, can be maintained in permanent motion consistently with the laws of gravitation. The principal hypotheses which present themselves are these—

    I.   The rings are solid bodies, regular or irregular.

    II.   The rings are fluid bodies, liquid or gaseous.

    III.   The rings are composed of loose materials.

The results of mathematical investigation applied to the first case are,—

1st.   That a uniform ring cannot have a permanent motion.

2nd.   That it is possible, by loading one side of the ring, to produce stability of motion, but that this loading must be very great compared with the whole mass of the rest of the ring, being as 82 to 18.

3rd. That this loading must not only be very great, but very nicely adjusted; because, if it were less than ·81, or more than ·83 of the whole, the motion would be unstable.

The mode in which such a system would be destroyed would be by the collision between the planet and the inside of the ring.

And it is evident that as no loading so enormous in comparison with the ring actually exists, we are forced to consider the rings as fluid, or at least not solid; and we find that, in the case of a fluid ring, waves would be generated, which would break it up into portions, the number of which would depend on the mass of Saturn directly, and on that of the ring inversely.

It appears, therefore, that the only constitution possible for such a ring is a series of disconnected masses, which may be fluid or solid, and need not be equal. The complicated internal motions of such a ring have been investigated, and found to consist of four series of waves, which, when combined together, will reproduce any form of original disturbance with all its consequences. The motion of one of these waves was exhibited to the Society by means of a small mechanical model made by Ramage of Aberdeen.

This theory of the rings, being indicated by the mechanical theory as the only one consistent with permanent motion, is further confirmed by recent observations on the inner obscure ring of Saturn. The limb of the planet is seen through the substance of this ring, not refracted, as it would be through a gas or fluid, but in its true position, as would be the case if the light passed through interstices between the separate particles composing the ring.

As the whole investigations are shortly to be published in a separate form, the mathematical methods employed were not laid before the Society.

XIX. *On the Stability of the motion of Saturn's Rings.*

[An Essay, which obtained the Adams Prize for the year 1856, in the University of Cambridge.]

ADVERTISEMENT.

THE Subject of the Prize was announced in the following terms :—

> The University having accepted a fund, raised by several members of St John's College, for the purpose of founding a Prize to be called the ADAMS PRIZE, for the best Essay on some subject of Pure Mathematics, Astronomy, or other branch of Natural Philosophy, the Prize to be given once in two years, and to be *open to the competition of all persons who have at any time been admitted to a degree in this University* :—

The Examiners give Notice, that the following is the subject for the Prize to be adjudged in 1857 :—

The Motions of Saturn's Rings.

** The problem may be treated on the supposition that the system of Rings is exactly or very approximately concentric with Saturn and symmetrically disposed about the plane of his Equator, and different hypotheses may be made respecting the physical constitution of the Rings. It may be supposed (1) that they are rigid : (2) that they are fluid, or in part aeriform : (3) that they consist of masses of matter not mutually coherent. The question will be considered to be answered by ascertaining on these hypotheses severally, whether the conditions of mechanical stability are satisfied by the mutual attractions and motions of the Planet and the Rings.

It is desirable that an attempt should also be made to determine on which of the above hypotheses the appearances both of the bright Rings and the recently discovered dark Ring may be most satisfactorily explained; and to indicate any causes to which a change of form, such as is supposed from a comparison of modern with the earlier observations to have taken place, may be attributed.

E. GUEST, *Vice-Chancellor.*
J. CHALLIS.
S. PARKINSON.
W. THOMSON.

March 23, 1855.

CONTENTS.

PART I.

ON THE MOTION OF A RIGID BODY OF ANY FORM ABOUT
A SPHERE.

PART II.

ON THE MOTION OF A RING, THE PARTS OF WHICH ARE NOT
RIGIDLY CONNECTED.

ON THE MUTUAL PERTURBATIONS OF TWO RINGS.

THERE are some questions in Astronomy, to which we are attracted rather on account of their peculiarity, as the possible illustration of some unknown principle, than from any direct advantage which their solution would afford to mankind. The theory of the Moon's inequalities, though in its first stages it presents theorems interesting to all students of mechanics, has been pursued into such intricacies of calculation as can be followed up only by those who make the improvement of the Lunar Tables the object of their lives. The value of the labours of these men is recognised by all who are aware of the importance of such tables in Practical Astronomy and Navigation. The methods by which the results are obtained are admitted to be sound, and we leave to professional astronomers the labour and the merit of developing them.

The questions which are suggested by the appearance of Saturn's Rings cannot, in the present state of Astronomy, call forth so great an amount of labour among mathematicians. I am not aware that any practical use has been made of Saturn's Rings, either in Astronomy or in Navigation. They are too distant, and too insignificant in mass, to produce any appreciable effect on the motion of other parts of the Solar system; and for this very reason it is difficult to determine those elements of their motion which we obtain so accurately in the case of bodies of greater mechanical importance.

But when we contemplate the Rings from a purely scientific point of view, they become the most remarkable bodies in the heavens, except, perhaps, those still less *useful* bodies—the spiral nebulæ. When we have actually seen that great arch swung over the equator of the planet without any visible connexion, we cannot bring our minds to rest. We cannot simply admit that such is the case, and describe it as one of the observed facts in nature, not admitting or requiring explanation. We must either explain its motion on the principles of mechanics, or admit that, in the Saturnian realms, there can be motion regulated by laws which we are unable to explain.

The arrangement of the rings is represented in the figure (1) on a scale of one inch to a hundred thousand miles. S is a section of Saturn through his equator, A, B and C are the three rings. A and B have been known for 200 years. They were mistaken by Galileo for protuberances on the planet itself, or perhaps satellites. Huyghens discovered that what he saw was a thin flat ring not touching the planet, and Ball discovered the division between A and B. Other divisions have been observed splitting these again into concentric rings, but these have not continued visible, the only well-established division being one in the middle of A. The third ring C was first detected by Mr Bond, at Cambridge U.S. on November 15, 1850; Mr Dawes, not aware of Mr Bond's discovery, observed it on November 29th, and Mr Lassel a few days later. It gives little light compared with the other rings, and is seen where it crosses the planet as an obscure belt, but it is so transparent that the limb of the planet is visible through it, and this without distortion, shewing that the rays of light have not passed through a transparent substance, but between the scattered particles of a discontinuous stream.

It is difficult to estimate the thickness of the system; according to the best estimates it is not more than 100 miles, the diameter of A being 176,418 miles; so that on the scale of our figure the thickness would be one thousandth of an inch.

Such is the scale on which this magnificent system of concentric rings is constructed; we have next to account for their continued existence, and to reconcile it with the known laws of motion and gravitation, so that by rejecting every hypothesis which leads to conclusions at variance with the facts, we may learn more of the nature of these distant bodies than the telescope can yet ascertain. We must account for the rings remaining suspended above the planet, concentric with Saturn and in his equatoreal plane; for the flattened figure of the section of each ring, for the transparency of the inner ring, and for the gradual approach of the inner edge of the ring to the body of Saturn as deduced from all the recorded observations by M. Otto Struvé (*Sur les dimensions des Anneaux de Saturne*—Recueil de Mémoires Astronomiques, Poulkowa, 15 Nov. 1851). For an account of the general appearance of the rings as seen from the planet, see Lardner on the Uranography of Saturn, *Mem. of the Astronomical Society*, 1853. See also the article "Saturn" in Nichol's *Cyclopædia of the Physical Sciences*.

Our curiosity with respect to these questions is rather stimulated than appeased by the investigations of Laplace. That great mathematician, though occupied with many questions which more imperiously demanded his attention, has devoted several chapters in various parts of his great work, to points connected with the Saturnian System.

He has investigated the law of attraction of a ring of small section on a point very near it (*Méc. Cél.* Liv. III. Chap. VI.), and from this he deduces the equation from which the ratio of the breadth to the thickness of each ring is to be found,

$$e = \frac{R^3}{3a^3}\frac{\rho}{\rho'} = \frac{\lambda(\lambda-1)}{(\lambda+1)(3\lambda^2+1)},$$

where R is the radius of Saturn, and ρ his density; a the radius of the ring, and ρ' its density; and λ the ratio of the breadth of the ring to its thickness. The equation for determining λ when e is given has one negative root which must be rejected, and two roots which are positive while $e < 0.0543$, and impossible when e has a greater value. At the critical value of e, $\lambda = 2.594$ nearly.

The fact that λ is impossible when e is above this value, shews that the ring cannot hold together if the ratio of the density of the planet to that of the ring exceeds a certain value. This value is estimated by Laplace at 1·3, assuming $a = 2R$.

We may easily follow the physical interpretation of this result, if we observe that the forces which act on the ring may be reduced to—

(1) The attraction of Saturn, varying inversely as the square of the distance from his centre.

(2) The centrifugal force of the particles of the ring, acting outwards, and varying directly as the distance from Saturn's polar axis.

(3) The attraction of the ring itself, depending on its form and density, and directed, roughly speaking, towards the centre of its section.

The first of these forces must balance the second somewhere near the mean distance of the ring. Beyond this distance their resultant will be outwards, within this distance it will act inwards.

If the attraction of the ring itself is not sufficient to balance these residual forces, the outer and inner portions of the ring will tend to separate, and the ring will be split up; and it appears from Laplace's result that this will be the case if the density of the ring is less than $\frac{10}{13}$ of that of the planet.

This condition applies to all rings whether broad or narrow, of which the parts are separable, and of which the outer and inner parts revolve with the same angular velocity.

Laplace has also shewn (Liv. v. Chap. III.), that on account of the oblateness of the figure of Saturn, the planes of the rings will follow that of Saturn's equator through every change of its position due to the disturbing action of other heavenly bodies.

Besides this, he proves most distinctly (Liv. III. Chap. VI.), that a solid uniform ring cannot possibly revolve about a central body in a permanent manner, for the slightest displacement of the centre of the ring from the centre of the planet would originate a motion which would never be checked, and would inevitably precipitate the ring upon the planet, not necessarily by breaking the ring, but by the inside of the ring falling on the equator of the planet.

He therefore infers that the rings are irregular solids, whose centres of gravity do not coincide with their centres of figure. We may draw the conclusion more formally as follows, " If the rings were solid and uniform, their motion would be unstable, and they would be destroyed. But they are not destroyed, and their motion is stable; therefore they are either not uniform or not solid."

I have not discovered * either in the works of Laplace or in those of more recent mathematicians, any investigation of the motion of a ring either not uniform or not solid. So that in the present state of mechanical science, we do not know whether an irregular solid ring, or a fluid or disconnected ring, can revolve permanently about a central body; and the Saturnian system still remains an unregarded witness in heaven to some necessary, but as yet unknown, development of the laws of the universe.

We know, since it has been demonstrated by Laplace, that a uniform solid ring cannot revolve permanently about a planet. We propose in this Essay to determine the amount and nature of the irregularity which would be required to make a permanent rotation possible. We shall find that the stability of the motion of the ring would be ensured by loading the ring at one point with a

---

* Since this was written, Prof. Challis has pointed out to me three important papers in Gould's *Astronomical Journal* :—Mr G. P. Bond *on the Rings of Saturn* (May 1851) and Prof. B. Pierce of Harvard University *on the Constitution of Saturn's Rings* (June 1851), and *on the Adams' Prize Problem* for 1856 (Sept. 1855). These American mathematicians have both considered the conditions of statical equilibrium of a transverse section of a ring, and have come to the conclusion that the rings, if they move each as a whole, must be very narrow compared with the observed rings, so that in reality there must be a great number of them, each revolving with its own velocity. They have also entered on the question of the fluidity of the rings, and Prof. Pierce has made an investigation as to the permanence of the motion of an irregular solid ring and of a fluid ring. The paper in which these questions are treated at large has not (so far as I am aware) been published, and the references to it in Gould's Journal are intended to give rather a popular account of the results, than an accurate outline of the methods employed. In treating of the attractions of an irregular ring, he makes admirable use of the theory of potentials, but his published investigation of the motion of such a body contains some oversights which are due perhaps rather to the imperfections of popular language than to any thing in the mathematical theory. The only part of the theory of a fluid ring which he has yet given an account of, is that in which he considers the form of the ring at any instant as an ellipse; corresponding to the case where $n = \omega$, and $m = 1$. As I had only a limited time for reading these papers, and as I could not ascertain the methods used in the original investigations, I am unable at present to state how far the results of this essay agree with or differ from those obtained by Prof. Pierce.

heavy satellite about $4\frac{1}{2}$ times the weight of the ring, but this load, besides being inconsistent with the observed appearance of the rings, must be far too artificially adjusted to agree with the natural arrangements observed elsewhere, for a very small error in excess or defect would render the ring again unstable.

We are therefore constrained to abandon the theory of a solid ring, and to consider the case of a ring, the parts of which are not rigidly connected, as in the case of a ring of independent satellites, or a fluid ring.

There is now no danger of the whole ring or any part of it being precipitated on the body of the planet. Every particle of the ring is now to be regarded as a satellite of Saturn, disturbed by the attraction of a ring of satellites at the same mean distance from the planet, each of which however is subject to slight displacements. The mutual action of the parts of the ring will be so small compared with the attraction of the planet, that no part of the ring can ever cease to move round Saturn as a satellite.

But the question now before us is altogether different from that relating to the solid ring. We have now to take account of variations in the form and arrangement of the parts of the ring, as well as its motion as a whole, and we have as yet no security that these variations may not accumulate till the ring entirely loses its original form, and collapses into one or more satellites, circulating round Saturn. In fact such a result is one of the leading doctrines of the "nebular theory" of the formation of planetary systems: and we are familiar with the actual breaking up of fluid rings under the action of "capillary" force, in the beautiful experiments of M. Plateau.

In this essay I have shewn that such a destructive tendency actually exists, but that by the revolution of the ring it is converted into the condition of dynamical stability. As the scientific interest of Saturn's Rings depends at present mainly on this question of their stability, I have considered their motion rather as an illustration of general principles, than as a subject for elaborate calculation, and therefore I have confined myself to those parts of the subject which bear upon the question of the permanence of a given form of motion.

There is a very general and very important problem in Dynamics, the solution of which would contain all the results of this Essay and a great deal more. It is this—

"Having found a particular solution of the equations of motion of any material system, to determine whether a slight disturbance of the motion indicated by the solution would cause a small periodic variation, or a total derangement of the motion."

The question may be made to depend upon the conditions of a maximum or a minimum of a function of many variables, but the theory of the tests for distinguishing maxima from minima by the Calculus of Variations becomes so intricate when applied to functions of several variables, that I think it doubtful whether the physical or the abstract problem will be first solved.

PART I.

ON THE MOTION OF A RIGID BODY OF ANY FORM ABOUT A SPHERE.

WE confine our attention for the present to the motion in the plane of reference, as the interest of our problem belongs to the character of this motion, and not to the librations, if any, from this plane.

Let S (Fig. 2) be the centre of gravity of the sphere, which we may call Saturn, and R that of the rigid body, which we may call the Ring. Join RS, and divide it in G so that

$$SG : GR :: R : S,$$

R and S being the masses of the Ring and Saturn respectively.

Then G will be the centre of gravity of the system, and its position will be unaffected by any mutual action between the parts of the system. Assume G as the point to which the motions of the system are to be referred. Draw GA in a direction fixed in space.

Let $\qquad\qquad AGR = \theta$, and $SR = r$,

then $\qquad\qquad GR = \dfrac{S}{S+R} r$, and $GS = \dfrac{R}{S+R} r$,

so that the positions of S and R are now determined.

Let BRB' be a straight line through R, *fixed with respect to the substance of the ring*, and let $BRK = \phi$.

This determines the angular position of the ring, so that from the values of r, θ, and ϕ the configuration of the system may be deduced, as far as relates to the plane of reference.

We have next to determine the forces which act between the ring and the sphere, and this we shall do by means of the *potential function* due to the ring, which we shall call V.

The value of V for any point of space S, depends on its position relatively to the ring, and it is found from the equation

$$V = \Sigma \left(\frac{dm}{r'} \right),$$

where dm is an element of the mass of the ring, and r' is the distance of that element from the given point, and the summation is extended over every element of mass belonging to the ring. V will then depend entirely upon the position of the point S relatively to the ring, and may be expressed as a function of r, the distance of S from R, the centre of gravity of the ring, and ϕ, the angle which the line SR makes with the line RB, fixed in the ring.

A particle P, placed at S, will, by the theory of potentials, experience a moving force $P \dfrac{dV}{dr}$ in the direction which tends to increase r, and $P \dfrac{1}{r} \dfrac{dV}{d\phi}$ in a tangential direction, tending to increase ϕ.

Now we know that the attraction of a sphere is the same as that of a particle of equal mass placed at its centre. The forces acting between the sphere and the ring are therefore $S \dfrac{dV}{dr}$ tending to increase r, and a tangential force $S \dfrac{1}{r} \dfrac{dV}{d\phi}$, applied at S tending to increase ϕ. In estimating the effect of this latter force on the ring, we must resolve it into a tangential force $S \dfrac{1}{r} \dfrac{dV}{d\phi}$ acting at R, and a couple $S \dfrac{dV}{d\phi}$ tending to increase ϕ.

We are now able to form the equations of motion for the planet and the ring.

For the planet

$$S \frac{d}{dt} \left\{ \left(\frac{Rr}{S+R} \right)^2 \frac{d\theta}{dt} \right\} = -\frac{R}{S+R} S \frac{dV}{d\phi} \quad \dots\dots\dots\dots (1),$$

$$S \frac{d^2}{dt^2} \left(\frac{Rr}{S+R} \right) - S \frac{Rr}{S+R} \left(\frac{d\theta}{dt} \right)^2 = S \frac{dV}{dr} \quad \dots\dots\dots\dots (2).$$

For the centre of gravity of the ring,

$$R \frac{d}{dt} \left\{ \left(\frac{Sr}{S+R} \right)^2 \frac{d\theta}{dt} \right\} = -\frac{S}{S+R} S \frac{dV}{d\phi} \quad \dots\dots\dots\dots (3),$$

$$R \frac{d^2}{dt^2} \left(\frac{Sr}{S+R} \right) - R \frac{Sr}{S+R} \left(\frac{d\theta}{dt} \right)^2 = S \frac{dV}{dr} \dots\dots\dots\dots (4).$$

For the rotation of the ring about its centre of gravity,

$$Rk^2 \frac{d^2}{dt^2} (\theta + \phi) = S \frac{dV}{d\phi} \quad \dots\dots\dots\dots\dots\dots (5),$$

where k is the radius of gyration of the ring about its centre of gravity.

Equation (3) and (4) are necessarily identical with (1) and (2), and shew that the orbit of the centre of gravity of the ring must be similar to that of the Planet. Equations (1) and (3) are equations of areas, (2) and (4) are those of the radius vector.

Equations (3), (4) and (5) may be thus written,

$$R \left\{ 2r \frac{dr}{dt} \frac{d\theta}{dt} + r^2 \frac{d^2\theta}{dt^2} \right\} + (R+S) \frac{dV}{d\phi} = 0 \quad \dots\dots\dots\dots (6),$$

$$R \left\{ \frac{d^2r}{dt^2} - r \left(\frac{d\theta}{dt} \right)^2 \right\} - (R+S) \frac{dV}{dr} \quad = 0 \quad \dots\dots\dots\dots (7),$$

$$Rk^2 \left(\frac{d^2\theta}{dt^2} + \frac{d^2\phi}{dt^2} \right) - S \frac{dV}{d\phi} \quad\quad = 0 \quad \dots\dots\dots\dots (8).$$

These are the necessary and sufficient data for determining the motion of the ring, the initial circumstances being given.

Prob. I.   To find the conditions under which a uniform motion of the ring is possible.

By a uniform motion is here meant a motion of uniform rotation, during which the position of the centre of the Planet with respect to the ring does not change.

In this case r and ϕ are constant, and therefore V and its differential coefficients are given. Equation (7) becomes,

$$Rr \left(\frac{d\theta}{dt}\right)^2 + (R+S)\frac{dV}{dr} = 0,$$

which shews that the angular velocity is constant, and that

$$\left(\frac{d\theta}{dt}\right)^2 = -\frac{R+S}{Rr}\frac{dV}{dr} = \omega^2, \text{ say} \quad\dots\dots\dots\dots\dots (9).$$

Hence, $\dfrac{d^2\theta}{dt^2} = 0$, and therefore by equation (8),

$$\frac{dV}{d\phi} = 0 \dots\dots\dots\dots\dots\dots\dots\dots\dots(10).$$

Equations (9) and (10) are the conditions under which the uniform motion is possible, and if they were exactly fulfilled, the uniform motion would go on for ever if not disturbed. But it does not follow that if these conditions were *nearly* fulfilled, or that if when accurately adjusted, the motion were *slightly* disturbed, the motion would go on for ever *nearly* uniform. The effect of the disturbance might be either to produce a periodic variation in the elements of the motion, the amplitude of the variation being small, or to produce a displacement which would increase indefinitely, and derange the system altogether. In the one case the motion would be *dynamically stable*, and in the other it would be *dynamically unstable*. The investigation of these displacements while still very small will form the next subject of inquiry.

PROB. II. To find the equations of the motion when slightly disturbed.

Let $r = r_0$, $\theta = \omega t$ and $\phi = \phi_0$ in the case of uniform motion, and let

$$r = r_0 + r_1,$$
$$\theta = \omega t + \theta_1,$$
$$\phi = \phi_0 + \phi_1,$$

when the motion is slightly disturbed, where r_1, θ_1, and ϕ_1 are to be treated as small quantities of the first order, and their powers and products are to be neglected. We may expand $\dfrac{dV}{dr}$ and $\dfrac{dV}{d\phi}$ by Taylor's Theorem,

$$\frac{dV}{dr} = \frac{dV}{dr} + \frac{d^2V}{dr^2}\ r_1 + \frac{d^2V}{drd\phi}\ \phi_1,$$

$$\frac{dV}{d\phi} = \frac{dV}{d\phi} + \frac{d^2V}{drd\phi}\ r_1 + \frac{d^2V}{d\phi^2}\ \phi_1,$$

where the values of the differential coefficients on the right-hand side of the equations are those in which r_0 stands for r, and ϕ_0 for ϕ.

Calling
$$\frac{d^2V}{dr^2} = L, \quad \frac{d^2V}{dr\,d\phi} = M, \quad \frac{d^2V}{d\phi^2} = N,$$

and taking account of equations (9) and (10), we may write these equations,

$$\frac{dV}{dr} = -\frac{Rr_0}{R+S}\omega^2 + Lr_1 + M\phi_1,$$

$$\frac{dV}{d\phi} = Mr_1 + N\phi_1.$$

Substituting these values in equations (6), (7), (8), and retaining all small quantities of the first order while omitting their powers and products, we have the following system of linear equations in r_1, θ_1, and ϕ_1,

$$R\left(2r_0\omega\frac{dr_1}{dt} + r_0^2\frac{d^2\theta_1}{dt^2}\right) + (R+S)(Mr_1 + N\phi_1) = 0 \ldots\ldots\ldots(11),$$

$$R\left(\frac{d^2r_1}{dt^2} - \omega^2 r_1 - 2r_0\omega\frac{d\theta_1}{dt}\right) - (R+S)(Lr_1 + M\phi_1) = 0 \ldots\ldots\ldots(12),$$

$$Rk^2\left(\frac{d^2\theta_1}{dt^2} + \frac{d\phi_1}{dt^2}\right) - S(Mr_1 + N\phi_1) = 0 \ldots\ldots\ldots(13).$$

PROB. III. To reduce the three simultaneous equations of motion to the form of a single linear equation.

Let us write n instead of the symbol $\dfrac{d}{dt}$, then arranging the equations in terms of r_1, θ_1, and ϕ_1, they may be written:

$$\{2R_0\omega n + (R+S)M\}\,r_1 + (Rr_0^2n^2)\,\theta_1 + (R+S)\,N\phi_1 = 0 \ldots\ldots(14),$$

$$\{Rn^2 - R\omega^2 - (R+S)\,L\}\,r_1 - (2Rr_0\omega n)\,\theta_1 - (R+S)\,M\phi_1 = 0 \ldots\ldots(15),$$

$$-(SM)\,r_1 + (Rk^2n^2)\,\theta_1 + (Rk^2n^2 - SN)\,\phi_1 = 0 \ldots\ldots(16).$$

Here we have three equations to determine three quantities r_1, θ_1, ϕ_1; but it is evident that only a relation can be determined between them, and that in the process for finding their absolute values, the three quantities will vanish together, and leave the following relation among the coefficients,

$$\left.\begin{array}{l} -\{2Rr_0\omega n + (R+S)\,M\}\,\{2Rr_0\omega n\}\,\{Rk^2n^2 - SN\} \\ +\{Rn^2 - R\omega^2 - (R+S)\,L\}\,\{Rk^2n^2\}\,\{(R+S)\,N\} \\ +(SM)\,(Rr_0^2n^2)\,(R+S)\,M - (SM)\,(2Rr_0\omega n)\,(R+S)\,N \\ +\{2Rr_0\omega n + (R+S)\,M\}\,\{Rk^2n^2\}\,\{(R+S)\,M\} \\ -\{Rn^2 - R\omega^2 - (R+S)\}\,\{Rr_0^2n^2\}\,\{Rk^2n^2 - SN\} \end{array}\right\} = 0 \;\ldots\ldots(17).$$

By multiplying up, and arranging by powers of n and dividing by Rn^2, this equation becomes

$$An^4 + Bn^2 + C = 0 \ldots\ldots\ldots\ldots\ldots\ldots\ldots (18),$$

where

$$\left.\begin{array}{l} A = R^2r_0^2k^2, \\ B = 3R^2r_0^2k^2\omega^2 - R\,(R+S)\,Lr_0^2k^2 - R\,\{(R+S)\,k^2 + Sr^2\}\,N \\ C = R\,\{(R+S)\,k^2 - 3Sr_0^2\}\,\omega^2 + (R+S)\,\{(R+S)\,k^2 + Sr_0^2\}\,(LN - M^2) \end{array}\right\} \ldots\ldots(19).$$

Here we have a biquadratic equation in n which may be treated as a quadratic in n^2, it being remembered that n stands for the operation $\dfrac{d}{dt}$.

PROB. IV. To determine whether the motion of the ring is stable or unstable, by means of the relations of the coefficients A, B, C.

The equations to determine the forms of r_1, θ_1, and ϕ_1 are all of the form

$$A\,\frac{d^4u}{dt^4} + B\,\frac{d^2u}{dt^2} + Cu = 0 \ldots\ldots\ldots\ldots\ldots\ldots (20),$$

and if n be one of the four roots of equation (18), then

$$u = De^{nt}$$

will be one of the four terms of the solution, and the values of r_1, θ_1, and ϕ_1 will differ only in the values of the coefficient D.

Let us inquire into the nature of the solution in different cases.

(1) If n be positive, this term would indicate a displacement which must increase indefinitely, so as to destroy the arrangement of the system.

(2) If n be negative, the disturbance which it belongs to would gradually die away.

(3) If n be a pure impossible quantity, of the form $\pm a\sqrt{-1}$, then there will be a term in the solution of the form $D\cos(at+a)$, and this would indicate a periodic variation, whose amplitude is D, and period $\dfrac{2\pi}{a}$.

(4) If n be of the form $b \pm \sqrt{-1}\,a$, the first term being positive and the second impossible, there will be a term in the solution of the form

$$D\epsilon^{bt} \cos (at + a),$$

which indicates a periodic disturbance, whose amplitude continually increases till it disarranges the system.

(5) If n be of the form $-b \pm \sqrt{-1}\,a$, a negative quantity and an impossible one, the corresponding term of the solution is

$$D\epsilon^{-bt} \cos (at + a),$$

which indicates a periodic disturbance whose amplitude is constantly diminishing.

It is manifest that the first and fourth cases are inconsistent with the permanent motion of the system. Now since equation (18) contains only even powers of n, it must have pairs of equal and opposite roots, so that every root coming under the second or fifth cases, implies the existence of another root belonging to the first or fourth. If such a root exists, some disturbance may occur to produce the kind of derangement corresponding to it, so that the system is not safe unless roots of the first and fourth kinds are altogether excluded. This cannot be done without excluding those of the second and fifth kinds, so that, to insure stability, all the four roots must be of the third kind, that is, pure impossible quantities.

That this may be the case, both values of n^2 must be real and negative, and the conditions of this are—

1st. That A, B, and C should be of the same sign,

2ndly. That $B^2 > 4AC$.

When these conditions are fulfilled, the disturbances will be periodic and consistent with stability. When they are not both fulfilled, a small disturbance may produce total derangement of the system.

PROB. V. To find the centre of gravity, the radius of gyration, and the variations of the potential near the centre of a circular ring of small but variable section.

Let a be the radius of the ring, and let θ be the angle subtended at the centre between the radius through the centre of gravity and the line through a given point in the ring. Then if μ be the mass of unit of length of the ring near the given point, μ will be a periodic function of θ, and may therefore be expanded by Fourier's theorem in the series,

$$\mu = \frac{R}{2\pi a}\{1 + 2f\cos\theta + \tfrac{2}{3}g\cos 2\theta + \tfrac{2}{3}h\sin 2\theta + 2i\cos(3\theta + a) + \&\text{c.}\}\dots\dots(21),$$

where f, g, h, &c. are arbitrary coefficients, and R is the mass of the ring.

(1) The moment of the ring about the diameter perpendicular to the prime radius is

$$Rr_0 = \int_0^{2\pi} \mu a^2 \cos\theta d\theta = Raf,$$

therefore the distance of the centre of gravity from the centre of the ring,

$$r_0 = af.$$

(2) The radius of gyration of the ring about its centre in its own plane is evidently the radius of the ring $= a$, but if k be that about the centre of gravity, we have

$$k^2 + r_0^2 = a^2;$$

$$\therefore k^2 = a^2(1 - f^2).$$

(3) The potential at any point is found by dividing the mass of each element by its distance from the given point, and integrating over the whole mass.

Let the given point be near the centre of the ring, and let its position be defined by the co-ordinates r' and ψ, of which r' is small compared with a.

The distance (ρ) between this point and a point in the ring is

$$\frac{1}{\rho} = \frac{1}{a}\{1 + \frac{r'}{a}\cos(\psi - \theta) + \tfrac{1}{4}\left(\frac{r'}{a}\right)^2 + \tfrac{3}{4}\left(\frac{r'}{a}\right)^2\cos 2(\psi - \theta) + \&\text{c.}\}.$$

The other terms contain powers of $\frac{r'}{a}$ higher than the second.

We have now to determine the value of the integral,

$$V = \int_0^{2\pi} \frac{\mu}{\rho} a d\theta;$$

and in multiplying the terms of (μ) by those of $\left(\frac{1}{\rho}\right)$, we need retain only those which contain constant quantities, for all those which contain sines or cosines of multiples of $(\psi - \theta)$ will vanish when integrated between the limits. In this way we find

$$V = \frac{R}{a} \left\{ 1 + f \frac{r'}{a} \cos \psi + \tfrac{1}{4} \frac{r'^2}{a^2} (1 + g \cos 2\psi + h \sin 2\psi) \right\} \dots\dots\dots (22).$$

The other terms containing higher powers of $\frac{r'}{a}$.

In order to express V in terms of r_1 and ϕ_1, as we have assumed in the former investigation, we must put

$$r' \cos \psi = -r_1 + \tfrac{1}{2} r_0 \phi_1^2,$$
$$r' \sin \psi = -r_0 \phi_1,$$

$$V = \frac{R}{a} \left\{ 1 - f \frac{r_1}{a} + \tfrac{1}{4} \frac{r_1^2}{a^2} (1 + g) + \tfrac{1}{2} \frac{h}{a} f r_1 \phi_1 + \tfrac{1}{4} f^2 \phi_1^2 (3 - g) \right\} \dots\dots\dots (23).$$

From which we find $\left(\dfrac{dV}{dr} \right)_0 = -\dfrac{R}{a^2} f,$

$$\left. \begin{aligned} \left(\frac{d^2V}{dr^2} \right)_0 &= L = \frac{R}{2a^3} (1 + g) \\[4pt] \left(\frac{d^2V}{dr\,d\phi} \right)_0 &= M = \frac{R}{2a^2} fh \\[4pt] \left(\frac{d^2V}{d\phi^2} \right)_0 &= N = \frac{R}{2a} f^2 (3 - g) \end{aligned} \right\} \dots\dots\dots\dots\dots\dots (24).$$

These 'results may be confirmed by the following considerations applicable to any circular ring, and not involving any expansion or integration. Let af be the distance of the centre of gravity from the centre of the ring, and let the ring revolve about its centre with velocity ω. Then the force necessary to keep the ring in that orbit will be $-Raf\omega^2$.

But let S be a mass fixed at the centre of the ring, then if

$$\omega^2 = \frac{S}{a^3},$$

every portion of the ring will be separately retained in its orbit by the attraction of S, so that the whole ring will be retained in its orbit. The resultant attraction must therefore pass through the centre of gravity, and be

$$-Raf\omega^2 = -RS \frac{f}{a^2};$$

therefore

$$\frac{dV}{dr} = -R \frac{f}{a^2}.$$

The equation
$$\frac{d^2V}{dx^2} + \frac{d^2V}{dy^2} + \frac{d^2V}{dz^2} + 4\pi\rho = 0$$

is true for any system of matter attracting according to the law of gravitation. If we bear in mind that the expression is identical in form with that which measures the total efflux of fluid from a differential element of volume, where $\frac{dV}{dx}$, $\frac{dV}{dy}$, $\frac{dV}{dz}$ are the rates at which the fluid passes through its sides, we may easily form the equation for any other case. Now let the position of a point in space be determined by the co-ordinates r, ϕ and z, where z is measured perpendicularly to the plane of the angle ϕ. Then by choosing the directions of the axes x, y, z, so as to coincide with those of the radius vector r, the perpendicular to it in the plane of ϕ, and the normal, we shall have

$$dx = dr, \qquad dy = rd\phi, \qquad dz = dz,$$

$$\frac{dV}{dx} = \frac{dV}{dr}, \quad \frac{dV}{dy} = \frac{1}{r}\frac{dV}{d\phi}, \quad \frac{dV}{dz} = \frac{dV}{dz}.$$

The quantities of fluid passing through an element of area in each direction are

$$\frac{dV}{dr}rd\phi dz, \quad \frac{dV}{d\phi}\frac{1}{r}drdz, \quad \frac{dV}{dz}rd\phi dr,$$

so that the expression for the whole efflux is

$$\frac{1}{r}\frac{dV}{dr} + \frac{d^2V}{dr^2} + \frac{1}{r^2}\frac{d^2V}{d\phi^2} + \frac{d^2V}{dz^2} \dots\dots\dots\dots\dots (25),$$

which is necessarily equivalent to the former expression.

Now at the centre of the ring $\frac{d^2V}{dz^2}$ may be found by considering the attraction on a point just above the centre at a distance z,

$$\frac{dV}{dz} = -R\frac{z}{(a^2+z^2)^{\frac{3}{2}}},$$

$$\frac{d^2V}{dz^2} = -\frac{R}{a^3}, \text{ when } z = 0.$$

Also we know $\qquad \dfrac{1}{r}\dfrac{dV}{dr} = -\dfrac{R}{a^3}$, and $r = af$,

so that in any circular ring $\qquad \dfrac{d^2V}{dr^2} + \dfrac{1}{a^2f^2}\dfrac{d^2V}{d\phi^2} = 2\dfrac{R}{a^3} \dots\dots\dots\dots\dots (26),$

an equation satisfied by the former values of L and N.

By referring to the original expression for the variable section of the ring, it appears that the effect of the coefficient f is to make the ring thicker on one side and thinner on the other in a uniformly graduated manner. The effect of g is to thicken the ring at two opposite sides, and diminish its section in the parts between. The coefficient h indicates an inequality of the same kind, only not symmetrically disposed about the diameter through the centre of gravity.

Other terms indicating inequalities recurring three or more times in the circumference of the ring, have no effect on the values of L, M and N. There is one remarkable case, however, in which the irregularity consists of a single heavy particle placed at a point on the circumference of the ring.

Let P be the mass of the particle, and Q that of the uniform ring on which it is fixed, then $R = P + Q$,

$$f = \frac{P}{R},$$

$$L = 2\,\frac{P}{a^3} + \frac{Q}{2a^3} = \frac{P+Q}{2a^3}\left(1 + 3\,\frac{P}{R}\right) = \frac{R}{2a^3}(1+g)\,;$$

$$\therefore\ g = \frac{3P}{R} = 3f \dots\dots\dots\dots\dots\dots\dots\dots\dots\dots\dots(27).$$

PROB. VI. To determine the conditions of stability of the motion in terms of the coefficients f, g, h, which indicate the distribution of mass in the ring.

The quantities which enter into the differential equation of motion (18) are R, S, k^2, r_0, ω^2, L, M, N. We must observe that S is very large compared with R, and therefore we neglect R in those terms in which it is added to S, and we put

$$S = a^3\omega^2,$$

$$k^2 = a^2\left(1 - f^2\right),$$

$$r_0 = af,$$

$$L = \frac{R}{2a^3}\left(1 + g\right),$$

$$M = \frac{R}{2a^2}fh,$$

$$N = \frac{R}{2a}f^2\left(3 - g\right).$$

Substituting these values in equation (18) and dividing by $R^2a^4f^2$, we obtain

$$(1-f^2)\,n^4+(1-\tfrac{5}{2}f^2+\tfrac{1}{2}f^2g)\,n^2\omega^2+(\tfrac{9}{4}-6f^2-\tfrac{1}{4}g^2-\tfrac{1}{4}h^2+2f^2g)\,\omega^4=0\ldots\ldots(28).$$

The condition of stability is that this equation shall give both values of n^2 negative, and this renders it necessary that all the coefficients should have the same sign, and that the square of the second should exceed four times the product of the first and third.

(1) Now if we suppose the ring to be uniform, f, g and h disappear, and the equation becomes

$$n^4+n^2\omega^2+\tfrac{9}{4}=0\ldots\ldots\ldots\ldots\ldots\ldots\ldots\ldots (29),$$

which gives impossible values to n^2 and indicates the instability of a uniform ring.

(2) If we make g and $h=0$, we have the case of a ring thicker at one side than the other, and varying in section according to the simple law of sines. We must remember, however, that f must be less than $\tfrac{1}{2}$, in order that the section of the ring at the thinnest part may be real. The equation becomes

$$(1-f^2)\,n^4+(1-\tfrac{5}{2}f^2)\,n^2\omega^2+(\tfrac{9}{4}-6f^2)\,\omega^4=0\ldots\ldots\ldots\ldots(30).$$

The condition that the third term should be positive gives

$$f^2<\cdot375.$$

The condition that n^2 should be real gives

$$71f^4-112f^2+32\text{ negative,}$$

which requires f^2 to be between $\cdot37445$ and $1\cdot2$.

The condition of stability is therefore that f^2 should lie between

$$\cdot37445\text{ and }\cdot375,$$

but the construction of the ring on this principle requires that f^2 should be less than $\cdot25$, so that it is impossible to reconcile this form of the ring with the conditions of stability.

(3) Let us next take the case of a uniform ring, loaded with a heavy particle at a point of its circumference. We have then $g=3f$, $h=0$, and the equation becomes

$$(1-f^2)\,n^4+(1-\tfrac{5}{2}f^2+\tfrac{3}{2}f^3)\,n^2\omega^2+(\tfrac{9}{4}-\tfrac{33}{4}f^2+6f^3)\,\omega^4=0\ldots\ldots\ldots(31).$$

39—2

Dividing each term by $1-f$, we get

$$(1+f)\, n^4 + (1+f-\tfrac{3}{2}f^2)\, n^2\omega^2 + \tfrac{3}{4}\{3\,(1+f) - 8f^2\}\, \omega^4 = 0 \ldots\ldots\ldots\ldots(32).$$

The first condition gives f less than $\cdot8279$.

The second condition gives f greater than $\cdot815865$.

Let us assume as a particular case between these limits $f = \cdot82$, which makes the ratio of the mass of the particle to that of the ring as 82 to 18, then the equation becomes

$$1\cdot82\, n^4 + \cdot8114\, n^2\omega^2 + \cdot9696\omega^4 = 0 \ldots\ldots\ldots\ldots\ldots (33),$$

which gives $\qquad\qquad \sqrt{-1}\,n = \pm\,\cdot5916\omega \ \text{ or } \pm\,\cdot3076\omega.$

These values of n indicate variations of r_1, θ_1, and ϕ_1, which are compounded of two simple periodic inequalities, the period of the one being $1\cdot69$ revolutions, and that of the other $3\cdot251$ revolutions of the ring. The relations between the phases and amplitudes of these inequalities must be deduced from equations (14), (15), (16), in order that the character of the motion may be completely determined.

Equations (14), (15), (16) may be written as follows:

$$(4n\omega + h\omega^2)\, \frac{r_1}{a} + 2fn^2\theta_1 + f(3-g)\, \omega^2\phi_1 = 0 \ldots\ldots\ldots\ldots (34),$$

$$\{n^2 - \tfrac{1}{2}\omega^2\,(3+g)\}\, \frac{r_1}{a} - 2f\omega n\theta_1 - \tfrac{1}{2}fh\omega^2\phi_1 = 0 \ldots\ldots\ldots\ldots (35),$$

$$-fh\omega^2\, \frac{r_1}{a} + 2\,(1-f^2)\, n^2\theta_1 + \{2\,(1-f^2)\, n^2 - f^2\,(3-g)\, \omega^2\}\, \phi_1 = 0 \ldots\ldots (36).$$

By eliminating one of the variables between any two of these equations, we may determine the relation between the two remaining variables. Assuming one of these to be a periodic function of t of the form $A\cos\nu t$, and remembering that n stands for the operation $\dfrac{d}{dt}$, we may find the form of the other.

Thus, eliminating θ_1 between the first and second equations,

$$\{n^3 + \tfrac{1}{2}n\omega^2\,(5-g) + h\omega^3\}\frac{r_1}{a} + f\omega^2\{(3-g)\,\omega - \tfrac{1}{2}hn\}\,\phi_1 = 0 \ldots\ldots\ldots\ldots (37).$$

Assuming $\dfrac{r_1}{a} = A \sin \nu t$, and $\phi_1 = Q \cos(\nu t - \beta)$,

$$\{-\nu^3 + \tfrac{1}{2}\nu\omega^2(5-g)\} A \cos \nu t + h\omega^3 A \sin \nu t + f\omega^3(3-g)Q\cos(\nu t - \beta) + \tfrac{1}{2}fh\omega^2\nu Q \sin(\nu t - \beta).$$

Equating νt to 0, and to $\dfrac{\pi}{2}$, we get the equations

$$\{\nu^3 - \tfrac{1}{2}\nu\omega^2(5-g)\} A = f\omega^2 Q \{(3-g)\omega \cos \beta - \tfrac{1}{2}h\nu \sin \beta\},$$

$$-h\omega^2 A = f\omega^2 Q \{(3-g)\omega \sin \beta + \tfrac{1}{2}h\nu \cos \beta\},$$

from which to determine Q and β.

In all cases in which the mass is disposed symmetrically about the diameter through the centre of gravity, $h = 0$ and the equations may be greatly simplified.

Let $\theta_1 = P \cos(\nu t - a)$, then the second equation becomes

$$\{\nu^2 + \tfrac{1}{2}\omega^2(3+g)\} A \sin \nu t = 2Pf\omega\nu \sin(\nu t - a),$$

whence $\qquad\qquad a = 0, \quad P = \dfrac{\nu^2 + \tfrac{1}{2}\omega^2(3+g)}{2f\omega\nu} A$ .... (38).

The first equation becomes

$$4A\omega\nu \cos \nu t - 2Pf\nu^2 \cos \nu t + Qf(3-g)\omega^2 \cos(\nu t - \beta) = 0,$$

whence $\qquad\qquad \beta = 0, \quad Q = \dfrac{\nu^3 - \tfrac{1}{2}\omega^2\nu(5-g)}{f(3-g)\omega^3} A$ .... (39).

In the numerical example in which a heavy particle was fixed to the circumference of the ring, we have, when $f = \cdot82$,

$$\frac{\nu}{\omega} = \begin{cases} \cdot5916 \\ \cdot3076 \end{cases}, \quad \frac{P}{A} = \begin{cases} 3\cdot21 \\ 5\cdot72 \end{cases}, \quad \frac{Q}{A} = \begin{cases} -1\cdot229 \\ -\ \cdot797 \end{cases},$$

so that if we put $\omega t = \theta_0 =$ the mean anomaly,

$$\frac{r_1}{a} = A \sin(\cdot5916\,\theta_0 - a) + B \sin(\cdot3076\,\theta_0 - \beta)$$ .... (40),

$$\theta_1 = 3\cdot21\,A \cos(\cdot5916\,\theta_0 - a) + 5\cdot72\,B \cos(\cdot3076\,\theta_0 - \beta)$$ ....(41),

$$\phi_1 = -1\cdot229\,A \cos(\cdot5916\,\theta_0 - a) - 5\cdot797\,B \cos(\cdot3076\,\theta_0 - \beta)$$ ... (42).

These three equations serve to determine r_1, θ_1 and ϕ_1 when the original motion is given. They contain four arbitrary constants A, B, a, β. Now since the original values r_1, θ_1, ϕ_1, and also their first differential coefficients with respect to t, are arbitrary, it would appear that six arbitrary constants ought to enter into the equation. The reason why they do not is that we assume r_0 and θ_0 as the *mean values* of r and θ in the *actual motion*. These quantities therefore depend on the original circumstances, and the two additional arbitrary constants enter into the values of r_0 and θ_0. In the analytical treatment of the problem the differential equation in n was originally of the sixth degree with a solution $n^2 = 0$, which implies the possibility of terms in the solution of the form $Ct + D$.

The existence of such terms depends on the previous equations, and we find that a term of this form may enter into the value of θ, and that r_1 may contain a constant term, but that in both cases these additions will be absorbed into the values of θ_0 and r_0.

PART II.

ON THE MOTION OF A RING, THE PARTS OF WHICH ARE NOT RIGIDLY CONNECTED.

1. In the case of the Ring of invariable form, we took advantage of the principle that the mutual actions of the parts of any system form at all times a system of forces in equilibrium, and we took no account of the attraction between one part of the ring and any other part, since no motion could result from this kind of action. But when we regard the different parts of the ring as capable of independent motion, we must take account of the attraction on each portion of the ring as affected by the irregularities of the other parts, and therefore we must begin by investigating the statical part of the problem in order to determine the forces that act on any portion of the ring, as depending on the instantaneous condition of the rest of the ring.

In order to bring the problem within the reach of our mathematical methods, we limit it to the case in which the ring is *nearly* circular and uniform, and has a transverse section very small compared with the radius of the ring. By analysing the difficulties of the theory of a linear ring, we shall be better able to appreciate those which occur in the theory of the actual rings.

The ring which we consider is therefore small in section, and very nearly circular and uniform, and revolving with nearly uniform velocity. The variations from circular form, uniform section, and uniform velocity must be expressed by a proper notation.

2. To express the position of an element of a variable ring at a given time in terms of the original position of the element in the ring.

Let S (fig. 3) be the central body, and SA a direction fixed in space.

Let SB be a radius, revolving with the mean angular velocity ω of the ring, so that $ASB = \omega t$.

Let π be an element of the ring in its actual position, and let P be the position it would have had if it had moved uniformly with the mean velocity ω and had not been displaced, then BSP is a constant angle $= s$, and the value of s enables us to identify any element of the ring.

The element may be removed from its mean position P in three different ways.

(1) By change of distance from S by a quantity $p\pi = \rho$.

(2) By change of angular position through a space $Pp = \sigma$.

(3) By displacement perpendicular to the plane of the paper by a quantity ζ.

ρ, σ and ζ are all functions of s and t. If we could calculate the attractions on any element as depending on the form of these functions, we might determine the motion of the ring for any given original disturbance. We cannot, however, make any calculations of this kind without knowing the form of the functions, and therefore we must adopt the following method of separating the original disturbance into others of simpler form, first given in Fourier's *Traité de Chaleur*.

3. Let U be a function of s, it is required to express U in a series of sines and cosines of multiples of s between the values $s = 0$ and $s = 2\pi$.

Assume $U = A_1 \cos s + A_2 \cos 2s + \&c. + A_m \cos ms + A_n \cos ns$

$+ B_1 \sin s + B_2 \cos 2s + \&c. + B_m \sin ms + B_n \sin ns.$

Multiply by $\cos ms\, ds$ and integrate, then all terms of the form

$$\int \cos ms \cos ns\, ds \text{ and } \int \cos ms \sin ns\, ds$$

will vanish, if we integrate from $s = 0$ to $s = 2\pi$, and there remains

$$\int_0^{2\pi} U \cos ms\, ds = \pi A_m, \qquad \int_0^{2\pi} U \sin ms\, ds = \pi B_m.$$

If we can determine the values of these integrals in the given case, we can find the proper coefficients A_m, B_m, &c., and the series will then represent the values of U from $s = 0$ to $s = 2\pi$, whether those values be continuous or discontinuous, and when none of those values are infinite the series will be convergent.

In this way we may separate the most complex disturbances of a ring into parts whose form is that of a circular function of s or its multiples. Each of these partial disturbances may be investigated separately, and its effect on the attractions of the ring ascertained either accurately or approximately.

4. To find the magnitude and direction of the attraction between two elements of a disturbed ring.

Let P and Q (fig. 4) be the two elements, and let their original positions be denoted by s_1 and s_2, the values of the arcs BP, BQ before displacement. The displacement consists in the angle BSP being increased by σ_1 and BSQ by σ_2, while the distance of P from the centre is increased by ρ_1 and that of Q by ρ_2. We have to determine the effect of these displacements on the distance PQ and the angle SPQ.

Let the radius of the ring be unity, and $s_2 - s_1 = 2\theta$, then the original value of PQ will be $2 \sin \theta$, and the increase due to displacement

$$= (\rho_2 + \rho_1) \sin \theta + (\sigma_2 - \sigma_1) \cos \theta.$$

We may write the complete value of PQ thus,

$$PQ = 2 \sin \theta \left\{ 1 + \tfrac{1}{2} (\rho_2 + \rho_1) + \tfrac{1}{2} (\sigma_2 - \sigma_1) \cot \theta \right\} \ldots\ldots\ldots\ldots (1).$$

The original value of the angle SPQ was $\dfrac{\pi}{2} - \theta$, and the increase due to displacement is

$$\tfrac{1}{2} (\rho_2 - \rho_1) \cot \theta - \tfrac{1}{2} (\sigma_2 - \sigma_1),$$

so that we may write the values of $\sin SPQ$ and $\cos SPQ$,

$$\sin SPQ = \cos \theta \left\{ 1 + \tfrac{1}{2}(\rho_2 - \rho_1) - \tfrac{1}{2}(\sigma_2 - \sigma_1)\tan\theta \right\} \quad \dots\dots\dots\dots (2),$$

$$\cos SPQ = \sin \theta \left\{ 1 - \tfrac{1}{2}(\rho_2 - \rho_1)\cot^2\theta + \tfrac{1}{2}(\sigma_2 - \sigma_1)\cot\theta \right\} \quad \dots\dots (3).$$

If we assume the masses of P and Q each equal to $\dfrac{1}{\mu}R$, where R is the mass of the ring, and μ the number of satellites of which it is composed, the accelerating effect of the radial force on P is

$$\frac{1}{\mu}R\frac{\cos SPQ}{PQ^2} = \frac{1}{\mu}\frac{R}{4\sin\theta}\left\{ 1 - (\rho_2 + \rho_1) - \tfrac{1}{2}(\rho_2 - \rho_1)\cot^2\theta - \tfrac{1}{2}(\sigma_2 - \sigma_1)\cot\theta \right\}\dots(4),$$

and the tangential force

$$\frac{1}{\mu}R\frac{\sin SPQ}{PQ^2} = \frac{1}{\mu}\frac{R\cos\theta}{4\sin^2\theta}\left\{ 1 - \tfrac{1}{2}\rho_2 - \tfrac{3}{2}\rho_1 - (\sigma_2 - \sigma_1)(\cot\theta + \tfrac{1}{2}\tan\theta) \right\}\dots\dots(5).$$

The normal force is $\dfrac{1}{\mu}R\dfrac{\zeta_2 - \zeta_1}{8\sin^3\theta}.$

5. Let us substitute for ρ, σ and ζ their values expressed in a series of sines and cosines of multiples of s, the terms involving ms being

$$\rho_1 = A\cos(ms + \alpha), \qquad \rho_2 = A\cos(ms + \alpha + 2\theta),$$

$$\sigma_1 = B\sin(ms + \beta), \qquad \sigma_2 = B\sin(ms + \beta + 2\theta),$$

$$\zeta_1 = C\cos(ms + \gamma), \qquad \zeta_2 = C\cos(ms + \gamma + 2\theta).$$

The radial force now becomes

$$\frac{1}{\mu}\frac{R}{4\sin\theta}\left\{ \begin{array}{l} 1 - A\cos(ms + \alpha)(1 + \cos 2m\theta) + A\sin(ms + \alpha)\sin 2m\theta \\ + \tfrac{1}{2}A\cos(ms + \alpha)(1 - \cos 2m\theta)\cot^2\theta - \tfrac{1}{2}A\sin(ms + \alpha)\sin 2m\theta\cot^2\theta \\ + \tfrac{1}{2}B\sin(ms + \beta)(1 - \cos 2m\theta)\cot\theta - \tfrac{1}{2}B\cos(ms + \beta)\sin 2m\theta\cot\theta \end{array} \right\} (6).$$

The radial component of the attraction of a corresponding particle on the other side of P may be found by changing the sign of θ. Adding the two together, we have for the effect of the pair

$$\frac{1}{\mu}\frac{R}{2\sin\theta}\left\{ 1 - A\cos(ms + \alpha)(2\cos^2 m\theta - \sin^2 m\theta\cot^2\theta) \right.$$

$$\left. - B\cos(ms + \beta)\tfrac{1}{2}\sin 2m\theta\cot\theta \right\} \dots\dots\dots\dots (7).$$

Let us put

$$L = \Sigma \left(\tfrac{1}{2} \frac{\sin^2 m\theta \cos^2 \theta}{\sin^3 \theta} - \frac{\cos^2 m\theta}{\sin \theta} \right)$$

$$M = \Sigma \left(\frac{\sin 2m\theta \cos \theta}{4 \sin^2 \theta} \right)$$

$$N = \Sigma \left(\frac{\sin^2 m\theta \cos^2 \theta}{\sin^3 \theta} + \tfrac{1}{2} \frac{\sin^2 m\theta}{\sin \theta} \right) \Bigg\} \quad \ldots\ldots\ldots\ldots\ldots\ldots (8)^*,$$

$$J = \Sigma \left(\frac{\sin^2 m\theta}{2 \sin^3 \theta} \right)$$

$$K = \Sigma \left(\frac{1}{2 \sin \theta} \right)$$

where the summation extends to all the satellites on the same side of P, that is, every value of θ of the form $\dfrac{x}{\mu} \pi$, where x is a whole number less than $\dfrac{\mu}{2}$.

The radial force may now be written

$$P = \frac{1}{\mu} R \{ K + LA \cos (ms + \alpha) - MB \cos (ms + \beta) \} \ldots\ldots\ldots\ldots (9).$$

* The following values of several quantities which enter into these investigations are calculated for a ring of 36 satellites.

$$K = 24\cdot5.$$

	$\Sigma \dfrac{\sin^2 m\theta \cos^2 \theta}{\sin^3 \theta}$	$\Sigma \dfrac{\cos^2 m\theta}{\sin \theta}$	L	M	N
$m = 0$	0	43	-43	0	0
$m = 1$	32	32	-16	16	37
$m = 2$	107	28	26	25	115
$m = 3$	212	25	81	28	221
$m = 4$	401	24	177	32	411
$m = 9$	975	20	468	30	986
$m = 18$	1569	18	767	0	1582

When μ is very great,

$$\left.\frac{\pi}{\mu}\right|^3 L = \cdot5259 \text{ when } m = \frac{\mu}{2},$$

$$= \cdot4342 \quad ,, \quad m = \frac{\mu}{3},$$

$$= \cdot3287 \quad ,, \quad m = \frac{\mu}{4}.$$

The tangential force may be calculated in the same way, it is

$$T = \frac{1}{\mu} R \{ MA \sin (ms + a) + NB \sin (ms + \beta) \} \ldots \ldots \ldots \ldots (10).$$

The normal force is

$$Z = -\frac{1}{\mu} RJC \cos (ms + \gamma) \ldots \ldots \ldots \ldots \ldots (11).$$

6. We have found the expressions for the forces which act upon each member of a system of equal satellites which originally formed a uniform ring, but are now affected with displacements depending on circular functions. If these displacements can be propagated round the ring in the form of waves with the velocity $\frac{m}{n}$, the quantities a, β, and γ will depend on t, and the complete expressions will be

$$\left. \begin{aligned} \rho &= A \cos (ms + nt + a) \\ \sigma &= B \sin (ms + nt + \beta) \\ \zeta &= C \cos (ms + n't + \gamma) \end{aligned} \right\} \ldots \ldots \ldots \ldots (12).$$

Let us find in what cases expressions such as these will be true, and what will be the result when they are not true.

Let the position of a satellite at any time be determined by the values of r, ϕ, and ζ, where r is the radius vector reduced to the plane of reference, ϕ the angle of position measured on that plane, and ζ the distance from it. The equations of motion will be

$$\left. \begin{aligned} r \left(\frac{d\phi}{dt} \right)^2 - \frac{d^2r}{dt^2} &= S \frac{1}{r^2} + P \\ 2 \frac{dr}{dt} \frac{d\phi}{dt} + r \frac{d^2\phi}{dt^2} &= T \\ \frac{d^2\zeta}{dt^2} &= -S \frac{\zeta}{r^3} + Z \end{aligned} \right\} \ldots \ldots \ldots \ldots (13).$$

If we substitute the value of ζ in the third equation and remember that r is nearly $= 1$, we find

$$n'^2 = S + \frac{1}{\mu} RJ \ldots \ldots \ldots \ldots (14).$$

As this expression is necessarily positive, the value of n' is always real, and the disturbances normal to the plane of the ring can always be propa-

40—2

gated as waves, and therefore can never be the cause of instability. We therefore confine our attention to the motion in the plane of the ring as deduced from the two former equations.

Putting $r = 1 + \rho$ and $\phi = \omega t + s + \sigma$, and omitting powers and products of ρ, σ and their differential coefficients,

$$\left. \begin{aligned} \omega^2 + \omega^2 \rho + 2\omega \frac{d\sigma}{dt} - \frac{d^2\rho}{dt^2} &= S - 2S\rho + P \\ 2\omega \frac{d\rho}{dt} + \frac{d^2\sigma}{dt^2} &= T \end{aligned} \right\} \quad \dots\dots\dots\dots (15).$$

Substituting the values of ρ and σ as given above, these equations become

$$\omega^2 - S - \frac{1}{\mu} RK + \left(\omega^2 + 2S - \frac{1}{\mu} RL + n^2 \right) A \cos(ms + nt + a)$$

$$+ \left(2\omega n + \frac{1}{\mu} RM \right) B \cos(ms + nt + \beta) = 0 \dots\dots\dots\dots (16),$$

$$\left(2\omega n + \frac{1}{\mu} RM \right) A \sin(ms + nt + a) + \left(n^2 + \frac{1}{\mu} RN \right) B \sin(ms + nt + \beta) = 0 \dots.(17).$$

Putting for $(ms + nt)$ any two different values, we find from the second equation (17)

$$a = \beta \dots\dots\dots\dots\dots\dots\dots\dots\dots\dots(18),$$

and
$$\left(2\omega n + \frac{1}{\mu} RM \right) A + \left(n^2 + \frac{1}{\mu} RN \right) B = 0 \dots\dots\dots\dots (19),$$

and from the first (16) $\left(\omega^2 + 2S - \frac{1}{\mu} RL + n^2 \right) A + \left(2\omega n + \frac{1}{\mu} RM \right) B = 0 \dots\dots (20),$

and
$$\omega^2 - S - \frac{1}{\mu} RK = 0 \dots\dots\dots\dots\dots\dots(21).$$

Eliminating A and B from these equations, we get

$$n^4 - \left\{ 3\omega^2 - 2S + \frac{1}{\mu} R(L - N) \right\} n^2$$

$$- 4\omega \frac{1}{\mu} RMn + \left(\omega^2 + 2S - \frac{1}{\mu} RL \right) \frac{1}{\mu} RN - \frac{1}{\mu^2} R^2 M^2 = 0 \dots\dots\dots\dots (22),$$

a biquadratic equation to determine n.

For every *real* value of n there are terms in the expressions for ρ and σ of the form

$$A \cos(ms + nt + a).$$

For every *pure impossible* root of the form $\pm\sqrt{-1}n'$ there are terms of the forms

$$A\epsilon^{\pm n't}\cos(ms+a).$$

Although the negative exponential coefficient indicates a continually diminishing displacement which is consistent with stability, the positive value which necessarily accompanies it indicates a continually increasing disturbance, which would completely derange the system in course of time.

For every mixed root of the form $\pm\sqrt{-1}n'+n$, there are terms of the form

$$A\epsilon^{\pm n't}\cos(ms+nt+a).$$

If we take the positive exponential, we have a series of m waves travelling with velocity $\dfrac{n}{m}$ and increasing in amplitude with the coefficient $\epsilon^{+n't}$. The negative exponential gives us a series of m waves gradually dying away, but the negative exponential cannot exist without the possibility of the positive one having a finite coefficient, so that it is necessary for the stability of the motion that the four values of n be all real, and none of them either impossible quantities or the sums of possible and impossible quantities.

We have therefore to determine the relations among the quantities K, L, M, N, R, S, that the equation

$$n^4-\left\{S+\frac{1}{\mu}R\left(3K+L-N\right)\right\}n^2$$

$$-4\omega\frac{1}{\mu}RMn+\left\{3S+\frac{1}{\mu}R\left(K-L\right)\right\}\frac{1}{\mu}RN-\frac{1}{\mu^2}R^2M^2=U=0$$

may have four real roots.

7. In the first place, U is positive, when n is a large enough quantity, whether positive or negative.

It is also positive when $n=0$, provided S be large, as it must be, compared with $\dfrac{1}{\mu}RL$, $\dfrac{1}{\mu}RM$ and $\dfrac{1}{\mu}RN$.

If we can now find a positive and a negative value of n for which U is negative, there must be four real values of n for which $U=0$, and the four roots will be real.

Now if we put $n = \pm\sqrt{\tfrac{1}{2}}\,\sqrt{S}$,

$$U = -\tfrac{1}{4}S^2 + \tfrac{1}{2}\,\frac{1}{\mu}\,R\,(7N \pm \mathfrak{t}\,\sqrt{2}M - L - 3K)\,S + \frac{1}{\mu^2}\,R^2\,(KN - LN - M^2),$$

which is negative if S be large compared to R.

So that a ring of satellites can always be rendered stable by increasing the mass of the central body and the angular velocity of the ring.

The values of L, M, and N depend on m, the number of undulations in the ring. When $m = \dfrac{\mu}{2}$, the values of L and N will be at their maximum and $M = 0$. If we determine the relation between S and R in this case so that the system may be stable, the stability of the system for every other displacement will be secured.

8. To find the mass which must be given to the central body in order that a ring of satellites may permanently revolve round it.

We have seen that when the attraction of the central body is sufficiently great compared with the forces arising from the mutual action of the satellites, a permanent ring is possible. Now the forces between the satellites depend on the manner in which the displacement of each satellite takes place. The conception of a perfectly arbitrary displacement of all the satellites may be rendered manageable by separating it into a number of partial displacements depending on periodic functions. The motions arising from these small displacements will take place independently, so that we have to consider only one at a time.

Of all these displacements, that which produces the greatest disturbing forces is that in which consecutive satellites are oppositely displaced, that is, when $m = \dfrac{\mu}{2}$, for then the nearest satellites are displaced so as to increase as much as possible the effects of the displacement of the satellite between them. If we make μ a large quantity, we shall have

$$\Sigma\,\frac{\sin^2 m\theta \cos^2\theta}{\sin^3\theta} = \frac{\mu^3}{\pi^3}\,(1 + 3^{-3} + 5^{-3} + \&\text{c.}) = \frac{\mu^3}{\pi^3}\,(1{\cdot}0518),$$

$$L = \frac{\mu^3}{\pi^3}\,{\cdot}5259, \qquad M = 0, \qquad N = 2L, \qquad K \text{ very small.}$$

Let $\frac{1}{\mu} RL = x$, then the equation of motion will be

$$n^4 - (S-x)\, n^2 + 2x\, (3S-x) = U = 0 \dots\dots\dots\dots\dots (23).$$

The conditions of this equation having real roots are

$$S > x \dots\dots\dots\dots\dots\dots\dots\dots\dots (24),$$

$$(S-x)^2 > 8x\, (3S-x) \dots\dots\dots\dots\dots (25).$$

The last condition gives the equation

$$S^2 - 26Sx + 9x^2 > 0,$$

whence $\qquad\qquad S > 26{\cdot}642x, \quad \text{or} \quad S < 0{\cdot}351x \dots\dots\dots\dots (26).$

The last solution is inadmissible because S must be greater than x, so that the true condition is $\qquad S > 25{\cdot}649x,$

$$> 25{\cdot}649\, \frac{1}{\mu}\, R\, \frac{\mu^3}{\pi^3}\, {\cdot}5259,$$

$$S > {\cdot}4352\mu^2 R \dots\dots\dots\dots\dots\dots (27).$$

So that if there were 100 satellites in the ring, then

$$S > 4352\, R$$

is the condition which must be fulfilled in order that the motion arising from every conceivable displacement may be periodic.

If this condition be not fulfilled, and if S be not sufficient to render the motion perfectly stable, then although the motion depending upon long undulations may remain stable, the short undulations will increase in amplitude till some of the neighbouring satellites are brought into collision.

9. To determine the nature of the motion when the system of satellites is of small mass compared with the central body.

The equation for the determination of n is

$$U = n^4 - \{\omega^2 + \frac{1}{\mu}\, R\, (2K + L - N)\}\, n^2 - 4\omega\, \frac{1}{\mu}\, RMn$$

$$+ \{3\omega^2 - \frac{1}{\mu}\, R\, (2K+L)\}\frac{1}{\mu}\, RN - \frac{1}{\mu^2}\, R^2 M^2 = 0 \dots\dots (28).$$

When R is very small we may approximate to the values of n by assuming that two of them are nearly $\pm\omega$, and that the other two are small.

If we put $n = \pm\omega$,

$$U = -\frac{1}{\mu} R\left(2K + L \pm 4M - 4N\right)\omega^2 + \&c.,$$

$$\frac{dU}{dn} = \pm 2\omega^3 + \&c.$$

Therefore the corrected values of n are

$$n = \pm\left\{\omega + \frac{1}{2\mu\omega} R\left(2K + L - 4N\right)\right\} + \frac{2}{\mu\omega} RM\ldots\ldots\ldots\ldots (29).$$

The small values of n are nearly $\pm\sqrt{3\frac{1}{\mu} RN}$: correcting them in the same way, we find the approximate values

$$n = \pm\sqrt{3\frac{1}{\mu}RN} - 2\frac{1}{\mu\omega} RM \ldots\ldots\ldots\ldots\ldots (30).$$

The four values of n are therefore

$$\left.\begin{array}{l} n_1 = -\omega - \dfrac{1}{2\mu\omega} R(2K + L - 4M - 4N) \\[2mm] n_2 = -\sqrt{3\dfrac{1}{\mu} RN} - \dfrac{2}{\mu\omega} RM \\[2mm] n_3 = +\sqrt{3\dfrac{1}{\mu} RN} - \dfrac{2}{\mu\omega} RM \\[2mm] n_4 = +\omega + \dfrac{1}{2\mu\omega} R(2K + L + 4M - 4N) \end{array}\right\} \ldots\ldots\ldots\ldots (31),$$

and the complete expression for ρ, so far as it depends on terms containing ms, is therefore

$$\rho = A_1 \cos\left(ms + n_1 t + a_1\right) + A_2 \cos\left(ms + n_2 t + a_2\right)$$
$$+ A_3 \cos\left(ms + n_3 t + a_3\right) + A_4 \cos\left(ms + n_4 t + a_4\right)\ldots\ldots\ldots\ldots(32),$$

and there will be other systems, of four terms each, for every value of m in the expansion of the original disturbance.

We are now able to determine the value of σ from equations (12), (20), by putting $\beta = a$, and

$$B = -\frac{2\omega n + \dfrac{1}{\mu} RM}{n^2 + \dfrac{1}{\mu} RN} A \ldots\ldots\ldots\ldots\ldots (33).$$

So that for every term of ρ of the form

$$\rho = A \cos\left(ms + nt + a\right) \dots\dots\dots\dots\dots\dots (34),$$

there is a corresponding term in σ,

$$\sigma = -\frac{2\omega n + \dfrac{1}{\mu} RM}{n^2 + \dfrac{1}{\mu} RN} A \sin\left(ms + nt + a\right) \dots\dots\dots\dots (35).$$

10. Let us now fix our attention on the motion of a single satellite, and determine its motion by tracing the changes of ρ and σ while t varies and s is constant, and equal to the value of s corresponding to the satellite in question.

We must recollect that ρ and σ are measured outwards and forwards from an imaginary point revolving at distance 1 and velocity ω, so that the motions we consider are not the absolute motions of the satellite, but its motions relative to a point fixed in a revolving plane. This being understood, we may describe the motion as elliptic, the major axis being in the tangential direction, and the ratio of the axes being nearly $2\dfrac{\omega}{n}$, which is nearly 2 for n_1 and n_4 and is very large for n_2 and n_3.

The time of revolution is $\dfrac{2\pi}{n}$, or if we take a revolution of the ring as the unit of time, the time of a revolution of the satellite about its mean position is $\dfrac{\omega}{n}$.

The *direction* of revolution of the satellite about its mean position is in every case opposite to that of the motion of the ring.

11. The absolute motion of a satellite may be found from its motion relative to the ring by writing

$$r = 1 + \rho = 1 + A \cos\left(ms + nt + a\right),$$

$$\theta = \omega t + s + \sigma = \omega t + s - 2\frac{\omega}{n} A \sin\left(ms + nt + a\right).$$

When n is nearly equal to $\pm \omega$, the motion of each satellite in space is nearly elliptic. The eccentricity is A, the longitude at epoch s, and the longitude when at the greatest distance from Saturn is for the negative value n_1

$$-\frac{1}{\mu\omega} R\left(2K+L-4M-4N\right) t + (m+1) s + a,$$

and for the positive value n_4

$$-\frac{1}{\mu\omega} R\left(2K+L+4M-4N\right) t - (m+1) s - a.$$

We must recollect that in all cases the quantity within brackets is negative, so that the major axis of the ellipse travels forwards in both cases. The chief difference between the two cases lies in the arrangement of the major axes of the ellipses of the different satellites. In the first case as we pass from one satellite to the next in front the axes of the two ellipses lie in the same order. In the second case the particle in front has its major axis behind that of the other. In the cases in which n is small the radius vector of each satellite increases and diminishes during a periodic time of several revolutions. This gives rise to an inequality, in which the tangential displacement far exceeds the radial, as in the case of the *annual equation* of the Moon.

12. Let us next examine the condition of the ring of satellites at a given instant. We must therefore fix on a particular value of t and trace the changes of ρ and σ for different values of s.

From the expression for ρ we learn that the satellites form a wavy line, which is furthest from the centre when $(ms + nt + a)$ is a multiple of 2π, and nearest to the centre for intermediate values.

From the expression for σ we learn that the satellites are sometimes in advance and sometimes in the rear of their mean position, so that there are places where the satellites are crowded together, and others where they are drawn asunder. When n is positive, B is of the opposite sign to A, and the crowding of the satellites takes place when they are furthest from the centre. When n is negative, the satellites are separated most when furthest from the centre, and crowded together when they approach it.

The form of the ring at any instant is therefore that of a string of beads forming a re-entering curve, nearly circular, but with a small variation of distance from the centre recurring m times, and forming m regular waves of transverse displacement at equal intervals round the circle. Besides these, there are waves of condensation and rarefaction, the effect of longitudinal displacement. When n is positive the points of greatest distance from the centre are points of greatest condensation, and when n is negative they are points of greatest rarefaction.

13. We have next to determine the velocity with which these waves of disturbance are propagated round the ring. We fixed our attention on a particular satellite by making s constant, and on a particular instant by making t constant, and thus we determined the motion of a satellite and the form of the ring. We must now fix our attention on a *phase* of the motion, and this we do by making ρ or σ constant. This implies

$$ms + nt + a = \text{constant},$$

$$\frac{ds}{dt} = -\frac{n}{m}.$$

So that the particular phase of the disturbance travels round the ring with an angular velocity $= -\dfrac{n}{m}$ relative to the ring itself. Now the ring is revolving in space with the velocity ω, so that the angular velocity of the wave in space is

$$\varpi = \omega - \frac{n}{m} \dots\dots\dots\dots\dots\dots\dots\dots\dots\dots\dots\dots(36).$$

Thus each satellite moves in an ellipse, while the general aspect of the ring is that of a curve of m waves revolving with velocity ϖ. This, however, is only the part of the whole motion, which depends on a single term of the solution. In order to understand the general solution we must shew how to determine the whole motion from the state of the ring at a given instant.

14. *Given the position and motion of every satellite at any one time, to calculate the position and motion of every satellite at any other time, provided that the condition of stability is fulfilled.*

The position of any satellite may be denoted by the values of ρ and σ for that satellite, and its velocity and direction of motion are then indicated by the values of $\dfrac{d\rho}{dt}$ and $\dfrac{d\sigma}{dt}$ at the given instant.

These four quantities may have for each satellite any four arbitrary values, as the position and motion of each satellite are independent of the rest, at the beginning of the motion.

Each of these quantities is therefore a perfectly arbitrary function of s, the mean angular position of the satellite in the ring.

But any function of s from $s = 0$ to $s = 2\pi$, however arbitrary or discontinuous, can be expanded in a series of terms of the form $A \cos(s + a) + A' \cos(2s + a') + \&c.$ See § 3.

Let each of the four quantities ρ, $\dfrac{d\rho}{dt}$, σ, $\dfrac{d\sigma}{dt}$ be expressed in terms of such a series, and let the terms in each involving ms be

$$\rho = E \cos(ms + e) \dots\dots\dots\dots\dots\dots\dots (37),$$

$$\frac{d\rho}{dt} = F \cos(ms + f) \dots\dots\dots\dots\dots\dots\dots (38),$$

$$\sigma = G \cos(ms + g) \dots\dots\dots\dots\dots\dots\dots (39),$$

$$\frac{d\sigma}{dt} = H \cos(ms + h) \dots\dots\dots\dots\dots\dots\dots (40).$$

These are the parts of the values of each of the four quantities which are capable of being expressed in the form of periodic functions of ms. It is evident that the eight quantities E, F, G, H, e, f, g, h, are all independent and arbitrary.

The next operation is to find the values of L, M, N, belonging to disturbances in the ring whose index is m [see equation (8)], to introduce these values into equation (28), and to determine the four values of n, (n_1, n_2, n_3, n_4).

This being done, the expression for ρ is that given in equation (32), which contains eight arbitrary quantities $(A_1, A_2, A_3, A_4, a_1, a_2, a_3, a_4)$.

Giving t its original value in this expression, and equating it to $E \cos(ms + e)$, we get an equation which is equivalent to two. For, putting $ms = 0$, we have

$$A_1 \cos a_1 + A_2 \cos a_2 + A_3 \cos a_3 + A_4 \cos a_4 = E \cos e \dots\dots\dots\dots (41).$$

And putting $ms = \dfrac{\pi}{2}$, we have another equation

$$A_1 \sin a_1 + A_2 \sin a_2 + A_3 \sin a_3 + A_4 \sin a_4 = E \sin e \dots\dots\dots\dots (42).$$

Differentiating (32) with respect to t, we get two other equations

$$-A_1 n_1 \sin \alpha - \&\text{c.} = F \cos f \dots\dots\dots\dots\dots\dots (43),$$

$$A_1 n_1 \cos \alpha + \&\text{c.} = F \sin f \dots\dots\dots\dots\dots (44).$$

Bearing in mind that B_1, B_2, &c. are connected with A_1, A_2, &c. by equation (33), and that B is therefore proportional to A, we may write $B = A\beta$, where

$$\beta = -\frac{2\omega n + \dfrac{1}{\mu} RM}{n^2 + \dfrac{1}{\mu} RN},$$

β being thus a function of n and a known quantity.

The value of σ then becomes at the epoch

$$\sigma = A_1 \beta_1 \sin(ms + \alpha_1) + \&\text{c.} = G \cos(ms + g),$$

from which we obtain the two equations

$$A_1 \beta_1 \sin \alpha_1 + \&\text{c.} = G \cos g \dots\dots\dots\dots\dots\dots(45),$$

$$A_1 \beta_1 \cos \alpha_1 + \&\text{c.} = -G \sin g \dots\dots\dots\dots\dots(46).$$

Differentiating with respect to t, we get the remaining equations

$$A_1 \beta_1 n_1 \cos \alpha_1 + \&\text{c.} = H \cos h \dots\dots\dots\dots\dots (47),$$

$$A_1 \beta_1 n_1 \sin \alpha_1 + \&\text{c.} = H \sin h \dots\dots\dots\dots\dots (48).$$

We have thus found eight equations to determine the eight quantities A_1, &c. and α_1, &c. To solve them, we may take the four in which $A_1 \cos \alpha_1$, &c. occur, and treat them as simple equations, so as to find $A_1 \cos \alpha_1$, &c. Then taking those in which $A_1 \sin \alpha_1$, &c. occur, and determining the values of those quantities, we can easily deduce the value of A_1 and α_1, &c. from these.

We now know the amplitude and phase of each of the four waves whose index is m. All other systems of waves belonging to any other index must be treated in the same way, and since the original disturbance, however arbitrary, can be broken up into periodic functions of the form of equations (37—40), our solution is perfectly general, and applicable to every possible disturbance of a ring fulfilling the condition of stability (27).

15. We come next to consider the effect of an external disturbing force, due either to the irregularities of the planet, the attraction of satellites, or the motion of waves in other rings.

All disturbing forces of this kind may be expressed in series of which the general term is

$$A \cos (vt + ms + a),$$

where v is an angular velocity and m a whole number.

Let $P \cos (ms + vt + p)$ be the central part of the force, acting inwards, and $Q \sin (ms + vt + q)$ the tangential part, acting forwards. Let $\rho = A \cos (ms + vt + a)$ and $\sigma = B \sin (ms + vt + \beta)$, be the terms of ρ and σ which depend on the external disturbing force. These will simply be added to the terms depending on the original disturbance which we have already investigated, so that the complete expressions for ρ and σ will be as general as before. In consequence of the additional forces and displacements, we must add to equations (16) and (17), respectively, the following terms:

$$\left\{3\omega^2 - \frac{1}{\mu} R \left(2K + L\right) + v^2\right\} A \cos (ms + vt + a)$$

$$+ \left(2\omega v + \frac{1}{\mu} RM\right) B \cos (ms + vt + \beta) - P \cos (ms + vt + p) = 0 \ldots\ldots(49).$$

$$\left(2\omega v + \frac{1}{\mu} RM\right) A \sin (ms + vt + a)$$

$$+ \left(v^2 + \frac{1}{\mu} RN\right) B \sin (ms + vt + \beta) + Q \sin (ms + vt + q) = 0 \ldots\ldots\ldots(50).$$

Making $ms + vt = 0$ in the first equation and $\frac{\pi}{2}$ in the second,

$$\left\{3\omega^2 - \frac{1}{\mu} R \left(2K + L\right) + v^2\right\} A \cos a + \left(2\omega v + \frac{1}{\mu} RM\right) B \cos \beta - P \cos p = 0 \ldots\ldots(51).$$

$$\left(2\omega v + \frac{1}{\mu} RM\right) A \cos a + \left(v^2 + \frac{1}{\mu} RN\right) B \cos \beta + Q \cos q = 0 \ldots\ldots(52).$$

Then if we put

$$U' = v^4 - \left\{\omega^2 + \frac{1}{\mu} R \left(2K + L - N\right)\right\} v^2 - 4 \frac{\omega}{\mu} RMv$$

$$+ \left\{3\omega^2 - \frac{1}{\mu} R \left(2K + L\right)\right\} \frac{1}{\mu} RN - \frac{1}{\mu^2} R^2 M^2 \ldots\ldots\ldots(53),$$

we shall find the value of $A \cos \alpha$ and $B \cos \beta$;

$$A \cos \alpha = \frac{v^2 + \frac{1}{\mu} RN}{U'} P \cos p + \frac{2\omega v + \frac{1}{\mu} RM}{U'} Q \cos q \ldots \ldots \ldots (54).$$

$$B \cos \beta = -\frac{2\omega v + \frac{1}{\mu} RM}{U'} P \cos p - \frac{v^2 + 3\omega^2 - \frac{1}{\mu} R(K+L)}{U'} Q \cos q \ldots \ldots (55).$$

Substituting sines for cosines in equations (51), (52), we may find the values of $A \sin \alpha$ and $B \sin \beta$.

Now U' is precisely the same function of v that U is of n, so that if v coincides with one of the four values of n, U' will vanish, the coefficients A and B will become infinite, and the ring will be destroyed. The disturbing force is supposed to arise from a revolving body, or an undulation of any kind which has an angular velocity $-\dfrac{v}{m}$ relatively to the ring, and therefore an absolute angular velocity $= \omega - \dfrac{v}{m}$.

If then the absolute angular velocity of the disturbing body is exactly or nearly equal to the absolute angular velocity of any of the free waves of the ring, that wave will increase till the ring be destroyed.

The velocities of the free waves are nearly

$$\omega \left(1 + \frac{1}{m}\right), \ \omega + \frac{1}{m}\sqrt{3\frac{1}{\mu}RN}, \ \omega - \frac{1}{m}\sqrt{3\frac{1}{\mu}RN}, \ \text{and} \ \omega\left(1 - \frac{1}{m}\right) \ldots \ldots (56).$$

When the angular velocity of the disturbing body is greater than that of the first wave, between those of the second and third, or less than that of the fourth, U' is positive. When it is between the first and second, or between the third and fourth, U' is negative.

Let us now simplify our conception of the disturbance by attending to the central force only, and let us put $p = 0$, so that P is a maximum when $ms + vt$ is a multiple of 2π. We find in this case $\alpha = 0$, and $\beta = 0$. Also

$$A = \frac{v^2 + \frac{1}{\mu} RN}{U'} P \ldots \ldots \ldots \ldots \ldots \ldots \ldots (57),$$

$$B = -\frac{2\omega v + \frac{1}{\mu} RM}{U'} P \ldots \ldots \ldots \ldots \ldots \ldots \ldots (58).$$

When U' is positive, A will be of the same sign as P, that is, the parts of the ring will be furthest from the centre where the disturbing force towards the centre is greatest.   When U' is negative, the contrary will be the case.

When v is positive, B will be of the opposite sign to A, and the parts of the ring furthest from the centre will be most crowded.   When v is negative, the contrary will be the case.

Let us now attend only to the tangential force, and let us put $q = 0$.   We find in this case also $\alpha = 0$, $\beta = 0$,

$$A = \frac{2\omega v + \dfrac{1}{\mu} RM}{U'}\, Q \dots\dots\dots\dots\dots\dots\dots\dots (59),$$

$$B = -\frac{v^2 + 3\omega^2 - \dfrac{1}{\mu} R(K+L)}{U'}\, Q \dots\dots\dots\dots (60).$$

The tangential displacement is here in the same or in the opposite direction to the tangential force, according as U' is negative or positive.   The crowding of satellites is at the points farthest from or nearest to Saturn according as v is positive or negative.

16.   The effect of any disturbing force is to be determined in the following manner.   The disturbing force, whether radial or tangential, acting on the ring may be conceived to vary from one satellite to another, and to be different at different times.   It is therefore a perfectly arbitrary function of s and t.

Let Fourier's method be applied to the general disturbing force so as to divide it up into terms depending on periodic functions of s, so that each term is of the form $F(t)\cos(ms + \alpha)$, where the function of t is still perfectly arbitrary.

But it appears from the general theory of the permanent motions of the heavenly bodies that they may all be expressed by periodic functions of t arranged in series.   Let vt be the argument of one of these terms, then the corresponding term of the disturbance will be of the form

$$P \cos(ms + vt + \alpha).$$

This term of the disturbing force indicates an alternately positive and negative action, disposed in m waves round the ring, completing its period relatively to each particle in the time $\dfrac{2\pi}{v}$, and travelling as a wave among the particles with an angular velocity $-\dfrac{v}{m}$, the angular velocity relative to fixed space being of course $\omega - \dfrac{v}{m}$. The whole disturbing force may be split up into terms of this kind.

17. Each of these elementary disturbances will produce its own wave in the ring, independent of those which belong to the ring itself. This new wave, due to external disturbance, and following different laws from the natural waves of the ring, is called the *forced wave*. The angular velocity of the forced wave is the same as that of the disturbing force, and its maxima and minima coincide with those of the force, but the extent of the disturbance and its direction depend on the comparative velocities of the forced wave and the four natural waves.

When the velocity of the forced wave lies between the velocities of the two middle free waves, or is greater than that of the swiftest, or less than that of the slowest, then the radial displacement due to a radial disturbing force is in the same direction as the force, but the tangential displacement due to a tangential disturbing force is in the opposite direction to the force.

The radial force therefore in this case produces a *positive forced wave*, and the tangential force a *negative forced wave*.

When the velocity of the forced wave is either between the velocities of the first and second free waves, or between those of the third and fourth, then the radial disturbance produces a forced wave in the contrary direction to that in which it acts, or a negative wave, and the tangential force produces a positive wave.

The coefficient of the forced wave changes sign whenever its velocity passes through the value of any of the velocities of the free waves, but it does so by becoming infinite, and not by vanishing, so that when the angular velocity very nearly coincides with that of a free wave, the forced wave becomes very great, and if the velocity of the disturbing force were made exactly equal to that of a free wave, the coefficient of the forced wave would become infinite. In such a case we should have to readjust our approximations, and to find whether such a coincidence might involve a physical impossibility.

The forced wave which we have just investigated is that which would maintain itself in the ring, supposing that it had been set agoing at the commencement of the motion. It is in fact the form of dynamical equilibrium of the ring under the influence of the given forces. In order to find the actual motion of the ring we must combine this forced wave with all the free waves, which go on independently of it, and in this way the solution of the problem becomes perfectly complete, and we can determine the whole motion under any given initial circumstances, as we did in the case where no disturbing force acted.

For instance, if the ring were perfectly uniform and circular at the instant when the disturbing force began to act, we should have to combine with the constant forced wave a system of four free waves so disposed, that at the given epoch, the displacements due to them should exactly neutralize those due to the forced wave. By the combined effect of these four free waves and the forced one the whole motion of the ring would be accounted for, beginning from its undisturbed state.

The disturbances which are of most importance in the theory of Saturn's rings are those which are produced in one ring by the action of attractive forces arising from waves belonging to another ring.

The effect of this kind of action is to produce in each ring, besides its own four free waves, four forced waves corresponding to the free waves of the other ring. There will thus be eight waves in each ring, and the corresponding waves in the two rings will act and react on each other, so that, strictly speaking, every one of the waves will be in some measure a forced wave, although the system of eight waves will be the free motion of the two rings taken together. The theory of the mutual disturbance and combined motion of two concentric rings of satellites requires special consideration.

18. On the motion of a ring of satellites when the conditions of stability are not fulfilled.

We have hitherto been occupied with the case of a ring of satellites, the stability of which was ensured by the smallness of mass of the satellites compared with that of the central body. We have seen that the statically unstable condition of each satellite between its two immediate neighbours may be compensated by the dynamical effect of its revolution round the planet, and a planet of sufficient mass can not only direct the motion of such satellites round its own body, but can likewise exercise an influence over their relations to each other, so as to overrule their natural tendency to crowd together, and distribute and preserve them in the form of a ring.

We have traced the motion of each satellite, the general shape of the disturbed ring, and the motion of the various waves of disturbance round the ring, and determined the laws both of the natural or free waves of the ring, and of the forced waves, due to extraneous disturbing forces.

We have now to consider the cases in which such a permanent motion of the ring is impossible, and to determine the mode in which a ring, originally regular, will break up, in the different cases of instability.

The equation from which we deduce the conditions of stability is—

$$U = n^4 - \left\{ \omega^2 + \frac{1}{\mu} R \left(2K + L - N \right) \right\} n^2 - 4\omega \frac{1}{\mu} RMn$$

$$+ \left\{ 3\omega^2 - \frac{1}{\mu} R \left(2K + L \right) \right\} \frac{1}{\mu} RN - \frac{1}{\mu^2} R^2 M^2 = 0.$$

The quantity, which, in the critical cases, determines the nature of the roots of this equation, is N. The quantity M in the third term is always small compared with L and N when m is large, that is, in the case of the dangerous short waves. We may therefore begin our study of the critical cases by leaving out the third term. The equation then becomes a quadratic in n^2, and in order that all the values of n may be real, both values of n^2 must be real and positive.

The condition of the values of n^2 being real is

$$\omega^4 + \omega^2 \frac{1}{\mu} R \left(4K + 2L - 14N \right) + \frac{1}{\mu^2} R^2 \left(2K + L + N \right)^2 > 0 \dots\dots\dots(61),$$

which shews that ω^2 must either be about 14 times at least smaller, or about 14 times at least greater, than quantities like $\frac{1}{\mu} RN$.

That both values of n^2 may be positive, we must have

$$\left. \begin{array}{l} \omega^2 + \frac{1}{\mu} R \left(2K + L - N \right) > 0 \\[2mm] \left\{ 3\omega^2 - \frac{1}{\mu} R \left(2K + L \right) \right\} \frac{1}{\mu} RN > 0 \end{array} \right\} \dots\dots\dots\dots\dots\dots (62).$$

42—2

We must therefore take the larger value of ω^2, and also add the condition that N be positive.

We may therefore state roughly, that, to ensure stability, $\dfrac{RN}{\mu}$, the coefficient of tangential attraction, must lie between zero and $\frac{1}{14}\omega^2$. If the quantity be negative, the two *small* values of n will become *pure impossible* quantities. If it exceed $\frac{1}{14}\omega^2$, *all* the values of n will take the form of *mixed impossible* quantities.

If we write x for $\dfrac{1}{\mu}RN$, and omit the other disturbing forces, the equation becomes
$$U = n^4 - (\omega^2 - x)\,n^2 + 3\omega^2 x = 0 \quad\dots\dots\dots\dots\dots (63),$$
whence
$$n^2 = \tfrac{1}{2}(\omega^2 - x) \pm \tfrac{1}{2}\sqrt{\omega^4 - 14\omega^2 x + x^2} \quad\dots\dots\dots\dots (64).$$

If x be small, two of the values of n are nearly $\pm\omega$, and the others are small quantities, real when x is positive and impossible when x is negative.

If x be greater than $(7 - \sqrt{48})\,\omega^2$, or $\dfrac{\omega^2}{14}$ nearly, the term under the radical becomes negative, and the value of n becomes
$$n = \pm\tfrac{1}{2}\sqrt{\sqrt{12\omega^2 x} + \omega^2 - x} \pm \tfrac{1}{2}\sqrt{-1}\sqrt{\sqrt{12\omega^2 x} - \omega^2 + x} \quad\dots\dots\dots (65),$$
where one of the terms is a real quantity, and the other impossible. Every solution may be put under the form
$$n = p \pm \sqrt{-1}\,q \quad\dots\dots\dots\dots\dots\dots\dots\dots (66),$$
where $q = 0$ for the case of stability, $p = 0$ for the pure impossible roots, and p and q finite for the mixed roots.

Let us now adopt this general solution of the equation for n, and determine its mechanical significance by substituting for the impossible circular functions their equivalent real exponential functions.

Substituting the general value of n in equations (34), (35),
$$\rho = A\left[\cos\{ms + (p + \sqrt{-1}q)\,t + a\} + \cos\{ms + (p - \sqrt{-1}q)\,t + a\}\right] \dots (67),$$
$$\left.\begin{aligned}
\sigma = &-A\,\frac{2\omega(p + \sqrt{-1}q)}{(p + \sqrt{-1}q)^2 + x}\sin\{ms + (p + \sqrt{-1}q)\,t + a\}\\
&-A\,\frac{2\omega(p - \sqrt{-1}q)}{(p - \sqrt{-1}q)^2 + x}\sin\{ms + (p - \sqrt{-1}q)\,t + a\}
\end{aligned}\right\} \dots\dots (68).$$

Introducing the exponential notation, these values become

$$\rho = A \left(\epsilon^{qt} + \epsilon^{-qt}\right) \cos\left(ms + pt + a\right) \quad\dots\dots\dots\dots\dots (69),$$

$$\sigma = -\frac{2\omega A}{\left(p^2 + q^2\right)^2 + 2\left(p^2 - q^2\right)x + x^2} \left\{ \begin{array}{l} p\left(p^2 + q^2 + x\right)\left(\epsilon^{qt} + \epsilon^{-qt}\right)\sin\left(ms + pt + a\right) \\ + q\left(p^2 + q^2 - x\right)\left(\epsilon^{qt} - \epsilon^{-qt}\right)\cos\left(ms + pt + a\right) \end{array} \right\} \dots (70).$$

We have now obtained a solution free from impossible quantities, and applicable to every case.

When $q = 0$, the case becomes that of real roots, which we have already discussed. When $p = 0$, we have the case of pure impossible roots arising from the negative values of n^2. The solutions corresponding to these roots are

$$\rho = A \left(\epsilon^{qt} + \epsilon^{-qt}\right) \cos\left(ms + a\right) \quad\dots\dots\dots\dots\dots (71).$$

$$\sigma = -\frac{2\omega q A}{q^2 - x} \left(\epsilon^{qt} - \epsilon^{-qt}\right) \cos\left(ms + a\right) \quad\dots\dots\dots\dots (72).$$

The part of the coefficient depending on ϵ^{-qt} diminishes indefinitely as the time increases, and produces no marked effect. The other part, depending on ϵ^{qt}, increases in a geometrical proportion as the time increases arithmetically, and so breaks up the ring. In the case of x being a small negative quantity, q^2 is nearly $3x$, so that the coefficient of σ becomes

$$-3\frac{\omega}{q}A.$$

It appears therefore that the motion of each particle is either outwards and backwards or inwards and forwards, but that the tangential part of the motion greatly exceeds the normal part.

It may seem paradoxical that a tangential force, acting *towards* a position of equilibrium, should produce instability, while a small tangential force *from* that position ensures stability, but it is easy to trace the destructive tendency of this apparently conservative force.

Suppose a particle slightly in front of a crowded part of the ring, then if x is negative there will be a tangential force pushing it forwards, and this force will cause its distance from the planet to increase, its angular velocity to diminish, and the particle itself to fall back on the crowded part, thereby increasing the irregularity of the ring, till the whole ring is broken up. In the same way it may be shewn that a particle *behind* a crowded part will be pushed into it. The only force which could preserve the ring from the effect of this action, is one which would prevent the particle from receding from the planet under the influence of the tangential force, or at least prevent the diminution of angular velocity. The transversal force of attraction of the ring is of this kind, and acts in the right direction, but it can never be of sufficient magnitude to have the required effect. In fact the thing to be done is to render the last term of the equation in n^2 positive when N is negative, which requires

$$\frac{1}{\mu} R \left(2K + L\right) > 3\omega^2,$$

and this condition is quite inconsistent with any constitution of the ring which fulfils the other condition of stability which we shall arrive at presently.

We may observe that the waves belonging to the two real values of n, $\pm \omega$, must be conceived to be travelling round the ring during the whole time of its breaking up, and conducting themselves like ordinary waves, till the excessive irregularities of the ring become inconsistent with their uniform propagation.

The irregularities which depend on the exponential solutions do not travel round the ring by propagation among the satellites, but remain among the same satellites which first began to move irregularly.

We have seen the fate of the ring when x is negative. When x is small we have two small and two large values of n, which indicate regular waves, as we have already shewn. As x increases, the small values of n increase, and the large values diminish, till they meet and form a pair of positive and a pair of negative equal roots, having values nearly $\pm \cdot 68\omega$. When x becomes greater than about $\frac{1}{14}\omega^2$, then all the values of n become impossible, of the form $p + \sqrt{-1}q$, q being small when x first begins to exceed its limits, and p being nearly $\pm \cdot 68\omega$.

The values of ρ and σ indicate periodic inequalities having the period $\dfrac{2\pi}{p}$, but increasing in amplitude at a rate depending on the exponential ϵ^{qt}. At the beginning of the motion the oscillations of the particles are in ellipses as in the case of stability, having the ratio of the axes about 1 in the normal direction to 3 in the tangential direction. As the motion continues, these ellipses increase in magnitude, and another motion depending on the second term of σ is combined with the former, so as to increase the ellipticity of the oscillations and to turn the major axis into an inclined position, so that its fore end points a little inwards, and its hinder end a little outwards. The oscillations of each particle round its mean position are therefore in ellipses, of which both axes increase continually while the eccentricity increases, and the major axis becomes slightly inclined to the tangent, and this goes on till the ring is destroyed. In the mean time the irregularities of the ring do not remain among the same set of particles as in the former case, but travel round the ring with a relative angular velocity $-\frac{p}{m}$. Of these waves there are four, two travelling forwards among the satellites, and two travelling backwards. One of each of these pairs depends on a negative value of q, and consists of a wave whose amplitude continually decreases. The other depends on a positive value of q, and is the destructive wave whose character we have just described.

19. We have taken the case of a ring composed of equal satellites, as that with which we may compare other cases in which the ring is constructed of loose materials differently arranged.

In the first place let us consider what will be the conditions of a ring composed of satellites of unequal mass. We shall find that the motion is of the same kind as when the satellites are equal.

For by arranging the satellites so that the smaller satellites are closer together than the larger ones, we may form a ring which will revolve uniformly about Saturn, the resultant force on each satellite being just sufficient to keep it in its orbit.

To determine the stability of this kind of motion, we must calculate the disturbing forces due to any given displacement of the ring. This calculation will be more complicated than in the former case, but will lead to results of the same general character. Placing these forces in the equations of motion, we shall find a solution of the same general character as in the former case, only instead of regular waves of displacement travelling round the ring, each wave will be split and reflected when it comes to irregularities in the chain of satellites. But if the condition of stability for every kind of wave be fulfilled, the motion of each satellite will consist of small oscillations about its position of dynamical equilibrium, and thus, on the whole, the ring will of itself assume the arrangement necessary for the continuance of its motion, if it be originally in a state not very different from that of equilibrium.

20. We now pass to the case of a ring of an entirely different construction. It is possible to conceive of a quantity of matter, either solid or liquid, not collected into a continuous mass, but scattered thinly over a great extent of space, and having its motion regulated by the gravitation of its parts to each other, or towards some dominant body. A shower of rain, hail, or cinders is a familiar illustration of a number of unconnected particles in motion; the visible stars, the milky way, and the resolved nebulæ, give us instances of a similar scattering of bodies on a larger scale. In the terrestrial instances we see the motion plainly, but it is governed by the attraction of the earth, and retarded by the resistance of the air, so that the mutual attraction of the parts is completely masked. In the celestial cases the distances are so enormous, and the time during which they have been observed so short, that we can perceive no motion at all. Still we are perfectly able to conceive of a collection of particles of small size compared with the distances between them, acting upon one another only by the attraction of gravitation, and revolving round a central body. The average density of such a system may be smaller than that of the rarest gas, while the particles themselves may be of great density; and the appearance from a distance will be that of a cloud of vapour, with this difference, that as the space between the particles is empty, the rays of light will pass through the system without being refracted, as they would have been if the system had been gaseous.

Such a system will have an *average density* which may be greater in some places than others. The resultant attraction will be towards places of greater average density, and thus the density of those places will be increased so as to increase the irregularities of density. The system will therefore be statically unstable, and nothing but motion of some kind can prevent the particles from forming agglomerations, and these uniting, till all are reduced to one solid mass.

We have already seen how dynamical stability can exist where there is statical instability in the case of a row of particles revolving round a central body. Let us now conceive a cloud of particles forming a ring of nearly uniform density revolving about a central body. There will be a primary effect of inequalities in density tending to draw particles towards the denser parts of the ring, and this will elicit a secondary effect, due to the motion of revolution, tending in the contrary direction, so as to restore the rings to uniformity. The relative magnitude of these two opposing forces determines the destruction or preservation of the ring.

To calculate these effects we must begin with the statical problem :—To determine the forces arising from the given displacements of the ring.

The longitudinal force arising from longitudinal displacements is that which has most effect in determining the stability of the ring. In order to estimate its limiting value we shall solve a problem of a simpler form.

21. An infinite mass, originally of uniform density k, has its particles displaced by a quantity ξ parallel to the axis of x, so that $\xi = A \cos mx$, to determine the attraction on each particle due to this displacement.

The density at any point will differ from the original density by a quantity k', so that

$$(k + k')\,(dx + d\xi) = k\,dx \dotfill (73),$$

$$k' = -k\,\frac{d\xi}{dx} = Akm \sin mx \dotfill (74).$$

The potential at any point will be $V + V'$, where V is the original potential, and V' depends on the displacement only, so that

$$\frac{d^2V'}{dx^2} + \frac{d^2V'}{dy^2} + \frac{d^2V'}{dz^2} + 4\pi k' = 0 \dotfill (75).$$

Now V' is a function of x only, and therefore,

$$V' = 4\pi A k\,\frac{1}{m}\sin mx \dotfill (76),$$

and the longitudinal force is found by differentiating V' with respect to x.

$$X = \frac{dV'}{dx} = 4\pi k A \cos mx = 4\pi k \xi \dotfill (77).$$

Now let us suppose this mass not of infinite extent, but of finite section parallel to the plane of yz. This change amounts to cutting off all portions of the mass beyond a certain boundary. Now the effect of the portion so cut off upon the longitudinal force depends on the value of m. When m is large, so that the wave-length is small, the effect of the external portion is insensible, so that the longitudinal force due to short waves is not diminished by cutting off a great portion of the mass.

22.   Applying this result to the case of a ring, and putting s for x, and σ for ξ we have

$$\sigma = A \cos ms, \text{ and } T = 4\pi kA \cos ms,$$

so that

$$\frac{1}{\mu} RN = 4\pi k,$$

when m is very large, and this is the greatest value of N.

The value of L has little effect on the condition of stability. If L and M are both neglected, that condition is

$$\omega^2 > 27 \cdot 856 \ (2\pi k) \dots\dots\dots\dots\dots\dots\dots\dots\dots (78),$$

and if L be as much as $\frac{1}{2}N$, then

$$\omega^2 > 25 \cdot 649 \ (2\pi k) \dots\dots\dots\dots\dots\dots\dots\dots(79),$$

so that it is not important whether we calculate the value of L or not.

The condition of stability is, that the average density must not exceed a certain value. Let us ascertain the relation between the maximum density of the ring and that of the planet.

Let b be the radius of the planet, that of the ring being unity, then the mass of Saturn is $\frac{4}{3}\pi b^3 k' = \omega^2$ if k' be the density of the planet. If we assume that the radius of the ring is twice that of the planet, as Laplace has done, then $b = \frac{1}{2}$ and

$$\frac{k'}{k} = 334 \cdot 2 \text{ to } 307 \cdot 7 \dots\dots\dots\dots\dots\dots\dots\dots (80),$$

so that the density of the ring cannot exceed $\frac{1}{300}$ of that of the planet. Now Laplace has shewn that if the outer and inner parts of the ring have the same angular velocity, the ring will not hold together if the ratio of the density of the planet to that of the ring exceeds $1 \cdot 3$, so that in the first place, our ring cannot have uniform angular velocity, and in the second place, Laplace's ring cannot preserve its form, if it is composed of loose materials acting on each other only by the attraction of gravitation, and moving with the same angular velocity throughout.

23.   On the forces arising from inequalities of thickness in a thin stratum of fluid of indefinite extent.

The forces which act on any portion of a continuous fluid are of two kinds, the pressures of contiguous portions of fluid, and the attractions of all portions of the fluid whether near or distant. In the case of a thin stratum of fluid, not acted on by any external forces, the pressures are due mainly to the component of the attraction which is perpendicular to the plane of the stratum. It is easy to shew that a fluid acted on by such a force will tend to assume a position of equilibrium, in which its free surface is plane; and that any irregularities will tend to equalise themselves, so that the plane surface will be one of stable equilibrium.

It is also evident, that if we consider only that part of the attraction which is parallel to the plane of the stratum, we shall find it always directed towards the thicker parts, so that the effect of this force is to draw the fluid from thinner to thicker parts, and so to increase irregularities and destroy equilibrium.

The normal attraction therefore tends to preserve the stability of equilibrium, while the tangential attraction tends to render equilibrium unstable.

According to the nature of the irregularities one or other of these forces will prevail, so that if the extent of the irregularities is small, the normal forces will ensure stability, while, if the inequalities cover much space, the tangential forces will render equilibrium unstable, and break up the stratum into beads.

To fix our ideas, let us conceive the irregularities of the stratum split up into the form of a number of systems of waves superposed on one another, then, by what we have just said, it appears, that very short waves will disappear of themselves, and be consistent with stability, while very long waves will tend to increase in height, and will destroy the form of the stratum.

In order to determine the law according to which these opposite effects take place, we must subject the case to mathematical investigation.

Let us suppose the fluid incompressible, and of the density k, and let it be originally contained between two parallel planes, at distances $+c$ and $-c$ from that of (xy), and extending to infinity. Let us next conceive a series of imaginary planes, parallel to the plane of (yz), to be plunged into the fluid stratum at infinitesimal distances from one another, so as to divide the fluid into imaginary slices perpendicular to the plane of the stratum.

Next let these planes be displaced parallel to the axis of x according to this law—that if x be the original distance of the plane from the origin, and ξ its displacement in the direction of x,

$$\xi = A \cos mx \quad\ldots\ldots\ldots\ldots\ldots\ldots\ldots\ldots (81).$$

43—2

According to this law of displacement, certain alterations will take place in the distances between consecutive planes; but since the fluid is incompressible, and of indefinite extent in the direction of y, the change of dimension must occur in the direction of z. The original thickness of the stratum was $2c$. Let its thickness at any point after displacement be $2c + 2\zeta$, then we must have

$$(2c + 2\zeta)\left(1 + \frac{d\xi}{dx}\right) = 2c \dots\dots\dots\dots\dots\dots (82),$$

or

$$\zeta = -c\,\frac{d\xi}{dx} = cmA \sin mx \dots\dots\dots\dots\dots (83).$$

Let us assume that the increase of thickness 2ζ is due to an increase of ζ at each surface; this is necessary for the equilibrium of the fluid between the imaginary planes.

We have now produced artificially, by means of these planes, a system of waves of longitudinal displacement whose length is $\dfrac{2\pi}{m}$ and amplitude A; and we have found that this has produced a system of waves of normal displacement on each surface, having the same length, with a height $= cmA$.

In order to determine the forces arising from these displacements, we must, in the first place, determine the potential function at any point of space, and this depends partly on the state of the fluid before displacement, and partly on the displacement itself.   We have, in all cases—

$$\frac{d^2V}{dx^2} + \frac{d^2V}{dy^2} + \frac{d^2V}{dz^2} = -4\pi\rho \dots\dots\dots\dots\dots (84).$$

Within the fluid, $\rho = k$; beyond it, $\rho = 0$.

Before displacement, the equation is reduced to

$$\frac{d^2V}{dz^2} = -4\pi\rho \dots\dots\dots\dots\dots\dots\dots (85).$$

Instead of assuming $V = 0$ at infinity, we shall assume $V = 0$ at the origin, and since in this case all is symmetrical, we have

within the fluid
$$V_1 = -2\pi k z^2; \quad \frac{dV_1}{dz} = -4\pi k z$$

at the bounding planes
$$V = -2\pi k c^2; \quad \frac{dV}{dz} = \mp 4\pi k c \qquad \left.\right\} \dots\dots\dots\dots (86);$$

beyond them
$$V_2 = 2\pi k c\,(\mp 2z \pm c); \quad \frac{dV}{dz} = \mp 4\pi k c$$

the upper sign being understood to refer to the boundary at distance $+c$, and the lower to the boundary at distance $-c$ from the origin.

Having ascertained the potential of the undisturbed stratum, we find that of the disturbance by calculating the effect of a stratum of density k and thickness ζ, spread over each surface according to the law of thickness already found. By supposing the coefficient A small enough, (as we may do in calculating the displacements on which stability depends), we may diminish the absolute thickness indefinitely, and reduce the case to that of a mere "superficial density," such as is treated of in the theory of electricity. We have here, too, to regard some parts as of *negative* density; but we must recollect that we are dealing with the *difference* between a disturbed and an undisturbed system, which may be positive or negative, though no real mass can be negative.

Let us for an instant conceive only one of these surfaces to exist, and let us transfer the origin to it. Then the law of thickness is

$$\zeta = mcA \sin mx \dots\dots\dots (83),$$

and we know that the normal component of attraction at the surface is the same as if the thickness had been uniform throughout, so that

$$\frac{dV}{dz} = -2\pi k\zeta,$$

on the positive side of the surface.

Also, the solution of the equation

$$\frac{d^2V}{dx^2} + \frac{d^2V}{dz^2} = 0,$$

consists of a series of terms of the form $C\epsilon^{iz} \sin ix$.

Of these the only one with which we have to do is that in which $i = -m$. Applying the condition as to the normal force at the surface, we get

$$V = 2\pi kc\epsilon^{-mz}A \sin mx \dots\dots\dots (87),$$

for the potential on the positive side of the surface, and

$$V = 2\pi kc\epsilon^{mz}A \sin mx \dots\dots\dots(88),$$

on the negative side.

Calculating the potentials of a pair of such surfaces at distances $+c$ and $-c$ from the plane of xy, and calling V' the sum of their potentials, we have for the space between these planes

$$V_1' = 2\pi kcA \sin mx\epsilon^{-mc}\left(\epsilon^{mz}+\epsilon^{-mz}\right)$$

beyond them

$$V_2' = 2\pi kcA \sin mx\epsilon^{\mp mz}\left(\epsilon^{mc}+\epsilon^{-mc}\right) \right\}\quad\cdots\cdots\cdots\cdots (89);$$

the upper or lower sign of the index being taken according as z is positive or negative.

These potentials must be added to those formerly obtained, to get the potential at any point after displacement.

We have next to calculate the pressure of the fluid at any point, on the supposition that the imaginary planes protect each slice of the fluid from the pressure of the adjacent slices, so that it is in equilibrium under the action of the forces of attraction, and the pressure of these planes on each side. Now in a fluid of density k, in equilibrium under forces whose potential is V, we have always—

$$\frac{dp}{dV}=k;$$

so that if we know that the value of p is p_0 where that of V is V_0, then at any other point

$$p = p_0 + k(V - V_0).$$

Now, at the free surface of the fluid, $p=0$, and the distance from the free surface of the disturbed fluid to the plane of the original surface is ζ, a small quantity. The attraction which acts on this stratum of fluid is, in the first place, that of the undisturbed stratum, and this is equal to $4\pi kc$, towards that stratum. The pressure due to this cause at the level of the original surface will be $4\pi k^2c\zeta$, and the pressure arising from the attractive forces due to the displacements upon this thin layer of fluid, will be small quantities of the second order, which we neglect. We thus find the pressure when $z=c$ to be,

$$p_0 = 4\pi k^2c^2mA \sin mx.$$

The potential of the undisturbed mass when $z=c$ is

$$V_0 = -2\pi kc^2,$$

and the potential of the disturbance itself for the same value of z, is

$$V_0' = 2\pi kcA \sin mx\left(1+\epsilon^{-2mc}\right).$$

So that we find the general value of p at any other point to be

$$p = 2\pi k^2 \left(c^2 - z^2\right) + 2\pi k^2 cA \sin mx \left\{2cm - 1 - \epsilon^{-2mc} + \epsilon^{mc} \left(\epsilon^{mz} + \epsilon^{-mz}\right)\right\} \dots (90).$$

This expression gives the pressure of the fluid at any point, as depending on the state of constraint produced by the displacement of the imaginary planes. The accelerating effect of these pressures on any particle, if it were allowed to move parallel to x, instead of being confined by the planes, would be

$$= \frac{1}{k}\frac{dp}{dx}.$$

The accelerating effect of the attractions in the same direction is

$$\frac{dV}{dx},$$

so that the whole acceleration parallel to x is

$$X = -2\pi kmcA \cos mx \left(2mc - \epsilon^{-2mc} - 1\right) \dots (91).$$

It is to be observed, that this quantity is independent of z, so that every particle in the slice, by the combined effect of pressure and attraction, is urged with the same force, and, if the imaginary planes were removed, each slice would move parallel to itself without distortion, as long as the absolute displacements remained small. We have now to consider the direction of the resultant force X, and its changes of magnitude.

We must remember that the original displacement is $A \cos mx$, if therefore $(2mc - \epsilon^{-2mc} - 1)$ be positive, X will be opposed to the displacement, and the equilibrium will be stable, whereas if that quantity be negative, X will act along with the displacement and increase it, and so constitute an unstable condition.

It may be seen that large values of mc give positive results and small ones negative. The sign changes when

$$2mc = 1\cdot147 \dots (92),$$

which corresponds to a wave-length

$$\lambda = 2c\,\frac{2\pi}{1\cdot147} = 2c\left(5\cdot471\right) \dots (93).$$

The length of the complete wave in the critical case is $5\cdot471$ times the thickness of the stratum. Waves shorter than this are stable, longer waves are unstable.

The quantity $\qquad$ $2mc\left(2mc - \epsilon^{-2mc} - 1\right),$

has a minimum when $\qquad$ $2mc = \cdot607$ ....(94),

and the wave-length is 10·353 times the thickness of the stratum.

In this case $\qquad$ $2mc\left(2mc - \epsilon^{-2mc} - 1\right) = -\cdot509$ ....(95),

and $\qquad$ $X = \cdot509\pi kA \cos mx$ .... (96).

24. Let us now conceive that the stratum of fluid, instead of being infinite in extent, is limited in breadth to about 100 times the thickness. The pressures and attractions will not be much altered by this removal of a distant part of the stratum. Let us also suppose that this thin but broad strip is bent round in its own plane into a circular ring whose radius is more than ten times the breadth of the strip, and that the waves, instead of being exactly parallel to each other, have their ridges in the direction of radii of the ring. We shall then have transformed our stratum into one of Saturn's Rings, if we suppose those rings to be liquid, and that a considerable breadth of the ring has the same angular velocity.

Let us now investigate the conditions of stability by putting

$$x = -2\pi kmc\left(2mc - \epsilon^{-2mc} - 1\right)$$

into the equation for n. We know that x must lie between 0 and $\dfrac{\omega^2}{13\cdot9}$ to ensure stability. Now the greatest value of x in the fluid stratum is $\cdot509\pi k$. Taking Laplace's ratio of the diameter of the ring to that of the planet, this gives 42·5 as the minimum value of the density of the planet divided by that of the fluid of the ring.

Now Laplace has shewn that any value of this ratio greater than 1·3 is inconsistent with the rotation of any considerable breadth of the fluid at the same angular velocity, so that our hypothesis of a broad ring with uniform velocity is untenable.

But the stability of such a ring is impossible for another reason, namely, that for waves in which $2mc > 1\cdot147$, x is negative, and the ring will be destroyed by these short waves in the manner described at page (333).

When the fluid ring is treated, not as a broad strip, but as a filament of circular or elliptic section, the mathematical difficulties are very much increased, but it may be shown that in this case also there will be a maximum value of x, which will require the density of the planet to be several times that of the ring, and that in all cases short waves will give rise to negative values of x, inconsistent with the stability of the ring.

It appears, therefore, that a ring composed of a continuous liquid mass cannot revolve about a central body without being broken up, but that the parts of such a broken ring may, under certain conditions, form a permanent ring of satellites.

On the Mutual Perturbations of Two Rings.

25. We shall assume that the difference of the mean radii of the rings is small compared with the radii themselves, but large compared with the distance of consecutive satellites of the same ring. We shall also assume that each ring separately satisfies the conditions of stability.

We have seen that the effect of a disturbing force on a ring is to produce a series of waves whose number and period correspond with those of the disturbing force which produces them, so that we have only to calculate the coefficient belonging to the wave from that of the disturbing force.

Hence in investigating the simultaneous motions of two rings, we may assume that the mutually disturbing waves travel with the same *absolute* angular velocity, and that a maximum in one corresponds either to a maximum or a minimum of the other, according as the coefficients have the same or opposite signs.

Since the motions of the particles of each ring are affected by the disturbance of the other ring, as well as of that to which they belong, the equations of motion of the two rings will be involved in each other, and the final equation for determining the wave-velocity will have eight roots instead of four. But as each of the rings has four *free* waves, we may suppose these to originate *forced* waves in the other ring, so that we may consider the eight waves of each ring as consisting of four free waves and four forced ones.

In strictness, however, the wave-velocity of the "free" waves will be affected by the existence of the forced waves which they produce in the other ring, so that none of the waves are really "free" in either ring independently, though the whole motion of the system of two rings as a whole is free.

We shall find, however, that it is best to consider the waves first as free, and then to determine the reaction of the other ring upon them, which is such as to alter the wave-velocity of both, as we shall see.

The forces due to the second ring may be separated into three parts.

1st. The constant attraction when both rings are at rest.

2nd. The variation of the attraction on the first ring, due to its own disturbances.

3rd. The variation of the attraction due to the disturbances of the second ring.

The first of these affects only the angular velocity. The second affects the waves of each ring independently, and the mutual action of the waves depends entirely on the third class of forces.

26. *To determine the attractions between two rings.*

Let R and a be the mass and radius of the exterior ring, R' and a' those of the interior, and let all quantities belonging to the interior ring be marked with accented letters. (Fig. 5.)

1st. *Attraction between the rings when at rest.*

Since the rings are at a distance small compared with their radii, we may calculate the attraction on a particle of the first ring as if the second were an infinite straight line at distance $a' - a$ from the first.

The mass of unit of length of the second ring is $\dfrac{R'}{2\pi a'}$, and the accelerating effect of the attraction of such a filament on an element of the first ring is

$$\frac{R'}{\pi a' (a - a')} \text{ inwards} \dots\dots\dots\dots\dots\dots\dots\dots\dots(97).$$

The attraction of the first ring on the second may be found by transposing accented and unaccented letters.

In consequence of these forces, the outer ring will revolve faster, and the inner ring slower than would otherwise be the case. These forces enter into the *constant terms* of the equations of motion, and may be included in the value of K.

2nd. *Variation due to disturbance of first ring.*

If we put $a(1+\rho)$ for a in the last expression, we get the attraction when the first ring is displaced. The part depending on ρ is

$$-\frac{R'a}{\pi a'(a-a')^2}\rho \text{ inwards} \dots\dots\dots\dots\dots(98).$$

This is the only variation of force arising from the displacement of the first ring. It affects the value of L in the equations of motion.

3rd. *Variation due to waves in the second ring.*

On account of the waves, the second ring varies in distance from the first, and also in mass of unit of length, and each of these alterations produces variations both in the radial and tangential force, so that there are four things to be calculated:

1st. Radial force due to radial displacement.

2nd. Radial force due to tangential displacement.

3rd. Tangential force due to radial displacement.

4th. Tangential force due to tangential displacement.

1st. Put $a'(1+\rho')$ for a', and we get the term in ρ'

$$\frac{R'}{\pi a'}\frac{(2a'-a)}{(a'-a)^2}\rho' \text{ inwards} = \lambda'\rho', \text{ say} \dots\dots\dots (99).$$

2nd. By the tangential displacement of the second ring the section is reduced in the proportion of 1 to $1-\dfrac{d\sigma'}{ds'}$, and therefore there is an alteration of the radial force equal to

$$-\frac{R'}{\pi a'(a-a')}\frac{d\sigma'}{ds'} \text{ inwards} = -\mu'\frac{d\sigma'}{ds'} \text{ say} \dots\dots\dots (100).$$

3rd. By the radial displacement of the second ring the direction of the filament near the part in question is altered, so that the attraction is no longer radial but forwards, and the tangential part of the force is

$$\frac{R'}{\pi a'(a-a')}\frac{d\rho'}{ds'} = +\mu'\frac{d\rho'}{ds'} \text{ forwards} \dots\dots\dots (101).$$

44—2

4th. By the tangential displacement of the second ring a tangential force arises, depending on the relation between the length of the waves and the distance between the rings.

If we make $m\,\dfrac{a-a'}{a'}=p$, and $m\displaystyle\int_{-\infty}^{+\infty}\dfrac{x\sin px}{(1+x^2)^{\frac{3}{2}}}\,dx=\Pi$,

the tangential force is $\qquad\dfrac{R'}{\pi a'\,(a-a')^2}\,\Pi\sigma'=\nu'\sigma'$ .... (102).

We may now write down the values of λ, μ, and ν by transposing accented and unaccented letters.

$$\lambda=\frac{R}{\pi a}\,\frac{(2a-a')}{(a-a')^2};\ \mu=\frac{R}{\pi a\,(a'-a)};\ \nu=\frac{R}{\pi a'\,(a-a')^2}\,\Pi \ \dots\dots\ (103).$$

Comparing these values with those of λ', μ', and ν', it will be seen that the following relations are approximately true when a is nearly equal to a':

$$\frac{\lambda'}{\lambda}=-\frac{\mu'}{\mu}=\frac{\nu'}{\nu}=\frac{R'a}{Ra'}\ \dots\dots\dots\dots\ (104).$$

27. To form the equations of motion.

*The original equations were

$$\omega^2+\omega^2\rho+2\omega\,\frac{d\sigma}{dt}-\frac{d^2\rho}{dt^2}=P=S+K-(2S-L)\,A\rho-MB\rho+\lambda'\rho'-\mu'\,\frac{d\sigma'}{ds'},$$

$$2\omega\,\frac{d\rho}{dt}+\frac{d^2\sigma}{dt^2}=Q=MA\sigma+NB\sigma+\mu'\,\frac{d\rho'}{ds'}+\nu'\sigma'.$$

Putting $\qquad\rho=A\cos(ms+nt),\ \ \sigma=B\sin(ms+nt),$
$$\rho'=A'\cos(ms+nt),\ \sigma'=B'\sin(ms+nt),$$
then $\qquad\qquad\qquad\omega^2=S+K$
$$\left.\begin{array}{l}(\omega^2+2S+n^2-L)\,A+(2\omega n+M)\,B-\lambda'A'+\mu'mB'=0\\(2\omega n+M)\,A+(n^2+N)\,B-\mu'mA'+\nu'B'=0\end{array}\right\}\dots\dots(105).$$

The corresponding equations for the second ring may be found by transposing accented and unaccented letters. We should then have four equations to determine the ratios of A, B, A', B', and a resultant equation of the eighth degree to determine n. But we may make use of a more convenient method, since λ', μ', and ν' are small. Eliminating B we find

$$\left.\begin{array}{l}An^4-A\,(\omega^2+2K+L-N)\,n^2-4A\omega Mn+AN\,(3\omega^2)\\(-\lambda'A'+\mu'mB')\,n^2+(\mu'mA'-\nu'B')\,2\omega n\end{array}\right\}=0\ \dots\dots\dots\ (106).$$

* [The analysis in this article is somewhat unsatisfactory, the equations of motion employed being those which were applicable in the case of a ring of radius unity. ED.]

Putting
$$B = \beta A, \quad A' = xA, \quad B' = \beta' A' = \beta' xA,$$

we have
$$\left. \begin{aligned} n^4 - \{\omega^2(+2K) + L - N\} n^2 - 4\omega Mn + 3\omega^2 N \\ + (-\lambda' + \mu' m\beta') n^2 x + (\mu' m - \nu' \beta') 2\omega nx \end{aligned} \right\} = U = 0 \dots (107).$$

$$\frac{dU}{dn} = 4n^3 - 2\omega^2 n + \&c. \dots (108),$$

$$\frac{dU}{dx} = -\lambda' n^2 + \mu' m\beta' n^2 + 2\mu' m\omega n - 2\nu' \beta' \omega n \dots (109),$$

whence
$$\frac{dn}{dx} = \frac{\lambda' n - \mu' m\beta' n - 2\mu' m\omega + 2\nu' \beta' \omega}{4n^2 - 2\omega^2} \dots (110).$$

28. If we were to solve the equation for n, leaving out the terms involving x, we should find the wave-velocities of the four free waves of the first ring, supposing the second ring to be prevented from being disturbed. But in reality the waves in the first ring produce a disturbance in the second, and these in turn react upon the first ring, so that the wave-velocity is somewhat different from that which it would be in the supposed case. Now if x be the ratio of the radial amplitude of displacement in the second ring to that in the first, and if $\bar{n}$ be a value of n supposing $x = 0$, then by Maclaurin's theorem,

$$n = +\bar{n} + \frac{dn}{dx} x \dots (111).$$

The wave-velocity relative to the ring is $-\dfrac{n}{m}$, and the absolute angular velocity of the wave in space is

$$\varpi = \omega - \frac{n}{m} = \omega - \frac{\bar{n}}{m} - \frac{1}{m} \frac{dn}{dx} x \dots (112),$$

$$= +p - qx \dots (113),$$

where $p = \omega - \dfrac{\bar{n}}{m}$, and $q = \dfrac{1}{m} \dfrac{dn}{dx}$.

Similarly in the second ring we should have

$$\varpi' = p' - q' \frac{1}{x} \dots (114) ;$$

and since the corresponding waves in the two rings must have the same absolute angular velocity,

$$\varpi = \varpi', \text{ or } p - qx = p' - q' \frac{1}{x} \dots (115).$$

This is a quadratic equation in x, the roots of which are real when

$$(p-p')^2 + 4qq'$$

is positive. When this condition is not fulfilled, the roots are impossible, and the general solution of the equations of motion will contain exponential factors, indicating destructive oscillations in the rings.

Since q and q' are small quantities, the solution is always real whenever p and p' are considerably different. The absolute angular velocities of the two pairs of reacting waves, are then nearly

$$p + \frac{qq'}{p-p'}, \text{ and } p' - \frac{qq'}{p-p'},$$

instead of p and p', as they would have been if there had been no reaction of the forced wave upon the free wave which produces it.

When p and p' are equal or nearly equal, the character of the solution will depend on the sign of qq'. We must therefore determine the signs of q and q' in such cases.

Putting $\beta' = \frac{2\omega'}{n'}$, we may write the values of q and q'

$$\left. \begin{array}{l} q = \frac{n}{m} \cdot \dfrac{\lambda' + 2\mu'm \left(\dfrac{\omega'}{n'} - \dfrac{\omega}{n}\right) - 4\nu' \dfrac{\omega'}{n'} \dfrac{\omega}{n}}{4n^2 - 2\omega^2} \\[4ex] q' = \frac{n'}{m'} \cdot \dfrac{\lambda + 2\mu m \left(\dfrac{\omega}{n} - \dfrac{\omega'}{n'}\right) - 4\nu \dfrac{\omega'}{n'} \dfrac{\omega}{n}}{4n'^2 - 2\omega'^2} \end{array} \right\} \quad \dots\dots\dots\dots\dots (116).$$

Referring to the values of the disturbing forces, we find that

$$\frac{\lambda'}{\lambda} = -\frac{\mu'}{\mu} = \frac{\nu'}{\nu} = \frac{R'a}{Ra'}.$$

Hence

$$\frac{q}{q'} = \frac{n}{n'} \frac{4n'^2 - 2\omega'^2}{4n^2 - 2\omega^2} \cdot \frac{R'a}{Ra'} \quad \dots\dots\dots\dots\dots\dots (117).$$

Since qq' is of the same sign as $\dfrac{q}{q'}$, we have only to determine whether $2n - \dfrac{\omega^2}{n}$, and $2n' - \dfrac{\omega'^2}{n}$, are of the same or of different signs. If these quantities are of the same sign, qq' is positive, if of different signs, qq' is negative.

Now there are four values of n, which give four corresponding values of $2n - \dfrac{\omega^2}{n}$:

$$n_1 = -\omega + \&c., \qquad 2n_1 - \frac{\omega^2}{n_1} \text{ is negative,}$$

$$n_2 = -\text{a small quantity,} \ 2n_2 - \frac{\omega^2}{n_2} \text{ is positive,}$$

$$n_3 = +\text{a small quantity,} \ 2n_3 - \frac{\omega^2}{n_3} \text{ is negative,}$$

$$n_4 = \omega - \&c., \qquad 2n_4 - \frac{\omega^2}{n_4} \text{ is positive.}$$

The quantity with which we have to do is therefore positive for the even orders of waves and negative for the odd ones, and the corresponding quantity in the other ring obeys the same law. Hence when the waves which act upon each other are either both of even or both of odd names, qq' will be positive, but when one belongs to an even series, and the other to an odd series, qq' is negative.

29. The values of p and p' are, roughly,

$$\left. \begin{array}{l} p_1 = \omega + \dfrac{\omega}{m} - \&c., \ p_2 = \omega + \&c., \ p_3 = \omega - \&c., \ p_4 = \omega - \dfrac{\omega}{m} + \&c. \\[2mm] p_1' = \omega' + \dfrac{\omega'}{m} - \&c., \ p_2' = \omega' + \&c., \ p_3' = \omega' - \&c., \ p_4' = \omega' - \dfrac{\omega'}{m} + \&c. \end{array} \right\} \dots\dots (118).$$

ω' is greater than ω, so that p_1' is the greatest, and p_4 the least of these values, and of those of the same order, the accented is greater than the unaccented. The following cases of equality are therefore possible under suitable circumstances :

$$
\begin{array}{ll}
p_1 = p_3', & p_1 = p_2', \\
p_2 = p_4', & p_1 = p_4', \\
p_4 = p_4' \ (\text{when } m = 1), & p_2 = p_3', \\
& p_3 = p_4',
\end{array}
$$

In the cases in the first column qq' will be positive, in those in the second column qq' will be negative.

30. Now each of the four values of p is a function of m, the number of undulations in the ring, and of a the radius of the ring, varying nearly as $a^{-\frac{3}{2}}$. Hence m being given, we may alter the radius of the ring till any one of the four values of p becomes equal to a given quantity, say a given value of p', so that if an indefinite number of rings coexisted, so as to form a sheet of rings, it would be always possible to discover instances of the equality of p and p' among them. If such a case of equality belongs to the first column given above, two constant waves will arise in both rings, one travelling a little faster, and the other a little slower than the free waves. If the case belongs to the second column, two waves will also arise in each ring, but the one pair will gradually die away, and the other pair will increase in amplitude indefinitely, the one wave strengthening the other till at last both rings are thrown into confusion.

The only way in which such an occurrence can be avoided is by placing the rings at such a distance that no value of m shall give coincident values of p and p'. For instance, if $\omega' > 2\omega$, but $\omega' < 3\omega$, no such coincidence is possible. For p_1 is always less than p_2', it is greater than p_4 when $m = 1$ or 2, and less than p_4 when m is 3 or a greater number. There are of course an infinite number of ways in which this noncoincidence might be secured, but it is plain that if a number of concentric rings were placed at small intervals from each other, such coincidences must occur accurately or approximately between some pairs of rings, and if the value of $(p - p')^2$ is brought lower than $-4qq'$, there will be destructive interference.

This investigation is applicable to any number of concentric rings, for, by the principle of superposition of small displacements, the reciprocal actions of any pair of rings are independent of all the rest.

31. *On the effect of long-continued disturbances on a system of rings.*

The result of our previous investigations has been to point out several ways in which disturbances may accumulate till collisions of the different particles of the rings take place. After such a collision the particles will still continue to revolve about the planet, but there will be a loss of energy in the system during the collision which can never be restored. Such collisions however will not affect what is called the Angular Momentum of the system about the planet, which will therefore remain constant.

Let M be the mass of the system of rings, and δm that of one ring whose radius is r, and angular velocity $\omega = S^{\frac{1}{2}} r^{-\frac{3}{2}}$. The angular momentum of the ring is

$$\omega r^2 \delta m = S^{\frac{1}{2}} r^{\frac{1}{2}} \delta m,$$

half its *vis viva* is

$$\tfrac{1}{2}\omega^2 r^2 \delta m = \tfrac{1}{2} S r^{-1} \delta m.$$

The potential energy due to Saturn's attraction on the ring is

$$- S r^{-1} \delta m.$$

The angular momentum of the whole system is invariable, and is

$$S^{\frac{1}{2}} \Sigma \left(r^{\frac{1}{2}} \delta m \right) = A \quad \dots\dots\dots\dots\dots\dots\dots\dots (119).$$

The whole energy of the system is the sum of half the *vis viva* and the potential energy, and is

$$- \tfrac{1}{2} S \Sigma \left(r^{-1} \delta m \right) = E \quad \dots\dots\dots\dots\dots\dots\dots (120).$$

A is invariable, while E necessarily diminishes. We shall find that as E diminishes, the distribution of the rings must be altered, some of the outer rings moving outwards, while the inner rings move inwards, so as either to spread out the whole system more, both on the outer and on the inner edge of the system, or, without affecting the extreme rings, to diminish the density or number of the rings at the mean distance, and increase it at or near the inner and outer edges.

Let us put $x = r^{\frac{1}{2}}$, then $A = S^{\frac{1}{2}} \Sigma (x dm)$ is constant.

Now let

$$x_1 = \frac{\Sigma (x dm)}{\Sigma (dm)},$$

and

$$x = x_1 + x',$$

then we may write

$$-\frac{2E}{S} = \Sigma \left(r^{-1} \delta m \right) = \Sigma \left(x^{-2} dm \right),$$

$$= \Sigma dm \left(x_1^{-2} - 2 \frac{x'}{x_1^3} + 3 \frac{x'^2}{x_1^4} - \&c. \right),$$

$$= \frac{1}{x_1^2} \Sigma (dm) - \frac{2}{x_1^3} \Sigma (x' dm) + \frac{3}{x_1^4} \Sigma (x'^2 \delta m) - \&c. \dots (121).$$

Now $\Sigma\,(dm) = M$ a constant, $\Sigma\,(x'dm) = 0$, and $\Sigma\,(x'^2\delta m)$ is a quantity which increases when the rings are spread out from the mean distance either way, x' being subject only to the restriction $\Sigma\,(x'dm) = 0$. But $\Sigma\,(x'^2 dm)$ may increase without the extreme values of x' being increased, provided some other values be increased.

32. In fact, if we consider the very innermost particle as moving in an ellipse, and at the further apse of its orbit encountering another particle belonging to a larger orbit, we know that the second particle, when at the same distance from the planet, moves the faster. The result is, that the interior satellite will receive a forward impulse at its further apse, and will move in a larger and less eccentric orbit than before. In the same way one of the outermost particles may receive a backward impulse at its nearer apse, and so be made to move in a smaller and less eccentric orbit than before. When we come to deal with collisions among bodies of unknown number, size, and shape, we can no longer trace the mathematical laws of their motion with any distinctness. All we can now do is to collect the results of our investigations and to make the best use we can of them in forming an opinion as to the constitution of the actual rings of Saturn which are still in existence and apparently in steady motion, whatever catastrophes may be indicated by the various theories we have attempted.

33. *To find the Loss of Energy due to internal friction in a broad Fluid Ring, the parts of which revolve about the Planet, each with the velocity of a satellite at the same distance.*

Conceive a fluid, the particles of which move parallel to the axis of x with a velocity u, u being a function of z, then there will be a tangential pressure on a plane parallel to xy

$$= \mu\frac{du}{dz} \text{ on unit of area}$$

due to the relative sliding of the parts of the fluid over each other.

In the case of the ring we have

$$\omega = S^{\frac{1}{2}} r^{-\frac{3}{2}}.$$

The absolute velocity of any particle is ωr. That of a particle at distance $(r + \delta r)$ is

$$\omega r + \frac{d}{dr}(\omega r)\,\delta r.$$

If the angular velocity had been uniform, there would have been no sliding, and the velocity would have been

$$\omega r + \omega \delta r.$$

The sliding is therefore

$$r \frac{d\omega}{dr} \delta r,$$

and the friction on unit of area perpendicular to r is $\mu r \dfrac{d\omega}{dr}$.

The loss of Energy, per unit of area, is the product of the sliding by the friction,

or,
$$\mu r^2 \left.\frac{d\omega}{dr}\right|^2 \delta r \text{ in unit of time.}$$

The loss of Energy in a part of the Ring whose radius is r, breadth δr, and thickness c, is

$$2\pi r^3 c \mu \left.\frac{d\omega}{dr}\right|^2 \delta r.$$

In the case before us it is $\qquad \frac{9}{2}\pi\mu S c r^{-2} \delta r.$

If the thickness of the ring is uniform between $r = a$ and $r = b$, the whole loss of Energy is

$$\frac{9}{2}\pi\mu S c \left(\frac{1}{b} - \frac{1}{a}\right),$$

in unit of time.

Now half the *vis viva* of an elementary ring is

$$\pi\rho c r \delta r \, r^2\omega^2 = \pi\rho c S \delta r,$$

and this between the limits $r = a$ and $r = b$ gives

$$\pi\rho c S (a - b).$$

The potential due to the attraction of S is twice this quantity with the sign changed, so that

$$E = -\pi\rho c S (a - b),$$

and
$$\frac{dE}{dt} = \frac{9}{2}\pi\mu S \left(\frac{1}{b} - \frac{1}{a}\right),$$

$$\frac{1}{E} \frac{dE}{dt} = -\frac{9}{2} \frac{\mu}{\rho} \frac{1}{ab}.$$

45—2

Now Professor Stokes finds $\sqrt{\dfrac{\mu}{\rho}} = 0 \cdot 0564$ for water,

and $= 0 \cdot 116$ for air,

taking the unit of space one English inch, and the unit of time one second. We may take $a = 88,209$ miles, and $b = 77,636$ for the ring A; and $a = 75,845$, and $b = 58,660$ for the ring B. We may also take one year as the unit of time. The quantity representing the ratio of the loss of energy in a year to the whole energy is

$$\frac{1}{E}\frac{dE}{dt} = \frac{1}{60,880,000,000,000} \text{ for the ring } A,$$

$$\text{and } \frac{1}{39,540,000,000,000} \text{ for the ring } B,$$

showing that the effect of internal friction in a ring of water moving with steady motion is inappreciably small. It cannot be from this cause therefore that any decay can take place in the motion of the ring, provided that no waves arise to disturb the motion.

Recapitulation of the Theory of the Motion of a Rigid Ring.

The position of the ring relative to Saturn at any given instant is defined by three variable quantities.

1st. The distance between the centre of gravity of Saturn and the centre of gravity of the ring. This distance we denote by r.

2nd. The angle which the line r makes with a fixed line in the plane of the motion of the ring. This angle is called θ.

3rd. The angle between the line r and a line fixed with respect to the ring so that it coincides with r when the ring is in its mean position. This is the angle ϕ.

The values of these three quantities determine the position of the ring so far as its motion in its own plane is concerned. They may be referred to as the *radius vector, longitude,* and *angle of libration* of the ring.

The forces which act between the ring and the planet depend entirely upon their relative positions. The method adopted above consists in determining the *potential* (V) of the ring at the centre of the planet in terms of r and ϕ. Then the *work done* by any displacement of the system is measured by the change of VS during that displacement. The attraction between the centre of gravity of the Ring and that of the planet is $-S\dfrac{dV}{dr}$, and the moment of the couple tending to turn the ring about its centre of gravity is $S\dfrac{dV}{d\phi}$.

It is proved in Problem V, that if a be the radius of a circular ring, $r_0 = af$ the distance of its centre of gravity from the centre of the circle, and R the mass of the ring, then, at the centre of the ring, $\dfrac{dV}{dr} = -\dfrac{R}{a^2}f$, $\dfrac{dV}{d\phi} = 0$.

It also appears that $\dfrac{d^2V}{dr^2} = \tfrac{1}{2}\dfrac{R}{a^3}(1+g)$, which is positive when $g > -1$, and that $\dfrac{d^2V}{d\phi^2} = \tfrac{1}{2}\dfrac{R}{a}f^2(3-g)$, which is positive when $g < 3$.

If $\dfrac{d^2V}{dr^2}$ is positive, then the attraction between the centres decreases as the distance increases, so that, if the two centres were kept at rest at a given distance by a constant force, the equilibrium would be unstable. If $\dfrac{d^2V}{d\phi^2}$ is positive, then the forces tend to increase the angle of libration, in whichever direction the libration takes place, so that if the ring were fixed by an axis through its centre of gravity, its equilibrium round that axis would be unstable.

In the case of the uniform ring with a heavy particle on its circumference whose weight $= \cdot 82$ of the whole, the direction of the whole attractive force of the ring near the centre will pass through a point lying in the same radius as the centre of gravity, but at a distance from the centre $= \tfrac{9}{8}a$. (Fig. 6.)

If we call this point O, the line SO will indicate the direction and position of the force acting on the ring, which we may call F.

It is evident that the force F, acting on the ring in the line OS, will tend to turn it round its centre of gravity R and to increase the angle of libration KRO. The direct action of this force can never reduce the angle of libration to zero again. To understand the indirect action of the force, we must recollect that the centre of gravity (R) of the ring is revolving about Saturn in the direction of the arrows, and that the ring is revolving about its centre of gravity with nearly the same velocity. If the angular velocity of the centre of gravity about Saturn were always equal to the rotatory velocity of the ring, there would be no libration.

Now suppose that the angle of rotation of the ring is in advance of the longitude of its centre of gravity, so that the line RO has got in advance of SRK by the angle of libration KRO. The attraction between the planet and the ring is a force F acting in SO. We resolve this force into a couple, whose moment is $F \cdot RN$, and a force F acting through R the centre of gravity of the ring.

The couple affects the rotation of the ring, but not the position of its centre of gravity, and the force RF acts on the centre of gravity without affecting the rotation.

Now the couple, in the case represented in the figure, acts in the positive direction, so as to *increase* the angular velocity of the ring, which was already greater than the velocity of revolution of R about S, so that the angle of libration would increase, and never be reduced to zero.

The force RF does not act in the direction of S, but behind it, so that it becomes a retarding force acting upon the centre of gravity of the ring. Now the effect of a retarding force is to cause the distance of the revolving body to decrease and the angular velocity to increase, so that a retarding force increases the angular velocity of R about S.

The effect of the attraction along SO in the case of the figure is, first, to increase the rate of rotation of the ring round R, and secondly, to increase the angular velocity of R about S. If the second effect is greater than the first, then, although the line RO increases its angular velocity, SR will increase its angular velocity more, and will overtake RO, and restore the ring to its original position, so that SRO will be made a straight line as at first. If this accelerating effect is not greater than the acceleration of rotation about R due to the couple, then no compensation will take place, and the motion will be essentially unstable.

If in the figure we had drawn ϕ negative instead of positive, then the couple would have been negative, the tangential force on R accelerative, r would have increased, and in the cases of stability the retardation of θ would be greater than that of $(\theta + \phi)$, and the normal position would be restored, as before.

The object of the investigation is to find the conditions under which this compensation is possible.

It is evident that when SRO becomes straight, there is still a difference of angular velocities between the rotation of the ring and the revolution of the centre of gravity, so that there will be an oscillation on the other side, and the motion will proceed by alternate oscillations without limit.

If we begin with r at its mean value, and ϕ negative, then the rotation of the ring will be retarded, r will be increased, the revolution of r will be more retarded, and thus ϕ will be reduced to zero. The next part of the motion will reduce r to its mean value, and bring ϕ to its greatest positive value. Then r will diminish to its least value, and ϕ will vanish. Lastly r will return to the mean value, and ϕ to the greatest negative value.

It appears from the calculations, that there are, in general, two different ways in which this kind of motion may take place, and that these may have different periods, phases, and amplitudes. The mental exertion required in following out the results of a combined motion of this kind, with all the variations of force and velocity during a complete cycle, would be very great in proportion to the additional knowledge we should derive from the exercise.

The result of this theory of a rigid ring shows not only that a perfectly uniform ring cannot revolve permanently about the planet, but that the irregularity of a permanently revolving ring must be a very observable quantity, the distance between the centre of the ring and the centre of gravity being between ·8158 and ·8279 of the radius. As there is no appearance about the rings justifying a belief in so great an irregularity, the theory of the solidity of the rings becomes very improbable.

When we come to consider the additional difficulty of the tendency of the fluid or loose parts of the ring to accumulate at the thicker parts, and thus to destroy that nice adjustment of the load on which stability depends, we have another powerful argument against solidity.

And when we consider the immense size of the rings, and their comparative thinness, the absurdity of treating them as rigid bodies becomes self-evident. An iron ring of such a size would be not only plastic but semifluid under the forces which it would experience, and we have no reason to believe these rings to be artificially strengthened with any material unknown on this earth.

Recapitulation of the Theory of a Ring of equal Satellites.

In attempting to conceive of the disturbed motion of a ring of unconnected satellites, we have, in the first place, to devise a method of identifying each satellite at any given time, and in the second place, to express the motion of every satellite under the same general formula, in order that the mathematical methods may embrace the whole system of bodies at once.

By conceiving the ring of satellites arranged regularly in a circle, we may easily identify any satellite, by stating the angular distance between it and a known satellite when so arranged. If the motion of the ring were undisturbed, this angle would remain unchanged during the motion, but, in reality, the satellite has its position altered in three ways: 1st, it may be further from or nearer to Saturn; 2ndly, it may be in advance or in the rear of the position it would have had if undisturbed; 3rdly, it may be on one side or other of the mean plane of the ring. Each of these displacements may vary in any way whatever as we pass from one satellite to another, so that it is impossible to assign beforehand the place of any satellite by knowing the places of the rest. § 2.

The formula, therefore, by which we are enabled to predict the place of every satellite at any given time, must be such as to allow the initial position of every satellite to be independent of the rest, and must express all future positions of that satellite by inserting the corresponding value of the quantity denoting time, and those of every other satellite by inserting the value of the angular distance of the given satellite from the point of reference. The three displacements of the satellite will therefore be functions of two variables—the angular position of the satellite, and the time. When the time alone is made to vary, we trace the complete motion of a single satellite; and when the time is made constant, and the angle is made to vary, we trace the form of the ring at a given time.

It is evident that the form of this function, in so far as it indicates the state of the whole ring at a given instant, must be wholly arbitrary, for the form of the ring and its motion at starting are limited only by the condition that the irregularities must be small. We have, however, the means of breaking up any function, however complicated, into a series of simple functions, so that the value of the function between certain limits may be accurately expressed as the sum of a series of sines and cosines of multiples of the variable. This method, due to Fourier, is peculiarly applicable to the case of a ring returning into itself, for the value of Fourier's series is necessarily periodic. We now regard the form of the disturbed ring at any instant as the result of the superposition of a number of separate disturbances, each of which is of the nature of a series of equal waves regularly arranged round the ring. Each of these elementary disturbances is characterised by the number of undulations in it, by their amplitude, and by the position of the first maximum in the ring. § 3.

When we know the form of each elementary disturbance, we may calculate the attraction of the disturbed ring on any given particle in terms of the constants belonging to that disturbance, so that as the actual displacement is the resultant of the elementary displacements, the actual attraction will be the resultant of the corresponding elementary attractions, and therefore the actual motion will be the resultant of all the motions arising from the elementary disturbances. We have therefore only to investigate the elementary disturbances one by one, and having established the theory of these, we calculate the actual motion by combining the series of motions so obtained.

Assuming the motion of the satellites in one of the elementary disturbances to be that of oscillation about a mean position, and the whole motion to be that of a uniformly revolving series of undulations, we find our supposition to be correct, provided a certain biquadratic equation is satisfied by the quantity denoting the rate of oscillation. § 6.

When the four roots of this equation are all real, the motion of each satellite is compounded of four different oscillations of different amplitudes and periods, and the motion of the whole ring consists of four series of undulations, travelling round the ring with different velocities. When any of these roots are impossible, the motion is no longer oscillatory, but tends to the rapid destruction of the ring.

To determine whether the motion of the ring is permanent, we must assure ourselves that the four roots of this equation are real, whatever be the number of undulations in the ring; for if any one of the possible elementary disturbances should lead to destructive oscillations, that disturbance might sooner or later commence, and the ring would be destroyed.

Now the number of undulations in the ring may be any whole number from one up to half the number of satellites. The forces from which danger is to be apprehended are greatest when the number of undulations is greatest, and by taking that number equal to half the number of satellites, we find the condition of stability to be

$$S > . 4352 \, \mu^2 R,$$

where S is the mass of the central body, R that of the ring, and μ the number of satellites of which it is composed. § 8. If the number of satellites be too great, destructive oscillations will commence, and finally some of the satellites will come into collision with each other and unite, so that the number of independent satellites will be reduced to that which the central body can retain and keep in discipline. When this has taken place, the satellites will not only be kept at the proper distance from the primary, but will be prevented by its preponderating mass from interfering with each other.

We next considered more carefully the case in which the mass of the ring is very small, so that the forces arising from the attraction of the ring are small compared with that due to the central body. In this case the values of the roots of the biquadratic are all real, and easily estimated. § 9.

If we consider the motion of any satellite about its mean position, as referred to axes fixed in the plane of the ring, we shall find that it describes an ellipse in the direction opposite to that of the revolution of the ring, the periodic time being to that of the ring as ω to n, and the tangential amplitude of oscillation being to the radial as 2ω to n. § 10.

The absolute motion of each satellite in space is nearly elliptic for the large values of n, the axis of the ellipse always advancing slowly in the direction of rotation. The path of a satellite corresponding to one of the small values of n is nearly circular, but the radius slowly increases and diminishes during a period of many revolutions. § 11.

The form of the ring at any instant is that of a re-entering curve, having m alternations of distance from the centre, symmetrically arranged, and m points of condensation, or crowding of the satellites, which coincide with the points of greatest distance when n is positive, and with the points nearest the centre when n is negative. § 12.

This system of undulations travels with an angular velocity $-\dfrac{n}{m}$ relative to the ring, and $\omega - \dfrac{n}{m}$ in space, so that during each oscillation of a satellite a complete wave passes over it. § 14.

To exhibit the movements of the satellites, I have made an arrangement by which 36 little ivory balls are made to go through the motions belonging to the first or fourth series of waves. (Figs. 7, 8.)

The instrument stands on a pillar A, in the upper part of which turns the cranked axle CC. On the parallel parts of this axle are placed two wheels, RR and TT, each of which has 36 holes at equal distances in a circle near its circumference. The two circles are connected by 36 small cranks of the form KK, the extremities of which turn in the corresponding holes of the two wheels. That axle of the crank K which passes through the hole in the wheel S is bored, so as to hold the end of the bent wire which carries the satellite S. This wire may be turned in the hole so as to place the bent part carrying the satellite at any angle with the crank. A pin P, which passes through the top of the pillar, serves to prevent the cranked axle from turning; and a pin Q, passing through the pillar horizontally, may be made to fix the wheel R, by inserting it in a hole in one of the spokes of that wheel. There is also a handle H, which is in one piece with the wheel T, and serves to turn the axle.

Now suppose the pin P taken out, so as to allow the cranked axle to turn, and the pin Q inserted in its hole, so as to prevent the wheel R from revolving; then if the crank C be turned by means of the handle H, the wheel T will have its centre carried round in a vertical circle, but will remain parallel to itself during the whole motion, so that every point in its plane will describe an equal circle, and all the cranks K will be made to revolve exactly as the large crank C does. Each satellite will therefore revolve in a small circular orbit, in the same time with the handle H, but the position of each satellite in that orbit may be arranged as we please, according as we turn the wire which supports it in the end of the crank.

In fig. 8, which gives a front view of the instrument, the satellites are so placed that each is turned 60^0 further round in its socket than the one behind it. As there are 36 satellites, this process will bring us back to our starting-point after six revolutions of the direction of the arm of the satellite; and therefore as we have gone round the ring once in the same direction, the arm of the satellite will have overtaken the radius of the ring five times.

Hence there will be five places where the satellites are beyond their mean distance from the centre of the ring, and five where they are within it, so that we have here a series of five undulations round the circumference of the ring. In this case the satellites are crowded together when nearest to the centre, so that the case is that of the *first* series of waves, when $m = 5$.

Now suppose the cranked axle C to be turned, and all the small cranks K to turn with it, as before explained, every satellite will then be carried round on its own arm in the same direction; but, since the direction of the arms of different satellites is different, their phases of revolution will preserve the same difference, and the system of satellites will still be arranged in five undulations, only the undulations will be propagated round the ring in the direction opposite to that of the revolution of the satellites.

To understand the motion better, let us conceive the centres of the orbits of the satellites to be arranged in a straight line instead of a circle, as in fig. 10. Each satellite is here represented in a different phase of its orbit, so that as we pass from one to another from left to right, we find the position of the satellite in its orbit altering in the direction opposite to that of the hands of a watch. The satellites all lie in a trochoidal curve, indicated by the line through them in the figure. Now conceive every satellite to move in its orbit through a certain angle in the direction of the arrows. The satellites will then lie in the dotted line, the form of which is the same as that of the former curve, only shifted in the direction of the large arrow. It appears, therefore, that as the satellites revolve, the undulation travels, so that any part of it reaches successively each satellite as it comes into the same phase of rotation. It therefore travels from those satellites which are most advanced in phase to those which are less so, and passes over a complete wave-length in the time of one revolution of a satellite.

Now if the satellites be arranged as in fig. 8, where each is more advanced in phase as we go round the ring in the direction of rotation, the wave will travel in the direction opposite to that of rotation, but if they are arranged as in fig. 12, where each satellite is less advanced in phase as we go round the ring, the wave will travel in the direction of rotation. Fig. 8 represents the *first* series of waves where $m = 5$, and fig. 12 represents the *fourth* series where $m = 7$. By arranging the satellites in their sockets before starting, we might make m equal to any whole number, from 1 to 18. If we chose any number above 18 the result would be the same as if we had taken a number as much below 18 and changed the arrangement from the first wave to the fourth.

In this way we can exhibit the motions of the satellites in the first and fourth waves. In reality they ought to move in ellipses, the major axes being twice the minor, whereas in the machine they move in circles: but the character of the motion is the same, though the form of the orbit is different.

We may now show these motions of the satellites among each other, combined with the motion of rotation of the whole ring. For this purpose we put in the pin P, so as to prevent the crank axle from turning, and take out the pin Q so as to allow the wheel R to turn. If we then turn the wheel T, all the small cranks will remain parallel to the fixed crank, and the wheel R will revolve at the same rate as T. The arm of each satellite will continue parallel to itself during the motion, so that the satellite will describe a circle whose centre is at a distance from the centre of R, equal to the arm of the satellite, and measured in the same direction. In our theory of real satellites, each moves in an ellipse, having the central body in its focus, but this motion in an eccentric circle is sufficiently near for illustration. The motion of the waves relative to the ring is the same as before. The waves of the first kind travel faster than the ring itself, and overtake the satellites, those of the fourth kind travel slower, and are overtaken by them.

In fig. 11 we have an exaggerated representation of a ring of twelve satellites affected by a wave of the fourth kind where $m = 2$. The satellites here lie in an ellipse at any given instant, and as each moves round in its circle about its mean position, the ellipse also moves round in the same direction with half their angular velocity. In the figure the dotted line represents the position of the ellipse when each satellite has moved forward into the position represented by a dot.

Fig. 13 represents a wave of the first kind where $m = 2$. The satellites at any instant lie in an epitrochoid, which, as the satellites revolve about their mean positions, revolves in the opposite direction with half their angular velocity, so that when the satellites come into the positions represented by the dots, the curve in which they lie turns round in the opposite direction and forms the dotted curve.

In fig. 9 we have the same case as in fig. 13, only that the absolute orbits of the satellites in space are given, instead of their orbits about their mean positions in the ring. Here each moves about the central body in an eccentric circle, which in strictness ought to be an ellipse not differing much from the circle.

As the satellites move in their orbits in the direction of the arrows, the curve which they form revolves in the same direction with a velocity $1\frac{1}{2}$ times that of the ring.

By considering these figures, and still more by watching the actual motion of the ivory balls in the model, we may form a distinct notion of the motions of the particles of a discontinuous ring, although the motions of the model are circular and not elliptic. The model, represented on a scale of one-third in figs. 7 and 8, was made in brass by Messrs. Smith and Ramage of Aberdeen.

We are now able to understand the mechanical principle, on account of which a massive central body is enabled to govern a numerous assemblage of satellites, and to space them out into a regular ring; while a smaller central body would allow disturbances to arise among the individual satellites, and collisions to take place.

When we calculated the attractions among the satellites composing the ring, we found that if any satellite be displaced tangentially, the resultant attraction will draw it away from its mean position, for the attraction of the satellites it approaches will increase, while that of those it recedes from will diminish, so that its equilibrium when in the mean position is unstable with respect to tangential displacements; and therefore, since every satellite of the ring is statically unstable between its neighbours, the slightest disturbance would tend to produce collisions among the satellites, and to break up the ring into groups of conglomerated satellites.

But if we consider the dynamics of the problem, we shall find that this effect need not necessarily take place, and that this very force which tends towards destruction may become the condition of the preservation of the ring. Suppose the whole ring to be revolving round a central body, and that one satellite gets in advance of its mean position. It will then be attracted forwards, its path will become less concave towards the attracting body, so that its distance from that body will increase. At this increased distance its angular velocity will be less, so that instead of overtaking those in front, it may by this means be made to fall back to its original position. Whether it does so or not must depend on the actual values of the attractive forces and on the angular velocity of the ring. When the angular velocity is great and the attractive forces small, the compensating process will go on vigorously, and the ring will be preserved. When the angular velocity is small and the attractive forces of the ring great, the dynamical effect will not compensate for the disturbing action of the forces and the ring will be destroyed.

If the satellite, instead of being displaced forwards, had been originally behind its mean position in the ring, the forces would have pulled it backwards, its path would have become more concave towards the centre, its distance from the centre would diminish, its angular velocity would increase, and it would gain upon the rest of the ring till it got in front of its mean position. This effect is of course dependent on the very same conditions as in the former case, and the actual effect on a disturbed satellite would be to make it describe an orbit about its mean position in the ring, so that if in advance of its mean position, it first recedes from the centre, then falls behind its mean position in the ring, then approaches the centre within the mean distance, then advances beyond its mean position, and, lastly, recedes from the centre till it reaches its starting-point, after which the process is repeated indefinitely, the orbit being always described in the direction opposite to that of the revolution of the ring.

We now understand what would happen to a disturbed satellite, if all the others were preserved from disturbance. But, since all the satellites are equally free, the motion of one will produce changes in the forces acting on the rest, and this will set them in motion, and this motion will be propagated from one satellite to another round the ring. Now propagated disturbances constitute waves, and all waves, however complicated, may be reduced to combinations of simple and regular waves; and therefore all the disturbances of the ring may be considered as the resultant of many series of waves, of different lengths, and travelling with different velocities. The investigation of the relation between the length and velocity of these waves forms the essential part of the problem, after which we have only to split up the original disturbance into its simple elements, to calculate the effect of each of these separately, and then to combine the results. The solution thus obtained will be perfectly general, and quite independent of the particular form of the ring, whether regular or irregular at starting. § 14.

We next investigated the effect upon the ring of an external disturbing force. Having split up the disturbing force into components of the same type with the waves of the ring (an operation which is always possible), we found that each term of the disturbing force generates a "forced wave" travelling with its own angular velocity. The magnitude of the forced wave depends not only on that of the disturbing force, but on the angular velocity with which the disturbance travels round the ring, being greater in proportion as this velocity more nearly coincides with that of one of the "free waves" of the ring. We also found that the displacement of the satellites was sometimes in the direction of the disturbing force, and sometimes in the opposite direction, according to the relative position of the forced wave among the four natural ones, producing in the one case positive, and in the other negative forced waves. In treating the problem generally, we must determine the forced waves belonging to every term of the disturbing force, and combine these with such a system of free waves as shall reproduce the initial state of the ring. The subsequent motion of the ring is that which would result from the free waves and forced waves together. The most important class of forced waves are those which are produced by waves in neighbouring rings. § 15.

We concluded the theory of a ring of satellites by tracing the process by which the ring would be destroyed if the conditions of stability were not fulfilled. We found two cases of instability, depending on the nature of the tangential force due to tangential displacement. If this force be in the direction opposite to the displacement, that is, if the parts of the ring are *statically stable*, the ring will be destroyed, the irregularities becoming larger and larger without being propagated round the ring. When the tangential force is in the direction of the tangential displacement, if it is below a certain value, the disturbances will be propagated round the ring without becoming larger, and we have the case of stability treated of at large. If the force exceed this value, the disturbances will still travel round the ring, but they will increase in amplitude continually till the ring falls into confusion. § 18.

We then proceeded to extend our method to the case of rings of different constitutions. The first case was that of a ring of satellites of unequal size. If the central body be of sufficient mass, such a ring will be spaced out, so that the larger satellites will be at wider intervals than the smaller ones, and the waves of disturbance will be propagated as before, except that there may be reflected waves when a wave reaches a part of the ring where there is a change in the average size of the satellites. § 19.

The next case was that of an annular cloud of meteoric stones, revolving uniformly about the planet. The *average density* of the space through which these small bodies are scattered will vary with every irregularity of the motion, and this variation of density will produce variations in the forces acting upon the other parts of the cloud, and so disturbances will be propagated in this ring, as in a ring of a finite number of satellites. The condition that such a ring should be free from destructive oscillations is, that the density of the planet should be more than three hundred times that of the ring. This would make the ring much rarer than common air, as regards its *average density*, though the density of the particles of which it is composed may be great. Comparing this result with Laplace's minimum density of a ring revolving as a whole, we find that such a ring cannot revolve as a whole, but that the inner parts must have a greater angular velocity than the outer parts. § 20.

We next took up the case of a flattened ring, composed of incompressible fluid, and moving with uniform angular velocity. The internal forces here arise partly from attraction and partly from fluid pressure. We began by taking the case of an infinite stratum of fluid affected by regular waves, and found the accurate values of the forces in this case. For long waves the resultant force is in the same direction as the displacement, reaching a maximum for waves whose length is about ten times the thickness of the stratum. For waves about five times as long as the stratum is thick there is no resultant force, and for shorter waves the force is in the opposite direction to the displacement. § 23.

Applying these results to the case of the ring, we find that it will be destroyed by the long waves unless the fluid is less than $\frac{1}{42}$ of the density of the planet, and that in all cases the short waves will break up the ring into small satellites.

Passing to the case of *narrow* rings, we should find a somewhat larger maximum density, but we should still find that very short waves produce forces in the direction opposite to the displacement, and that therefore, as already explained (page 333), these short undulations would increase in magnitude without being propagated along the ring, till they had broken up the fluid filament into drops. These drops may or may not fulfil the condition formerly given for the stability of a ring of equal satellites. If they fulfil it, they will move as a permanent ring. If they do not, short waves will arise and be propagated among the satellites, with ever increasing magnitude, till a sufficient number of drops have been brought into collision, so as to unite and form a smaller number of larger drops, which may be capable of revolving as a permanent ring.

We have already investigated the disturbances produced by an external force independent of the ring; but the special case of the mutual perturbations of two concentric rings is considerably more complex, because the existence of a double system of waves changes the character of both, and the waves produced react on those that produced them.

We determined the attraction of a ring upon a particle of a concentric ring, first, when both rings are in their undisturbed state; secondly, when the particle is disturbed; and, thirdly, when the attracting ring is disturbed by a series of waves.   § 26.

We then formed the equations of motion of one of the rings, taking in the disturbing forces arising from the existence of a wave in the other ring, and found the small variation of the velocity of a wave in the first ring as dependent on the magnitude of the wave in the second ring, which travels with it.   § 27.

The forced wave in the second ring must have the same absolute angular velocity as the free wave of the first which produces it, but this velocity of the free wave is slightly altered by the reaction of the forced wave upon it. We find that if a free wave of the first ring has an absolute angular velocity not very different from that of a free wave of the second ring, then if both free waves be of even orders (that is, of the second or fourth varieties of waves), or both of odd orders (that is, of the first or third), then the swifter of the two free waves has its velocity increased by the forced wave which it produces, and the slower free wave is rendered still slower by its forced wave; and even when the two free waves have the same angular velocity, their mutual action will make them both split into two, one wave in each ring travelling faster, and the other wave in each ring travelling slower, than the rate with which they would move if they had not acted on each other.

But if one of the free waves be of an even order and the other of an odd order, the swifter free wave will travel slower, and the slower free wave will travel swifter, on account of the reaction of their respective forced waves. If the two free waves have naturally a certain small difference of velocities, they will be made to travel together, but if the difference is less than this, they will again split into two pairs of waves, one pair continually increasing in magnitude without limit, and the other continually diminishing, so that one of the waves in each ring will increase in violence till it has thrown the ring into a state of confusion.

There are four cases in which this may happen. The first wave of the outer ring may conspire with the second or the fourth of the inner ring, the second of the outer with the third of the inner, or the third of the outer with the fourth of the inner. That two rings may revolve permanently, their distances must be arranged so that none of these conspiracies may arise between odd and even waves, whatever be the value of m. The number of conditions to be fulfilled is therefore very great, especially when the rings are near together and have nearly the same angular velocity, because then there are a greater number of dangerous values of m to be provided for.

In the case of a large number of concentric rings, the stability of each pair must be investigated separately, and if in the case of any two, whether consecutive rings or not, there are a pair of conspiring waves, those two rings will be agitated more and more, till waves of that kind are rendered impossible by the breaking up of those rings into some different arrangement. The presence of the other rings cannot prevent the mutual destruction of any pair which bear such relations to each other.

It appears, therefore, that in a system of many concentric rings there will be continually new cases of mutual interference between different pairs of rings. The forces which excite these disturbances being very small, they will be slow of growth, and it is possible that by the irregularities of each of the rings the waves may be so broken and confused (see § 19), as to be incapable of mounting up to the height at which they would begin to destroy the arrangement of the ring. In this way it may be conceived to be possible that the gradual disarrangement of the system may be retarded or indefinitely postponed.

But supposing that these waves mount up so as to produce collisions among the particles, then we may deduce the result upon the system from general dynamical principles. There will be a tendency among the exterior rings to remove further from the planet, and among the interior rings to approach the planet, and this either by the extreme interior and exterior rings diverging from each other, or by intermediate parts of the system moving away from the mean ring. If the interior rings are observed to approach the planet, while it is known that none of the other rings have expanded, then the cause of the change cannot be the mutual action of the parts of the system, but the resistance of some medium in which the rings revolve.   § 31.

There is another cause which would gradually act upon a broad fluid ring of which the parts revolve each with the angular velocity due to its distance from the planet, namely, the internal friction produced by the slipping of the concentric rings with different angular velocities.   It appears, however (§ 33), that the effect of fluid friction would be insensible if the motion were regular.

Let us now gather together the conclusions we have been able to draw from the mathematical theory of various kinds of conceivable rings.

We found that the stability of the motion of a solid ring depended on so delicate an adjustment, and at the same time so unsymmetrical a distribution of mass, that even if the exact condition were fulfilled, it could scarcely last long, and if it did, the immense preponderance of one side of the ring would be easily observed, contrary to experience.   These considerations, with others derived from the mechanical structure of so vast a body, compel us to abandon any theory of solid rings.

We next examined the motion of a ring of equal satellites, and found that if the mass of the planet is sufficient, any disturbances produced in the arrangement of the ring will be propagated round it in the form of waves, and will not introduce dangerous confusion.   If the satellites are unequal, the propagation of the waves will no longer be regular, but disturbances of the ring will in this, as in the former case, produce only waves, and not growing confusion.   Supposing the ring to consist, not of a single row of large satellites, but of a cloud of evenly distributed unconnected particles, we found that such a cloud must have a very small density in order to be permanent, and that this is inconsistent with its outer and inner parts moving with the same angular velocity.   Supposing the ring to be fluid and continuous, we found that it will be necessarily broken up into small portions.

We conclude, therefore, that the rings must consist of disconnected particles; these may be either solid or liquid, but they must be independent.   The entire system of rings must therefore consist either of a series of many concentric rings, each moving with its own velocity, and having its own systems of waves, or else of a confused multitude of revolving particles, not arranged in rings, and continually coming into collision with each other.

Taking the first case, we found that in an indefinite number of possible cases the mutual perturbations of two rings, stable in themselves, might mount up in time to a destructive magnitude, and that such cases must continually occur in an extensive system like that of Saturn, the only retarding cause being the possible irregularity of the rings.

The result of long-continued disturbance was found to be the spreading out of the rings in breadth, the outer rings pressing outwards, while the inner rings press inwards.

The final result, therefore, of the mechanical theory is, that the only system of rings which can exist is one composed of an indefinite number of unconnected particles, revolving round the planet with different velocities according to their respective distances. These particles may be arranged in series of narrow rings, or they may move through each other irregularly. In the first case the destruction of the system will be very slow, in the second case it will be more rapid, but there may be a tendency towards an arrangement in narrow rings, which may retard the process.

We are not able to ascertain by observation the constitution of the two outer divisions of the system of rings, but the inner ring is certainly transparent, for the limb of Saturn has been observed through it. It is also certain, that though the space occupied by the ring is transparent, it is not through the material parts of it that Saturn was seen, for his limb was observed without distortion; which shows that there was no refraction, and therefore that the rays did not pass through a medium at all, but between the solid or liquid particles of which the ring is composed. Here then we have an optical argument in favour of the theory of independent particles as the material of the rings. The two outer rings may be of the same nature, but not so exceedingly rare that a ray of light can pass through their whole thickness without encountering one of the particles.

Finally, the two outer rings have been observed for 200 years, and it appears, from the careful analysis of all the observations by M. Struvé, that the second ring is broader than when first observed, and that its inner edge is nearer the planet than formerly. The inner ring also is suspected to be approaching the planet ever since its discovery in 1850. These appearances seem to indicate the same slow progress of the rings towards separation which we found to be the result of theory, and the remark, that the inner edge of the inner ring is most distinct, seems to indicate that the approach towards the planet is less rapid near the edge, as we had reason to conjecture. As to the apparent unchangeableness of the exterior diameter of the outer ring, we must remember that the outer rings are certainly far more dense than the inner one, and that a small change in the outer rings must balance a great change in the inner one. It is possible, however, that some of the observed changes may be due to the existence of a resisting medium. If the changes already suspected should be confirmed by repeated observations with the same instruments, it will be worth while to investigate more carefully whether Saturn's Rings are permanent or transitionary elements of the Solar System, and whether in that part of the heavens we see celestial immutability, or terrestrial corruption and generation, and the old order giving place to new before our own eyes.

APPENDIX.

On the Stability of the Steady Motion of a Rigid Body about a Fixed Centre of Force.
By PROFESSOR W. THOMSON *(communicated in a letter).*

THE body will be supposed to be symmetrical on the two sides of a certain plane containing the centre of force, and no motion except that of parts of the body parallel to the plane will be considered. Taking it as the plane of construction, let G (fig. 14) be the centre of gravity of the body, and O a point at which the resultant attraction of the body is in the line OG towards G. Then if the body be placed with O coinciding with the centre of force, and set in a state of rotation about that point as an axis, with an angular velocity equal to $\sqrt{\dfrac{fS}{aM}}$, (where f denotes the attraction of the body on a unit of matter at O, S the amount of matter in the central body, M the mass of the revolving body, and a the distance OG), it will continue, provided it be perfectly undisturbed, to revolve uniformly at this rate, and the attraction Sf on the moving body will be constantly balanced by the centrifugal force $\omega^2 aM$ of its motion.

Let us now suppose the motion to be slightly disturbed, and let it be required to investigate the consequences. Let X, S, Y, be rectangular axes of reference revolving uniformly with the angular velocity ω, round S, the fixed attracting point. Let $\bar{x}$, $\bar{y}$, be the co-ordinates of G with reference to these axes, and let XS, YS denote the components of the whole force of attraction of S on the rigid body. Then since this force is in the line through S, its moment round G is

$$SY\bar{x} - SX\bar{y};$$

the components of the forces on the moving body being reckoned as *positive* when they tend to *diminish* $\bar{x}$ and $\bar{y}$ respectively. Hence if k denote the radius of gyration of the body round G, and if ϕ denote the angle which OG makes with SX (*i.e.* the angle GOK), the equations of motion are,

$$M\left(\frac{d^2\bar{x}}{dt^2} - 2\omega\frac{d\bar{y}}{dt} - \omega^2\bar{x}\right) + SX = 0,$$

$$M\left(\frac{d^2\bar{y}}{dt^2} + 2\omega\frac{d\bar{x}}{dt} - \omega^2\bar{y}\right) + SY = 0,$$

$$Mk^2\frac{d^2\phi}{dt^2} - S(Y\bar{x} - X\bar{y}) = 0.$$

In the first place we see that one integral of these equations is

$$M\left(\bar{x}\frac{d\bar{y}}{dt} - \bar{y}\frac{d\bar{x}}{dt}\right) + M\omega(\bar{x}^2 + \bar{y}^2) + Mk^2\frac{d\phi}{dt} = H.$$

This is the "equation of angular momentum."

In considering whether the motion round S with velocity ω when O coincides with S is stable or unstable, we must find whether every possible motion with the same "angular momentum" round S is such that it will never bring O to more than an infinitely small distance from S: that is to say, we must find whether, for every possible solution in which $H = M(a^2 + k^2)\omega$, and for which the co-ordinates of O are infinitely small at one time, these co-ordinates remain infinitely small. Let these values at time t be denoted thus: $SN = \xi$, and $NO = \eta$; let OG be at first infinitely nearly parallel to OX, *i.e.* let ϕ be infinitely small (the full solution will tell us whether or not ϕ remains infinitely small); then, as long as ϕ is infinitely small, we have

$$\bar{x} = a + \xi, \quad \bar{y} = \eta + a\phi,$$

and the equations of motion have the forms

$$M\left\{\frac{d^2\xi}{dt^2} - 2\omega\left(\frac{d\eta}{dt} + a\frac{d\phi}{dt}\right) - \omega^2(a + \xi)\right\} + SX = 0,$$

$$M\left\{\frac{d^2\eta}{dt^2} + a\frac{d^2\phi}{dt^2} + 2\omega\frac{d\xi}{dt} - \omega^2(\eta + a\phi)\right\} + SY = 0,$$

and we may write the equation of angular momentum instead of the third equation,

$$M\left\{(a + \xi)\left(\frac{d\eta}{dt} + a\frac{d\phi}{dt}\right) - (\eta + a\phi)\left(\frac{d\xi}{dt}\right) + \omega(a + \xi)^2 + \omega(\eta + a\phi)^2 + k^2\frac{d\phi}{dt}\right\} = H.$$

If now we suppose ξ and η to be infinitely small, the last of these equations becomes

$$(a^2 + k^2)\frac{d\phi}{dt} + 2\omega a\xi + a\frac{d\eta}{dt} = 0. \quad\dots\dots\dots\dots\dots\dots\dots(a).$$

If p and q denote the components parallel and perpendicular to OG of the attraction of the body on a unit of matter at S, we have

$$X = p \cos \phi - q \sin \phi = p, \text{ and } Y = p \sin \phi + q \cos \phi = p\phi + q,$$

since q and ϕ are each infinitely small; and if we put $V =$ potential at S, and

$$\alpha = \frac{d^2 V}{d\xi^2}, \quad \beta = \frac{d^2 V}{d\eta^2}, \quad \gamma = \frac{d^2 V}{d\xi d\eta},$$

then

$$p = f - \alpha\xi - \gamma\eta, \quad q = -\beta\eta - \gamma\xi,$$
$$X = f - \alpha\xi - \gamma\eta, \quad Y = f\phi - \beta\eta - \gamma\xi.$$

If we make these substitutions for X and Y, and take into account that

$$f = \omega^2 a \frac{M}{S} \quad\dotfill\quad (b),$$

the first and second equations of motion become

$$\frac{d^2\xi}{dt^2} - 2\omega \frac{d\eta}{dt} - \omega^2\xi - 2\omega a \frac{d\phi}{dt} - \frac{S}{M}(\alpha\xi + \gamma\eta) = 0 \quad\dotfill\quad (c),$$

$$\frac{d^2\eta}{dt^2} + 2\omega \frac{d\xi}{dt} - \omega^2\eta + a \frac{d^2\phi}{dt^2} - \frac{S}{M}(\beta\eta + \gamma\xi) = 0 \quad\dotfill\quad (d).$$

Combining equations (a), (c), and (d), by the same method as that adopted in the text, we find that the differential equation in ξ, η, or ϕ, is of the form

$$A \frac{d^4 u}{dt^4} + B \frac{d^2 u}{dt^2} + Cu = 0,$$

where $A = k^2,$

$$B = \omega^2 (2k^2 - a^2) - \frac{S}{M}\{k^2\alpha + (a^2 + k^2)\beta\},$$

$$C = \omega^4 (k^2 - 3a^2) + \omega^2 \frac{S}{M}\{(a^2 + k^2)(\alpha + \beta) - 4a^2\beta\} + (a^2 + k^2)\frac{S^2}{M^2}(\alpha\beta - \gamma).$$

In comparing this result with that obtained in the Essay, we must put

$$r_0 \text{ for } a,$$
$$R \text{ for } M,$$
$$R + S \text{ for } S,$$
$$L \text{ for } \alpha,$$
$$N r_0^2 \text{ for } \beta,$$
$$M r_0 \text{ for } \gamma.$$

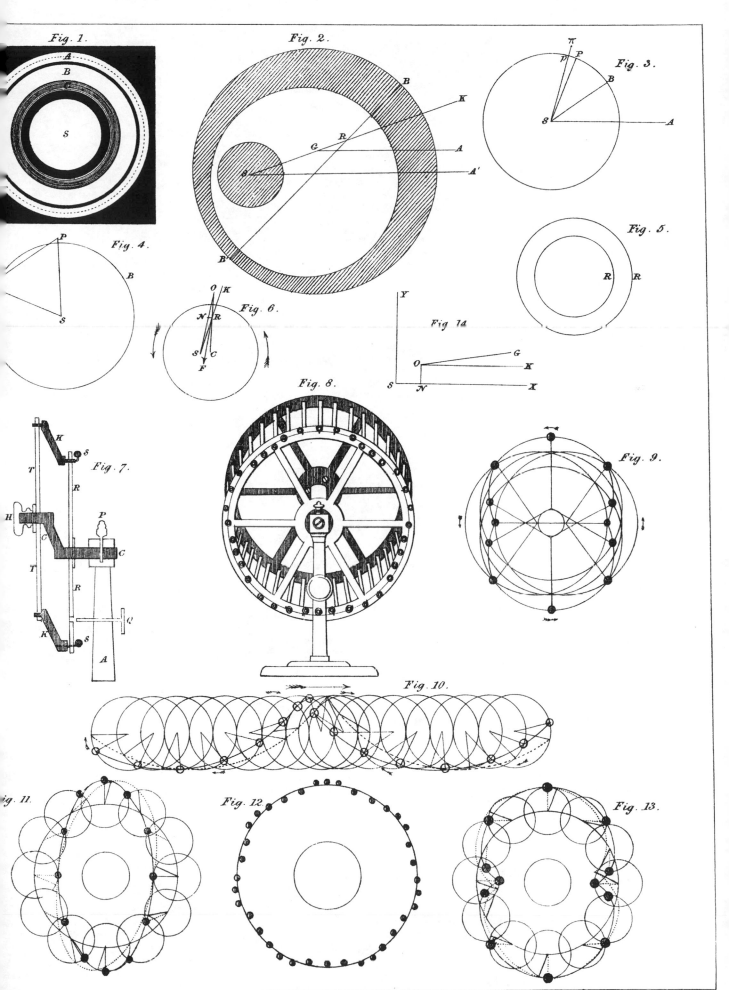

Fig. 1.

Fig. 2.

Fig. 3.

Fig. 4.

Fig. 5.

Fig. 6.

Fig. 7.

Fig. 8.

Fig. 9.

Fig. 10.

Fig. 11.

Fig. 12.

Fig. 13.

Fig. 14.

[From the *Philosophical Magazine* for January and July, 1860.]

XX. *Illustrations of the Dynamical Theory of Gases*[*].

PART I.

ON THE MOTIONS AND COLLISIONS OF PERFECTLY ELASTIC SPHERES.

So many of the properties of matter, especially when in the gaseous form, can be deduced from the hypothesis that their minute parts are in rapid motion, the velocity increasing with the temperature, that the precise nature of this motion becomes a subject of rational curiosity. Daniel Bernouilli, Herapath, Joule, Krönig, Clausius, &c. have shewn that the relations between pressure, temperature, and density in a perfect gas can be explained by supposing the particles to move with uniform velocity in straight lines, striking against the sides of the containing vessel and thus producing pressure. It is not necessary to suppose each particle to travel to any great distance in the same straight line; for the effect in producing pressure will be the same if the particles strike against each other; so that the straight line described may be very short. M. Clausius has determined the mean length of path in terms of the average distance of the particles, and the distance between the centres of two particles when collision takes place. We have at present no means of ascertaining either of these distances; but certain phenomena, such as the internal friction of gases, the conduction of heat through a gas, and the diffusion of one gas through another, seem to indicate the possibility of determining accurately the mean length of path which a particle describes between two successive collisions. In order to lay the foundation of such investigations on strict mechanical principles, I shall demonstrate the laws of motion of an indefinite number of small, hard, and perfectly elastic spheres acting on one another only during impact.

* Read at the Meeting of the British Association at Aberdeen, September 21, 1859.

If the properties of such a system of bodies are found to correspond to those of gases, an important physical analogy will be established, which may lead to more accurate knowledge of the properties of matter. If experiments on gases are inconsistent with the hypothesis of these propositions, then our theory, though consistent with itself, is proved to be incapable of explaining the phenomena of gases. In either case it is necessary to follow out the consequences of the hypothesis.

Instead of saying that the particles are hard, spherical, and elastic, we may if we please say that the particles are centres of force, of which the action is insensible except at a certain small distance, when it suddenly appears as a repulsive force of very great intensity. It is evident that either assumption will lead to the same results. For the sake of avoiding the repetition of a long phrase about these repulsive forces, I shall proceed upon the assumption of perfectly elastic spherical bodies. If we suppose those aggregate molecules which move together to have a bounding surface which is not spherical, then the rotatory motion of the system will store up a certain proportion of the whole *vis viva*, as has been shewn by Clausius, and in this way we may account for the value of the specific heat being greater than on the more simple hypothesis.

On the Motion and Collision of Perfectly Elastic Spheres.

Prop. I. Two spheres moving in opposite directions with velocities inversely as their masses strike one another; to determine their motions after impact.

Let P and Q be the position of the centres at impact; AP, BQ the directions and magnitudes of the velocities before impact; Pa, Qb the same after impact; then, resolving the velocities parallel and perpendicular to PQ the line of centres, we find that the velocities parallel to the line of centres are exactly reversed, while those perpendicular to that line are unchanged. Compounding these velocities again, we find that the velocity of each ball is the same before and after impact, and that the directions before and after impact lie in the same plane with the line of centres, and make equal angles with it.

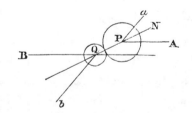

Prop. II. To find the probability of the direction of the velocity after impact lying between given limits.

In order that a collision may take place, the line of motion of one of the balls must pass the centre of the other at a distance less than the sum of their radii; that is, it must pass through a circle whose centre is that of the other ball, and radius (s) the sum of the radii of the balls. Within this circle every position is equally probable, and therefore the probability of the distance from the centre being between r and $r+dr$ is

$$\frac{2rdr}{s^2}.$$

Now let ϕ be the angle APa between the original direction and the direction after impact, then $APN = \frac{1}{2}\phi$, and $r = s \sin \frac{1}{2}\phi$, and the probability becomes

$$\tfrac{1}{2} \sin \phi d\phi.$$

The area of a spherical zone between the angles of polar distance ϕ and $\phi + d\phi$ is

$$2\pi \sin \phi d\phi;$$

therefore if ω be any small area on the surface of a sphere, radius unity, the probability of the direction of rebound passing through this area is

$$\frac{\omega}{4\pi};$$

so that the probability is independent of ϕ, that is, all directions of rebound are equally likely.

Prop. III. Given the direction and magnitude of the velocities of two spheres before impact, and the line of centres at impact; to find the velocities after impact.

Let OA, OB represent the velocities before impact, so that if there had been no action between the bodies they would have been at A and B at the end of a second. Join AB, and let G be their centre of gravity, the position of which is not affected by their mutual action. Draw GN parallel to the line of centres at impact (not necessarily in the plane AOB). Draw aGb 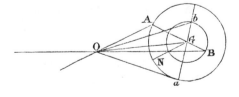 in the plane AGN, making $NGa = NGA$, and $Ga = GA$ and $Gb = GB$; then by

48—2

Prop. I. Ga and Gb will be the velocities relative to G; and compounding these with OG, we have Oa and Ob for the true velocities after impact.

By Prop. II. all directions of the line aGb are equally probable. It appears therefore that the velocity after impact is compounded of the velocity of the centre of gravity, and of a velocity equal to the velocity of the sphere relative to the centre of gravity, which may with equal probability be in any direction whatever.

If a great many equal spherical particles were in motion in a perfectly elastic vessel, collisions would take place among the particles, and their velocities would be altered at every collision; so that after a certain time the *vis viva* will be divided among the particles according to some regular law, the average number of particles whose velocity lies between certain limits being ascertainable, though the velocity of each particle changes at every collision.

Prop. IV. To find the average number of particles whose velocities lie between given limits, after a great number of collisions among a great number of equal particles.

Let N be the whole number of particles. Let x, y, z be the components of the velocity of each particle in three rectangular directions, and let the number of particles for which x lies between x and $x+dx$, be $Nf(x)dx$, where $f(x)$ is a function of x to be determined.

The number of particles for which y lies between y and $y+dy$ will be $Nf(y)dy$; and the number for which z lies between z and $z+dz$ will be $Nf(z)dz$, where f always stands for the same function.

Now the existence of the velocity x does not in any way affect that of the velocities y or z, since these are all at right angles to each other and independent, so that the number of particles whose velocity lies between x and $x+dx$, and also between y and $y+dy$, and also between z and $z+dz$, is

$$Nf(x)f(y)f(z)\,dx\,dy\,dz.$$

If we suppose the N particles to start from the origin at the same instant, then this will be the number in the element of volume $(dx\,dy\,dz)$ after unit of time, and the number referred to unit of volume will be

$$Nf(x)f(y)f(z).$$

But the directions of the coordinates are perfectly arbitrary, and therefore this number must depend on the distance from the origin alone, that is

$$f(x)f(y)f(z) = \phi(x^2 + y^2 + z^2).$$

Solving this functional equation, we find

$$f(x) = Ce^{Ax^2}, \qquad \phi(r^2) = C^3 e^{Ar^2}.$$

If we make A positive, the number of particles will increase with the velocity, and we should find the whole number of particles infinite. We therefore make A negative and equal to $-\dfrac{1}{a^2}$, so that the number between x and $x + dx$ is

$$NCe^{-\frac{x^2}{a^2}}dx.$$

Integrating from $x = -\infty$ to $x = +\infty$, we find the whole number of particles,

$$NC\sqrt{\pi}a = N, \quad \therefore \quad C = \frac{1}{a\sqrt{\pi}},$$

$f(x)$ is therefore

$$\frac{1}{a\sqrt{\pi}}e^{-\frac{x^2}{a^2}}.$$

Whence we may draw the following conclusions :—

1st.   The number of particles whose velocity, resolved in a certain direction, lies between x and $x + dx$ is

$$N\frac{1}{a\sqrt{\pi}}e^{-\frac{x^2}{a^2}}dx \dots\dots\dots\dots\dots\dots\dots\dots (1).$$

2nd.   The number whose actual velocity lies between v and $v + dv$ is

$$N\frac{4}{a^3\sqrt{\pi}}v^2e^{-\frac{v^2}{a^2}}dv\dots\dots\dots\dots\dots\dots\dots\dots(2).$$

3rd.   To find the mean value of v, add the velocities of all the particles together and divide by the number of particles; the result is

$$\text{mean velocity} = \frac{2a}{\sqrt{\pi}} \dots\dots\dots\dots\dots\dots\dots\dots(3).$$

4th.   To find the mean value of v^2, add all the values together and divide by N,

$$\text{mean value of } v^2 = \tfrac{3}{2}a^2\dots\dots\dots\dots\dots\dots\dots(4).$$

This is greater than the square of the mean velocity, as it ought to be.

It appears from this proposition that the velocities are distributed among the particles according to the same law as the errors are distributed among the observations in the theory of the "method of least squares." The velocities range from 0 to ∞, but the number of those having great velocities is comparatively small. In addition to these velocities, which are in all directions equally, there may be a general motion of translation of the entire system of particles which must be compounded with the motion of the particles relatively to one another. We may call the one the motion of translation, and the other the motion of agitation.

PROP. V. Two systems of particles move each according to the law stated in Prop. IV.; to find the number of pairs of particles, one of each system, whose relative velocity lies between given limits.

Let there be N particles of the first system, and N' of the second, then NN' is the whole number of such pairs. Let us consider the velocities in the direction of x only; then by Prop. IV. the number of the first kind, whose velocities are between x and $x+dx$, is

$$N \frac{1}{a\sqrt{\pi}} e^{-\frac{x^2}{a^2}} dx.$$

The number of the second kind, whose velocity is between $x+y$ and $x+y+dy$, is

$$N' \frac{1}{\beta\sqrt{\pi}} e^{-\frac{(x+y)^2}{\beta^2}} dy,$$

where β is the value of a for the second system.

The number of pairs which fulfil both conditions is

$$NN' \frac{1}{a\beta\pi} e^{-\left(\frac{x^2}{a^2}+\frac{(x+y)^2}{\beta^2}\right)} dx\, dy.$$

Now x may have any value from $-\infty$ to $+\infty$ consistently with the difference of velocities being between y and $y+dy$; therefore integrating between these limits, we find

$$NN' \frac{1}{\sqrt{a^2+\beta^2}\sqrt{\pi}} e^{-\frac{y^2}{a^2+\beta^2}} dy \dots\dots\dots\dots\dots\dots (5)$$

for the whole number of pairs whose difference of velocity lies between y and $y+dy$.

This expression, which is of the same form with (1) if we put NN' for N, $a^2 + \beta^2$ for a^2, and y for x, shews that the distribution of relative velocities is regulated by the same law as that of the velocities themselves, and that the mean relative velocity is the square root of the sum of the squares of the mean velocities of the two systems.

Since the direction of motion of every particle in one of the systems may be reversed without changing the distribution of velocities, it follows that the velocities compounded of the velocities of two particles, one in each system, are distributed according to the same formula (5) as the relative velocities.

PROP. VI. Two systems of particles move in the same vessel; to prove that the mean *vis viva* of each particle will become the same in the two systems.

Let P be the mass of each particle of the first system, Q that of each particle of the second. Let p, q be the mean velocities in the two systems before impact, and let p', q' be the mean velocities after one impact. Let $OA = p$ and $OB = q$, and let AOB be a right angle; then, by Prop. V., AB will be the mean relative velocity, OG will be the mean velocity of the centre of gravity; and drawing aGb at right angles to OG, and making $aG = AG$ and $bG = BG$, then Oa will be the mean velocity of P after impact, compounded of OG and Ga, and Ob will be that of Q after impact.

Now
$$ AB = \sqrt{p^2 + q^2}, \; AG = \frac{Q}{P+Q}\sqrt{p^2 + q^2}, \; BG = \frac{P}{P+Q}\sqrt{p^2 + q^2}, $$

$$ OG = \frac{\sqrt{P^2 p^2 + Q^2 q^2}}{P + Q}, $$

therefore
$$ p' = Oa = \frac{\sqrt{Q^2(p^2 + q^2) + P^2 p^2 + Q^2 q^2}}{P + Q}, $$

and
$$ q' = Ob = \frac{\sqrt{P^2(p^2 + q^2) + P^2 p^2 + Q^2 q^2}}{P + Q}, $$

and
$$ Pp'^2 - Qq'^2 = \left(\frac{P-Q}{P+Q}\right)^2 (Pp^2 - Qq^2) \quad \dots\dots\dots\dots\dots\dots (6). $$

It appears therefore that the quantity $Pp^2 - Qq^2$ is diminished at every impact in the same ratio, so that after many impacts it will vanish, and then

$$Pp^2 = Qq^2.$$

Now the mean *vis viva* is $\frac{3}{2}Pa^2 = \frac{3\pi}{8}Pp^2$ for P, and $\frac{3\pi}{8}Qq^2$ for Q; and it is manifest that these quantities will be equal when $Pp^2 = Qq^2$.

If any number of different kinds of particles, having masses P, Q, R and velocities p, q, r respectively, move in the same vessel, then after many impacts

$$Pp^2 = Qq^2 = Rr^2, \&c. \dots\dots\dots\dots\dots\dots\dots\dots(7).$$

PROP. VII.   A particle moves with velocity r relatively to a number of particles of which there are N in unit of volume; to find the number of these which it approaches within a distance s in unit of time.

If we describe a tubular surface of which the axis is the path of the particle, and the radius the distance s, the content of this surface generated in unit of time will be $\pi r s^2$, and the number of particles included in it will be

$$N\pi r s^2 \dots\dots\dots\dots\dots\dots\dots\dots\dots\dots\dots\dots(8),$$

which is the number of particles to which the moving particle approaches within a distance s.

PROP. VIII.   A particle moves with velocity v in a system moving according to the law of Prop. IV.; to find the number of particles which have a velocity relative to the moving particle between r and $r + dr$.

Let u be the actual velocity of a particle of the system, v that of the original particle, and r their relative velocity, and θ the angle between v and r, then

$$u^2 = v^2 + r^2 - 2vr \cos \theta.$$

If we suppose, as in Prop. IV., all the particles to start from the origin at once, then after unit of time the "density" or number of particles to unit of volume at distance u will be

$$N \frac{1}{a^3 \pi^{\frac{3}{2}}} e^{-\frac{u^2}{a^2}}.$$

From this we have to deduce the number of particles in a shell whose centre is at distance v, radius $= r$, and thickness $= dr$,

$$N \frac{1}{a\sqrt{\pi}} \frac{r}{v} \left\{ e^{-\frac{(r-v)^2}{a^2}} - e^{-\frac{(r+v)^2}{a^2}} \right\} dr \dots\dots\dots\dots\dots\dots (9),$$

which is the number required.

Cor. It is evident that if we integrate this expression from $r = 0$ to $r = \infty$, we ought to get the whole number of particles $= N$, whence the following mathematical result,

$$\int_0^\infty dx \cdot x \left(e^{-\frac{(x-a)^2}{a^2}} - e^{-\frac{(x+a)^2}{a^2}} \right) = \sqrt{\pi} a a \dots\dots\dots\dots\dots (10).$$

Prop. IX. Two sets of particles move as in Prop. V.; to find the number of pairs which approach within a distance s in unit of time.

The number of the second kind which have a velocity between v and $v + dv$ is

$$N' \frac{4}{\beta^3 \sqrt{\pi}} v^2 e^{-\frac{v^2}{\beta^2}} dv = n'.$$

The number of the first kind whose velocity relative to these is between r and $r + dr$ is

$$N \frac{1}{a\sqrt{\pi}} \frac{r}{v} \left(e^{-\frac{(r-v)^2}{a^2}} - e^{-\frac{(r+v)^2}{a^2}} \right) dr = n,$$

and the number of pairs which approach within distance s in unit of time is

$$nn'\pi r s^2,$$

$$= NN' \frac{4}{a\beta^3} s^2 r^2 v e^{-\frac{v^2}{\beta}} \left\{ e^{-\frac{(v-r)^2}{a^2}} - e^{-\frac{(v+r)^2}{a^2}} \right\} dr\, dv.$$

By the last proposition we are able to integrate with respect to v, and get

$$NN' \frac{4\sqrt{\pi}}{(a^2+\beta^2)^{\frac{3}{2}}} s^2 r^3 e^{-\frac{r^2}{a^2+\beta^2}} dr.$$

Integrating this again from $r = 0$ to $r = \infty$,

$$2NN' \sqrt{\pi} \sqrt{a^2 + \beta^2} s^2 \dots\dots\dots\dots\dots\dots (11)$$

is the number of collisions in unit of time which take place in unit of volume between particles of different kinds, s being the distance of centres at collision.

The number of collisions between two particles of the first kind, s_1 being the striking distance, is

$$2N^2 \sqrt{\pi} \sqrt{2a^2 s_1^2} \, ;$$

and for the second system it is

$$2N'^2 \sqrt{\pi} \sqrt{2\beta^2 s_2^2}.$$

The mean velocities in the two systems are $\dfrac{2a}{\sqrt{\pi}}$ and $\dfrac{2\beta}{\sqrt{\pi}}$; so that if l_1 and l_2 be the mean distances travelled by particles of the first and second systems between each collision, then

$$\frac{1}{l_1} = \pi N_1 \sqrt{2} s_1^2 + \pi N_2 \frac{\sqrt{a^2 + \beta^2}}{a} s^2,$$

$$\frac{1}{l_2} = \pi N_1 \frac{\sqrt{a^2 + \beta^2}}{\beta} s^2 + \pi N_2 \sqrt{2} s_2^2.$$

Prop. X. To find the probability of a particle reaching a given distance before striking any other.

Let us suppose that the probability of a particle being stopped while passing through a distance dx, is adx; that is, if N particles arrived at a distance x, $Nadx$ of them would be stopped before getting to a distance $x + dx$. Putting this mathematically,

$$\frac{dN}{dx} = -Na, \text{ or } N = Ce^{-ax}.$$

Putting $N = 1$ when $x = 0$, we find e^{-ax} for the probability of a particle not striking another before it reaches a distance x.

The *mean distance* travelled by each particle before striking is $\dfrac{1}{a} = l$. The probability of a particle reaching a distance $= nl$ without being struck is e^{-n}. (See a paper by M. Clausius, *Philosophical Magazine*, February 1859.)

If all the particles are at rest but one, then the value of a is

$$a = \pi s^2 N,$$

where s is the distance between the centres at collision, and N is the number of particles in unit of volume. If v be the velocity of the moving particle relatively to the rest, then the number of collisions in unit of time will be

$$v \pi s^2 N \, ;$$

and if v_1 be the actual velocity, then the number will be $v_1 a$; therefore

$$a = \frac{v}{v_1} \pi s^2 N,$$

where v_1 is the actual velocity of the striking particle, and v its velocity relatively to those it strikes. If v_2 be the actual velocity of the other particles, then $v = \sqrt{v_1^2 + v_2^2}$. If $v_1 = v_2$, then $v = \sqrt{2} v_1$, and

$$a = \sqrt{2} \pi s^2 N.$$

Note *. M. Clausius makes $a = \frac{4}{3} \pi s^2 N$.

PROP. XI. In a mixture of particles of two different kinds, to find the mean path of each particle.

Let there be N_1 of the first, and N_2 of the second in unit of volume. Let s_1 be the distance of centres for a collision between two particles of the first set, s_2 for the second set, and s' for collision between one of each kind. Let v_1 and v_2 be the coefficients of velocity, M_1, M_2 the mass of each particle.

The probability of a particle M_1 not being struck till after reaching a distance x_1 by another particle of the same kind is

$$e^{-\sqrt{2} \pi s_1^2 N_1 x}.$$

---

* [In the *Philosophical Magazine* of 1860, Vol. I. pp. 434—6 Clausius explains the method by which he found his value of the mean relative velocity. It is briefly as follows: If u, v be the velocities of two particles their relative velocity is $\sqrt{u^2 + v^2 - 2uv \cos \theta}$ and the mean of this as regards direction only, all directions of v being equally probable, is shewn to be

$$v + \frac{1}{3} \frac{u^2}{v} \text{ when } u < v, \text{ and } u + \frac{1}{3} \frac{v^2}{u} \text{ when } u > v.$$

If $v = u$ these expressions coincide. Clausius in applying this result and putting u, v for the mean velocities assumes that the mean relative velocity is given by expressions of the same form, so that when the mean velocities are each equal to u the mean relative velocity would be $\frac{4}{3} u$. This step is, however, open to objection, and in fact if we take the expressions given above for the mean velocity, treating u and v as the velocities of two particles which may have any values between 0 and ∞, to calculate the mean relative velocity we should proceed as follows: Since the number of particles with velocities between u and $u + du$ is $N \frac{4}{a^3 \sqrt{\pi}} u^2 e^{-\frac{u^2}{a^2}} du$, the mean relative velocity is

$$\frac{16}{a^3 \beta^3 \pi} \int_0^\infty \int_v^\infty u^2 v^2 e^{-\left(\frac{u^2}{a^2} + \frac{v^2}{\beta^2}\right)} \left(u + \frac{1}{3} \frac{v^2}{u}\right) du\, dv + \frac{16}{a^3 \beta^3 \pi} \int_0^\infty \int_0^v u^2 v^2 e^{-\left(\frac{u^2}{a^2} + \frac{v^2}{\beta^2}\right)} \left(v + \frac{1}{3} \frac{u^2}{v}\right) du\, dv.$$

This expression, when reduced, leads to $\frac{2}{\sqrt{\pi}} \sqrt{a^2 + \beta^2}$, which is the result in the text. Ed.]

The probability of not being struck by a particle of the other kind in the same distance is

$$e^{-\sqrt{1+\frac{v_2^2}{v_1^2}}\,\pi s'^2 N_2 x}.$$

Therefore the probability of not being struck by any particle before reaching a distance x is

$$e^{-\pi\left(\sqrt{2}s_1^2 N_1 + \sqrt{1+\frac{v_2^2}{v_1^2}}\,s'^2 N_2\right)x};$$

and if l_1 be the *mean distance* for a particle of the first kind,

$$\frac{1}{l_1} = \sqrt{2}\,\pi s_1^2 N_1 + \pi\sqrt{1+\frac{v_2^2}{v_1^2}}\,s'^2 N_2 \dots\dots\dots\dots\dots (12).$$

Similarly, if l_2 be the mean distance for a particle of the second kind,

$$\frac{1}{l_2} = \sqrt{2}\,\pi s_2^2 N_2 + \pi\sqrt{1+\frac{v_1^2}{v_2^2}}\,s'^2 N_1 \dots\dots\dots\dots\dots(13).$$

The mean density of the particles of the first kind is $N_1 M_1 = \rho_1$, and that of the second $N_2 M_2 = \rho_2$. If we put

$$A = \sqrt{2}\,\frac{\pi s_1^2}{M_1}, \quad B = \pi\sqrt{1+\frac{v_2^2}{v_1^2}}\,\frac{s'^2}{M_2}, \quad C = \pi\sqrt{1+\frac{v_1^2}{v_2^2}}\,\frac{s'^2}{M_1}, \quad D = \sqrt{2}\,\frac{\pi s_2^2}{M_2}\dots\dots(14),$$

$$\frac{1}{l_1} = A\rho_1 + B\rho_2, \quad \frac{1}{l_2} = C\rho_1 + D\rho_2\dots\dots\dots\dots\dots(15),$$

and

$$\frac{B}{C} = \frac{M_1 v_2}{M_2 v_1} = \frac{v_2^3}{v_1^3}\dots\dots\dots\dots\dots\dots(16).$$

Prop. XII. To find the pressure on unit of area of the side of the vessel due to the impact of the particles upon it.

Let N = number of particles in unit of volume;

M = mass of each particle;

v = velocity of each particle;

l = mean path of each particle;

then the number of particles in unit of area of a stratum dz thick is

$$Ndz \dots\dots\dots\dots\dots\dots\dots (17).$$

The number of collisions of these particles in unit of time is

$$Ndz\,\frac{v}{l}\dots\dots\dots\dots\dots\dots(18).$$

The number of particles which after collision reach a distance between nl and $(n+dn)\,l$ is

$$N\,\frac{v}{l}\,e^{-n}\,dz\,dn \dots\dots\dots\dots\dots\dots\dots (19).$$

The proportion of these which strike on unit of area at distance z is

$$\frac{nl-z}{2nl} \dots\dots\dots\dots\dots\dots\dots\dots\dots\dots(20)\,;$$

the mean velocity of these in the direction of z is

$$v\,\frac{nl+z}{2nl} \dots\dots\dots\dots\dots\dots\dots\dots (21).$$

Multiplying together (19), (20), and (21), and M, we find the momentum at impact

$$MN\,\frac{v^2}{4n^2l^3}\,(n^2l^2-z^2)\,e^{-n}\,dz\,dn.$$

Integrating with respect to z from 0 to nl, we get

$$\tfrac{1}{6}MNv^2\,ne^{-n}\,dn.$$

Integrating with respect to n from 0 to ∞, we get

$$\tfrac{1}{6}MNv^2$$

for the momentum in the direction of z of the striking particles; for the momentum of the particles after impact is the same, but in the opposite direction; so that the whole pressure on unit of area is twice this quantity, or

$$p=\tfrac{1}{3}MNv^2.$$

This value of p is independent of l the length of path. In applying this result to the theory of gases, we put $MN=\rho$, and $v^2=3k$, and then

$$p=k\rho,$$

which is Boyle and Mariotte's law.   By (4) we have

$$v^2=\tfrac{3}{2}a^2, \quad \therefore\ a^2=2k \dots\dots\dots\dots\dots\dots (23).$$

We have seen that, on the hypothesis of elastic particles moving in straight lines, the pressure of a gas can be explained by the assumption that the square of the velocity is proportional directly to the absolute temperature, and inversely to the specific gravity of the gas at constant temperature, so that at the same pressure and temperature the value of NMv^2 is the same for all gases. But we found in Prop. VI. that when two sets of particles communicate agitation to one another, the value of Mv^2 is the same in each. From this it appears that N, the number of particles in unit of volume, is the same for all gases at the same pressure and temperature. This result agrees with the chemical law, that equal volumes of gases are chemically equivalent.

We have next to determine the value of l, the mean length of the path of a particle between consecutive collisions. The most direct method of doing this depends upon the fact, that when different strata of a gas slide upon one another with different velocities, they act upon one another with a tangential force tending to prevent this sliding, and similar in its results to the friction between two solid surfaces sliding over each other in the same way. The explanation of gaseous friction, according to our hypothesis, is, that particles having the mean velocity of translation belonging to one layer of the gas, pass out of it into another layer having a different velocity of translation; and by striking against the particles of the second layer, exert upon it a tangential force which constitutes the internal friction of the gas. The whole friction between two portions of gas separated by a plane surface, depends upon the total action between all the layers on the one side of that surface upon all the layers on the other side.

PROP. XIII. To find the internal friction in a system of moving particles.

Let the system be divided into layers parallel to the plane of xy, and let the motion of translation of each layer be u in the direction of x, and let $u = A + Bz$. We have to consider the mutual action between the layers on the positive and negative sides of the plane xy. Let us first determine the action between two layers dz and dz', at distances z and $-z'$ on opposite sides of this plane, each unit of area. The number of particles which, starting from dz in unit of time, reach a distance between nl and $(n + dn) l$ is by (19),

$$N \frac{v}{l} e^{-n} \, dz \, dn.$$

The number of these which have the ends of their paths in the layer dz' is

$$N \frac{v}{2nl^2} e^{-n} \, dz \, dz' \, dn.$$

The mean velocity in the direction of x which each of these has before impact is $A + Bz$, and after impact $A + Bz'$; and its mass is M, so that a mean momentum $= MB(z-z')$ is communicated by each particle. The whole action due to these collisions is therefore

$$NMB\frac{v}{2nl^2}(z-z')e^{-n}\,dz\,dz'\,dn.$$

We must first integrate with respect to z' between $z'=0$ and $z'=z-nl$; this gives

$$\tfrac{1}{2}NMB\frac{v}{2nl^2}(n^2l^2-z^2)e^{-n}\,dz\,dn$$

for the action between the layer dz and all the layers below the plane xy. Then integrate from $z=0$ to $z=nl$,

$$\tfrac{1}{6}MNBlvn^2e^{-n}\,dn.$$

Integrate from $n=0$ to $n=\infty$, and we find the whole friction between unit of area above and below the plane to be

$$F=\tfrac{1}{3}MNlvB=\tfrac{1}{3}\rho lv\,\frac{du}{dz}=\mu\frac{du}{dz},$$

where μ is the ordinary coefficient of internal friction,

$$\mu=\tfrac{1}{3}\rho lv=\frac{1}{3\sqrt{2}}\frac{Mv}{\pi s^2}\;\dotfill\;(24),$$

where ρ is the density, l the mean length of path of a particle, and v the mean velocity $v=\dfrac{2a}{\sqrt{\pi}}=2\sqrt{\dfrac{2k}{\pi}}$,

$$l=\tfrac{3}{2}\frac{\mu}{\rho}\sqrt{\frac{\pi}{2k}}\;\dotfill\;(25).$$

Now Professor Stokes finds by experiments on air,

$$\sqrt{\frac{\mu}{\rho}}=\cdot116.$$

If we suppose $\sqrt{k}=930$ feet per second for air at 60°, and therefore the mean velocity $v=1505$ feet per second, then the value of l, the mean distance travelled over by a particle between consecutive collisions, $=\frac{1}{447000}$th of an inch, and each particle makes 8,077,200,000 collisions per second.

A remarkable result here presented to us in equation (24), is that if this explanation of gaseous friction be true, the coefficient of friction is independent of the density. Such a consequence of a mathematical theory is very startling, and the only experiment I have met with on the subject does not seem to confirm it. We must next compare our theory with what is known of the diffusion of gases, and the conduction of heat through a gas.

PART II.

* On the Process of Diffusion of two or more kinds of moving particles among one another.

We have shewn, in the first part of this paper, that the motions of a system of many small elastic particles are of two kinds : one, a general motion of translation of the whole system, which may be called the motion in mass; and the other a motion of agitation, or molecular motion, in virtue of which velocities in all directions are distributed among the particles according to a certain law. In the cases we are considering, the collisions are so frequent that the law of distribution of the molecular velocities, if disturbed in any way, will be re-established in an inappreciably short time; so that the motion will always consist of this definite motion of agitation, combined with the general motion of translation.

When two gases are in communication, streams of the two gases might run freely in opposite directions, if it were not for the collisions which take place between the particles. The rate at which they actually interpenetrate each other must be investigated. The diffusion is due partly to the spreading of the particles by the molecular agitation, and partly to the actual motion of the two opposite currents in mass, produced by the pressure behind, and resisted

---

* [The methods and results of this paper have been criticised by Clausius in a memoir published in Poggendorff's *Annalen*, Vol. cxv., and in the *Philosophical Magazine*, Vol. xxiii. His main objection is that the various circumstances of the strata, discussed in the paper, have not been sufficiently represented in the equations. In particular, if there be a series of strata at different temperatures perpendicular to the axis of x, then the proportion of molecules whose directions form with the axis of x angles whose cosines lie between μ and $\mu + d\mu$ is not $\frac{1}{2}d\mu$ as has been assumed by Maxwell throughout his work, but $\frac{1}{2}Hd\mu$ where H is a factor to be determined. In discussing the steady conduction of heat through a gas Clausius assumes that, in addition to the velocity attributed to the molecule according to Maxwell's theory, we must also suppose a velocity normal to the stratum and depending on the temperature of the stratum. On this assumption the factor H is investigated along with other modifications, and an expression for the assumed velocity is determined from the consideration that when the flow of heat is steady there is no movement of the mass. Clausius combining his own results with those of Maxwell points out that the expression contained in (28) of the paper involves as a result the motion of the gas. He also disputes the accuracy of expression (59) for the Conduction of Heat. In the introduction to the memoir published in the *Phil. Trans.*, 1866, it will be found that Maxwell expresses dissatisfaction with his former theory of the Diffusion of Gases, and admits the force of the objections made by Clausius to his expression for the Conduction of Heat. Ed.]

by the collisions of the opposite stream. When the densities are equal, the diffusions due to these two causes respectively are as 2 to 3.

PROP. XIV. *In a system of particles whose density, velocity, &c. are functions of* x, *to find the quantity of matter transferred across the plane of* yz, *due to the motion of agitation alone.*

If the number of particles, their velocity, or their length of path is greater on one side of this plane than on the other, then more particles will cross the plane in one direction than in the other; and there will be a transference of matter across the plane, the amount of which may be calculated.

Let there be taken a stratum whose thickness is dx, and area unity, at a distance x from the origin. The number of collisions taking place in this stratum in unit of time will be

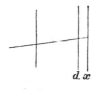

$$N \frac{v}{l} dx.$$

The proportion of these which reach a distance between nl and $(n+dn)l$ before they strike another particle is

$$e^{-n} dn.$$

The proportion of these which pass through the plane yz is

$$\frac{nl+x}{2nl} \text{ when } x \text{ is between } -nl \text{ and } 0,$$

and $\qquad -\dfrac{nl-x}{2nl} \text{ when } x \text{ is between } 0 \text{ and } +nl;$

the sign being negative in the latter case, because the particles cross the plane in the negative direction. The mass of each particle is M; so that the quantity of matter which is projected from the stratum dx, crosses the plane yz in a positive direction, and strikes other particles at distances between nl and $(n+dn)l$ is

$$\frac{MNv\,(x \mp nl)}{2nl^{2}} dx\, e^{-n} dn \dots\dots\dots\dots\dots\dots(26),$$

where x must be between $\pm nl$, and the upper or lower sign is to be taken according as x is positive or negative.

In integrating this expression, we must remember that N, v, and l are functions of x, not vanishing with x, and of which the variations are very small between the limits $x = -nl$ and $x = +nl$.

As we may have occasion to perform similar integrations, we may state here, to save trouble, that if U and r are functions of x not vanishing with x, whose variations are very small between the limits $x = +r$ and $x = -r$,

$$\int_{-r}^{+r} \pm U x^m dx = \frac{2}{m+2} \frac{d}{dx} (U r^{m+2}) \dotfill (27).$$

When m is an odd number, the upper sign only is to be considered; when m is even or zero, the upper sign is to be taken with positive values of x, and the lower with negative values. Applying this to the case before us,

$$\int_{-nl}^{+nl} \frac{MNvx}{2nl^2} dx = \tfrac{1}{3} \frac{d}{dx} (MNvn^2l),$$

$$\int_{-nl}^{+nl} \mp \frac{MNv}{2l} dx = -\tfrac{1}{2} \frac{d}{dx} (MNvn^2l).$$

We have now to integrate

$$\int_0^\infty -\tfrac{1}{6} \frac{d}{dx} (MNvl)\, n^2 e^{-n} dn,$$

n being taken from 0 to ∞. We thus find for the quantity of matter transferred across unit of area by the motion of agitation in unit of time,

$$q = -\tfrac{1}{3} \frac{d}{dx} (\rho v l) \dotfill (28),$$

where $\rho = MN$ is the density, v the mean velocity of agitation, and l the mean length of path.

PROP. XV. The quantity transferred, in consequence of a mean motion of translation V, would obviously be

$$Q = V\rho \dotfill (29).$$

PROP. XVI. *To find the resultant dynamical effect of all the collisions which take place in a given stratum.*

Suppose the density and velocity of the particles to be functions of x, then more particles will be thrown into the given stratum from that side on which the density is greatest; and those particles which have greatest velocity will have the greatest effect, so that the stratum will not be generally in equilibrium, and the dynamical measure of the force exerted on the stratum will be the resultant momentum of all the particles which lodge in it during unit of time. We shall first take the case in which there is no mean motion of translation, and then consider the effect of such motion separately.

Let a stratum whose thickness is a (a small quantity compared with l), and area unity, be taken at the origin, perpendicular to the axis of x; and let another stratum, of thickness dx, and area unity, be taken at a distance x from the first.

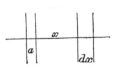

If M_1 be the mass of a particle, N the number in unit of volume, v the velocity of agitation, l the mean length of path, then the number of collisions which take place in the stratum dx is

$$N \frac{v}{l} \, dx.$$

The proportion of these which reach a distance between nl and $(n+dn)l$ is

$$e^{-n} \, dn.$$

The proportion of these which have the extremities of their paths in the stratum a is

$$\frac{a}{2nl}.$$

The velocity of these particles, resolved in the direction of x, is

$$-\frac{vx}{nl},$$

and the mass is M; so that multiplying all these terms together, we get

$$\frac{NMv^2ax}{2n^2l^3} e^{-n} \, dx \, dn \dotfill (30)$$

for the momentum of the particles fulfilling the above conditions.

To get the whole momentum, we must first integrate with respect to x from $x = -nl$ to $x = +nl$, remembering that l may be a function of x, and is a very small quantity. The result is

$$\frac{d}{dx} \left(\frac{NMv^2}{3} \right) ane^{-n} \, dn.$$

Integrating with respect to n from $n = 0$ to $n = \infty$, the result is

$$- a \, \frac{d}{dx} \left(\frac{NMv^2}{3} \right) = aX\rho \quad\ldots\ldots\ldots\ldots\ldots\ldots\ldots (31)$$

as the whole resultant force on the stratum a arising from these collisions. Now $\dfrac{NMv^2}{3} = p$ by Prop. XII., and therefore we may write the equation

$$- \frac{dp}{dx} = X\rho \quad\ldots\ldots\ldots\ldots\ldots\ldots\ldots\ldots\ldots (32),$$

the ordinary hydrodynamical equation.

PROP. XVII. *To find the resultant effect of the collisions upon each of several different systems of particles mixed together.*

Let M_1, M_2, &c. be the masses of the different kinds of particles, N_1, N_2, &c. the number of each kind in unit of volume, v_1, v_2, &c. their velocities of agitation, l_1, l_2 their mean paths, p_1, p_2, &c. the pressures due to each system of particles; then

$$\left. \begin{aligned} \frac{1}{l_1} &= A\rho_1 + B\rho_2 + \text{\&c.} \\ \frac{1}{l_2} &= C\rho_1 + D\rho_2 + \text{\&c.} \end{aligned} \right\} \quad\ldots\ldots\ldots\ldots\ldots\ldots (33).$$

The number of collisions of the first kind of particles with each other in unit of time will be

$$N_1 v_1 A \rho_1.$$

The number of collisions between particles of the first and second kinds will be

$$N_1 v_1 B \rho_2, \quad \text{or} \quad N_2 v_2 C \rho_1, \quad \text{because} \quad v_1^3 B = v_2^3 C.$$

The number of collisions between particles of the second kind will be $N_2 v_2 D \rho_2$, and so on, if there are more kinds of particles.

Let us now consider a thin stratum of the mixture whose volume is unity.

The resultant momentum of the particles of the first kind which lodge in it during unit of time is

$$- \frac{dp_1}{dx}.$$

The proportion of these which strike particles of the first kind is

$$A\rho_1 l_1.$$

The whole momentum of these remains among the particles of the first kind. The proportion which strike particles of the second kind is

$$B\rho_2 l_1.$$

The momentum of these is divided between the striking particles in the ratio of their masses; so that $\dfrac{M_1}{M_1+M_2}$ of the whole goes to particles of the first kind, and $\dfrac{M_2}{M_1+M_2}$ to particles of the second kind.

The effect of these collisions is therefore to produce a force

$$-\frac{dp_1}{dx}\left(A\rho_1 l_1 + B\rho_2 l_1 \frac{M_1}{M_1+M_2}\right)$$

on particles of the first system, and

$$-\frac{dp_1}{dx} B\rho_2 l_1 \frac{M_2}{M_1+M_2}$$

on particles of the second system.

The effect of the collisions of those particles of the second system which strike into the stratum, is to produce a force

$$-\frac{dp_2}{dx} C\rho_1 l_2 \frac{M_1}{M_1+M_2}$$

on the first system, and

$$-\frac{dp_2}{dx}\left(C\rho_1 l_2 \frac{M_2}{M_1+M_2} + D\rho_2 l_2\right)$$

on the second.

The whole effect of these collisions is therefore to produce a resultant force

$$-\frac{dp_1}{dx}\left(A\rho_1 l_1 + B\rho_2 l_1 \frac{M_1}{M_1+M_2}\right) - \frac{dp_2}{dx} C\rho_1 l_2 \frac{M_1}{M_1+M_2} + \&c. \ldots\ldots\ldots(34)$$

on the first system,

$$-\frac{dp_1}{dx} B\rho_2 l_1 \frac{M_2}{M_1+M_2} - \frac{dp_2}{dx}\left(C\rho_1 l_2 \frac{M_2}{M_1+M_2} + D\rho_2 l_2\right) + \&c. \ldots\ldots(35)$$

on the second, and so on.

PROP. XVIII. *To find the mechanical effect of a difference in the mean velocity of translation of two systems of moving particles.*

Let V_1, V_2 be the mean velocities of translation of the two systems respectively, then $\dfrac{M_1 M_2}{M_1 + M_2}(V_1 - V_2)$ is the mean momentum lost by a particle of the first, and gained by a particle of the second at collision. The number of such collisions in unit of volume is

$$N_1 B \rho_2 v_1, \text{ or } N_2 C \rho_1 v_2;$$

therefore the whole effect of the collisions is to produce a force

$$= - N_1 B \rho_2 v_1 \frac{M_1 M_2}{M_1 + M_2}(V_1 - V_2) \dots\dots\dots\dots\dots\dots(36)$$

on the first system, and an equal and opposite force

$$= + N_2 C \rho_1 v_2 \frac{M_1 M_2}{M_1 + M_2}(V_1 - V_2) \dots\dots\dots\dots\dots (37)$$

on unit of volume of the second system.

PROP. XIX. *To find the law of diffusion in the case of two gases diffusing into each other through a plug made of a porous material, as in the case of the experiments of Graham.*

The pressure on each side of the plug being equal, it was found by Graham that the quantities of the gases which passed in opposite directions through the plug in the same time were directly as the square roots of their specific gravities.

We may suppose the action of the porous material to be similar to that of a number of particles fixed in space, and obstructing the motion of the particles of the moving systems. If L_1 is the mean distance a particle of the first kind would have to go before striking a fixed particle, and L_2 the distance for a particle of the second kind, then the mean paths of particles of each kind will be given by the equations

$$\frac{1}{l_1} = A\rho_1 + B\rho_2 + \frac{1}{L_1}, \quad \frac{1}{l_2} = C\rho_1 + D\rho_2 + \frac{1}{L_2} \dots\dots\dots\dots (38).$$

The mechanical effect upon the plug of the pressures of the gases on each side, and of the percolation of the gases through it, may be found by Props. XVII. and XVIII. to be

$$\frac{M_1 N_1 v_1 V_1}{L_1} + \frac{M_2 N_2 v_2 V_2}{L_2} - \frac{dp_1}{dx}\frac{l_1}{L_1} - \frac{dp_2}{dx}\frac{l_2}{L_2} = 0 \dots\dots\dots\dots(39);$$

and this must be zero, if the pressures are equal on each side of the plug. Now if Q_1, Q_2 be the quantities transferred through the plug by the mean motion of translation, $Q_1 = \rho_1 V_1 = M_1 N_1 V_1$; and since by Graham's law

$$\frac{Q_1}{Q_2} = -\sqrt{\frac{M_1}{M_2}} = -\frac{v_2}{v_1},$$

we shall have

$$M_1 N_1 v_1 V_1 = -M_2 N_2 v_2 V_2 = U \text{ suppose};$$

and since the pressures on the two sides are equal, $\frac{dp_2}{dx} = -\frac{dp_1}{dx}$, and the only way in which the equation of equilibrium of the plug can generally subsist is when $L_1 = L_2$ and $l_1 = l_2$. This implies that $A = C$ and $B = D$. Now we know that $v_1^3 B = v_2^3 C$. Let $K = 3\dfrac{A}{v_1^3}$, then we shall have

$$A = C = \tfrac{1}{3} K v_1^3, \quad B = D = \tfrac{1}{3} K v_2^3 \quad\text{.....................} (40),$$

and

$$\frac{1}{l_1} = \frac{1}{l_2} = K\,(v_1 p_1 + v_2 p_2) + \frac{1}{L} \quad\text{............................}(41).$$

The diffusion is due partly to the motion of translation, and partly to that of agitation. Let us find the part due to the motion of translation.

The equation of motion of one of the gases through the plug is found by adding the forces due to pressures to those due to resistances, and equating these to the moving force, which in the case of slow motions may be neglected altogether. The result for the first is

$$\frac{dp_1}{dx}\left(A\rho_1 l_1 + B\rho_2 l_1 \frac{M_1}{M_1 + M_2}\right) + \frac{dp_2}{dx}\,C\rho_1 l_2 \frac{M_1}{M_1 + M_2}$$

$$+ N_1 B \rho_2 v_1 \frac{M_1 M_2}{M_1 + M_2}\,(V_1 - V_2) + \frac{\rho_1 v_1 V_1}{L} = 0 \quad\text{......} (42).$$

Making use of the simplifications we have just discovered, this becomes

$$\frac{dp}{dx}\frac{Kl}{v_1^2 + v_2^2}\,(v_1^3 p_1 + v_2^3 p_2) + K\frac{v_1 v_2}{v_1^2 + v_2^2}\,(p_1 v_2 + p_2 v_1)\,U + \frac{1}{L}\,U \quad\text{.........}(43),$$

whence

$$U = -\frac{dp}{dx}\frac{Kl\,(v_1^3 p_1 + v_2^3 p_2)}{Kv_1 v_2\,(p_1 v_2 + p_2 v_1) + \dfrac{v_1^2 + v_2^2}{L}} \quad\text{..................} (44);$$

whence the rate of diffusion due to the motion of translation may be found; for

$$Q_1 = \frac{U}{v_1}, \text{ and } Q_2 = -\frac{U}{v_2} \quad\dots\dots\dots\dots\dots\dots (45).$$

To find the diffusion due to the motion of agitation, we must find the value of q_1.

$$q_1 = -\tfrac{1}{3}\frac{d}{dx}(\rho_1 v_1 l_1),$$

$$= -\frac{L}{v_1}\frac{d}{dx}\frac{p_1}{1 + KL(v_1 p_1 + v_2 p_2)},$$

$$q_1 = -\frac{l^2}{v_1 L}\frac{dp}{dx}\{1 + KLv_2(p_1 + p_2)\} \quad\dots\dots\dots\dots (46).$$

Similarly,
$$q_2 = +\frac{l^2}{v_2 L}\frac{dp}{dx}\{1 + KLv_1(p_1 + p_2)\} \quad\dots\dots\dots\dots (47).$$

The whole diffusions are $Q_1 + q_1$ and $Q_2 + q_2$. The values of q_1 and q_2 have a term not following Graham's law of the square roots of the specific gravities, but following the law of equal volumes. The closer the material of the plug, the less will this term affect the result.

Our assumptions that the porous plug acts like a system of fixed particles, and that Graham's law is fulfilled more accurately the more compact the material of the plug, are scarcely sufficiently well verified for the foundation of a theory of gases; and even if we admit the original assumption that they are systems of moving elastic particles, we have not very good evidence as yet for the relation among the quantities A, B, C, and D.

Prop. XX. *To find the rate of diffusion between two vessels connected by a tube.*

When diffusion takes place through a large opening, such as a tube connecting two vessels, the question is simplified by the absence of the porous diffusion plug; and since the pressure is constant throughout the apparatus, the volumes of the two gases passing opposite ways through the tube at the same time must be equal. Now the quantity of gas which passes through the tube is due partly to the motion of agitation as in Prop. XIV., and partly to the mean motion of translation as in Prop. XV.

Let us suppose the volumes of the two vessels to be a and b, and the length of the tube between them c, and its transverse section s. Let a be filled with the first gas, and b with the second at the commencement of the experiment, and let the pressure throughout the apparatus be P.

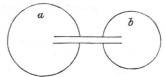

Let a volume y of the first gas pass from a to b, and a volume y' of the second pass from b to a; then if p_1 and p_2 represent the pressures in a due to the first and second kinds of gas, and p'_1 and p'_2 the same in the vessel b,

$$p_1 = \frac{a-y}{a}\,P, \quad p_2 = \frac{y'}{a}\,P, \quad p'_1 = \frac{y}{b}\,P, \quad p'_2 = \frac{b-y'}{b}\,P \dots\dots\dots\dots(48).$$

Since there is still equilibrium,

$$p_1 + p_2 = p'_1 + p'_2,$$

which gives

$$y = y' \text{ and } p_1 + p_2 = P = p'_1 + p'_2 \dots\dots\dots\dots\dots(49).$$

The rate of diffusion will be $+\dfrac{dy}{dt}$ for the one gas, and $-\dfrac{dy}{dt}$ for the other, measured in volume of gas at pressure P.

Now the rate of diffusion of the first gas will be

$$\frac{dy}{dt} = s\,\frac{k_1 q_1 + p_1 V_1}{P} = s\,\frac{-\frac{1}{3}v_1\dfrac{d}{dx}(p_1 l_1) + p_1 V_1}{P} \dots\dots\dots\dots(50);$$

and that of the second,

$$-\frac{dy}{dt} = s\,\frac{-\frac{1}{3}v_2\dfrac{d}{dx}(p_2 l_2) + p_2 V_2}{P} \dots\dots\dots\dots(51).$$

We have also the equation, derived from Props. XVI. and XVII.,

$$\frac{dp_1}{dx}\{A\rho_1 l_1 (M_1 + M_2) + B\rho_2 l_1 M_1 - C\rho_1 l_2 M_1\} + B\rho_1 \rho_2 v_1 M_2 (V_1 - V_2) = 0 \dots(52).$$

From these three equations we can eliminate V_1 and V_2, and find $\dfrac{dy}{dt}$ in terms of p and $\dfrac{dp}{dx}$, so that we may write

$$\frac{dy}{dt} = f\left(p_1,\ \frac{dp_1}{dx}\right) \dots\dots\dots\dots\dots\dots(53).$$

Since the capacity of the tube is small compared with that of the vessels, we may consider $\dfrac{dy}{dt}$ constant through the whole length of the tube. We may then solve the differential equation in p and x; and then making $p = p_1$ when $x = 0$, and $p = p'_1$ when $x = c$, and substituting for p_1 and p'_1 their values in terms of y, we shall have a differential equation in y and t, which being solved, will give the amount of gas diffused in a given time.

The solution of these equations would be difficult unless we assume relations among the quantities A, B, C, D, which are not yet sufficiently established in the case of gases of different density. Let us suppose that in a particular case the two gases have the same density, and that the four quantities A, B, C, D are all equal.

The volume diffused, owing to the motion of agitation of the particles, is then

$$-\tfrac{1}{3}\frac{s}{P}\frac{dp}{dx}\,vl,$$

and that due to the motion of translation, or the interpenetration of the two gases in opposite streams, is

$$-\frac{s}{P}\frac{dp}{dx}\frac{kl}{v}.$$

The values of v are distributed according to the law of Prop. IV., so that the mean value of v is $\dfrac{2a}{\sqrt{\pi}}$, and that of $\dfrac{1}{v}$ is $\dfrac{2}{\sqrt{\pi a}}$, that of k being $\tfrac{1}{2}a^2$. The diffusions due to these two causes are therefore in the ratio of 2 to 3, and their sum is

$$\frac{dy}{dt} = -\tfrac{4}{3}\sqrt{\frac{2k}{\pi}}\,\frac{sl}{P}\frac{dp}{dx}\ \dots\dots\dots\dots\dots\dots\dots (54).$$

If we suppose $\dfrac{dy}{dt}$ constant throughout the tube, or, in other words, if we regard the motion as *steady* for a short time, then $\dfrac{dp}{dx}$ will be constant and equal to $\dfrac{p'_1 - p_1}{c}$; or substituting from (48),

$$\frac{dy}{dt} = -\tfrac{4}{3}\sqrt{\frac{2k}{\pi}}\,\frac{sl}{abc}\{(a+b)\,y - ab\}\ \dots\dots\dots\dots (55);$$

whence
$$y = \frac{ab}{a+b}\left(1 - e^{-\tfrac{4}{3}\sqrt{\frac{2k}{\pi}}\frac{sl}{abc}(a+b)t}\right)\dots\dots\dots\dots\dots\dots(56).$$

By choosing pairs of gases of equal density, and ascertaining the amount of diffusion in a given time, we might determine the value of l in this expression. The diffusion of nitrogen into carbonic oxide or of deutoxide of nitrogen into carbonic acid, would be suitable cases for experiment. The only existing experiment which approximately fulfils the conditions is one by Graham, quoted by Herapath from Brande's *Quarterly Journal of Science*, Vol. XVIII. p. 76.

A tube 9 inches long and 0·9 inch diameter, communicated with the atmosphere by a tube 2 inches long and 0·12 inch diameter; 152 parts of olefiant gas being placed in the tube, the quantity remaining after four hours was 99 parts.

In this case there is not much difference of specific gravity between the gases, and we have $a = 9 \times (0\cdot9)^2 \frac{\pi}{4}$ cubic inches, $b = \infty$, $c = 2$ inches, and $s = (0\cdot12)^2 \frac{\pi}{4}$ square inches;

$$l = \sqrt{\frac{\pi}{2k}} \cdot \frac{3}{4} \frac{ac}{s} \log_e 10 \cdot \frac{1}{t} \cdot \log_{10} \left(\frac{a}{a-y} \right) \dots\dots\dots\dots (57);$$

$$\therefore \ l = 0\cdot00000256 \text{ inch} = \tfrac{1}{389000} \text{ inch} \dots\dots\dots\dots (58).$$

PROP. XXI. *To find the amount of energy which crosses unit of area in unit of time when the velocity of agitation is greater on one side of the area than on the other.*

The energy of a single particle is composed of two parts,—the *vis viva* of the centre of gravity, and the *vis viva* of the various motions of rotation round that centre, or, if the particle be capable of internal motions, the *vis viva* of these. We shall suppose that the whole *vis viva* bears a constant proportion to that due to the motion of the centre of gravity, or

$$E = \tfrac{1}{2}\beta M v^2,$$

where β is a coefficient, the experimental value of which is 1·634. Substituting E for M in Prop. XIV., we get for the transference of energy across unit of area in unit of time,

$$Jq = -\tfrac{1}{3} \frac{d}{dx} \left(\tfrac{1}{2}\beta M v^2 N v l \right),$$

where J is the mechanical equivalent of heat in foot-pounds, and q is the transfer of heat in thermal units.

Now $MN = \rho$, and $l = \dfrac{1}{A\rho}$, so that $MNl = \dfrac{1}{A}$;

$$\therefore Jq = -\tfrac{1}{2}\frac{\beta v^2}{A}\frac{dv}{dx} \quad\dotfill (59).$$

Also, if T is the absolute temperature,

$$\frac{1}{T}\frac{dT}{dx} = \frac{2}{v}\frac{dv}{dx};$$

$$\therefore Jq = -\tfrac{3}{4}\beta plv\,\frac{1}{T}\frac{dT}{dx} \dotfill (60),$$

where p must be measured in dynamical units of force.

Let $J = 772$ foot-pounds, $p = 2116$ pounds to square foot, $l = \frac{1}{400000}$ inch, $v = 1505$ feet per second, $T = 522$ or $62°$ Fahrenheit ; then

$$q = \frac{T' - T}{40000x} \dotfill (61),$$

where q is the flow of heat in thermal units per square foot of area ; and T' and T are the temperatures at the two sides of a stratum of air x *inches* thick.

In Prof. Rankine's work on the Steam-engine, p. 259, values of the *thermal resistance*, or the reciprocal of the *conductivity*, are given for various substances as computed from a Table of conductivities deduced by M. Peclet from experiments by M. Despretz :—

	Resistance.
Gold, Platinum, Silver	0·0036
Copper	0·0040
Iron	0·0096
Lead	0·0198
Brick	0·3306

Air by our calculation............40000

It appears, therefore, that the resistance of a stratum of air to the conduction of heat is about 10,000,000 times greater than that of a stratum of copper of equal thickness. It would be almost impossible to establish the value of the conductivity of a gas by direct experiment, as the heat radiated from the sides of the vessel would be far greater than the heat conducted through the air, even if currents could be entirely prevented*.

PART III.

ON THE COLLISION OF PERFECTLY ELASTIC BODIES OF ANY FORM.

When two perfectly smooth spheres strike each other, the force which acts between them always passes through their centres of gravity; and therefore their motions of rotation, if they have any, are not affected by the collision, and do not enter into our calculations. But, when the bodies are not spherical, the force of compact will not, in general, be in the line joining their centres of gravity; and therefore the force of impact will depend both on the motion of the centres and the motions of rotation before impact, and it will affect both these motions after impact.

In this way the velocities of the centres and the velocities of rotation will act and react on each other, so that finally there will be some relation established between them; and since the rotations of the particles about their three axes are quantities related to each other in the same way as the three velocities of their centres, the reasoning of Prop. IV. will apply to rotation as well as velocity, and both will be distributed according to the law

$$\frac{dN}{dx} = N \frac{1}{a\sqrt{\pi}} e^{-\frac{x}{a^2}}.$$

* [Clausius, in the memoir cited in the last foot-note, has pointed out two oversights in this calculation. In the first place the numbers have not been properly reduced to English measure, and have still to be multiplied by ·4356, the ratio of the English pound to the kilogramme. The numbers have, further, been calculated with one hour as the unit of time, whereas Maxwell has used them as if a second had been the unit. Taking account of these circumstances and using his own expression for the conduction which differs from (59) only in having $\frac{5}{12}$ in place of $\frac{1}{2}$ on the right-hand side, Clausius finds that the resistance of a stratum of air to the conduction of heat is 1400 times greater than that of a stratum of lead of the same thickness, or about 7000 times greater than that of copper. Ed.]

Also, by Prop. V., if x be the average velocity of one set of particles, and y that of another, then the average value of the sum or difference of the velocities is

$$\sqrt{x^2 + y^2};$$

from which it is easy to see that, if in each individual case

$$u = ax + by + cz,$$

where x, y, z are independent quantities distributed according to the law above stated, then the *average values* of these quantities will be connected by the equation

$$u^2 = a^2 x^2 + b^2 y^2 + c^2 z^2.$$

PROP. XXII.   *Two perfectly elastic bodies of any form strike each other : given their motions before impact, and the line of impact, to find their motions after impact.*

Let M_1 and M_2 be the centres of gravity of the two bodies. $M_1 X_1$, $M_1 Y_1$, and $M_1 Z_1$ the principal axes of the first ; and $M_2 X_2$, $M_2 Y_2$ and $M_2 Z_2$ those of the second. Let I be the point of impact, and $R_1 I R_2$ the line of impact.

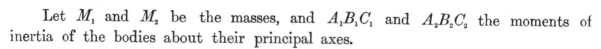

Let the co-ordinates of I with respect to M_1 be $x_1 y_1 z_1$, and with respect to M_2 let them be $x_2 y_2 z_2$.

Let the direction-cosines of the line of impact $R_1 I R_2$ be $l_1 m_1 n_1$ with respect to M_1, and $l_2 m_2 n_2$ with respect to M_2.

Let M_1 and M_2 be the masses, and $A_1 B_1 C_1$ and $A_2 B_2 C_2$ the moments of inertia of the bodies about their principal axes.

Let the velocities of the centres of gravity, resolved in the direction of the principal axes of each body, be

$$U_1, \ V_1, \ W_1, \ \text{and} \ U_2, \ V_2, \ W_2, \ \text{before impact,}$$
and
$$U'_1, \ V'_1, \ W'_1, \ \text{and} \ U'_2, \ V'_2, \ W'_2, \ \text{after impact.}$$

Let the angular velocities round the same axes be

$$p_1, \ q_1, \ r_1, \ \text{and} \ p_2, \ q_2, \ r_2, \ \text{before impact,}$$
and
$$p'_1, \ q'_1, \ r'_1, \ \text{and} \ p'_2, \ q'_2, \ r'_2, \ \text{after impact.}$$

Let R be the impulsive force between the bodies, measured by the momentum it produces in each.

Then, for the velocities of the centres of gravity, we have the following equations :

$$U'_1 = U_1 + \frac{Rl_1}{M_1}, \quad U'_2 = U_2 - \frac{Rl_2}{M_2} \dots\dots\dots\dots\dots (62),$$

with two other pairs of equations in V and W.

The equations for the angular velocities are

$$p'_1 = p_1 + \frac{R}{A_1}(y_1 n_1 - z_1 m_1), \quad p'_2 = p_2 - \frac{R}{A_2}(y_2 n_2 - z_2 m_2)\dots\dots\dots(63),$$

with two other pairs of equations for q and r.

The condition of perfect elasticity is that the whole *vis viva* shall be the same after impact as before, which gives the equation

$$M_1(U'^2_1 - U^2_1) + M_2(U'^2_2 - U^2_2) + A_1(p'^2_1 - p^2_1) + A_2(p'^2_2 - p^2_2) + \&c. = 0\dots.(64).$$

The terms relating to the axis of x are here given; those relating to y and z may be easily written down.

Substituting the values of these terms, as given by equations (62) and (63), and dividing by R, we find

$$l_1(U'_1 + U_1) - l_2(U'_2 + U_2) + (y_1 n_1 - z_1 m_1)(p'_1 + p_1) - (y_2 n_2 - z_2 m_2)(p'_2 + p_2) + \&c. = 0\dots(65).$$

Now if v_1 be the velocity of the striking-point of the first body before impact, resolved along the line of impact,

$$v_1 = l_1 U_1 + (y_1 n_1 - z_1 m_1) p_1 + \&c.;$$

and if we put v_2 for the velocity of the other striking-point resolved along the same line, and v'_1 and v'_2 the same quantities after impact, we may write, equation (65),

$$v_1 + v'_1 - v_2 - v'_2 = 0 \dots\dots\dots\dots\dots\dots (66),$$

or

$$v_1 - v_2 = v'_2 - v'_1 \dots\dots\dots\dots\dots\dots\dots (67),$$

which shows that the velocity of separation of the striking-points resolved in the line of impact is equal to that of approach.

Substituting the values of the accented quantities in equation (65) by means of equations (63) and (64), and transposing terms in R, we find

$$2\left\{U_1 l_1 - U_2 l_2 + p_1\left(y_1 n_1 - z_1 m_1\right) - p_2\left(y_2 n_2 - z_2 m_2\right)\right\} + \&\text{c.}$$

$$= -R\left\{\frac{l_1^2}{M_1} + \frac{l_2^2}{M_2} + \frac{\left(y_1 n_1 - z_1 m_1\right)^2}{A_1} + \frac{\left(y_2 n_2 - z_2 m_2\right)^2}{A_2} + \&\text{c.} \dots\dots\dots (68),$$

the other terms being related to y and z as these are to x. From this equation we may find the value of R; and by substituting this in equations (63), (64), we may obtain the values of all the velocities after impact.

We may, for example, find the value of U'_1 from the equation

$$
\left.
\begin{aligned}
U'_1 &\left\{\frac{l_1^2}{M_1} + \frac{l_2^2}{M_2} + \frac{\left(y_1 n_1 - z_1 m_1\right)^2}{A_1} + \frac{\left(y_2 n_2 - z_2 m_2\right)^2}{A_2} + \&\text{c.}\right\}\frac{M_1}{l_1} \\
= U_1 &\left\{-\frac{l_1^2}{M_1} + \frac{l_2^2}{M_2} + \frac{\left(y_1 n_1 - z_1 m_1\right)^2}{A_1} + \frac{\left(y_2 n_2 - z_2 m_2\right)^2}{A_2} + \&\text{c.}\right\}\frac{M_1}{l_1} \\
&+ 2U_2 l_2 - 2p_1\left(y_1 n_1 - z_1 m_1\right) + 2p_2\left(y_2 n_2 - z_2 m_2\right) - \&\text{c.}
\end{aligned}
\right\} \dots\dots\dots (69).
$$

PROP. XXIII. *To find the relations between the average velocities of translation and rotation after many collisions among many bodies.*

Taking equation (69), which applies to an individual collision, we see that U'_1 is expressed as a linear function of U_1, U_2, p_1, p_2, &c., all of which are quantities of which the values are distributed among the different particles according to the law of Prop. IV. It follows from Prop. V., that if we square every term of the equation, we shall have a new equation between the *average values* of the different quantities. It is plain that, as soon as the required relations have been established, they will remain the same after collision, so that we may put $U_1'^2 = U_1^2$ in the equation of averages. The equation between the average values may then be written

$$\left(M_1 U_1^2 - M_2 U_2^2\right)\frac{l_2^2}{M_2} + \left(M_1 U_1^2 - A_1 p_1^2\right)\frac{\left(y_1 n_1 - z_1 m_1\right)^2}{A_1} + \left(M_1 U_1^2 - A_2 p_2^2\right)\frac{\left(y_2 n_2 - z_2 m_2\right)^2}{A_2} + \&\text{c.} = 0.$$

Now since there are collisions in every possible way, so that the values of l, m, n, &c. and x, y, z, &c. are infinitely varied, this equation cannot subsist unless

$$M_1 U_1^2 = M_2 U_2^2 = A_1 p_1^2 = A_2 p_2^2 = \&\text{c.}$$

The final state, therefore, of any number of systems of moving particles of any form is that in which the average *vis viva* of translation along each of the three axes is the same in all the systems, and equal to the average *vis viva* of rotation about each of the three principal axes of each particle.

Adding the *vires vivæ* with respect to the other axes, we find that the whole *vis viva* of translation is equal to that of rotation in each system of particles, and is also the same for different systems, as was proved in Prop. VI.

This result (which is true, however nearly the bodies approach the spherical form, provided the motion of rotation is at all affected by the collisions) seems decisive against the unqualified acceptation of the hypothesis that gases are such systems of hard elastic particles. For the ascertained fact that γ, the ratio of the specific heat at constant pressure to that at constant volume, is equal to 1·408, requires that the ratio of the whole *vis viva* to the *vis viva* of translation should be

$$\beta = \frac{2}{3\,(\gamma - 1)} = 1\text{·}634\,;$$

whereas, according to our hypothesis, $\beta = 2$.

We have now followed the mathematical theory of the collisions of hard elastic particles through various cases, in which there seems to be an analogy with the phenomena of gases. We have deduced, as others have done already, the relations of pressure, temperature, and density of a single gas. We have also proved that when two different gases act freely on each other (that is, when at the same temperature), the mass of the single particles of each is inversely proportional to the square of the molecular velocity; and therefore, at equal temperature and pressure, *the number of particles in unit of volume is the same.*

We then offered an explanation of the internal friction of gases, and deduced from experiments a value of the mean length of path of a particle between successive collisions.

We have applied the theory to the law of diffusion of gases, and, from an experiment on olefiant gas, we have deduced a value of the length of path not very different from that deduced from experiments on friction.

Using this value of the length of path between collisions, we found that the resistance of air to the conduction of heat is 10,000,000 times that of copper, a result in accordance with experience.

Finally, by establishing a necessary relation between the motions of translation and rotation of all particles not spherical, we proved that a system of such particles could not possibly satisfy the known relation between the two specific heats of all gases.

[From the *Philosophical Transactions*, MDCCCLX.]

XXI. *On the Theory of Compound Colours, and the Relations of the Colours of the Spectrum.* Communicated by Professor STOKES, Sec. R.S.

(Received January 5,—Read March 22, 1860.)

§ I. *Introduction.*

ACCORDING to Newton's analysis of light[*], every colour in nature is produced by the mixture, in various proportions, of the different kinds of light into which white light is divided by refraction. By means of a prism we may analyse any coloured light, and determine the proportions in which the different homogeneous rays enter into it; and by means of a lens we may recombine these rays, and reproduce the original coloured light.

Newton has also shewn[†] how to combine the different rays of the spectrum so as to form a single beam of light, and how to alter the proportions of the different colours so as to exhibit the result of combining them in any arbitrary manner.

The number of different kinds of homogeneous light being infinite, and the proportion in which each may be combined being also variable indefinitely, the results of such combinations could not be appreciated by the eye, unless the chromatic effect of every mixture, however complicated, could be expressed in some simpler form. Colours, as seen by the human eye of the normal type, can all be reduced to a few classes, and expressed by a few well-known names; and even those colours which have different names have obvious relations among themselves. Every colour, except purple, is similar to some colour of the spectrum[‡],

[*] *Optics*, Book I. Part 2, Prop. 7.
[†] *Lectiones Opticæ*, Part 2, § 1, pp. 100 to 105; and *Optics*, Book I. Part 2, Prop. 11.
[‡] *Optics*, Book I. Part 2, Prop. 4.

although less intense; and all purples may be compounded of blue and red, and diluted with white to any required tint. Brown colours, which at first sight seem different, are merely red, orange or yellow of feeble intensity, more or less diluted with white.

It appears therefore that the result of any mixture of colours, however complicated, may be defined by its relation to a certain small number of well-known colours. Having selected our standard colours, and determined the relations of a given colour to these, we have defined that colour completely as to its appearance. Any colour which has the same relation to the standard colours, will be identical in appearance, though its optical constitution, as revealed by the prism, may be very different.

We may express this by saying that two compound colours may be *chromatically* identical, but *optically* different. The *optical* properties of light are those which have reference to its origin and propagation through media, till it falls on the sensitive organ of vision; the *chromatical* properties of light are those which have reference to its power of exciting certain sensations of *colour*, perceived through the organ of vision.

The investigation of the chromatic relations of the rays of the spectrum must therefore be founded upon observations of the apparent identity of compound colours, as seen by an eye either of the normal or of some abnormal type; and the results to which the investigation leads must be regarded as partaking of a physiological, as well as of a physical character, and as indicating certain laws of sensation, depending on the constitution of the organ of vision, which may be different in different individuals. We have to determine the laws of the composition of colours in general, to reduce the number of standard colours to the smallest possible, to discover, if we can, what they are, and to ascertain the relation which the homogeneous light of different parts of the spectrum bears to the standard colours.

§ II. *History of the Theory of Compound Colours.*

The foundation of the theory of the composition of colours was laid by Newton*. He first shews that, by the mixture of homogeneal light, colours may be produced which are "like to the colours of homogeneal light as to the appearance of colour, but not as to the immutability of colour and consti-

---

* *Optics,* Book I. Part 2, Props. 4, 5, 6.

tution of light." Red and yellow give an orange colour, which is chromatically similar to the orange of the spectrum, but optically different, because it is resolved into its component colours by a prism, while the orange of the spectrum remains unchanged. When the colours to be mixed lie at a distance from one another in the spectrum, the resultant appears paler than that intermediate colour of the spectrum which it most resembles; and when several are mixed, the resultant may appear white. Newton* is always careful, however, not to call any mixture white, unless it agrees with common white light in its optical as well as its chromatical properties, and is a mixture of *all* the homogeneal colours. The theory of compound colours is first presented in a mathematical form in Prop. 6, "*In a mixture of primary colours, the quantity and quality of each being given, to know the colour of the compound.*" He divides the circumference of a circle into seven parts, proportional to the seven musical intervals, in accordance with his opinion about the proportions of the colours in the spectrum. At the centre of gravity of each of these arcs he places a little circle, whose area is proportional to the number of rays of the corresponding colour which enter into the given mixture. The position of the centre of gravity of all these circles indicates the nature of the resultant colour. A radius drawn through it points out that colour of the spectrum which it most resembles, and the distance from the centre determines the fulness of its colour.

With respect to this construction, Newton says, "This rule I conceive accurate enough for practice, though not mathematically accurate." He gives no reasons for the different parts of his rule, but we shall find that his method of finding the centre of gravity of the component colours is completely confirmed by my observations, and that it involves mathematically the theory of three elements of colour; but that the disposition of the colours on the circumference of a circle was only a provisional arrangement, and that the true relations of the colours of the spectrum can only be determined by direct observation.

Young† appears to have originated the theory, that the three elements of colour are determined as much by the constitution of the sense of sight as by anything external to us. He conceives that three different sensations may be excited by light, but that the proportion in which each of the three is excited depends on the nature of the light. He conjectures that these primary sensa-

---

* 7th and 8th Letters to Oldenburg.

† Young's *Lectures on Natural Philosophy*, Kelland's Edition, p. 345, or Quarto, 1807, Vol. I. p. 441 ; see also Young in *Philosophical Transactions*, 1801, or Works in Quarto, Vol. II. p. 617.

tions correspond to red, green, and violet. A blue ray, for example, though homogeneous in itself, he conceives capable of exciting both the green and the violet sensation, and therefore he would call blue a compound colour, though the colour of a simple kind of light. The *quality* of any colour depends, according to this theory, on the *ratios* of the intensities of the three sensations which it excites, and its *brightness* depends on the *sum* of these three intensities.

Sir David Brewster, in his paper entitled " On a New Analysis of Solar Light, indicating three Primary Colours, forming Coincident Spectra of equal length*," regards the actual colours of the spectrum as arising from the intermixture, in various proportions, of three primary kinds of light, red, yellow, and blue, each of which is variable in intensity, but uniform in colour, from one end of the spectrum to the other; so that every colour in the spectrum is really compound, and might be shewn to be so if we had the means of separating its elements.

Sir David Brewster, in his researches, employed coloured media, which, according to him, absorb the three elements of a single prismatic colour in different degrees, and change their proportions, so as to alter the colour of the light, without altering its refrangibility.

In this paper I shall not enter into the very important questions affecting the physical theory of light, which can only be settled by a careful inquiry into the phenomena of absorption. The physiological facts, that we have a threefold sensation of colour, and that the three elements of this sensation are affected in different proportions by light of different refrangibilities, are equally true, whether we adopt the physical theory that there are three kinds of light corresponding to these three colour-sensations, or whether we regard light of definite refrangibility as an undulation of known length, and therefore variable only in intensity, but capable of producing different chemical actions on different substances, of being absorbed in different degrees by different media, and of exciting in different degrees the three different colour-sensations of the human eye.

Sir David Brewster has given a diagram of three curves, in which the base-line represents the length of the spectrum, and the ordinates of the curves represent, by estimation, the intensities of the three kinds of light at each point of the spectrum. I have employed a diagram of the same kind to express the

---

* *Transactions of the Royal Society of Edinburgh*, Vol. XII. p. 123.

results arrived at in this paper, the ordinates being made to represent the intensities of each of the three elements of colour, as calculated from the experiments.

The most complete series of experiments on the mixture of the colours of the spectrum, is that of Professor Helmholtz*, of Königsberg. By using two slits at right angles to one another, he formed two pure spectra, the fixed lines of which were seen crossing one another when viewed in the ordinary way by means of a telescope. The colours of these spectra were thus combined in every possible way, and the effect of the combination of any two could be seen separately by drawing the eye back from the eye-piece of the telescope, when the compound colour was seen by itself at the eye-hole. The proportion of the components was altered by turning the combined slits round in their own plane.

One result of these experiments was, that a colour, chromatically identical with white, could be formed by combining yellow with indigo. M. Helmholtz was not then able to produce white with any other pair of simple colours, and considered that three simple colours were required in general to produce white, one from each of the three portions into which the spectrum is divided by the yellow and indigo.

Professor Grassmann† shewed that Newton's theory of compound colours implies that there are an infinite number of pairs of complementary colours in the spectrum, and pointed out the means of finding them. He also shewed how colours may be represented by lines, and combined by the method of the parallelogram.

In a second memoir‡, M. Helmholtz describes his method of ascertaining these pairs of complementary colours. He formed a pure spectrum by means of a slit, a prism, and a lens; and in this spectrum he placed an apparatus having two parallel slits which were capable of adjustment both in position and breadth, so as to let through any two portions of the spectrum, in any proportions. Behind this slit, these rays were united in an image of the prism, which was received on paper. By arranging the slits, the colour of this image may be reduced to white, and made identical with that of paper illuminated with white light. The wave-lengths of the component colours were then measured by observing the angle of diffraction through a grating. It was found that the

---

* Poggendorff's *Annalen*, Band LXXXVII. (*Philosophical Magazine*, 1852, December).

† *Ibid*. Band LXXXIX. (*Philosophical Magazine*, 1854, April).   ‡ *Ibid*. Band XCIV.

colours from red to green-yellow ($\lambda = 2082$) were complementary to colours ranging from green-blue ($\lambda = 1818$) to violet, and that the colours between green-yellow and green-blue have no homogeneous complementaries, but must be neutralized by mixtures of red and violet.

M. Helmholtz also gives a provisional diagram of the curve formed by the spectrum on Newton's diagram, for which his experiments did not furnish him with the complete data.

Accounts of experiments by myself on the mixture of artificial colours by rapid rotation, may be found in the *Transactions of the Royal Society of Edinburgh*, Vol. XXI. Pt. 2 (1855); in an appendix to Professor George Wilson's work on Colour-Blindness; in the *Report of the British Association* for 1856, p. 12; and in the *Philosophical Magazine*, July 1857, p. 40. These experiments shew that, for the normal eye, there are three, and only three, elements of colour, and that in the colour-blind one of these is absent. They also prove that chromatic observations may be made, both by normal and abnormal eyes, with such accuracy, as to warrant the employment of the results in the calculation of colour-equations, and in laying down colour-diagrams by Newton's rule.

The first instrument which I made (in 1852) to examine the mixtures of the colours of the spectrum was similar to that which I now use, but smaller, and it had no constant light for a term of comparison. The second was $6\frac{1}{2}$ feet long, made in 1855, and shewed *two* combinations of colour side by side. I have now succeeded in making the mixture much more perfect, and the comparisons more exact, by using white reflected light, instead of the second compound colour. An apparatus in which the light passes through the prisms, and is reflected back again in nearly the same path by a concave mirror, was shewn by me to the British Association in 1856. It has the advantage of being portable, and need not be more than half the length of the other, in order to produce a spectrum of equal length. I am so well satisfied with the working of this form of the instrument, that I intend to make use of it in obtaining equations from a greater variety of observers than I could meet with when I was obliged to use the more bulky instrument. It is difficult at first to get the observer to believe that the compound light can ever be so adjusted as to appear to his eyes identical with the white light in contact with it. He has to learn what adjustments are necessary to produce the requisite alteration under all circumstances, and he must never be satisfied till the two parts of the field are identical in colour and illumination. To do this thoroughly, implies not merely good eyes, but a power of judging as to the exact nature of the difference between two very pale and nearly identical tints, whether they differ in the amount of red, green, or blue, or in brightness of illumination.

In the following paper I shall first lay down the mathematical theory of Newton's diagram, with its relation to Young's theory of the colour-sensation. I shall then describe the experimental method of mixing the colours of the spectrum, and determining the wave-lengths of the colours mixed. The results of my experiments will then be given, and the chromatic relations of the spectrum exhibited in a system of colour-equations, in Newton's diagram, and in three curves of intensity, as in Brewster's diagram. The differences between the results of two observers will then be discussed, shewing on what they depend, and in what way such differences may affect the vision of persons otherwise free from defects of sight.

§ III. *Mathematical Theory of Newton's Diagram of Colours.*

Newton's diagram is a plane figure, designed to exhibit the relations of colours to each other.

Every point in the diagram represents a colour, simple or compound, and we may conceive the diagram itself so painted, that every colour is found at its corresponding point. Any colour, differing only in quantity of illumination from one of the colours of the diagram, is referred to it as a unit, and is measured by the ratio of the illumination of the given colour to that of the corresponding colour in the diagram. In this way the *quantity* of a colour is estimated. The resultant of mixing any two colours of the diagram is found by dividing the line joining them inversely as the quantity of each; then, if the sum of these quantities is unity, the resultant will have the illumination as well as the colour of the point so found; but if the sum of the components is different from unity, the *quantity* of the resultant will be measured by the sum of the components.

This method of determining the position of the resultant colour is mathematically identical with that of finding the centre of gravity of two weights, and placing a weight equal to their sum at the point so found. We shall therefore speak of the resultant tint as the sum of its components placed at their centre of gravity.

By compounding this resultant tint with some other colour, we may find the position of a mixture of three colours, at the centre of gravity of its components; and by taking these components in different proportions, we may obtain colours corresponding to every part of the triangle of which they are the angular points. In this way, by taking any three colours we should be able to construct a triangular portion of Newton's diagram by painting it with mixtures of the three colours. Of course these mixtures must be made to correspond with optical mixtures of light, not with mechanical mixtures of pigments.

Let us now take any colour belonging to a point of the diagram outside this triangle. To make the centre of gravity of the three weights coincide with this point, one or more of the weights must be made negative. This, though following from mathematical principles, is not capable of direct physical interpretation, as we cannot exhibit a negative colour.

The equation between the three selected colours, x, y, z, and the new colour u, may in the first case be written

$$u = x + y + z \dots\dots\dots\dots\dots\dots\dots\dots\dots\dots\dots(1),$$

x, y, z being the quantities of colour required to produce u. In the second case suppose that z must be made negative,

$$u = x + y - z \dots\dots\dots\dots\dots\dots\dots\dots\dots\dots\dots(2).$$

As we cannot realize the term $-z$ as a negative colour, we transpose it to the other side of the equation, which then becomes

$$u + z = x + y \dots\dots\dots\dots\dots\dots\dots\dots\dots\dots\dots(3),$$

which may be interpreted to mean, that the resultant tint, $u + z$, is identical with the resultant, $x + y$. We thus find a mixture of the new colour with one of the selected colours, which is chromatically equivalent to a mixture of the other two selected colours.

When the equation takes the form

$$u = x - y - z \dots\dots\dots\dots\dots\dots\dots\dots\dots\dots\dots(4),$$

two of the components being negative, we must transpose them thus,

$$u + y + z = x \dots\dots\dots\dots\dots\dots\dots\dots\dots\dots\dots(5),$$

which means that a mixture of certain proportions of the new colour and two of the three selected, is chromatically equivalent to the third. We may thus in all cases find the relation between any three colours and a fourth, and exhibit this relation in a form capable of experimental verification; and by proceeding in this way we may map out the positions of all colours upon Newton's diagram. Every colour in nature will then be defined by the position of the corresponding colour in the diagram, and by the ratio of its illumination to that of the colour in the diagram.

§ IV. *Method of representing Colours by Straight Lines drawn from a Point.*

To extend our ideas of the relations of colours, let us form a new geometrical conception by the aid of solid geometry.

Let us take as origin any point not in the plane of the diagram, and let us draw lines through this point to the different points of the diagram; then the direction of any of these lines will depend upon the position of the point of the diagram through which it passes, so that we may take this line as the representative of the corresponding colour on the diagram.

In order to indicate the *quantity* of this colour, let it be produced beyond the plane of the diagram in the same ratio as the given colour exceeds in illumination the colour on the diagram. In this way every colour in nature will be represented by a line drawn through the origin, whose *direction* indicates the *quality* of the colour, while its *length* indicates its *quantity*.

Let us find the resultant of two colours by this method. Let O be the origin and AB be a section of the plane of the diagram by that of the paper. Let OP, OQ be lines representing colours, A, B the 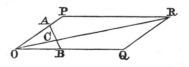 corresponding points in the diagram; then the quantity of P will be $\dfrac{OP}{OA}=p$,

and that of Q will be $\dfrac{OQ}{OB}=q$. The resultant of these will be represented in the diagram by the point C, where $AC : CB :: q : p$, and the quantity of the resultant will be $p+q$, so that if we produce OC to R, so that $OR=(p+q)OC$, the line OR will represent the resultant of OP and OQ in direction and magnitude. It is easy to prove, from this construction, that OR is the diagonal of the parallelogram of which OP and OQ are two sides. It appears therefore that if colours are represented in quantity and quality by the magnitude and direction of straight lines, the rule for the composition of colours is identical with that for the composition of forces in mechanics. This analogy has been well brought out by Professor Grassmann in Poggendorff's *Annalen*, Bd. LXXXIX.

We may conceive an arrangement of actual colours in space founded upon this construction. Suppose each of these radiating lines representing a given colour to be itself illuminated with that colour, the brightness increasing from zero at the origin to unity, where it cuts the plane of the diagram, and becoming continually more intense in proportion to the distance from the origin. In this way every colour in nature may be matched, both in quality and quantity, by some point in this coloured space.

If we take any three lines through the origin as axes, we may, by co-ordinates parallel to these lines, express the position of any point in space. That point will correspond to a colour which is the resultant of the three colours represented by the three co-ordinates.

This system of co-ordinates is an illustration of the resolution of a colour into three components. According to the theory of Young, the human eye is capable of three distinct primitive sensations of colour, which by their composition in various proportions, produce the sensations of actual colour in all their varieties. Whether any kinds of light have the power of exciting these primitive sensations separately, has not yet been determined.

If colours corresponding to the three primitive sensations can be exhibited, then all colours, whether produced by light, disease, or imagination, are compounded of these, and have their places within the triangle formed by joining the three primaries. If the colours of the pure spectrum, as laid down on the diagram, form a triangle, the colours at the angles *may* correspond to the primitive sensations. If the curve of the spectrum does not reach the angles of the circumscribing triangle, then no colour in the spectrum, and therefore no colour in nature, corresponds to any of the three primary sensations.

The only data at present existing for determining the primary colours, are derived from the comparison of observations of colour-equations by colour-blind, and by normal eyes. The colour-blind equations differ from the others by the non-existence of one of the elements of colour, the relation of which to known colours can be ascertained. It appears, from observations made for me by two colour-blind persons*, that the elementary sensation which they do not possess is a red approaching to crimson, lying beyond both vermilion and carmine. These

---

* *Transactions of the Royal Society of Edinburgh*, Vol. XXI. Pt. 2, p. 286.

observations are confirmed by those of Mr Pole, and by others which I have obtained since. I have hopes of being able to procure a set of colour-blind equations between the colours of the spectrum, which will indicate the missing primary in a more exact manner.

The experiments which I am going to describe have for their object the determination of the position of the colours of the spectrum upon Newton's diagram, from actual observations of the mixtures of those colours. They were conducted in such a way, that in every observation the judgment of the observer was exercised upon two parts of an illuminated field, one of which was so adjusted as to be chromatically identical with the other, which, during the whole series of observations, remained of one constant intensity of white. In this way the effects of subjective colours were entirely got rid of, and all the observations were of the same kind, and therefore may claim to be equally accurate; which is not the case when comparisons are made between bright colours of different kinds.

The chart of the spectrum, deduced from these observations, exhibits the colours arranged very exactly along two sides of a triangle, the extreme red and violet forming doubtful portions of the third side. This result greatly simplifies the theory of colour, if it does not actually point out the three primary colours themselves.

§ V. *Description of an Instrument for making definite Mixtures of the Colours of the Spectrum.*

The experimental method which I have used consists in forming a combination of three colours belonging to different portions of the spectrum, the quantity of each being so adjusted that the mixture shall be white, and equal in intensity to a given white. Fig. 1, Plate VI. p. 444, represents the instrument for making the observations. It consists of two tubes, or long boxes, of deal, of rectangular section, joined together at an angle of about 100°.

The part AK is about five feet long, seven inches broad, and four deep; KN is about two feet long, five inches broad, and four deep; BD is a partition parallel to the side of the long box. The whole of the inside of the instrument is painted black, and the only openings are at the end AC, and at E. At the angle there is a lid, which is opened when the optical parts have to be adjusted or cleaned.

At E is a fine vertical slit; L is a lens; at P there are two equilateral prisms. The slit E, the lens L, and the prisms P are so adjusted, that when light is admitted at E a *pure spectrum* is formed at AB, the extremity of the long box. A mirror at M is also adjusted so as to reflect the light from E along the narrow compartment of the long box to BC. See Fig. 3.

At AB is placed the contrivance shewn in Fig. 2, Plate I. $A'B'$ is a rectangular frame of brass, having a rectangular aperture of 6×1 inches. On this frame are placed six brass sliders, X, Y, Z. Each of these carries a knife-edge of brass in the plane of the surface of the frame.

These six moveable knife-edges form three slits, X, Y, Z, which may be so adjusted as to coincide with any three portions of the pure spectrum formed by light from E. The intervals behind the sliders are closed by hinged shutters, which allow the sliders to move without letting light pass between them.

The inner edge of the brass frame is graduated to twentieths of an inch, so that the position of any slit can be read off. The breadth of the slit is ascertained by means of a wedge-shaped piece of metal, six inches long, and tapering to a point from a breadth of half an inch. This is gently inserted into each slit, and the breadth is determined by the distance to which it enters, the divisions on the wedge corresponding to the 200th of an inch difference in breadth, so that the unit of breadth is ·005 inch.

Now suppose light to enter at E, to pass through the lens, and to be refracted by the two prisms at P; a pure spectrum, shewing Fraunhofer's lines, is formed at AB, but only that part is allowed to pass which falls on the three slits X, Y, Z. The rest is stopped by the shutters. Suppose that the portion falling on X belongs to the red part of the spectrum; then, of the white light entering at E, only the red will come through the slit X. If we were to admit red light at X it would be refracted to E, by the principle in Optics, that the course of any ray may be reversed. If, instead of red light, we were to admit white light at X, still only red light would come to E; for all other light would be either more or less refracted, and would not reach the slit at E. Applying the eye at the slit E, we should see the prism P uniformly illuminated with red light, of the kind corresponding to the part of the spectrum which falls on the slit X when light is admitted at E.

Let the slit Y correspond to another portion of the spectrum, say the green; then, if white light is admitted at Y, the prism, as seen by an eye at E, will be uniformly illuminated with green light; and if white light be admitted at X

and Y simultaneously, the colour seen at E will be a compound of red and green, the proportions depending on the breadth of the slits and the intensity of the light which enters them. The third slit Z, enables us to combine any three kinds of light in any given proportions, so that an eye at E shall see the face of the prism at P uniformly illuminated with the colour resulting from the combination of the three. The position of these three rays in the spectrum is found by admitting the light at E, and comparing the position of the slits with the position of the principal fixed lines ; and the breadth of the slits is determined by means of the wedge.

At the same time white light is admitted through BC to the mirror of black glass at M, whence it is reflected to E, past the edge of the prism at P, so that the eye at E sees through the lens a field consisting of two portions, separated by the edge of the prism; that on the left hand being compounded of three colours of the spectrum refracted by the prism, while that on the right hand is white light reflected from the mirror. By adjusting the slits properly, these two portions of the field may be made equal, both in colour and brightness, so that the edge of the prism becomes almost invisible.

In making experiments, the instrument was placed on a table in a room moderately lighted, with the end AB turned towards a large board covered with white paper, and placed in the open air, so as to be uniformly illuminated by the sun. In this way the three slits and the mirror M were all illuminated with white light of the same intensity, and all were affected in the same ratio by any change of illumination; so that if the two halves of the field were rendered equal when the sun was under a cloud, they were found nearly correct when the sun again appeared. No experiments, however, were considered good unless the sun remained uniformly bright during the whole series of experiments.

After each set of experiments light was admitted at E, and the position of the fixed lines D and F of the spectrum was read off on the scale at AB. It was found that after the instrument had been some time in use these positions were invariable, shewing that the eye-hole, the prisms, and the scale might be considered as rigidly connected.

§ VI. *Method of determining the Wave-length corresponding to any point of the Spectrum on the Scale* AB.

Two plane surfaces of glass were kept apart by two parallel strips of gold-beaters' leaf, so as to enclose a stratum of air of nearly uniform thickness. Light reflected from this stratum of air was admitted at *E*, and the spectrun formed by it was examined at *AB* by means of a lens. This spectrum consists of a large number of bright bands, separated by dark spaces at nearly uniform intervals, these intervals, however, being considerably larger as we approach the violet end of the spectrum.

The reason of these alternations of brightness is easily explained. By the theory of Newton's rings, the light reflected from a stratum of air consists of two parts, one of which has traversed a path longer than that of the other, by an interval depending on the thickness of the stratum and the angle of incidence. Whenever the interval of retardation is an exact multiple of a wave-length, these two portions of light destroy each other by interference; and when the interval is an odd number of half wave-lengths, the resultant light is a maximum.

In the ordinary case of Newton's rings, these alternations depend upon the varying thickness of the stratum; while in this case a pencil of rays of different wave-lengths, but all experiencing the same retardation, is analysed into a spectrum, in which the rays are arranged in order of their respective wave-lengths. Every ray whose wave-length is an exact submultiple of the retardation will be destroyed by interference, and its place will appear dark in the spectrum; and there will be as many dark bands seen as there are rays whose wave-lengths fulfil this condition.

If, then, we observe the positions of the dark bands on the scale *AB*, the wave-lengths corresponding to these positions will be a series of submultiples of the retardation.

Let us call the first dark band visible on the red side of the spectrum zero, and let us number them in order 1, 2, 3, &c. towards the violet end. Let N be the number of undulations corresponding to the band zero which are contained in the retardation R; then if n be the number of any other band, $N+n$ will be the number of the corresponding wave-lengths in the retardation, or in symbols,

$$R = (N+n)\lambda \dots\dots\dots(6).$$

Now observe the position of two of Fraunhofer's fixed lines with respect to the dark bands, and let n_1, n_2 be their positions expressed in the number of bands, whole or fractional, reckoning from zero. Let λ_1, λ_2 be the wave-lengths of these fixed lines as determined by Fraunhofer, then

$$R = (N + n_1)\,\lambda_1 = (N + n_2)\,\lambda_2, \quad\dots\dots\dots\dots\dots (7);$$

whence

$$N = \frac{n_2\lambda_2 - n_1\lambda_1}{\lambda_1 - \lambda_2} = \frac{(n_2 - n_1)}{\lambda_1 - \lambda_2}\,\lambda_2 - n_1 \quad\dots\dots\dots\dots (8),$$

and

$$R = \frac{n_2 - n_1}{\lambda_1 - \lambda_2}\,\lambda_1\lambda_2 \quad\dots\dots\dots\dots\dots\dots (9).$$

Having thus found N and R, we may find the wave-length corresponding to the dark band n from the formula

$$\lambda = \frac{R}{N + n} \quad\dots\dots\dots\dots\dots\dots\dots (10).$$

In my experiments the line D corresponded with the seventh dark band, and F was between the 15th and 16th, so that $n_2 = 15\cdot7$. Here then for D,

and for F, $\quad\quad\quad \begin{matrix} n_1 = 7, & \lambda_1 = 2175 \\ n_2 = 15\cdot7, & \lambda_2 = 1794 \end{matrix} \Bigg\}$ in Fraunhofer's measure $\dots\dots\dots$ (11),

whence we find $\quad\quad\quad N = 34, \ R = 89175 \dots\dots\dots\dots\dots\dots\dots (12)$.

There were 22 bands visible, corresponding to 22 different positions on the scale AB, as determined 4th August, 1859.

TABLE I.

Band.	Scale.		Band.	Scale.		Band.	Scale.
$n=$ 1	17		$n=$ 9	36		$n=$ 16	57
2	19		10	39		17	61
3	$21\frac{1}{4}$		11	42		18	65
4	$23\frac{1}{2}$		12	45		19	69
5	26		13	48		20	73
6	$28\frac{1}{2}$		14	51		21	77
7	31		15	54		22	82
8	$33\frac{1}{2}$						

Sixteen equidistant points on the scale were chosen for standard colours in the experiments to be described. The following Table gives the reading on the scale AB, the value of $N+n$, and the calculated wave-length for each of these :—

Table II.

Scale.	(N + n).	Wave-length.	Colour.
20	36·4	2450	Red.
24	38·3	2328	Scarlet.
28	39·8	2240	Orange.
32	41·4	2154	Yellow.
36	42·9	2078	Yellow-Green.
40	44·3	2013	Green.
44	45·7	1951	Green.
48	47·0	1879	Bluish green.
52	48·3	1846	Blue-green.
56	49·6	1797	Greenish blue.
60	50·8	1755	Blue.
64	51·8	1721	Blue.
68	52·8	1688	Blue.
72	53·7	1660	Indigo.
76	54·7	1630	Indigo.
80	55·6	1604	Indigo.

Having thus selected sixteen distinct points of the spectrum on which to operate, and determined their wave-lengths and apparent colours, I proceeded to ascertain the mathematical relations between these colours in order to lay them down on Newton's diagram. For this purpose I selected three of these as points of reference, namely, those at 24, 44, and 68 of the scale. I chose these points because they are well separated from each other on the scale, and because the colour of the spectrum at these points does not appear to the eye to vary very rapidly, either in hue or brightness, in passing from one point to another. Hence a small error of position will not make so serious an alteration of colour at these points, as if we had taken them at places of rapid variation; and we may regard the amount of the illumination produced by the light entering through the slits in these positions as sensibly proportional to the breadth of the slits.

(24) corresponds to a bright scarlet about one-third of the distance from C to D; (44) is a green very near the line E; and (68) is a blue about one-third of the distance from F to G.

§ VII. *Method of Observation.*

The instrument is turned with the end AB towards a board, covered with white paper, and illuminated by sunlight. The operator sits at the end AB, to move the sliders, and adjust the slits; and the observer sits at the end E, which is shaded from any bright light. The operator then places the slits so that their centres correspond to the three standard colours, and adjusts their breadths till the observer sees the prism illuminated with pure white light of the same intensity with that reflected by the mirror M. In order to do this, the observer must tell the operator what difference he observes in the two halves of the illuminated field, and the operator must alter the breadth of the slits accordingly, always keeping the centre of each slit at the proper point of the scale. The observer may call for more or less red, blue or green; and then the operator must increase or diminish the width of the slits X, Y, and Z respectively. If the variable field is darker or lighter than the constant field, the operator must widen or narrow all the slits in the same proportion. When the variable part of the field is nearly adjusted, it often happens that the constant white light from the mirror appears tinged with the complementary colour. This is an indication of what is required to make the resemblance of the two parts of the field of view perfect. When no difference can be detected between the two parts of the field, either in colour or in brightness, the observer must look away for some time, to relieve the strain on the eye, and then look again. If the eye thus refreshed still judges the two parts of the field to be equal, the observation may be considered complete, and the operator must measure the breadth of each slit by means of the wedge, as before described, and write down the result as a colour-equation, thus—

Oct. 18, J. $\qquad$ $18 \cdot 5 \, (24) + 27 \, (44) + 37 \, (68) = \mathrm{W} *$ .... (13).

This equation means that on the 18th of October the observer J. (myself) made an observation in which the breadth of the slit X was $18 \cdot 5$, as measured by the wedge, while its centre was at the division (24) of the scale; that the breadths of Y and Z were 27 and 37, and their positions (44) and (68); and that the illumination produced by these slits was exactly equal, in my estimation as an observer, to the constant white W.

The position of the slit X was then shifted from (24) to (28), and when the proper adjustments were made, I found a second colour-equation of this form—

Oct. 18, J. $\qquad 16\,(28)+21\,(44)+37\,(68)=\mathrm{W} \dots\dots\dots\dots (14).$

Subtracting one equation from the other and remembering that the figures in brackets are merely symbols of position, not of magnitude, we find

$$16\,(28)=18\cdot5\,(24)+6\,(44) \dots\dots\dots\dots\dots (15),$$

shewing that (28) can be made up of (24) and (44), in the proportion of 18·5 to 6.

In this way, by combining each colour with two standard colours, we may produce a white equal to the constant white. The red and yellow colours from (20) to (32) must be combined with green and blue, the greens from (36) to (52) with red and blue, and the blues from (56) to (80) with red and green.

The following is a specimen of an actual series of observations made in this way by another observer (K.) :—

TABLE III.

Oct. 13, 1859.   Observer (K.).

$(X)\qquad (Y)\qquad (Z)$

$18\tfrac{1}{2}(24)+32\tfrac{1}{2}(44)+32\ (68)=\mathrm{W}*.$
$17\tfrac{1}{2}(24)+32\tfrac{1}{2}(44)+63\ (80)=\mathrm{W}.$
$18\ (24)+32\tfrac{1}{2}(44)+35\ (72)=\mathrm{W}.$
$19\ (24)+32\ (44)+31\tfrac{1}{2}(68)=\mathrm{W}*.$
$19\ (24)+30\tfrac{1}{2}(44)+35\ (64)=\mathrm{W}.$
$20\ (24)+23\ (44)+39\ (60)=\mathrm{W}.$
$21\ (24)+14\ (44)+58\ (56)=\mathrm{W}.$
$22\ (24)+62\ (52)+11\ (68)=\mathrm{W}.$
$22\ (24)+42\ (48)+29\tfrac{1}{2}(68)=\mathrm{W}.$
$19\ (24)+31\tfrac{1}{2}(44)+33\ (68)=\mathrm{W}*.$
$16\ (24)+28\ (40)+32\tfrac{1}{2}(68)=\mathrm{W}.$
$6\ (24)+27\ (36)+32\tfrac{1}{2}(68)=\mathrm{W}.$
$23\ (32)+11\tfrac{1}{2}(44)+32\tfrac{1}{2}(68)=\mathrm{W}.$
$17\ (28)+26\ (44)+32\tfrac{1}{2}(68)=\mathrm{W}.$
$20\ (24)+33\tfrac{1}{2}(44)+32\tfrac{1}{2}(68)=\mathrm{W}*.$
$46\ (20)+33\ (44)+30\ (68)=\mathrm{W}.$

The equations marked with an asterisk (*) are those which involve the three standard colours, and since every other equation must be compared with them, they must be often repeated.

The following Table contains the *means* of four sets of observations by the same observer (K.):—

TABLE IV. (K.)

$$44 \cdot 3 \,(20) + 31 \cdot 0 \,(44) + 27 \cdot 7 \,(68) = W.$$
$$16 \cdot 1 \,(28) + 25 \cdot 6 \,(44) + 30 \cdot 6 \,(68) = W.$$
$$22 \cdot 0 \,(32) + 12 \cdot 1 \,(44) + 30 \cdot 6 \,(68) = W.$$
$$6 \cdot 4 \,(24) + 25 \cdot 2 \,(36) + 31 \cdot 3 \,(68) = W.$$
$$15 \cdot 3 \,(24) + 26 \cdot 0 \,(40) + 30 \cdot 7 \,(68) = W.$$
$$19 \cdot 8 \,(24) + 35 \cdot 0 \,(46) + 30 \cdot 2 \,(68) = W.$$
$$21 \cdot 2 \,(24) + 41 \cdot 4 \,(48) + 27 \cdot 0 \,(68) = W.$$
$$22 \cdot 0 \,(24) + 62 \cdot 0 \,(52) + 13 \cdot 0 \,(68) = W.$$
$$21 \cdot 7 \,(24) + 10 \cdot 4 \,(44) + 61 \cdot 7 \,(56) = W.$$
$$20 \cdot 5 \,(24) + 23 \cdot 7 \,(44) + 40 \cdot 5 \,(60) = W.$$
$$19 \cdot 7 \,(24) + 30 \cdot 3 \,(44) + 33 \cdot 7 \,(64) = W.$$
$$18 \cdot 0 \,(24) + 31 \cdot 2 \,(44) + 32 \cdot 3 \,(72) = W.$$
$$17 \cdot 5 \,(24) + 30 \cdot 7 \,(44) + 44 \cdot 0 \,(76) = W.$$
$$18 \cdot 3 \,(24) + 33 \cdot 2 \,(44) + 63 \cdot 7 \,(80) = W.$$

§ VIII. *Determination of the Average Error in Observations of different kinds.*

In order to estimate the degree of accuracy of these observations, I have taken the differences between the values of the three standard colours as originally observed, and their means as given by the above Table. The sum of all the errors of the red (24) from the means, was 31·1, and the number of observations was 42, which gives the average error ·74.

The sum of errors in green (44) was 48·0, and the number of observations 31, giving a mean error 1·55.

The sum of the errors in blue (68) was 46·9, and the number of observations 35, giving a mean error 1·16.

It appears therefore that in the observations generally, the average error does not exceed 1·5; and therefore the results, if confirmed by several observations, may safely be trusted to that degree of accuracy.

The equation between the three standard colours was repeatedly observed, in order to detect any alteration in the character of the light, or any other change of condition which would prevent the observations from being comparable with one another; and also because this equation is used in the reduction of all the others, and therefore requires to be carefully observed. There are twenty observations of this equation, the mean of which gives

$$18 \cdot 6 \,(24) + 31 \cdot 4 \,(44) + 30 \cdot 5 \,(68) = W^* \quad \dots\dots\dots\dots\dots(16)$$

as the standard equation.

We may use the twenty observations of this equation as a means of determining the relations between the errors in the different colours, and thus of estimating the accuracy of the observer in distinguishing colours.

The following Table gives the result of these operations, where R stands for (24), G for (44), and B for (68) :—

TABLE V.—Mean Errors in the Standard Equation.

$(R) = \cdot54$	$(G - B) = \cdot99$	$(G + B) = 2 \cdot 31$	$\sqrt{G^2 + B^2} = 1 \cdot 67$
$(G) = 1 \cdot 22$	$(B - R) = \cdot 85$	$(B + R) = 1 \cdot 59$	$\sqrt{B^2 + R^2} = 1 \cdot 26$
$(B) = 1 \cdot 15$	$(R - G) = \cdot 86$	$(R + G) = 1 \cdot 57$	$\sqrt{R^2 + G^2} = 1 \cdot 33$

$$(R + G + B) = 2 \cdot 67 \qquad \sqrt{R^2 + G^2 + B^2} = 1 \cdot 76$$

The first column gives the mean difference between the observed value of each of the colours and the mean of all the observations. The second column shews the average error of the observed *differences* between the values of the standards, from the mean value of those differences. The third column shews the average error of the *sums* of two standards, from the mean of such sums. The fourth column gives the square root of the sum of the squares of the quantities in the first column. I have also given the average error of the sum of R, G and B, from its mean value, and the value of $\sqrt{R^2 + G^2 + B^2}$.

It appears from the first column that the red is more accurately observed than the green and blue.

§ IX. *Relative Accuracy in Observations of Colour and of Brightness.*

If the errors in the different colours occurred perfectly independent of each other, then the probable mean error in the sum or difference of any two colours would be the square root of the sum of their squares, as given in the fourth column. It will be seen, however, that the number in the second column is always less, and that in the third always greater, than that in the fourth; shewing that the errors are not independent of each other, but that positive errors in any colour coincide more often with positive than with negative errors in another colour. Now the *hue* of the resultant depends on the *ratios* of the components, while its *brightness* depends on their sum. Since, therefore, the difference of two colours is always more accurately observed than their sum, variations of *colour* are more easily detected than variations in *brightness*, and the eye appears to be a more accurate judge of the identity of colour of the two parts of the field than of their equal illumination. The same conclusion may be drawn from the value of the mean error of the sum of the three standards, which is 2·67, while the square root of the sum of the squares of the errors is 1·76.

§ X. *Reduction of the Observations.*

By eliminating W from the equations of page 428 by means of the standard equation, we obtain equations involving each of the fourteen selected colours of the spectrum, along with the three standard colours; and by transposing the selected colour to one side of the equation, we obtain its value in terms of the three standards. If any of the terms of these equations are negative, the equation has no physical interpretation as it stands, but by transposing the negative term to the other side it becomes positive, and then the equation may be verified.

The following Table contains the values of the fourteen selected tints in terms of the standards. To avoid repetition, the symbols of the standard colours are placed at the head of each column.

TABLE VI.

Observer (K.).	(24.)	(44.)	(68.)
44·3 (20) =	18·6	+ 0·4	+ 2·8
16·1 (28) =	18·6	+ 5·8	− 0·1
22·0 (32) =	18·6	+19·3	− 0·1
25·2 (36) =	12·2	+31·4	− 0·8
26·0 (40) =	3·3	+31·4	− 0·2
35·0 (46) = −	1·2	+31·4	+ 0·3
41·4 (48) = −	2·6	+31·4	+ 3·5
62·0 (52) = −	3·4	+31·4	+17·5
61·7 (56) = −	3·1	+21·0	+30·5
40·5 (60) = −	1·9	+ 7·7	+30·5
33·7 (64) = −	1·1	+ 1·1	+30·5
32·3 (72) = +	0·6	+ 0·2	+30·5
44·0 (76) = +	1·1	+ 0·7	+30·5
63·7 (80) = +	0·3	− 1·8	+30·5

From these equations we may lay down a chart of the spectrum on Newton's diagram by the following method:—Take any three points, A, B, C, and let A represent the standard colour (24), B (44), and C (68). Then, to find the position of any other colour, say (20), divide AC in P so that $(18\cdot6)\,AP = (2\cdot8)\,PC$, and then divide BP in Q so that $(18\cdot6 + 2\cdot8)\,PQ = (0\cdot4)\,QB$. At the point Q the colour corresponding to (20) must be placed. In this way the diagram of fig. 4, Plate VI., p. 444, has been constructed from the observations of all the colours.

§ XI. *The Spectrum as laid down on Newton's Diagram.*

The curve on which these points lie has this striking feature, that two portions of it are nearly, if not quite, straight lines. One of these portions extends from (24) to (46), and the other from (48) to (64). The colour (20) and those beyond (64), are not far from the line joining (24) and (68). The spectrum, therefore, as exhibited in Newton's diagram, forms two sides of a triangle, with doubtful fragments of the third side. Now if three colours in Newton's diagram lie in a straight line, the middle one is a compound of the two others. Hence all the colours of the spectrum may be compounded of those which lie at the angles of this triangle. These correspond to the following colours :—

TABLE VII.

		Scale.	Wave-length.	Index in water.	Wave-length in water.
R	Scarlet . .	24	2328	1·332	1·747
G	Green . . .	$46\frac{3}{4}$	1914	1·334	1·435
B	Blue . . .	$64\frac{1}{2}$	1717	1·339	1·282

All the other colours of the spectrum may be produced by combinations of these; and since all natural colours are compounded of the colours of the spectrum, they may be compounded of these three primary colours. I have strong reason to believe that these are the three primary colours corresponding to three modes of sensation in the organ of vision, on which the whole system of colour, as seen by the normal eye, depends.

§ XII. *Results found by a second Observer.*

We may now consider the results of three series of observations made by myself (J.) as observer, in order to determine the relation of one observer to

another in the perception of colour.   The standard colours are connected by the
following equation, as determined by six observations :—

$$18 \cdot 1 \,(24) + 27 \cdot 5 \,(44) + 37 \,(68) = W^* \quad \ldots \ldots \ldots \ldots \ldots (17).$$

The average errors in these observations were—

TABLE VIII.

R, ·28	G + B, ·83	G − B, ·83	
G, ·83	B + R, ·42	B − R, ·28	R + G + B, ·95
B, ·16	R + G, ·95	R − G, ·72	

shewing that in this case, also, the power of distinguishing *colour* is more to be
depended on than that of distinguishing degrees of *illumination.*

The average error in the other observations from the means was ·64 for red,
·76 for green, and 1·02 for blue.

TABLE IX.

Observations by J., October 1859.

	(24.)	(44.)	(68.)
44·3 (20) =	18·1	− 2·5	+ 2·3
16·0 (28) =	18·1	+ 6·2	− 0·7
21·5 (32) =	18·1	+ 25·2	− 0·7
19·3 (36) =	8·1	+ 27·5	− 0·3
20·7 (40) =	2·1	+ 27·5	− 0·5
52·3 (48) = −	1·4	+ 27·5	+ 10·7
95·0 (52) = −	2·4	+ 27·5	+ 37·0
51·7 (56) = −	2·2	+ 4·8	+ 37·0
37·2 (60) = −	1·2	+ 0·8	+ 37·0
36·7 (64) = −	0·2	+ 0·8	+ 37·0
35·0 (72) = +	0·6	− 0·2	+ 37·0
40·0 (76) = +	0·9	+ 0·5	+ 37·0
51·0 (80) = +	1·1	+ 0·5	+ 37·0

§ XIII.   *Comparison of Results by Newton's Diagram.*

The relations of the colours, as given by these observations, are laid down
in fig. 5, Plate VI., p. 444.   It appears from this diagram, that the positions of
the colours lie nearly in a straight line from (24) to (44), and from (48) to (60).
The colours beyond (60) are crowded together, as in the other diagram, and
the observations are not yet sufficiently accurate to distinguish their relative
positions accurately.   The colour (20) at the red end of the spectrum is further

from the line joining (24) and (68) than in the other diagram, but I have not obtained satisfactory observations of these extreme colours. It will be observed that (32), (36), and (40) are placed further to the right in fig. 5 than in fig. 4, shewing that the second observer (J.) sees more green in these colours than the first (K.), also that (48), (52), (56), and (60) are much further up in fig. 5, shewing that to the second observer they appear more blue and less green. These differences were well seen in making an observation. When the instrument was adjusted to suit the first observer (K.), then, if the selected colour were (32), (36), or (40), the second (J.), on looking into the instrument, saw it too green; but if (48), (52), (56), or (60) were the selected colour, then, if right to the first observer, it appeared too blue to the second. If the instrument were adjusted to suit the second observer, then, in the first case, the other saw red, and in the second green; shewing that there was a real difference in the eyes of these two individuals, producing constant and measurable differences in the apparent colour of objects.

§ XIV. *Comparison by Curves of Intensity of the Primaries.*

Figs. 6 and 7, Plate VI. p. 444, are intended to indicate the intensities of the three standard colours at different points of the spectrum. The curve marked (R) indicates the intensity of the red or (24), (G) that of green or (44), and (B) that of blue or (68). The curve marked (S) has its ordinates equal to the sum of the ordinates of the other three curves. The intensities are found by dividing every colour-equation by the coefficient of the colour on the left-hand side. Fig. 6 represents the results of observations by K., and fig. 7 represents those of J. It will be observed that the ordinates in fig. 7 are smaller between (48) and (56) than in fig. 6. This indicates the feeble intensity of certain kinds of light as seen by the eyes of J., which made it impossible to get observations of the colour (52) at all without making the slit so wide as to include all between (48) and (56).

This blindness of my eyes to the parts of the spectrum between the fixed lines E and F appears to be confined to the region surrounding the axis of vision, as the field of view, when adjusted for my eyes looking directly at the colour, is decidedly out of adjustment when I view it by indirect vision, turning the axis of my eye towards some other point. The prism then appears greener and brighter than the mirror, shewing that the parts of my eye at a distance from the axis are more sensitive to this blue-green light than the parts close to the axis.

It is to be noticed that this insensibility is not to all light of a green or blue colour, but to light of a definite refrangibility. If I had a species of colour-blindness rendering me totally or partially insensible to that element of colour which most nearly corresponds with the light in question, then the light from the mirror, as well as that from the prism, would appear to me deficient in that colour, and I should still consider them chromatically identical; or if there were any difference, it would be the same for all colours nearly the same in appearance, such as those just beyond the line F, which appear to me quite bright.

We must also observe that the peculiarity is confined to a certain portion of the retina, which is known to be of a yellow colour, and which is the seat of several ocular phenomena observed by Purkinje and Wheatstone, and of the sheaf or brushes seen by Haidinger in polarized light; and also that though, of the two observers whose results are given here, one is much more affected with this peculiarity than the other, both are less sensible to the light between E and F than to that on either side; and other observers, whose results are not here given, confirm this.

§ XV. *Explanation of the Differences between the two Observers.*

I think, therefore, that the yellow spot at the foramen centrale of Soemmering will be found to be the cause of this phenomenon, and that it absorbs the rays between E and F, and would, if placed in the path of the incident light, produce a corresponding dark band in the spectrum formed by a prism.

The reason why white light does not appear yellow in consequence, is that this absorbing action is constant, and we reckon as white the *mean* of all the colours we are accustomed to see. This may be proved by wearing spectacles of any strong colour for some time, when we shall find that we judge white objects to be white, in spite of the rays which enter the eye being coloured.

Now ordinary white light is a mixture of all kinds of light, including that between E and F, which is partially absorbed. If, therefore, we compound an artificial white containing the absorbed ray as one of its three components, it will be much more altered by the absorption than the ordinary light, which contains many rays of nearly the same colour, which are not absorbed. On the other hand, if the artificial light do not contain the absorbed ray, it will be less altered than the ordinary light which contains it. Hence the greater the absorption the less green will those colours appear which are near the absorbed part, such as (48), (52), (56), and the more green will the colours appear which are not near it, such as (32), (36), (40). And these are the chief differences between fig. 4 and fig. 5.

I first observed this peculiarity of my eyes when observing the spectrum formed by a very long vertical slit. I saw an elongated dark spot running up and down in the blue, as if confined in a groove, and following the motion of the eye as it moved up or down the spectrum, but refusing to pass out of the blue into other colours. By increasing the breadth of the spectrum, the dark portion was found to correspond to the *foramen centrale*, and to be visible only when the eye is turned towards the blue-green between E and F. The spot may be well seen by first looking at a yellow paper, and then at a blue one, when the spot will be distinctly seen for a short time, but it soon disappears when the eye gets accustomed to the blue*.

I have been the more careful in stating this peculiarity of my eyes, as I have reason to believe that it affects most persons, especially those who can see Haidinger's brushes easily. Such persons, in comparing their vision with that of others, may be led to think themselves affected with partial colour-blindness, whereas their colour-vision may be of the ordinary kind, but the rays which reach their sense of sight may be more or less altered in their proportions by passing through the media of the eye. The existence of real, though partial colour-blindness will make itself apparent, in a series of observations, by the discrepancy between the observed values and the means being greater in certain colours than in others.

§ XVI. *General Conclusions.*

Neither of the observers whose results are given here shew any indications of colour-blindness, and when the differences arising from the absorption of the rays between E and F are put out of account, they agree in proving that there are three colours in the spectrum, red, green, and blue, by the mixtures of

---

* See the *Report of the British Association* for 1856, p. 12.

which colours chromatically identical with the other colours of the spectrum may be produced. The exact position of the red and blue is not yet ascertained; that of the green is $\frac{1}{4}$ from E towards F.

The orange and yellow of the spectrum are chromatically equivalent to mixtures of red and green. They are neither richer nor paler than the corresponding mixtures, and the only difference is that the mixture may be resolved by a prism, whereas the colour in the spectrum cannot be so resolved. This result seems to put an end to the pretension of yellow to be considered a primary element of colour.

In the same way the colours from the primary green to blue are chromatically identical with mixtures of these; and the extreme ends of the spectrum are probably equivalent to mixtures of red and blue, but they are so feeble in illumination that experiments on the same plan with the rest can give no result, but they must be examined by some special method. When observations have been obtained from a greater number of individuals, including those whose vision is dichromatic, the chart of the spectrum may be laid down independently of accidental differences, and a more complete discussion of the laws of the sensation of colour attempted.

POSTSCRIPT.

[Received May 8,—Read May 24, 1860.]

Since sending the above paper to the Royal Society, I have obtained some observations of the colour of the spectrum by persons whose vision is "dichromic," and who are therefore said to be "colour-blind."

The instrument used in making these observations was similar in principle to that formerly described, except that, in order to render it portable, the rays are reflected back through the prisms, nearly in their original direction; thus rendering one of the limbs of the instrument unnecessary, and allowing the other to be shortened considerably on account of the greater angular dispersion. The principle of reflecting light, so as to pass twice through the same prism, was employed by me in an instrument for combining colours made in 1856, and a reflecting instrument for observing the spectrum has been constructed independently by M. Porro.

Light from a sheet of paper illuminated by sunlight is admitted at the slits X, Y, Z (fig. 8, Plate VII. p. 444), falls on the prisms P and P' (angles $= 45°$), then on a concave silvered glass, S, radius 34 inches. The light, after reflexion, passes again through the prisms P' and P, and is reflected by a small mirror, e, to the slit E, where the eye is placed to receive the light compounded of the colours corresponding to the positions and breadths of the slits X, Y, and Z.

At the same time, another portion of the light from the illuminated paper enters the instrument at BC, is reflected at the mirror M, passes through the lens L, is reflected at the mirror M', passes close to the edge of the prism P, and is reflected along with the coloured light at e, to the eye-slit at E.

In this way the compound colour is compared with a constant white light in optical juxtaposition with it. The mirror M is made of silvered glass, that at M' is made of glass roughened and blackened at the back, to reduce the intensity of the constant light to a convenient value for the experiments.

This instrument gives a spectrum in which the lines are very distinct, and the length of the spectrum from A to H is 3·6 inches. The outside measure of the box is 3 feet 6 inches, by 11 inches by 4 inches, and it can be carried about, and set up in any position, without readjustment. It was made by Messrs Smith and Ramage of Aberdeen.

In obtaining observations from colour-blind persons, two slits only are required to produce a mixture chromatically equivalent to white; and at one point of the spectrum the colour of the pure rays appears identical with white. This point is near the line F, a little on the less refrangible side. From this point to the more refrangible end of the spectrum appears to them "blue." The colours on the less refrangible side appear to them all of the same quality, but of different degrees of brightness; and when any of them are made sufficiently bright, they are called "yellow." It is convenient to use the term "yellow" in speaking of the colours from red to green inclusive, since it will be found that a dichromic person in speaking of red, green, orange, and brown, refers to different degrees of brightness or purity of a single colour, and not to different colours perceived by him. This colour we may agree to call "yellow," though it is not probable that the sensation of it is like that of yellow as perceived by us.

Of the three standard colours which I formerly assumed, the red appears to them "yellow," but so feeble that there is not enough in the whole red division of the spectrum to form an equivalent to make up the standard white.

The green at E appears a good "yellow," and the blue at $\frac{2}{3}$ from F towards G appears a good "blue." I have therefore taken these as standard colours for reducing dichromic observations. The three standard colours will be referred to as (104), (88), and (68), these being the positions of the red, green, and blue on the scale of the new instrument.

Mr James Simpson, formerly student of Natural Philosophy in my class, has furnished me with thirty-three observations taken in good sunlight. Ten of these were between the two standard colours, and give the following result :—

$$33 \cdot 7\,(88) + 33 \cdot 1\,(68) = W \dots\dots\dots\dots\dots(1).$$

The mean errors of these observations were as follows :—

Error of $(88) = 2 \cdot 5$; of $(68) = 2 \cdot 3$; of $(88) + (68) = 4 \cdot 8$; of $(88) - (68) = 1 \cdot 3$.

The fact that the mean error of the sum was so much greater than the mean error of the difference indicates that in this case, as in all others that I have examined, observations of equality of tint can be depended on much more than observations of equality of illumination or brightness.

From six observations of my own, made at the same time, I have deduced the "trichromic" equation

$$22 \cdot 6\,(104) + 26\,(88) + 37 \cdot 4\,(68) = W \dots\dots\dots\dots(2).$$

If we suppose that the light which reached the organ of vision was the same in both cases, we may combine these equations by subtraction, and so find

$$22 \cdot 6\,(104) - 7 \cdot 7\,(88) + 4 \cdot 3\,(68) = D \dots\dots\dots\dots(3),$$

where D is that colour, the absence of the sensation of which constitutes the defect of the dichromic eye. The sensation which I have in addition to those of the dichromic eye is therefore similar to the full red (104), but different from it, in that the red (104) has $7 \cdot 7$ of green (88) in it which must be removed, and $4 \cdot 3$ of blue (68) substituted. This agrees pretty well with the colour which Mr Pole* describes as neutral to him, though crimson to others. It must be remembered, however, that different persons of ordinary vision require different proportions of the standard colours, probably owing to differences in the absorptive powers of the media of the eye, and that the above equation (2), if observed by K., would have been

$$23\,(104) + 32\,(88) + 31\,(68) = W \dots\dots\dots\dots(4).$$

* *Philosophical Transactions*, 1859, Part I. p. 329.

and the value of D, as deduced from these observers, would have been

$$23 (104) - 1\cdot 7 (88) - 1\cdot 1 (68) = D \dots\dots\dots\dots\dots\dots(5),$$

in which the defective sensation is much nearer to the red of the spectrum. It is probably a colour to which the extreme red of the spectrum tends, and which differs from the extreme red only in not containing that small proportion of "yellow" light which renders it visible to the colour-blind.

From other observations by Mr Simpson the following results have been deduced :—

TABLE a.

	(88.)	(68.)			(88.)	(68.)
$(99\cdot2 +) =$	33·7	1·9				
$31\cdot3 (96) =$	33·7	2·1		$100 (96) =$	108	7
$28 \;\; (92) =$	33·7	1·4		$100 (92) =$	120	5
$33\cdot7 (88) =$	33·7	0		$100 (88) =$	100	0
$54\cdot7 (84) =$	33·7	6·1		$100 (84) =$	61	11
$71 \;\; (82) =$	33·7	15·1		$100 (82) =$	47	21
$99 \;\; (80) =$	33·7	33·1		$100 (80) =$	34	33
$70 \;\; (78) =$	15·7	33·1		$100 (78) =$	22	47
$56 \;\; (76) =$	5·7	33·1		$100 (76) =$	10	59
$36 \;\; (72) = -$	0·3	33·1		$100 (72) = -$	1	92
$33\cdot1 (68) =$	0	33·1		$100 (68) =$	0	100
$40 \;\; (64) =$	0·2	33·1		$100 (64) =$	0	83
$55\cdot5 (60) =$	1·7	33·1		$100 (60)$	3	60
$(57 -) \;\; = -$	0·3	33·1				

In the Table on the left side $(99\cdot2 +)$ means the whole of the spectrum beyond $(99\cdot2)$ on the scale, and $(57 -)$ means the whole beyond (57) on the scale. The position of the fixed lines with reference to the scale was as follows :—

A, 116; a, 112; B, 110; C, 106; D, 98·3; E, 88; F, 79; G, 61; H, 44.

The values of the standard colours in different parts of the spectrum are given on the right side of the above Table, and are represented by the curves of fig. 9, Plate VII. p. 444, where the left-hand curve represents the intensity of the "yellow" element, and the right-hand curve that of the "blue" element of colour as it appears to the colour-blind.

The appearance of the spectrum to the colour-blind is as follows :—

From A to E the colour is pure "yellow" very faint up to D, and reaching a maximum between D and E. From E to one-third beyond F towards

G the colour is mixed, varying from "yellow" to "blue," and becoming neutral or "white" at a point near F. In this part of the spectrum, the total intensity, as given by the dotted line, is decidedly less than on either side of it, and near the line F, the retina close to the "yellow spot" is less sensible to light than the parts further from the axis of the eye. This peculiarity of the light near F is even more marked in the colour-blind than in the ordinary eye. Beyond F the "blue" element comes to a maximum between F and G, and then diminishes towards H; the spectrum from this maximum to the end being pure "blue."

In fig. 10, Plate VII. p. 444, these results are represented in a different manner. The point D, corresponding to the sensation wanting in the colour-blind, is taken as the origin of coordinates, the "yellow" element of colour is represented by distances measured horizontally to the right from D, and the "blue" element by distances measured vertically from the horizontal line through D. The numerals indicate the different colours of the spectrum according to the scale shewn in fig. 9, and the coordinates of each point indicate the composition of the corresponding colour. The triangle of colours is reduced, in the case of dichromic vision, to a straight line "B" "Y," and the proportions of "blue" and "yellow" in each colour are indicated by the ratios in which this line is cut by the line from D passing through the position of that colour.

The results given above were all obtained with the light of white paper, placed in clear sunshine. I have obtained similar results, when the sun was hidden, by using the light of uniformly illuminated clouds, but I do not consider these observations sufficiently free from disturbing circumstances to be employed in calculation. It is easy, however, by means of such observations, to verify the most remarkable phenomena of colour-blindness, as for instance, that the colours from red to green appear to differ only in brightness, and that the brightness may be made identical by changing the width of the slit; that the colour near F is a neutral tint, and that the eye in viewing it sees a dark spot in the direction of the axis of vision; that the colours beyond are all blue of different intensities, and that any "blue" may be combined with any "yellow" in such proportions as to form "white." These results I have verified by the observations of another colour-blind gentleman, who did not obtain sunlight for his observations; and as I have now the means of carrying the requisite apparatus easily, I hope to meet with other colour-blind observers, and to obtain their observations under more favourable circumstances.

On the Comparison of Colour-blind with ordinary Vision by means of Observations with Coloured Papers.

In March 1859 I obtained a set of observations by Mr Simpson, of the relations between six coloured papers as seen by him. The experiments were made with the colour-top in the manner described in my paper in the *Transactions of the Royal Society of Edinburgh*, Vol. XXI. pt. 2, p. 286; and the colour-equations were arranged so as to be equated to zero, as in those given in the *Philosophical Magazine*, July, 1857. The colours were—Vermilion (V), ultramarine (U), emerald-green (G), ivory-black (B), snow-white (W), and pale chrome-yellow (Y). These six colours afford fifteen colour-blind equations, since four colours enter into each equation. Fourteen of these were observed by Mr Simpson, and from these I deduced three equations, giving the relation of the three standards (V), (U), (G) to the other colours, according to his kind of vision. From these three equations I then deduced fifteen equations, admitting of comparison with the observed equations, and necessarily consistent in themselves.

The comparison of these equations furnishes a test of the truth of the theory that the colour-blind see by means of two colour-sensations, and that therefore every colour may be expressed in terms of *two* given colours, just as in ordinary vision it may be expressed in terms of three given colours. The one set of equations are each the result of a single observation; the other set are deduced from three equations in accordance with this theory, and the two sets agree to within an average error = 2·1.

TABLE *b.*

		V.	U.	G.	B.	W.	Y.	
1.	Observed ...	0	0	− 100	+ 45	+ 22	+ 33	= 0.
	Calculated ..	0	0	− 100	+ 37·5	+ 26·5	+ 36	= 0.
2.	Observed ...	0	+ 58	0	− 69	− 31	− 42	= 0.
	Calculated ..	0	+ 58·3	0	− 67·3	− 32·7	+ 41·7	= 0.
3.	Observed ...	0	+ 32	− 100	0	+ 12	+ 56	= 0.
	Calculated ..	0	+ 32·3	− 100	0	+ 8·3	+ 59·4	= 0.
4.	Observed ...	0	+ 38	− 89	− 11	0	+ 62	= 0.
	Calculated ..	0	+ 40	− 85	− 15	0	+ 60	= 0.
5.	Observed ...	0	+ 32	+ 68	− 60	− 40	0	= 0.
	Calculated ..	0	+ 34	+ 66	− 63·5	− 36·5	0	= 0.

TABLE *b* (*continued*).

		V.	U.	G.	B.	W.	Y.	
6.	Observed ...	− 100	0	0	+ 82	+ 5	+ 13	= 0.
	Calculated ..	− 100	0	0	+ 83·9	+ 4·5	+ 11·6	= 0.
7.	Observed ...	+ 47	0	− 100	0	+ 22	+ 31	= 0.
	Calculated ..	+ 44·7	0	− 100	0	+ 24·5	+ 30·8	= 0.
8.	Observed ...	− 100	0	+ 20	+ 77	0	+ 3	= 0.
	Calculated ..	− 100	0	+ 17	+ 77·5	0	+ 5·5	= 0.
9.	Not Observed.							
	Calculated ..	+ 96	0	− 31	− 69	+ 4	0	= 0.
10.	Observed ...	− 70	+ 53	0	0	− 30	+ 47	= 0.
	Calculated ..	− 73·5	+ 53	0	0	− 26·5	+ 47	= 0.
11.	Observed ...	− 100	+ 8	0	+ 71	0	+ 21	= 0.
	Calculated ..	− 100	+ 8	0	+ 74·5	0	+ 17·5	= 0.
12.	Observed ...	+ 85	+ 15	0	− 88	− 12	0	= 0.
	Calculated ..	+ 86	+ 14	0	− 88·5	− 11·5	0	= 0.
13.	Observed ...	− 20	+ 39	− 80	0	0	+ 61	= 0.
	Calculated ..	− 19	+ 40	− 81	0	0	+ 60	= 0.
14.	Observed ...	− 66	+ 30	+ 70	0	− 34	0	= 0.
	Calculated ..	− 70	+ 27	+ 73	0	− 30	0	= 0.
15.	Observed ...	+ 100	− 2	− 27	− 71	0	0	= 0.
	Calculated ..	+ 96	+ 4	− 24	− 76	0	0	= 0.

But, according to our theory, colour-blind vision is not only dichromic, but the two elements of colour are identical with two of the three elements of colour as seen by the ordinary eye; so that it differs from ordinary vision only in not perceiving a particular colour, the relation of which to known colours may be numerically defined. This colour may be expressed under the form

$$aV + bU + cG = D \dots\dots\dots\dots\dots\dots\dots (16),$$

where V, U, and G are the standard colours used in the experiments, and D is the colour which is visible to the ordinary eye, but invisible to the colour-blind. If we know the value of D, we may always change an ordinary colour-equation into a colour-blind equation by subtracting from it nD (n being chosen so that one of the standard colours is eliminated), and adding n of black.

In September 1856 I deduced, from thirty-six observations of my own, the chromatic relations of the same set of six coloured papers. These observations, with a comparison of them with the trichromic theory of vision, are to be found in the *Philosophical Magazine* for July 1857. The relations of the six colours may be deduced from two equations, of which the most convenient form is

V.	U.	G.	B.	W.	Y.	
$+39\cdot7$	$+26\cdot6$	$+33\cdot7$	$-22\cdot7$	$-77\cdot3$	0	$=0$.........(17).
$-62\cdot4$	$+18\cdot6$	$-37\cdot6$	0	$+45\cdot7$	$+35\cdot7$	$=0$.........(18).

The value of D, as deduced from a comparison of these equations with the colour-blind equations, is

$$1\cdot198\,V + 0\cdot078\,U - 0\cdot276\,G = D \quad\ldots\ldots\ldots\ldots\ldots (19).$$

By making D the same thing as black (B), and eliminating W and Y respectively from the two ordinary colour-equations by means of D, we obtain three colour-blind equations, calculated from the ordinary equations and consistent with them, supposing that the colour (D) is black to the colour-blind.

The following Table is a comparison of the colour-blind equations deduced from Mr Simpson's observations alone, with those deduced from my observations and the value of D.

TABLE c.

	V.	U.	G.	B.	W.	Y.
(15) Calculated . .	$+96$	$+4$	-24	-76	0	0
By (19)	$+93\cdot9$	$+6\cdot1$	$-21\cdot7$	$-78\cdot3$	0	0
(14) Calculated . .	-70	$+27$	$+73$	0	-30	0
By (17) and (19) . .	-70	$+27\cdot2$	$-72\cdot8$	0	-30	0
(13) Calculated . .	-19	$+40$	-81	0	0	$+60$
By (18) and (19) . .	$-13\cdot6$	$+38\cdot5$	$-86\cdot4$	0	0	$+61\cdot5$

The average error here is 1·9, smaller than the average error of the individual colour-blind observations, shewing that the theory of colour-blindness being the want of a certain colour-sensation which is one of the three ordinary colour-sensations, agrees with observation to within the limits of error.

In fig. 11, Plate VII. p. 444, I have laid down the chromatic relations of these colours according to Newton's method. V (vermilion), U (ultramarine), and G (emerald-green) are assumed as standard colours, and placed at the angles of an equilateral triangle. The position of W (white) and Y (pale chrome-yellow) with respect to these are laid down from equations (17) and (18), deduced from my own observations. The positions of the defective colour, of white, and of yellow, as deduced from Mr Simpson's equations alone, are given at "d," "w," and "y." The positions of these points, as deduced from a combination

56—2

of these equations with my own, are given at "D," "W," and "Y." The difference of these positions from those of "*d*," "*w*," and "*y*," shews the amount of discrepancy between observation and theory.

It will be observed that D is situated near V (vermilion), but that a line from D to W cuts UV at C near to V. D is therefore a red colour, not scarlet, but further from yellow. It may be called crimson, and may be *imitated* by a mixture of 86 vermilion and 14 ultramarine. This compound colour will be of the same *hue* as D; but since C lies between D and W, C must be regarded as D diluted with a certain amount of white; and therefore D must be imagined to be like C in hue, but without the intermixture of white which is unavoidable in actual pigments, and which reduces the purity of the tint.

Lines drawn from D through "W" and "Y," the colour-blind positions of white and yellow, pass through W and Y, their positions in ordinary vision. The reason why they do not coincide with W and Y, is that the white and yellow papers are much brighter than the colours corresponding to the points W and Y of the triangle V, U, G; and therefore lines from D, which represent them in intensity as well as in quality, must be longer than DW and DY in the proportion of their brightness.

On Compound Colours.

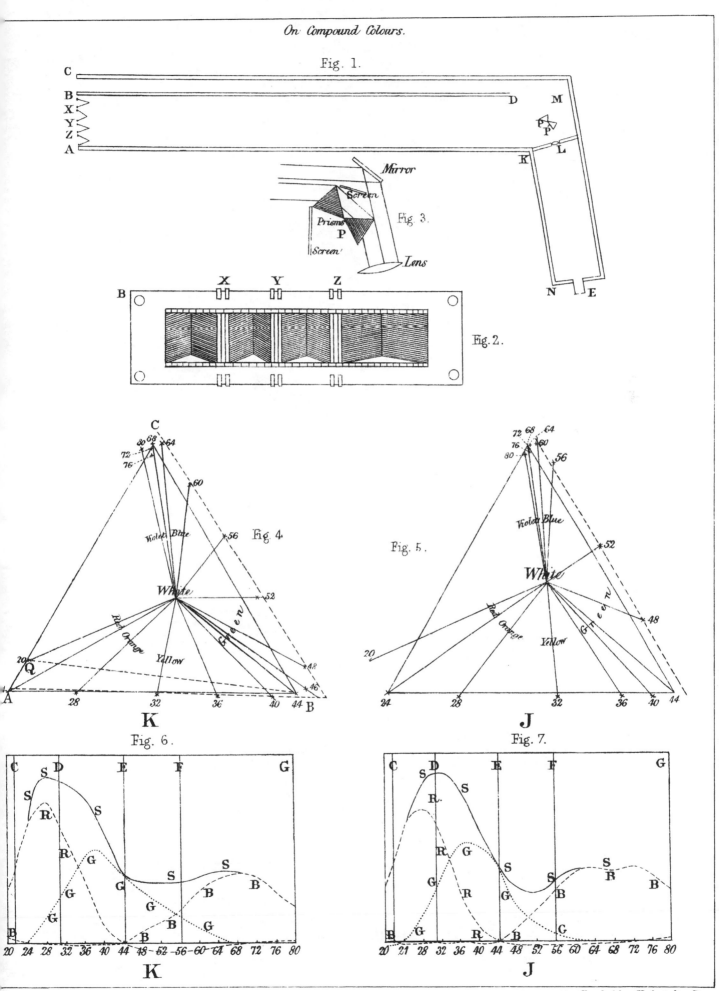

Fig. 1.

Fig. 2.

Fig. 3.

Fig. 4.

Fig. 5.

Fig. 6.

Fig. 7.

Cambridge University Press

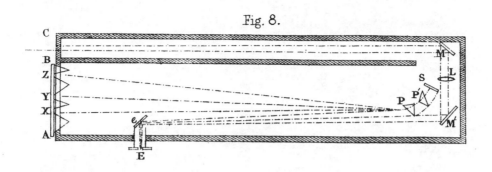

Fig. 8.

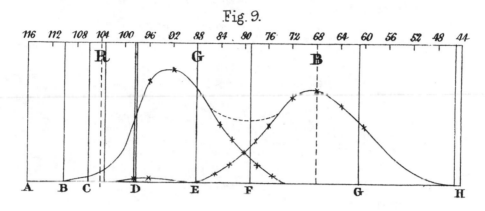

Fig. 9.

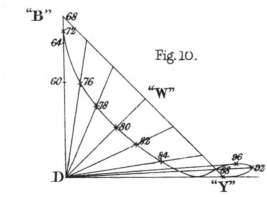

Fig. 10.

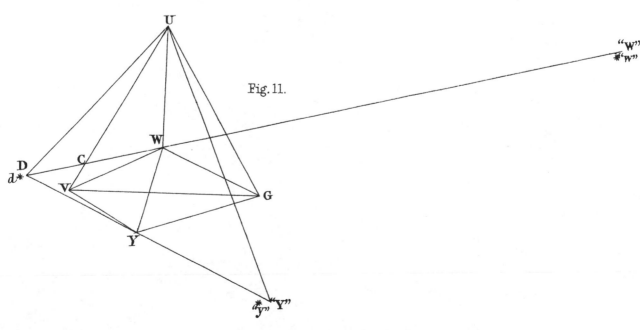

Fig. 11.

[Lecture at the Royal Institution of Great Britain. May 17, 1861.]

XXII. *On the Theory of Three Primary Colours.*

THE speaker commenced by shewing that our power of vision depends entirely on our being able to distinguish the intensity and quality of colours. The forms of visible objects are indicated to us only by differences in colour or brightness between them and surrounding objects. To classify and arrange these colours, to ascertain the physical conditions on which the differences of coloured rays depend, and to trace, as far as we are able, the physiological process by which these different rays excite in us various sensations of colour, we must avail ourselves of the united experience of painters, opticians, and physiologists. The speaker then proceeded to state the results obtained by these three classes of inquirers, to explain their apparent inconsistency by means of Young's Theory of Primary Colours, and to describe the tests to which he had subjected that theory.

Painters have studied the relations of colours, in order to imitate them by means of pigments. As there are only a limited number of coloured substances adapted for painting, while the number of tints in nature is infinite, painters are obliged to produce the tints they require by mixing their pigments in proper proportions. This leads them to regard these tints as actually compounded of other colours, corresponding to the pure pigments in the mixture. It is found, that by using three pigments only, we can produce all colours lying within certain limits of intensity and purity. For instance, if we take carmine (red), chrome yellow, and ultramarine (blue), we get by mixing the carmine and the chrome, all varieties of orange, passing through scarlet to crimson on the one side, and to yellow on the other; by mixing chrome and ultramarine we get all hues of green; and by mixing ultramarine with carmine, we get all hues of purple, from violet to mauve and crimson. Now these are all the strong colours that we ever see or can imagine: all others are like these, only less pure in tint. Our three colours can be mixed so as to form a neutral grey; and if this grey be mixed with any of the hues produced by mixing two colours only, all the tints of that hue will be exhibited, from the pure colour to neutral grey. If we could assume that the colour of a mixture of different kinds of paint is a true mixture of the colours of the pigments, and in the same proportion, then an analysis of colour might be made with the same ease as a chemical analysis of a mixture of substances.

The colour of a mixture of pigments, however, is often very different from a true mixture of the colours of the pure pigments. It is found to depend on the size of the particles, a finely ground pigment producing more effect than one coarsely ground. It has also been shewn by Professor Helmholtz, that when light falls on a mixture of pigments, part of it is acted on by one pigment only, and part of it by another; while a third portion is acted on by both pigments in succession before it is sent back to the eye. The two parts reflected directly from the pure pigments enter the eye together, and form a true mixture of colours; but the third portion, which has suffered absorption from both pigments, is often so considerable as to give its own character to the resulting tint. This is the explanation of the green tint produced by mixing most blue and yellow pigments.

In studying the mixture of colours, we must avoid these sources of error, either by mixing the rays of light themselves, or by combining the impressions of colours within the eye by the rotation of coloured papers on a disc.

The speaker then stated what the opticians had discovered about colour. White light, according to Newton, consists of a great number of different kinds of coloured light which can be separated by a prism. Newton divided these into seven classes, but we now recognize many thousand distinct kinds of light in the spectrum, none of which can be shewn to be a compound of more elementary rays. If we accept the theory that light is an undulation, then, as there are undulations of every different period from the one end of the spectrum to the other, there are an *infinite* number of possible kinds of light, no one of which can be regarded as compounded of any others.

Physical optics does not lead us to any theory of three primary colours, but leaves us in possession of an infinite number of pure rays with an infinitely more infinite number of compound beams of light, each containing any proportions of any number of the pure rays.

These beams of light, passing through the transparent parts of the eye, fall on a sensitive membrane, and we become aware of various colours. We know that the colour we see depends on the nature of the light; but the opticians say there are an infinite number of kinds of light; while the painters, and all who pay attention to what they see, tell us that they can account for all actual colours by supposing them mixtures of three primary colours.

The speaker then next drew attention to the physiological difficulties in accounting for the perception of colour. Some have supposed that the different kinds of light are distinguished by the time of their vibration. There are about 447 billions of vibrations of red light in a second; and 577 billions of vibrations of green light in the same time. It is certainly not by any mental process of which we are conscious that we distinguish between these infinitesimal portions of time, and it is difficult to conceive any mechanism by which the vibrations could be counted so that we should become conscious of the results, especially when many rays of different periods of vibration act on the same part of the eye at once.

Besides, all the evidence we have on the nature of nervous action goes to prove that whatever be the nature of the agent which excites a nerve, the sensation will differ only in being more or less acute. By acting on a nerve in various ways, we may produce the faintest sensation or the most violent pain; but if the intensity of the sensation is the same, its quality must be the same.

Now, we may perceive by our eyes a faint red light which may be made stronger and stronger till our eyes are dazzled. We may then perform the same experiment with a green light or a blue light. We shall thus see that our sensation of colour may differ in other ways, besides in being stronger or fainter. The sensation of colour, therefore, cannot be due to one nerve only.

The speaker then proceeded to state the theory of Dr Thomas Young, as the only theory which completely reconciles these difficulties in accounting for the perception of colour.

Young supposes that the eye is provided with three distinct sets of nervous fibres, each set extending over the whole sensitive surface of the eye. Each of these three systems of nerves, when excited, gives us a different sensation. One of them, which gives us the sensation we call red, is excited most by the red rays, but also by the orange and yellow, and slightly by the violet; another is acted on by the green rays, but also by the orange and yellow and part of the blue; while the third is acted on by the blue and violet rays.

If we could excite one of these sets of nerves without acting on the others, we should have the pure sensation corresponding to that set of nerves. This would be truly a primary colour, whether the nerve were excited by pure or by compound light, or even by the action of pressure or disease.

If such experiments could be made, we should be able to see the primary colours separately, and to describe their appearance by reference to the scale of colours in the spectrum.

But we have no direct consciousness of the contrivances of our own bodies, and we never feel any sensation which is not infinitely complex, so that we can never know directly how many sensations are combined when we see a colour. Still less can we isolate one or more sensations by artificial means, so that in general when a ray enters the eye, though it should be one of the pure rays of the spectrum, it may excite more than one of the three sets of nerves, and thus produce a compound sensation.

The terms simple and compound, therefore, as applied to colour-sensation, have by no means the same meaning as they have when applied to a ray of light.

The speaker then stated some of the consequences of Young's theory, and described the tests to which he had subjected it :—

1st.   There are three primary colours.

2nd.   Every colour is either a primary colour, or a mixture of primary colours.

3rd.   Four colours may always be arranged in one of two ways.   Either one of them is a mixture of the other three, or a mixture of two of them can be found, identical with a mixture of the other two.

4th.   These results may be stated in the form of colour-equations, giving the numerical value of the amount of each colour entering into any mixture. By means of the Colour Top*, such equations can be obtained for coloured papers, and they may be obtained with a degree of accuracy shewing that the colour-judgment of the eye may be rendered very perfect.

The speaker had tested in this way more than 100 different pigments and mixtures, and had found the results agree with the theory of three primaries

* Described in the *Trans. of the Royal Society of Edinburgh*, Vol. XXI., and in the *Phil. Mag.*

in every case. He had also examined all the colours of the spectrum with the same result.

The experiments with pigments do not indicate what colours are to be considered as primary; but experiments on the prismatic spectrum shew that all the colours of the spectrum, and therefore all the colours in nature, are equivalent to mixtures of three colours of the spectrum itself, namely, red, green (near the line E), and blue (near the line G). Yellow was found to be a mixture of red and green.

The speaker, assuming red, green, and blue as primary colours, then exhibited them on a screen by means of three magic lanterns, before which were placed glass troughs containing respectively sulphocyanide of iron, chloride of copper, and ammoniated copper.

A triangle was thus illuminated, so that the pure colours appeared at its angles, while the rest of the triangle contained the various mixtures of the colours as in Young's triangle of colour.

The graduated intensity of the primary colours in different parts of the spectrum was exhibited by three coloured images, which, when superposed on the screen, gave an artificial representation of the spectrum.

Three photographs of a coloured ribbon taken through the three coloured solutions respectively, were introduced into the camera, giving images representing the red, the green, and the blue parts separately, as they would be seen by each of Young's three sets of nerves separately. When these were superposed, a coloured image was seen, which, if the red and green images had been as fully photographed as the blue, would have been a truly-coloured image of the ribbon. By finding photographic materials more sensitive to the less refrangible rays, the representation of the colours of objects might be greatly improved.

The speaker then proceeded to exhibit mixtures of the colours of the pure spectrum. Light from the electric lamp was passed through a narrow slit, a lens and a prism, so as to throw a pure spectrum on a screen containing three moveable slits, through which three distinct portions of the spectrum were suffered to pass. These portions were concentrated by a lens on a screen at a distance, forming a large, uniformly coloured image of the prism.

When the whole spectrum was allowed to pass, this image was white, as in Newton's experiment of combining the rays of the spectrum. When portions of the spectrum were allowed to pass through the moveable slits, the image was uniformly illuminated with a mixture of the corresponding colours. In order to see these colours separately, another lens was placed between the moveable slits and the screen. A magnified image of the slits was thus thrown on the screen, each slit shewing, by its colour and its breadth, the quality and quantity of the colour which it suffered to pass. Several colours were thus exhibited, first separately, and then in combination. Red and blue, for instance, produced purple; red and green produced yellow; blue and yellow produced a pale pink; red, blue, and green produced white; and red and a bluish green near the line F produced a colour which appears very different to different eyes.

The speaker concluded by stating the peculiarities of colour-blind vision, and by shewing that the investigation into the theory of colour is truly a physiological inquiry, and that it requires the observations and testimony of persons of every kind in order to discover and explain the various peculiarities of vision.

[From the *Philosophical Magazine,* Vol. XXI.]

XXIII. *On Physical Lines of Force.*

PART I.

The Theory of Molecular Vortices applied to Magnetic Phenomena.

In all phenomena involving attractions or repulsions, or any forces depending on the relative position of bodies, we have to determine the *magnitude* and *direction* of the force which would act on a given body, if placed in a given position.

In the case of a body acted on by the gravitation of a sphere, this force is inversely as the square of the distance, and in a straight line to the centre of the sphere. In the case of two attracting spheres, or of a body not spherical, the magnitude and direction of the force vary according to more complicated laws. In electric and magnetic phenomena, the magnitude and direction of the resultant force at any point is the main subject of investigation. Suppose that the direction of the force at any point is known, then, if we draw a line so that in every part of its course it coincides in direction with the force at that point, this line may be called a *line of force,* since it indicates the direction of the force in every part of its course.

By drawing a sufficient number of lines of force, we may indicate the direction of the force in every part of the space in which it acts.

Thus if we strew iron filings on paper near a magnet, each filing will be magnetized by induction, and the consecutive filings will unite by their opposite poles, so as to form fibres, and these fibres will *indicate* the direction of the lines of force. The beautiful illustration of the presence of magnetic force afforded by this experiment, naturally tends to make us think of the lines of force as something real, and as indicating something more than the mere resultant of two forces, whose seat of action is at a distance, and which do not exist there

57—2

at all until a magnet is placed in that part of the field. We are dissatisfied with the explanation founded on the hypothesis of attractive and repellent forces directed towards the magnetic poles, even though we may have satisfied ourselves that the phenomenon is in strict accordance with that hypothesis, and we cannot help thinking that in every place where we find these lines of force, some physical state or action must exist in sufficient energy to produce the actual phenomena.

My object in this paper is to clear the way for speculation in this direction, by investigating the mechanical results of certain states of tension and motion in a medium, and comparing these with the observed phenomena of magnetism and electricity. By pointing out the mechanical consequences of such hypotheses, I hope to be of some use to those who consider the phenomena as due to the action of a medium, but are in doubt as to the relation of this hypothesis to the experimental laws already established, which have generally been expressed in the language of other hypotheses.

I have in a former paper* endeavoured to lay before the mind of the geometer a clear conception of the relation of the lines of force to the space in which they are traced. By making use of the conception of currents in a fluid, I shewed how to draw lines of force, which should indicate by their number the amount of force, so that each line may be called a unit-line of force (see Faraday's *Researches*, 3122); and I have investigated the path of the lines where they pass from one medium to another.

In the same paper I have found the geometrical significance of the "Electrotonic State," and have shewn how to deduce the mathematical relations between the electrotonic state, magnetism, electric currents, and the electromotive force, using mechanical illustrations to assist the imagination, but not to account for the phenomena.

I propose now to examine magnetic phenomena from a mechanical point of view, and to determine what tensions in, or motions of, a medium are capable of producing the mechanical phenomena observed. If, by the same hypothesis, we can connect the phenomena of magnetic attraction with electromagnetic phenomena and with those of induced currents, we shall have found a theory which, if not true, can only be proved to be erroneous by experiments which will greatly enlarge our knowledge of this part of physics.

---

* See a paper "On Faraday's Lines of Force," *Cambridge Philosophical Transactions*, Vol. x. Part i. Page 155 of this volume.

The mechanical conditions of a medium under magnetic influence have been variously conceived of, as currents, undulations, or states of displacement or strain, or of pressure or stress.

Currents, issuing from the north pole and entering the south pole of a magnet, or circulating round an electric current, have the advantage of representing correctly the geometrical arrangement of the lines of force, if we could account on mechanical principles for the phenomena of attraction, or for the currents themselves, or explain their continued existence

Undulations issuing from a centre would, according to the calculations of Professor Challis, produce an effect similar to attraction in the direction of the centre; but admitting this to be true, we know that two series of undulations traversing the same space do not combine into one resultant as two attractions do, but produce an effect depending on relations of *phase* as well as intensity, and if allowed to proceed, they diverge from each other without any mutual action. In fact the mathematical laws of attractions are not analogous in any respect to those of undulations, while they have remarkable analogies with those of currents, of the conduction of heat and electricity, and of elastic bodies.

In the *Cambridge and Dublin Mathematical Journal* for January 1847, Professor William Thomson has given a "Mechanical Representation of Electric, Magnetic, and Galvanic Forces," by means of the displacements of the particles of an elastic solid in a state of strain. In this representation we must make the angular displacement at every point of the solid proportional to the magnetic force at the corresponding point of the magnetic field, the direction of the axis of rotation of the displacement corresponding to the direction of the magnetic force. The absolute displacement of any particle will then correspond in magnitude and direction to that which I have identified with the electrotonic state; and the relative displacement of any particle, considered with reference to the particle in its immediate neighbourhood, will correspond in magnitude and direction to the quantity of electric current passing through the corresponding point of the magneto-electric field. The author of this method of representation does not attempt to explain the origin of the observed forces by the effects due to these strains in the elastic solid, but makes use of the mathematical analogies of the two problems to assist the imagination in the study of both.

We come now to consider the magnetic influence as existing in the form of some kind of pressure or tension, or, more generally, of *stress* in the medium.

Stress is action and reaction between the consecutive parts of a body, and consists in general of pressures or tensions different in different directions at the same point of the medium.

The necessary relations among these forces have been investigated by mathematicians; and it has been shewn that the most general type of a stress consists of a combination of three principal pressures or tensions, in directions at right angles to each other.

When two of the principal pressures are equal, the third becomes an axis of symmetry, either of greatest or least pressure, the pressures at right angles to this axis being all equal.

When the three principal pressures are equal, the pressure is equal in every direction, and there results a stress having no determinate axis of direction, of which we have an example in simple hydrostatic pressure.

The general type of a stress is not suitable as a representation of a magnetic force, because a line of magnetic force has direction and intensity, but has no third quality indicating any difference between the *sides* of the line, which would be analogous to that observed in the case of polarized light*.

We must therefore represent the magnetic force at a point by a stress having a single axis of greatest or least pressure, and all the pressures at right angles to this axis equal. It may be objected that it is inconsistent to represent a line of force, which is essentially dipolar, by an axis of stress, which is necessarily isotropic; but we know that *every* phenomenon of action and reaction is isotropic in its *results*, because the effects of the force on the bodies between which it acts are equal and opposite, while the nature and origin of the force may be dipolar, as in the attraction between a north and a south pole.

Let us next consider the mechanical effect of a state of stress symmetrical about an axis. We may resolve it, in all cases, into a simple hydrostatic pressure, combined with a simple pressure or tension along the axis. When the axis is that of greatest pressure, the force along the axis will be a pressure. When the axis is that of least pressure, the force along the axis will be a tension.

If we observe the lines of force between two magnets, as indicated by iron filings, we shall see that whenever the lines of force pass from one pole to another, there is *attraction* between those poles; and where the lines of force from the poles avoid each other and are dispersed into space, the poles *repel*

---

* See Faraday's *Researches*, 3252.

each other, so that in both cases they are drawn in the direction of the resultant of the lines of force.

It appears therefore that the stress in the axis of a line of magnetic force is a *tension*, like that of a rope.

If we calculate the lines of force in the neighbourhood of two gravitating bodies, we shall find them the same in direction as those near two magnetic poles of the same name; but we know that the mechanical effect is that of attraction instead of repulsion. The lines of force in this case do not run between the bodies, but avoid each other, and are dispersed over space. In order to produce the effect of attraction, the stress along the lines of gravitating force must be a *pressure*.

Let us now suppose that the phenomena of magnetism depend on the existence of a tension in the direction of the lines of force, combined with a hydrostatic pressure; or in other words, a pressure greater in the equatorial than in the axial direction: the next question is, what mechanical explanation can we give of this inequality of pressures in a fluid or mobile medium? The explanation which most readily occurs to the mind is that the excess of pressure in the equatorial direction arises from the centrifugal force of vortices or eddies in the medium having their axes in directions parallel to the lines of force.

This explanation of the cause of the inequality of pressures at once suggests the means of representing the dipolar character of the line of force. Every vortex is essentially dipolar, the two extremities of its axis being distinguished by the direction of its revolution as observed from those points.

We also know that when electricity circulates in a conductor, it produces lines of magnetic force passing through the circuit, the direction of the lines depending on the direction of the circulation. Let us suppose that the direction of revolution of our vortices is that in which vitreous electricity must revolve in order to produce lines of force whose direction within the circuit is the same as that of the given lines of force.

We shall suppose at present that all the vortices in any one part of the field are revolving in the same direction about axes nearly parallel, but that in passing from one part of the field to another, the direction of the axes, the velocity of rotation, and the density of the substance of the vortices are subject to change. We shall investigate the resultant mechanical effect upon an element of the medium, and from the mathematical expression of this resultant we shall deduce the physical character of its different component parts.

PROP. I.—If in two fluid systems geometrically similar the velocities and densities at corresponding points are proportional, then the differences of pressure at corresponding points due to the motion will vary in the duplicate ratio of the velocities and the simple ratio of the densities.

Let l be the ratio of the linear dimensions, m that of the velocities, n that of the densities, and p that of the pressures due to the motion. Then the ratio of the *masses* of corresponding portions will be l^3n, and the ratio of the velocities acquired in traversing similar parts of the systems will be m; so that l^3mn is the ratio of the momenta acquired by similar portions in traversing similar parts of their paths.

The ratio of the surfaces is l^2, that of the forces acting on them is l^2p, and that of the times during which they act is $\dfrac{l}{m}$; so that the ratio of the impulse of the forces is $\dfrac{l^3p}{m}$, and we have now

$$l^3mn = \frac{l^3p}{m},$$

or $$m^2n = p\ ;$$

that is, the ratio of the pressures due to the motion (p) is compounded of the ratio of the densities (n) and the duplicate ratio of the velocities (m^2), and does not depend on the linear dimensions of the moving systems.

In a circular vortex, revolving with uniform angular velocity, if the pressure at the axis is p_0, that at the circumference will be $p_1 = p_0 + \frac{1}{2}\rho v^2$, where ρ is the density and v the velocity at the circumference. The *mean pressure* parallel to the axis will be

$$p_0 + \tfrac{1}{4}\rho v^2 = p_2.$$

If a number of such vortices were placed together side by side with their axes parallel, they would form a medium in which there would be a pressure p_2 parallel to the axes, and a pressure p_1 in any perpendicular direction. If the vortices are circular, and have uniform angular velocity and density throughout, then

$$p_1 - p_2 = \tfrac{1}{4}\rho v^2.$$

If the vortices are not circular, and if the angular velocity and the density are not uniform, but vary according to the same law for all the vortices,

$$p_1 - p_2 = C\rho v^2,$$

where ρ is the mean density, and C is a numerical quantity depending on the distribution of angular velocity and density in the vortex. In future we shall write $\frac{\mu}{4\pi}$ instead of $C\rho$, so that

$$p_1 - p_2 = \frac{1}{4\pi}\mu v^2 \quad \dots \dots \dots \dots \dots \dots \dots \dots \dots \dots \dots (1),$$

where μ is a quantity bearing a constant ratio to the density, and v is the linear velocity at the circumference of each vortex.

A medium of this kind, filled with molecular vortices having their axes parallel, differs from an ordinary fluid in having different pressures in different directions. If not prevented by properly arranged pressures, it would tend to expand laterally. In so doing, it would allow the diameter of each vortex to expand and its velocity to diminish in the same proportion. In order that a medium having these inequalities of pressure in different directions should be in equilibrium, certain conditions must be fulfilled, which we must investigate.

PROP. II.—If the direction-cosines of the axes of the vortices with respect to the axes of x, y, and z be l, m, and n, to find the normal and tangential stresses on the co-ordinate planes.

The actual stress may be resolved into a simple hydrostatic pressure p_1 acting in all directions, and a simple tension $p_1 - p_2$, or $\frac{1}{4\pi}\mu v^2$, acting along the axis of stress.

Hence if p_{xx}, p_{yy}, and p_{zz} be the normal stresses parallel to the three axes, considered positive when they tend to increase those axes; and if p_{yz}, p_{zx}, and p_{xy} be the tangential stresses in the three co-ordinate planes, considered positive when they tend to increase simultaneously the symbols subscribed, then by the resolution of stresses*,

$$p_{xx} = \frac{1}{4\pi}\mu v^2 l^2 - p_1,$$

$$p_{yy} = \frac{1}{4\pi}\mu v^2 m^2 - p_1,$$

$$p_{zz} = \frac{1}{4\pi}\mu v^2 n^2 - p_1,$$

---

* Rankine's *Applied Mechanics*, Art. 106.

$$p_{yz} = \frac{1}{4\pi} \mu v^2 mn,$$

$$p_{zx} = \frac{1}{4\pi} \mu v^2 nl,$$

$$p_{xy} = \frac{1}{4\pi} \mu v^2 lm.$$

If we write $\qquad\qquad a = vl, \quad \beta = vm, \text{ and } \gamma = vn,$

then

$$\left. \begin{array}{ll} p_{xx} = \dfrac{1}{4\pi} \mu a^2 - p_1, & p_{yz} = \dfrac{1}{4\pi} \mu \beta \gamma \\[2mm] p_{yy} = \dfrac{1}{4\pi} \mu \beta^2 - p_1, & p_{zx} = \dfrac{1}{4\pi} \mu \gamma a \\[2mm] p_{zz} = \dfrac{1}{4\pi} \mu \gamma^2 - p_1, & p_{xy} = \dfrac{1}{4\pi} \mu a \beta \end{array} \right\} \quad \ldots \ldots \ldots \ldots \ldots (2).$$

Prop. III.—To find the resultant force on an element of the medium, arising from the variation of internal stress.

We have in general, for the force in the direction of x per unit of volume by the law of equilibrium of stresses[*],

$$X = \frac{d}{dx} p_{xx} + \frac{d}{dy} p_{xy} + \frac{d}{dz} p_{xz} \ldots \ldots \ldots \ldots \ldots \ldots (3).$$

In this case the expression may be written

$$X = \frac{1}{4\pi} \left\{ \frac{d(\mu a)}{dx} a + \mu a \frac{da}{dx} - 4\pi \frac{dp_1}{dx} + \frac{d(\mu \beta)}{dy} a + \mu \beta \frac{da}{dy} + \frac{d(\mu \gamma)}{dz} a + \mu \gamma \frac{da}{dz} \right\} \ldots (4).$$

Remembering that $a \dfrac{da}{dx} + \beta \dfrac{d\beta}{dx} + \gamma \dfrac{d\gamma}{dx} = \dfrac{1}{2} \dfrac{d}{dx} (a^2 + \beta^2 + \gamma^2)$, this becomes

$$X = a \frac{1}{4\pi} \left\{ \frac{d}{dx} (\mu a) + \frac{d}{dy} (\mu \beta) + \frac{d}{dz} (\mu \gamma) \right\} + \frac{1}{8\pi} \mu \frac{d}{dx} (a^2 + \beta^2 + \gamma^2)$$

$$- \mu \beta \frac{1}{4\pi} \left(\frac{d\beta}{dx} - \frac{da}{dy} \right) + \mu \gamma \frac{1}{4\pi} \left(\frac{da}{dz} - \frac{d\gamma}{dx} \right) - \frac{dp_1}{dx} \ldots \ldots \ldots (5).$$

The expressions for the forces parallel to the axes of y and z may be written down from analogy.

[*] Rankine's *Applied Mechanics*, Art. 116.

We have now to interpret the meaning of each term of this expression.

We suppose a, β, γ to be the components of the force which would act upon that end of a unit magnetic bar which points to the north.

μ represents the magnetic inductive capacity of the medium at any point referred to air as a standard. μa, $\mu\beta$, $\mu\gamma$ represent the quantity of magnetic induction through unit of area perpendicular to the three axes of x, y z respectively.

The total amount of magnetic induction through a closed surface surrounding the pole of a magnet, depends entirely on the strength of that pole; so that if $dx\,dy\,dz$ be an element, then

$$\left(\frac{d}{dx}\mu a + \frac{d}{dy}\mu\beta + \frac{d}{dz}\mu\gamma\right) dx\,dy\,dz = 4\pi m\,dx\,dy\,dz \ldots\ldots\ldots\ldots (6),$$

which represents the total amount of magnetic induction outwards through the surface of the element $dx\,dy\,dz$, represents the amount of "imaginary magnetic matter" within the element, of the kind which points north.

The *first term* of the value of X, therefore,

$$a\,\frac{1}{4\pi}\left(\frac{d}{dx}\mu a + \frac{d}{dy}\mu\beta + \frac{d}{dz}\mu\gamma\right) \ldots\ldots\ldots\ldots\ldots\ldots (7),$$

may be written

$$am \ldots\ldots\ldots\ldots\ldots\ldots\ldots\ldots\ldots\ldots\ldots (8),$$

where a is the intensity of the magnetic force, and m is the amount of magnetic matter pointing north in unit of volume.

The physical interpretation of this term is, that the force urging a north pole in the positive direction of x is the product of the intensity of the magnetic force resolved in that direction, and the strength of the north pole of the magnet.

Let the parallel lines from left to right in fig. 1 represent a field of magnetic force such as that of the earth, sn being the direction from south to north. The vortices, according to our hypothesis, will be in the direction shewn by the arrows in fig. 3, that is, in a plane perpendicular to the lines of force, and revolving in the direction of the hands of a watch when observed from s looking towards n. The parts of the vortices above the plane of the paper will be moving towards e, and the parts below that plane towards w.

We shall always mark by an arrow-head the direction in which we must
look in order to see the vortices rotating in the
direction of the hands of a watch. The arrow-head
will then indicate the *northward* direction in the
magnetic field, that is, the direction in which that
end of a magnet which points to the north would
set itself in the field.

Fig. 1.

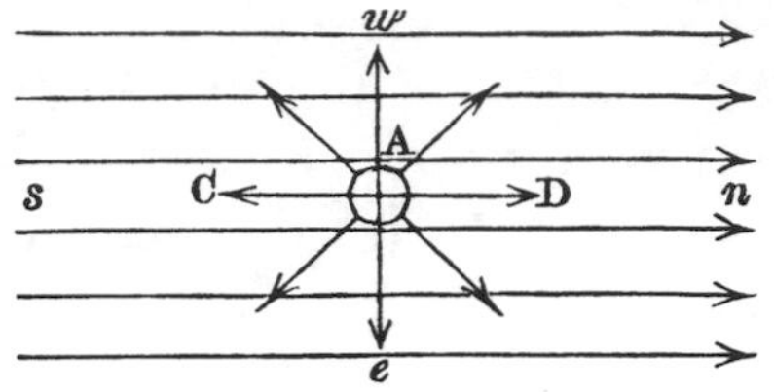

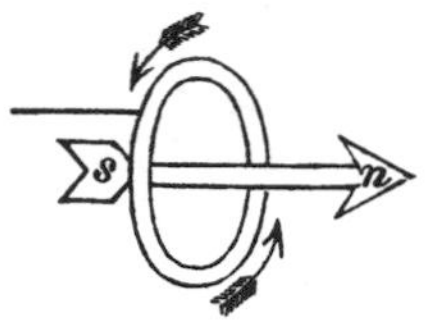

Now let *A* be the end of a magnet which
points north. Since it repels the north ends of
other magnets, the lines of force will be directed
from A outwards in all directions. On the north
side the line *AD* will be in the *same* direction with
the lines of the magnetic field, and the velocity of
the vortices will be *increased*. On the south side
the line *AC* will be in the opposite direction, and
the velocity of the vortices will be diminished, so
that the lines of force are more powerful on the
north side of *A* than on the south side.

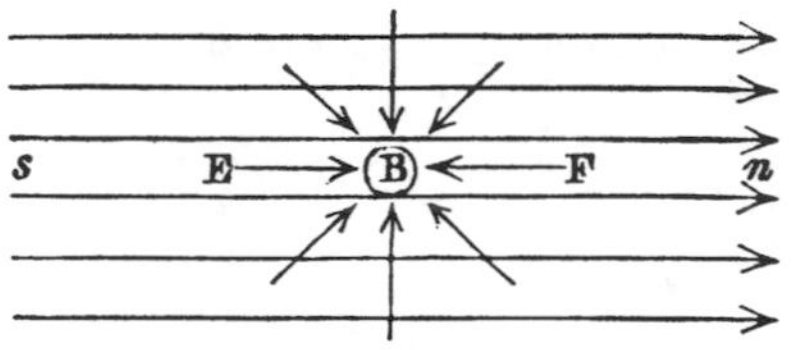

Fig. 2.

We have seen that the mechanical effect of the
vortices is to produce a tension along their axes,
so that the resultant effect on *A* will be to pull

Fig. 3.

it more powerfully towards *D* than towards *C*; that is, *A* will tend to move
to the north.

Let *B* in fig. 2 represent a south pole. The lines of force belonging to *B*
will tend *towards B*, and we shall find that the lines of force are rendered
stronger towards *E* than towards *F*, so that the effect in this case is to urge *B*
towards the south.

It appears therefore that, on the hypothesis of molecular vortices, our first
term gives a mechanical explanation of the force acting on a north or south
pole in the magnetic field.

We now proceed to examine the second term,

$$\frac{1}{8\pi}\,\mu\,\frac{d}{dx}\,(\alpha^2+\beta^2+\gamma^2).$$

Here $\alpha^2+\beta^2+\gamma^2$ is the square of the intensity at any part of the field, and
μ is the magnetic inductive capacity at the same place. Any body therefore

placed in the field will be urged *towards places of stronger magnetic intensity* with a force depending partly on its own capacity for magnetic induction, and partly on the rate at which the square of the intensity increases.

If the body be placed in a fluid medium, then the medium, as well as the body, will be urged towards places of greater intensity, so that its hydrostatic pressure will be increased in that direction. The resultant effect on a body placed in the medium will be the *difference* of the actions on the body and on the portion of the medium which it displaces, so that the body will tend to or from places of greatest magnetic intensity, according as it has a greater or less capacity for magnetic induction than the surrounding medium.

In fig. 4 the lines of force are represented as converging and becoming more powerful towards the right, so that the magnetic tension at B is stronger than at A, and the body AB will be urged to the right. If the capacity for magnetic induction is greater in the body than in the surrounding medium, it will move to the right, but if less it will move to the left.

Fig. 4.   Fig. 5.

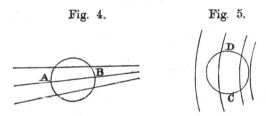

We may suppose in this case that the lines of force are converging to a magnetic pole, either north or south, on the right hand.

In fig. 5 the lines of force are represented as vertical, and becoming more numerous towards the right. It may be shewn that if the force increases towards the right, the lines of force will be curved towards the right. The effect of the magnetic tensions will then be to draw any body towards the right with a force depending on the excess of its inductive capacity over that of the surrounding medium.

We may suppose that in this figure the lines of force are those surrounding an electric current perpendicular to the plane of the paper and on the right hand of the figure.

These two illustrations will shew the mechanical effect on a paramagnetic or diamagnetic body placed in a field of varying magnetic force, whether the increase of force takes place along the lines or transverse to them. The form of the second term of our equation indicates the general law, which is quite independent of the direction of the lines of force, and depends solely on the manner in which the force *varies* from one part of the field to another.

We come now to the third term of the value of X,

$$-\mu\beta\,\frac{1}{4\pi}\left(\frac{d\beta}{dx}-\frac{da}{dy}\right).$$

Here $\mu\beta$ is, as before, the quantity of magnetic induction through unit of area perpendicular to the axis of y, and $\dfrac{d\beta}{dx}-\dfrac{da}{dy}$ is a quantity which would disappear if $adx+\beta dy+\gamma dz$ were a complete differential, that is, if the force acting on a unit north pole were subject to the condition that no work can be done upon the pole in passing round any closed curve. The quantity represents the work done on a north pole in travelling round unit of area in the direction from $+x$ to $+y$ parallel to the plane of xy. Now if an electric current whose strength is r is traversing the axis of z, which, we may suppose, points vertically upwards, then, if the axis of x is east and that of y north, a unit north pole will be urged round the axis of z in the direction from x to y, so that in one revolution the work done will be $=4\pi r$. Hence $\dfrac{1}{4\pi}\left(\dfrac{d\beta}{dx}-\dfrac{da}{dy}\right)$ represents the *strength of an electric current parallel to* z through unit of area; and if we write

$$\frac{1}{4\pi}\left(\frac{d\gamma}{dy}-\frac{d\beta}{dz}\right)=p,\quad \frac{1}{4\pi}\left(\frac{da}{dz}-\frac{d\gamma}{dx}\right)=q,\quad \frac{1}{4\pi}\left(\frac{d\beta}{dx}-\frac{da}{dy}\right)=r\ldots\ldots\ldots(9),$$

then p, q, r will be the quantity of electric current per unit of area perpendicular to the axes of x, y, and z respectively.

The physical interpretation of the third term of X, $-\mu\beta r$, is that if $\mu\beta$ is the quantity of magnetic induction parallel to y, and r the quantity of electricity flowing in the direction of z, the element will be urged in the direction of $-x$, transversely to the direction of the current and of the lines of force; that is, an *ascending* current in a field of force magnetized towards the *north* would tend to move *west*.

To illustrate the action of the molecular vortices, let sn be the direction of magnetic force in the field, and let C be the section of an ascending magnetic current perpendicular to the paper. The lines of force due to this current will be circles drawn in the opposite direction from that of the hands of a watch; that is, in the direction *nwse*. At *e* the lines of force will be the sum of those of the field and of the current, and at *w* they will be the difference of the two sets of lines; so that the vortices on the east side of the current will be more powerful than those on the west side. Both sets of vortices have their equatorial parts turned towards *C*, so that they tend to expand towards *C*, but those on the east side have the greatest effect, so that the resultant effect on the current is to urge it towards the *west*.

Fig. 6.

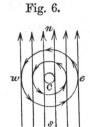

The fourth term,

$$+ \mu\gamma \frac{1}{4\pi} \left(\frac{da}{dz} - \frac{d\gamma}{dx} \right), \text{ or } + \mu\gamma q \quad \dots\dots\dots\dots (10),$$

may be interpreted in the same way, and indicates that a current *q* in the direction of *y*, that is, to the north, placed in a magnetic field in which the lines are vertically upwards in the direction of *z*, will be urged towards the *east*.

The fifth term,

$$- \frac{dp_1}{dx} \quad \dots\dots\dots\dots\dots\dots\dots\dots (11),$$

merely implies that the element will be urged in the direction in which the hydrostatic pressure p_1 diminishes.

We may now write down the expressions for the components of the resultant force on an element of the medium per unit of volume, thus:

$$X = am + \frac{1}{8\pi} \mu \frac{d}{dx} (v^2) - \mu\beta r + \mu\gamma q - \frac{dp_1}{dx} \quad \dots\dots\dots\dots (12),$$

$$Y = \beta m + \frac{1}{8\pi} \mu \frac{d}{dy} (v^2) - \mu\gamma p + \mu ar - \frac{dp_1}{dy} \quad \dots\dots\dots\dots (13),$$

$$Z = \gamma m + \frac{1}{8\pi} \mu \frac{d}{dz} (v^2) - \mu aq + \mu\beta p - \frac{dp_1}{dz} \quad \dots\dots\dots\dots (14).$$

The first term of each expression refers to the force acting on magnetic poles.

The second term to the action on bodies capable of magnetism by induction.

The third and fourth terms to the force acting on electric currents.

And the fifth to the effect of simple pressure.

Before going further in the general investigation, we shall consider equations (12, 13, 14), in particular cases, corresponding to those simplified cases of the actual phenomena which we seek to obtain in order to determine their laws by experiment.

We have found that the quantities p, q, and r represent the resolved parts of an electric current in the three co-ordinate directions. Let us suppose in the first instance that there is *no* electric current, or that p, q, and r vanish. We have then by (9),

$$\frac{d\gamma}{dy} - \frac{d\beta}{dz} = 0, \quad \frac{da}{dz} - \frac{d\gamma}{dx} = 0, \quad \frac{d\beta}{dx} - \frac{da}{dy} = 0 \dots \dots \dots (15),$$

whence we learn that $\quad a\,dx + \beta\,dy + \gamma\,dz = d\phi \dots \dots \dots \dots \dots (16),$

is an exact differential of ϕ, so that

$$a = \frac{d\phi}{dx}, \quad \beta = \frac{d\phi}{dy}, \quad \gamma = \frac{d\phi}{dz} \dots \dots \dots \dots (17):$$

μ is proportional to the density of the vortices, and represents the "capacity for magnetic induction" in the medium. It is equal to 1 in air, or in whatever medium the experiments were made which determined the powers of the magnets, the strengths of the electric currents, &c.

Let us suppose μ constant, then

$$m = \frac{1}{4\pi}\left\{\frac{d}{dx}(\mu a) + \frac{d}{dy}(\mu\beta) + \frac{d}{dz}(\mu\gamma)\right\} = \frac{1}{4\pi}\mu\left(\frac{d^2\phi}{dx^2} + \frac{d^2\phi}{dy^2} + \frac{d^2\phi}{dz^2}\right) \dots \dots (18)$$

represents the amount of imaginary magnetic matter in unit of volume. That there may be no resultant force on that unit of volume arising from the action represented by the first term of equations (12, 13, 14), we must have $m = 0$, or

$$\frac{d^2\phi}{dx^2} + \frac{d^2\phi}{dy^2} + \frac{d^2\phi}{dz^2} = 0 \dots \dots \dots \dots \dots (19).$$

Now it may be shewn that equation (19), if true within a given space, implies that the forces acting within that space are such as would result from a distribution of centres of force beyond that space, attracting or repelling inversely as the square of the distance.

Hence the lines of force in a part of space where μ is uniform, and where there are no electric currents, must be such as would result from the theory of "imaginary matter" acting at a distance. The assumptions of that theory are unlike those of ours, but the results are identical.

Let us first take the case of a single magnetic pole, that is, one end of a long magnet, so long that its other end is too far off to have a perceptible influence on the part of the field we are considering. The conditions then are, that equation (18) must be fulfilled at the magnetic pole, and (19) everywhere else. The only solution under these conditions is

$$\phi = -\frac{m}{\mu}\frac{1}{r} \dots\dots\dots\dots\dots\dots (20),$$

where r is the distance from the pole, and m the strength of the pole.

The repulsion at any point on a unit pole of the same kind is

$$\frac{d\phi}{dr} = \frac{m}{\mu}\frac{1}{r^2} \dots\dots\dots\dots\dots\dots (21).$$

In the standard medium $\mu = 1$; so that the repulsion is simply $\frac{m}{r^2}$ in that medium, as has been shewn by Coulomb.

In a medium having a greater value of μ (such as oxygen, solutions of salts of iron, &c.) the attraction, on our theory, ought to be *less* than in air, and in diamagnetic media (such as water, melted bismuth, &c.) the attraction between the same magnetic poles ought to be *greater* than in air.

The experiments necessary to demonstrate the difference of attraction of two magnets according to the magnetic or diamagnetic character of the medium in which they are placed, would require great precision, on account of the limited range of magnetic capacity in the fluid media known to us, and the small amount of the difference sought for as compared with the whole attraction.

Let us next take the case of an electric current whose quantity is C, flowing through a cylindrical conductor whose radius is R, and whose length is infinite as compared with the size of the field of force considered.

Let the axis of the cylinder be that of z, and the direction of the current positive, then within the conductor the quantity of current per unit of area is

$$r = \frac{C}{\pi R^2} = \frac{1}{4\pi}\left(\frac{d\beta}{dx} - \frac{d\alpha}{dy}\right)\dots\dots\dots\dots(22);$$

so that within the conductor

$$\alpha = -2\frac{C}{R^2}y, \quad \beta = 2\frac{C}{R^2}x, \quad \gamma = 0\dots\dots\dots\dots(23).$$

Beyond the conductor, in the space round it,

$$\phi = 2C \tan^{-1} \frac{y}{x} \quad\dots\dots\dots\dots\dots (24),$$

$$a = \frac{d\phi}{dx} = -2C\frac{y}{x^2+y^2}, \quad \beta = \frac{d\phi}{dy} = 2C\frac{x}{x^2+y^2}, \quad \gamma = \frac{d\phi}{dz} = 0 \dots\dots (25).$$

If $\rho = \sqrt{x^2+y^2}$ is the perpendicular distance of any point from the axis of the conductor, a unit north pole will experience a force $= \dfrac{2C}{\rho}$, tending to move it round the conductor in the direction of the hands of a watch, if the observer view it in the direction of the current.

Let us now consider a current running parallel to the axis of z in the plane of xz at a distance ρ. Let the quantity of the current be c', and let the length of the part considered be l, and its section s, so that $\dfrac{c'}{s}$ is its strength per unit of section. Putting this quantity for ρ in equations (12, 13, 14), we find

$$X = -\mu\beta\frac{c'}{s}$$

per unit of volume; and multiplying by ls, the volume of the conductor considered, we find

$$X = -\mu\beta c' l$$
$$= -2\mu\frac{Cc'l}{\rho} \quad\dots\dots\dots\dots\dots (26),$$

shewing that the second conductor will be attracted towards the first with a force inversely as the distance.

We find in this case also that the amount of attraction depends on the value of μ, but that it varies directly instead of inversely as μ; so that the attraction between two conducting wires will be greater in oxygen than in air, and greater in air than in water.

We shall next consider the nature of electric currents and electromotive forces in connexion with the theory of molecular vortices.

PART II.

THE THEORY OF MOLECULAR VORTICES APPLIED TO ELECTRIC CURRENTS.

We have already shewn that all the forces acting between magnets, substances capable of magnetic induction, and electric currents, may be mechanically accounted for on the supposition that the surrounding medium is put into such a state that at every point the pressures are different in different directions, the direction of least pressure being that of the observed lines of force, and the difference of greatest and least pressures being proportional to the square of the intensity of the force at that point.

Such a state of stress, if assumed to exist in the medium, and to be arranged according to the known laws regulating lines of force, will act upon the magnets, currents, &c. in the field with precisely the same resultant forces as those calculated on the ordinary hypothesis of direct action at a distance. This is true independently of any particular theory as to the *cause* of this state of stress, or the mode in which it can be sustained in the medium. We have therefore a satisfactory answer to the question, "Is there any mechanical hypothesis as to the condition of the medium indicated by lines of force, by which the observed resultant forces may be accounted for?" The answer is, the lines of force indicate the direction of *minimum pressure* at every point of the medium.

The second question must be, "What is the mechanical cause of this difference of pressure in different directions?" We have supposed, in the first part of this paper, that this difference of pressures is caused by molecular vortices, having their axes parallel to the lines of force.

We also assumed, perfectly arbitrarily, that the direction of these vortices is such that, on looking along a line of force from south to north, we should see the vortices revolving in the direction of the hands of a watch.

We found that the velocity of the circumference of each vortex must be proportional to the intensity of the magnetic force, and that the density of the substance of the vortex must be proportional to the capacity of the medium for magnetic induction.

We have as yet given no answers to the questions, "How are these vortices set in rotation?" and "Why are they arranged according to the known laws of lines of force about magnets and currents?" These questions are certainly of a higher order of difficulty than either of the former; and I wish to separate the suggestions I may offer by way of provisional answer to them, from the mechanical deductions which resolved the first question, and the hypothesis of vortices which gave a probable answer to the second.

We have, in fact, now come to inquire into the physical connexion of these vortices with electric currents, while we are still in doubt as to the nature of electricity, whether it is one substance, two substances, or not a substance at all, or in what way it differs from matter, and how it is connected with it.

We know that the lines of force are affected by electric currents, and we know the distribution of those lines about a current; so that from the force we can determine the amount of the current. Assuming that our explanation of the lines of force by molecular vortices is correct, why does a particular distribution of vortices indicate an electric current? A satisfactory answer to this question would lead us a long way towards that of a very important one, "What is an electric current?"

I have found great difficulty in conceiving of the existence of vortices in a medium, side by side, revolving in the same direction about parallel axes. The contiguous portions of consecutive vortices must be moving in opposite directions; and it is difficult to understand how the motion of one part of the medium can coexist with, and even produce, an opposite motion of a part in contact with it.

The only conception which has at all aided me in conceiving of this kind of motion is that of the vortices being separated by a layer of particles, revolving each on its own axis in the opposite direction to that of the vortices, so that the contiguous surfaces of the particles and of the vortices have the same motion.

In mechanism, when two wheels are intended to revolve in the same direction, a wheel is placed between them so as to be in gear with both, and this wheel is called an "idle wheel." The hypothesis about the vortices which I have to suggest is that a layer of particles, acting as idle wheels, is interposed between each vortex and the next, so that each vortex has a tendency to make the neighbouring vortices revolve in the same direction with itself.

In mechanism, the idle wheel is generally made to rotate about a *fixed* axle; but in epicyclic trains and other contrivances, as, for instance, in Siemens's governor for steam-engines *, we find idle wheels whose centres are capable of motion. In all these cases the motion of the centre is the half sum of the motions of the circumferences of the wheels between which it is placed. Let us examine the relations which must subsist between the motions of our vortices and those of the layer of particles interposed as idle wheels between them.

PROP. IV.—To determine the motion of a layer of particles separating two vortices.

Let the circumferential velocity of a vortex, multiplied by the three direction-cosines of its axis respectively, be α, β, γ, as in Prop. II. Let l, m, n be the direction-cosines of the normal to any part of the surface of this ·vortex, the outside of the surface being regarded positive. Then the components of the velocity of the particles of the vortex at this part of its surface will be

$$n\beta - m\gamma \quad \text{parallel to } x,$$
$$l\gamma - n\alpha \quad \text{parallel to } y,$$
$$m\alpha - l\beta \quad \text{parallel to } z.$$

If this portion of the surface be in contact with another vortex whose velocities are α', β', γ', then a layer of very small particles placed between them will have a velocity which will be the mean of the superficial velocities of the vortices which they separate, so that if u is the velocity of the particles in the direction of x,

$$u = \tfrac{1}{2}m\,(\gamma' - \gamma) - \tfrac{1}{2}n\,(\beta' - \beta) \quad \dots\dots\dots\dots\dots\dots(27),$$

since the normal to the second vortex is in the opposite direction to that of the first.

PROP. V.—To determine the whole amount of particles transferred across unit of area in the direction of x in unit of time.

Let x_1, y_1, z_1 be the co-ordinates of the centre of the first vortex, x_2, y_2, z_2 those of the second, and so on. Let V_1, V_2, &c. be the volumes of the first, second, &c. vortices, and $\overline{V}$ the sum of their volumes. Let dS be an element of the surface separating the first and second vortices, and x, y, z its co-ordinates. Let ρ be the quantity of particles on every unit of surface. Then if p be the whole quantity of particles transferred across unit of area in unit of time in

* See Goodeve's *Elements of Mechanism*, p. 118.

the direction of x, the whole momentum parallel to x of the particles within the space whose volume is $\overline{V}$ will be $\overline{V}p$, and we shall have

$$\overline{V}p = \Sigma u\rho dS \quad\dots\dots\dots\dots\dots\dots\dots\dots\dots(28),$$

the summation being extended to every surface separating any two vortices within the volume $\overline{V}$.

Let us consider the surface separating the first and second vortices. Let an element of this surface be dS, and let its direction-cosines be l_1, m_1, n_1 with respect to the first vortex, and l_2, m_2, n_2 with respect to the second; then we know that

$$l_1 + l_2 = 0, \quad m_1 + m_2 = 0, \quad n_1 + n_2 = 0\dots\dots\dots\dots\dots(29).$$

The values of a, β, γ vary with the position of the centre of the vortex; so that we may write

$$a_2 = a_1 + \frac{da}{dx}(x_2 - x_1) + \frac{da}{dy}(y_2 - y_1) + \frac{da}{dz}(z_2 - z_1) \dots\dots\dots(30),$$

with similar equations for β and γ.

The value of u may be written :—

$$u = \tfrac{1}{2}\frac{d\gamma}{dx}\{m_1(x - x_1) + m_2(x - x_2)\}$$

$$+ \tfrac{1}{2}\frac{d\gamma}{dy}\{m_1(y - y_1) + m_2(y - y_2)\} + \tfrac{1}{2}\frac{d\gamma}{dz}\{m_1(z - z_1) + m_2(z - z_2)\}$$

$$- \tfrac{1}{2}\frac{d\beta}{dx}\{n_1(x - x_1) + n_2(x - x_2)\} - \tfrac{1}{2}\frac{d\beta}{dy}\{n_1(y - y_1) + n_2(y - y_2)\}$$

$$- \tfrac{1}{2}\frac{d\beta}{dz}\{n_1(z - z_1) + n_1(z - z_2)\}\dots\dots\dots\dots\dots\dots\dots\dots\dots\dots(31).$$

In effecting the summation of $\Sigma u\rho dS$, we must remember that round any closed surface $\Sigma l dS$ and all similar terms vanish; also that terms of the form $\Sigma l y dS$, where l and y are measured in different directions, also vanish; but that terms of the form $\Sigma l x dS$, where l and x refer to the same axis of co-ordinates, do not vanish, but are equal to the volume enclosed by the surface. The result is

$$\overline{V}p = \tfrac{1}{2}\rho\left(\frac{d\gamma}{dy} - \frac{d\beta}{dz}\right)(V_1 + V_2 + \&c.)\dots\dots\dots\dots\dots(32);$$

or dividing by $\overline{V} = V_1 + V_2 + \&c.$,

$$p = \tfrac{1}{2}\rho \left(\frac{d\gamma}{dy} - \frac{d\beta}{dz} \right) \dots\dots\dots\dots\dots\dots (33).$$

If we make

$$\rho = \frac{1}{2\pi} \dots\dots\dots\dots\dots\dots\dots (34),$$

then equation (33) will be identical with the first of equations (9), which give the relation between the quantity of an electric current and the intensity of the lines of force surrounding it.

It appears therefore that, according to our hypothesis, an electric current is represented by the transference of the moveable particles interposed between the neighbouring vortices. We may conceive that these particles are very small compared with the size of a vortex, and that the mass of all the particles together is inappreciable compared with that of the vortices, and that a great many vortices, with their surrounding particles, are contained in a single complete molecule of the medium. The particles must be conceived to roll without sliding between the vortices which they separate, and not to touch each other, so that, as long as they remain within the same complete molecule, there is no loss of energy by resistance. When, however, there is a general transference of particles in one direction, they must pass from one molecule to another, and in doing so, may experience resistance, so as to waste electrical energy and generate heat.

Now let us suppose the vortices arranged in a medium in any arbitrary manner. The quantities $\frac{d\gamma}{dy} - \frac{d\beta}{dz}$, &c. will then in general have values, so that there will at first be electrical currents in the medium. These will be opposed by the electrical resistance of the medium; so that, unless they are kept up by a continuous supply of force, they will quickly disappear, and we shall then have $\frac{d\gamma}{dy} - \frac{d\beta}{dz} = 0$, &c.; that is, $\alpha dx + \beta dy + \gamma dz$ will be a complete differential (see equations (15) and (16)); so that our hypothesis accounts for the distribution of the lines of force.

In Plate VIII. p. 488, fig. 1, let the vertical circle EE represent an electric current flowing from copper C to zinc Z through the conductor EE', as shewn by the arrows.

Let the horizontal circle MM' represent a line of magnetic force embracing the electric circuit, the north and south directions being indicated by the lines SN and NS.

Let the vertical circles V and V' represent the molecular vortices of which the line of magnetic force is the axis. V revolves as the hands of a watch, and V' the opposite way.

It will appear from this diagram, that if V and V' were contiguous vortices, particles placed between them would move downwards; and that if the particles were forced downwards by any cause, they would make the vortices revolve as in the figure. We have thus obtained a point of view from which we may regard the relation of an electric current to its lines of force as analogous to the relation of a toothed wheel or rack to wheels which it drives.

In the first part of the paper we investigated the relations of the statical forces of the system. We have now considered the connexion of the motions of the parts considered as a system of mechanism. It remains that we should investigate the dynamics of the system, and determine the forces necessary to produce given changes in the motions of the different parts.

PROP. VI.—To determine the actual energy of a portion of a medium due to the motion of the vortices within it.

Let a, β, γ be the components of the circumferential velocity, as in Prop. II., then the actual energy of the vortices in unit of volume will be proportional to the density and to the square of the velocity. As we do not know the distribution of density and velocity in each vortex, we cannot determine the numerical value of the energy directly; but since μ also bears a constant though unknown ratio to the mean density, let us assume that the energy in unit of volume is

$$E = C\mu\,(a^2 + \beta^2 + \gamma^2),$$

where C is a constant to be determined.

Let us take the case in which

$$a = \frac{d\phi}{dx}, \quad \beta = \frac{d\phi}{dy}, \quad \gamma = \frac{d\phi}{dz} \quad\dots\dots\dots\dots\dots\dots (35).$$

Let
$$\phi = \phi_1 + \phi_2 \quad\dots\dots\dots\dots\dots\dots\dots\dots (36),$$

and let $\quad \dfrac{\mu}{4\pi}\left(\dfrac{d^2\phi_1}{dx^2} + \dfrac{d^2\phi_1}{dy^2} + \dfrac{d^2\phi_1}{dz^2}\right) = m_1,$ and $\dfrac{\mu}{4\pi}\left(\dfrac{d^2\phi_2}{dx^2} + \dfrac{d^2\phi_2}{dy^2} + \dfrac{d^2\phi_2}{dz^2}\right) = m_2 \dots(37);$

then ϕ_1 is the potential at any point due to the magnetic system m_1, and ϕ_2 that due to the distribution of magnetism represented by m_2. The actual energy of all the vortices is

$$E = \Sigma C\mu\,(\alpha^2 + \beta^2 + \gamma^2)\,dV \quad\dots\dots\dots\dots\dots\dots\dots (38),$$

the integration being performed over all space.

This may be shewn by integration by parts (see Green's 'Essay on Electricity,' p. 10) to be equal to

$$E = -4\pi C\Sigma\,(\phi_1 m_1 + \phi_2 m_2 + \phi_1 m_2 + \phi_2 m_1)\,dV \dots\dots\dots\dots (39).$$

Or since it has been proved (Green's 'Essay,' p. 10) that

$$\Sigma \phi_1 m_2\, dV = \Sigma \phi_2 m_1\, dV,$$

$$E = -4\pi C\,(\phi_1 m_1 + \phi_2 m_2 + 2\phi_1 m_2)\,dV \dots\dots\dots\dots (40).$$

Now let the magnetic system m_1 remain at rest, and let m_2 be moved parallel to itself in the direction of x through a space δx; then, since ϕ_1 depends on m_1 only, it will remain as before, so that $\phi_1 m_1$ will be constant; and since ϕ_2 depends on m_2 only, the distribution of ϕ_2 about m_2 will remain the same, so that $\phi_2 m_2$ will be the same as before the change. The only part of E that will be altered is that depending on $2\phi_1 m_2$, because ϕ_1 becomes $\phi_1 + \dfrac{d\phi_1}{dx}\,\delta x$ on account of the displacement. The variation of actual energy due to the displacement is therefore

$$\delta E = -4\pi C\Sigma\left(2\frac{d\phi_1}{dx}\,m_2\right)dV\delta x \dots\dots\dots\dots\dots (41).$$

But by equation (12) the work done by the mechanical forces on m_2 during the motion is

$$\delta W = \Sigma\left(\frac{d\phi_1}{dx}\,m_2 dV\right)\delta x \dots\dots\dots\dots\dots\dots (42);$$

and since our hypothesis is a purely mechanical one, we must have by the conservation of force,

$$\delta E + \delta W = 0 \dots\dots\dots\dots\dots\dots\dots\dots (43);$$

that is, the loss of energy of the vortices must be made up by work done in moving magnets, so that

$$-4\pi C\Sigma\left(2\frac{d\phi_1}{dx}\,m_2 dV\right)\delta x + \Sigma\left(\frac{d\phi_1}{dx}\,m_2 dV\right)\delta x = 0,$$

or

$$C = \frac{1}{8\pi} \dots\dots\dots\dots\dots\dots\dots\dots\dots (44);$$

so that the energy of the vortices in unit of volume is

$$\frac{1}{8\pi}\mu\left(\alpha^2+\beta^2+\gamma^2\right)\ldots\ldots\ldots\ldots\ldots\ldots\ldots\ldots(45);$$

and that of a vortex whose volume is V is

$$\frac{1}{8\pi}\mu\left(\alpha^2+\beta^2+\gamma^2\right)V\ldots\ldots\ldots\ldots\ldots\ldots(46).$$

In order to produce or destroy this energy, work must be expended on, or received from, the vortex, either by the tangential action of the layer of particles in contact with it, or by change of form in the vortex. We shall first investigate the tangential action between the vortices and the layer of particles in contact with them.

PROP. VII.—To find the energy spent upon a vortex in unit of time by the layer of particles which surrounds it.

Let P, Q, R be the forces acting on unity of the particles in the three co-ordinate directions, these quantities being functions of x, y, and z. Since each particle touches two vortices at the extremities of a diameter, the reaction of the particle on the vortices will be equally divided, and will be

$$-\tfrac{1}{2}P,\quad-\tfrac{1}{2}Q,\quad-\tfrac{1}{2}R$$

on each vortex for unity of the particles; but since the superficial density of the particles is $\frac{1}{2\pi}$ (see equation (34)), the forces on unit of surface of a vortex will be

$$-\frac{1}{4\pi}P,\quad-\frac{1}{4\pi}Q,\quad-\frac{1}{4\pi}R.$$

Now let dS be an element of the surface of a vortex. Let the direction-cosines of the normal be l, m, n. Let the co-ordinates of the element be x, y, z. Let the component velocities of the surface be u, v, w. Then the work expended on that element of surface will be

$$\frac{dE}{dt}=-\frac{1}{4\pi}\left(Pu+Qv+Rw\right)dS\ldots\ldots\ldots\ldots\ldots(47).$$

Let us begin with the first term, $Pu\,dS$. P may be written

$$P_0+\frac{dP}{dx}x+\frac{dP}{dy}y+\frac{dP}{dz}z\ldots\ldots\ldots\ldots\ldots\ldots(48),$$

and

$$u=n\beta-m\gamma.$$

Remembering that the surface of the vortex is a closed one, so that

$$\Sigma nxdS = \Sigma mxdS = \Sigma mydS = \Sigma mzdS = 0,$$

and

$$\Sigma mydS = \Sigma nzdS = V,$$

we find

$$\Sigma PudS = \left(\frac{dP}{dz}\beta - \frac{dP}{dy}\gamma\right) V \dots\dots\dots\dots\dots (49),$$

and the whole work done on the vortex in unit of time will be

$$\frac{dE}{dt} = -\frac{1}{4\pi}\Sigma (Pu + Qv + Rw)\, dS$$

$$= \frac{1}{4\pi}\left\{ a\left(\frac{dQ}{dz} - \frac{dR}{dy}\right) + \beta\left(\frac{dR}{dx} - \frac{dP}{dz}\right) + \gamma\left(\frac{dP}{dy} - \frac{dQ}{dx}\right)\right\} V \dots\dots (50).$$

PROP. VIII.—To find the relations between the alterations of motion of the vortices, and the forces P, Q, R which they exert on the layer of particles between them.

Let V be the volume of a vortex, then by (46) its energy is

$$E = \frac{1}{8\pi}\mu\left(a^2 + \beta^2 + \gamma^2\right) V \dots\dots\dots\dots\dots (51),$$

and

$$\frac{dE}{dt} = \frac{1}{4\pi}\mu V\left(a\frac{da}{dt} + \beta\frac{d\beta}{dt} + \gamma\frac{d\gamma}{dt}\right) \dots\dots\dots\dots\dots (52).$$

Comparing this value with that given in equation (50), we find

$$a\left(\frac{dQ}{dz} - \frac{dR}{dy} - \mu\frac{da}{dt}\right) + \beta\left(\frac{dR}{dx} - \frac{dP}{dz} - \mu\frac{d\beta}{dt}\right) + \gamma\left(\frac{dP}{dy} - \frac{dQ}{dx} - \mu\frac{d\gamma}{dt}\right) = 0 \dots\dots (53).$$

This equation being true for all values of a, β, and γ, first let β and γ vanish, and divide by a. We find

$$\left.\begin{array}{l} \dfrac{dQ}{dz} - \dfrac{dR}{dy} = \mu\dfrac{da}{dt} \\[2ex] \dfrac{dR}{dx} - \dfrac{dP}{dz} = \mu\dfrac{d\beta}{dt} \\[2ex] \dfrac{dP}{dy} - \dfrac{dQ}{dx} = \mu\dfrac{d\gamma}{dt} \end{array}\right\} \dots\dots\dots\dots\dots\dots\dots (54).$$

Similarly,

and

From these equations we may determine the relation between the alterations of motion $\dfrac{da}{dt}$, &c. and the forces exerted on the layers of particles between the vortices, or, in the language of our hypothesis, the relation between changes in the state of the magnetic field and the electromotive forces thereby brought into play.

In a memoir "On the Dynamical Theory of Diffraction" (*Cambridge Philosophical Transactions*, Vol. IX. Part 1, section 6), Professor Stokes has given a method by which we may solve equations (54), and find P, Q, and R in terms of the quantities on the right hand of those equations. I have pointed out* the application of this method to questions in electricity and magnetism.

Let us then find three quantities F, G, H from the equations

$$
\left.
\begin{aligned}
\frac{dG}{dz} - \frac{dH}{dy} &= \mu a \\[4pt]
\frac{dH}{dx} - \frac{dF}{dz} &= \mu \beta \\[4pt]
\frac{dF}{dy} - \frac{dG}{dx} &= \mu \gamma
\end{aligned}
\right\} \quad \dots\dots\dots\dots\dots\dots (55),
$$

with the conditions
$$
\frac{1}{4\pi}\left(\frac{d}{dx}\mu a + \frac{d}{dy}\mu\beta + \frac{d}{dz}\mu\gamma\right) = m = 0 \dots\dots\dots\dots (56),
$$

and
$$
\frac{dF}{dx} + \frac{dG}{dy} + \frac{dH}{dz} = 0 \dots\dots\dots\dots\dots (57).
$$

Differentiating (55) with respect to t, and comparing with (54), we find

$$
P = \frac{dF}{dt}, \quad Q = \frac{dG}{dt}, \quad R = \frac{dH}{dt} \dots\dots\dots\dots\dots (58).
$$

We have thus determined three quantities, F, G, H, from which we can find P, Q, and R by considering these latter quantities as the rates at which the former ones vary. In the paper already referred to, I have given reasons for considering the quantities F, G, H as the resolved parts of that which Faraday has conjectured to exist, and has called the *electrotonic state*. In that paper I have stated the mathematical relations between this electrotonic state and the lines of magnetic force as expressed in equations (55), and also between the electrotonic state and electromotive force as expressed in equations (58). We must now endeavour to interpret them from a mechanical point of view in connexion with our hypothesis.

---

* *Cambridge Philosophical Transactions*, Vol. X. Part I. Art. 3. "On Faraday's Lines of Force," pp. 205—209 of this vol.

We shall in the first place examine the process by which the lines of force are produced by an electric current.

Let *AB*, Plate VIII., p. 488, fig. 2, represent a current of electricity in the direction from *A* to *B*. Let the large spaces above and below *AB* represent the vortices, and let the small circles separating the vortices represent the layers of particles placed between them, which in our hypothesis represent electricity.

Now let an electric current from left to right commence in *AB*. The row of vortices *gh* above *AB* will be set in motion in the opposite direction to that of a watch. (We shall call this direction +, and that of a watch −.) We shall suppose the row of vortices *kl* still at rest, then the layer of particles between these rows will be acted on by the row *gh* on their lower sides, and will be at rest above. If they are free to move, they will rotate in the negative direction, and will at the same time move from right to left, or in the opposite direction from the current, and so form an *induced* electric current.

If this current is checked by the electrical resistance of the medium, the rotating particles will act upon the row of vortices *kl*, and make them revolve in the positive direction till they arrive at such a velocity that the motion of the particles is reduced to that of rotation, and the induced current disappears. If, now, the primary current *AB* be stopped, the vortices in the row *gh* will be checked, while those of the row *kl* still continue in rapid motion. The momentum of the vortices beyond the layer of particles *pq* will tend to move them from left to right, that is, in the direction of the primary current; but if this motion is resisted by the medium, the motion of the vortices beyond *pq* will be gradually destroyed.

It appears therefore that the phenomena of induced currents are part of the process of communicating the rotatory velocity of the vortices from one part of the field to another.

As an example of the action of the vortices in producing induced currents, let us take the following case:—Let *B*, Plate VIII., p. 488, fig. 3, be a circular ring, of uniform section, lapped uniformly with covered wire. It may be shewn that if an electric current is passed through this wire, a magnet placed within the coil of wire will be strongly affected, but no magnetic effect will be produced on any external point. The effect will be that of a magnet bent round till its two poles are in contact.

If the coil is properly made, no effect on a magnet placed outside it can be discovered, whether the current is kept constant or made to vary in strength; but if a conducting wire C be made to *embrace* the ring any number of times, an electromotive force will act on that wire whenever the current in the coil is made to vary; and if the circuit be *closed*, there will be an actual current in the wire C.

This experiment shews that, in order to produce the electromotive force, it is not necessary that the conducting wire should be placed in a field of magnetic force, or that lines of magnetic force should pass through the substance of the wire or near it. All that is required is that lines of force should pass through the circuit of the conductor, and that these lines of force should vary in quantity during the experiment.

In this case the vortices, of which we suppose the lines of magnetic force to consist, are all within the hollow of the ring, and outside the ring all is at rest. If there is no conducting circuit embracing the ring, then, when the primary current is made or broken, there is no action outside the ring, except an instantaneous pressure between the particles and the vortices which they separate. If there is a continuous conducting circuit embracing the ring, then, when the primary current is made, there will be a current in the opposite direction through C; and when it is broken, there will be a current through C in the same direction as the primary current.

We may now perceive that induced currents are produced when the electricity yields to the electromotive force,—this force, however, still existing when the formation of a sensible current is prevented by the resistance of the circuit.

The electromotive force, of which the components are P, Q, R, arises from the action between the vortices and the interposed particles, when the velocity of rotation is altered in any part of the field. It corresponds to the pressure on the axle of a wheel in a machine when the velocity of the driving wheel is increased or diminished.

The electrotonic state, whose components are F, G, H, is what the electromotive force would be if the currents, &c. to which the lines of force are due, instead of arriving at their actual state by degrees, had started instantaneously from rest with their actual values. It corresponds to the *impulse* which would act on the axle of a wheel in a machine if the actual velocity were suddenly given to the driving wheel, the machine being previously at rest.

If the machine were suddenly stopped by stopping the driving wheel, each wheel would receive an impulse equal and opposite to that which it received when the machine was set in motion.

This impulse may be calculated for any part of a system of mechanism, and may be called the *reduced momentum* of the machine for that point. In the varied motion of the machine, the actual force on any part arising from the variation of motion may be found by differentiating the reduced momentum with respect to the time, just as we have found that the electromotive force may be deduced from the electrotonic state by the same process.

Having found the relation between the velocities of the vortices and the electromotive forces when the centres of the vortices are at rest, we must extend our theory to the case of a fluid medium containing vortices, and subject to all the varieties of fluid motion. If we fix our attention on any one elementary portion of a fluid, we shall find that it not only travels from one place to another, but also changes its form and position, so as to be elongated in certain directions and compressed in others, and at the same time (in the most general case) turned round by a displacement of rotation.

These changes of form and position produce changes in the velocity of the molecular vortices, which we must now examine.

The alteration of form and position may always be reduced to three simple extensions or compressions in the direction of three rectangular axes, together with three angular rotations about any set of three axes. We shall first consider the effect of three simple extensions or compressions.

PROP. IX.—To find the variations of α, β, γ in the parallelopiped x, y, z when x becomes $x + \delta x$; y, $y + \delta y$; and z, $z + \delta z$; the volume of the figure remaining the same.

By Prop. II. we find for the work done by the vortices against pressure,

$$\delta W = p_1 \delta (xyz) - \frac{\mu}{4\pi} (\alpha^2 yz \delta x + \beta^2 zx \delta y + \gamma^2 xy \delta z) \dots \dots \dots \dots (59);$$

and by Prop. VI. we find for the variation of energy,

$$\delta E = \frac{\mu}{4\pi} (\alpha \delta \alpha + \beta \delta \beta + \gamma \delta \gamma) xyz \dots \dots \dots \dots \dots (60).$$

The sum $\delta W + \delta E$ must be zero by the conservation of energy, and $\delta(xyz) = 0$, since xyz is constant; so that

$$\alpha\left(\delta\alpha - \alpha\,\frac{\delta x}{x}\right) + \beta\left(\delta\beta - \beta\,\frac{\delta y}{y}\right) + \gamma\left(\delta\gamma - \gamma\,\frac{\delta z}{z}\right) = 0 \ldots\ldots\ldots (61).$$

In order that this should be true independently of any relations between α, β, and γ, we must have

$$\delta\alpha = \alpha\,\frac{\delta x}{x}, \quad \delta\beta = \beta\,\frac{\delta y}{y}, \quad \delta\gamma = \gamma\,\frac{\delta z}{z} \ldots\ldots\ldots\ldots\ldots (62).$$

Prop. X.—To find the variations of α, β, γ due to a rotation θ_1 about the axis of x from y to z, a rotation θ_2 about the axis of y from z to x, and a rotation θ_3 about the axis of z from x to y.

The axis of β will move away from the axis of x by an angle θ_3; so that β resolved in the direction of x changes from 0 to $-\beta\theta_3$.

The axis of γ approaches that of x by an angle θ_2; so that the resolved part of γ in direction x changes from 0 to $\gamma\theta_2$.

The resolved part of α in the direction of x changes by a quantity depending on the second power of the rotations, which may be neglected. The variations of α, β, γ from this cause are therefore

$$\delta\alpha = \gamma\theta_2 - \beta\theta_3, \quad \delta\beta = \alpha\theta_3 - \gamma\theta_1, \quad \delta\gamma = \beta\theta_1 - \alpha\theta_2 \ldots\ldots\ldots\ldots (63).$$

The most general expressions for the distortion of an element produced by the displacement of its different parts depend on the nine quantities

$$\frac{d}{dx}\delta x, \quad \frac{d}{dy}\delta x, \quad \frac{d}{dz}\delta x; \quad \frac{d}{dx}\delta y, \quad \frac{d}{dy}\delta y, \quad \frac{d}{dz}\delta y; \quad \frac{d}{dx}\delta z, \quad \frac{d}{dy}\delta z, \quad \frac{d}{dz}\delta z;$$

and these may always be expressed in terms of nine other quantities, namely, three simple extensions or compressions,

$$\frac{\delta x'}{x'}, \quad \frac{\delta y'}{y'}, \quad \frac{\delta z'}{z'}$$

along three axes properly chosen, x', y', z', the nine direction-cosines of these axes with their six connecting equations, which are equivalent to three independent quantities, and the three rotations θ_1, θ_2, θ_3 about the axes of x, y, z.

Let the direction-cosines of x' with respect to x, y, z be l_1, m_1, n_1, those of y', l_2, m_2, n_2, and those of z', l_3, m_3, n_3; then we find

$$\left.\begin{array}{l}\dfrac{d}{dx}\,\delta x = l_1{}^2\dfrac{\delta x'}{x'} + l_2{}^2\dfrac{\delta y'}{y'} + l_3{}^2\dfrac{\delta z'}{z'} \\[2mm] \dfrac{d}{dy}\,\delta x = l_1 m_1\dfrac{\delta x'}{x'} + l_2 m_2\dfrac{\delta y'}{y'} + l_3 m_3\dfrac{\delta z'}{z'} - \theta_3 \\[2mm] \dfrac{d}{dz}\,\delta x = l_1 n_1\dfrac{\delta x'}{x'} + l_2 n_2\dfrac{\delta y'}{y'} + l_3 n_3\dfrac{dz'}{z'} + \theta_2\end{array}\right\} \quad \dots\dots\dots\dots\dots (64),$$

with similar equations for quantities involving δy and δz.

Let a', β', γ' be the values of a, β, γ referred to the axes x', y', z'; then

$$\left.\begin{array}{l}a' = l_1 a + m_1\beta + n_1\gamma \\ \beta' = l_2 a + m_2\beta + n_2\gamma \\ \gamma' = l_3 a + m_3\beta + n_3\gamma\end{array}\right\} \quad \dots\dots\dots\dots\dots\dots (65).$$

We shall then have
$$\delta a = l_1\delta a' + l_2\delta\beta' + l_3\delta\gamma' + \gamma\theta_2 - \beta\theta_3 \dots\dots\dots\dots\dots (66),$$

$$= l_1 a'\dfrac{\delta x'}{x'} + l_2\beta'\dfrac{\delta y'}{y'} + l_3\gamma'\dfrac{\delta z'}{z'} + \gamma\theta_2 - \beta\theta_3 \dots\dots\dots (67).$$

By substituting the values of a', β', γ', and comparing with equations (64), we find

$$\delta a = a\dfrac{d}{dx}\,\delta x + \beta\dfrac{d}{dy}\,\delta x + \gamma\dfrac{d}{dz}\,\delta x \dots\dots\dots\dots\dots (68)$$

as the variation of a due to the change of form and position of the element. The variations of β and γ have similar expressions.

PROP. XI.—To find the electromotive forces in a moving body.

The variation of the velocity of the vortices in a moving element is due to two causes—the action of the electromotive forces, and the change of form and position of the element. The whole variation of a is therefore

$$\delta a = \dfrac{1}{\mu}\left(\dfrac{dQ}{dz} - \dfrac{dR}{dy}\right)\delta t + a\dfrac{d}{dx}\,\delta x + \beta\dfrac{d}{dy}\,\delta x + \gamma\dfrac{d}{dz}\,\delta x \dots\dots\dots\dots (69).$$

But since a is a function of x, y, z and t, the variation of a may be also written

$$\delta a = \dfrac{da}{dx}\,\delta x + \dfrac{da}{dy}\,\delta y + \dfrac{da}{dz}\,\delta z + \dfrac{da}{dt}\,\delta t \dots\dots\dots\dots\dots (70).$$

Equating the two values of δa and dividing by δt, and remembering that in the motion of an incompressible medium

$$\dfrac{d}{dx}\dfrac{dx}{dt} + \dfrac{d}{dy}\dfrac{dy}{dt} + \dfrac{d}{dz}\dfrac{dz}{dt} = 0 \dots\dots\dots\dots\dots (71),$$

and that in the absence of free magnetism

$$\frac{d\alpha}{dx} + \frac{d\beta}{dy} + \frac{d\gamma}{dz} = 0 \dots\dots\dots\dots\dots\dots\dots\dots(72),$$

we find

$$\frac{1}{\mu}\left(\frac{dQ}{dz} - \frac{dR}{dy}\right) + \gamma\frac{d}{dz}\frac{dx}{dt} - \alpha\frac{d}{dz}\frac{dz}{dt} - \alpha\frac{d}{dy}\frac{dy}{dt} + \beta\frac{d}{dy}\frac{dx}{dt}$$

$$+ \frac{d\gamma}{dz}\frac{dx}{dt} - \frac{d\alpha}{dz}\frac{dz}{dt} - \frac{d\alpha}{dy}\frac{dy}{dt} + \frac{d\beta}{dy}\frac{dx}{dt} - \frac{d\alpha}{dt} = 0 \dots\dots\dots (73).$$

Putting

$$\alpha = \frac{1}{\mu}\left(\frac{dG}{dz} - \frac{dH}{dy}\right) \dots\dots\dots\dots\dots\dots\dots (74),$$

and

$$\frac{d\alpha}{dt} = \frac{1}{\mu}\left(\frac{d^2G}{dz\,dt} - \frac{d^2H}{dy\,dt}\right) \dots\dots\dots\dots\dots\dots(75),$$

where F, G, and H are the values of the electrotonic components for a fixed point of space, our equation becomes

$$\frac{d}{dz}\left(Q + \mu\gamma\frac{dx}{dt} - \mu\alpha\frac{dz}{dt} - \frac{dG}{dt}\right) - \frac{d}{dy}\left(R + \mu\alpha\frac{dy}{dt} - \mu\beta\frac{dx}{dt} - \frac{dH}{dt}\right) = 0 \dots\dots(76).$$

The expressions for the variations of β and γ give us two other equations which may be written down from symmetry. The complete solution of the three equations is

$$\left.\begin{aligned}
P &= \mu\gamma\frac{dy}{dt} - \mu\beta\frac{dz}{dt} + \frac{dF}{dt} - \frac{d\Psi}{dx} \\[4pt]
Q &= \mu\alpha\frac{dz}{dt} - \mu\gamma\frac{dx}{dt} + \frac{dG}{dt} - \frac{d\Psi}{dy} \\[4pt]
R &= \mu\beta\frac{dx}{dt} - \mu\alpha\frac{dy}{dt} + \frac{dH}{dt} - \frac{d\Psi}{dz}
\end{aligned}\right\} \dots\dots\dots\dots\dots\dots (77).$$

The first and second terms of each equation indicate the effect of the motion of any body in the magnetic field, the third term refers to changes in the electrotonic state produced by alterations of position or intensity of magnets or currents in the field, and Ψ is a function of x, y, z, and t, which is indeterminate as far as regards the solution of the original equations, but which may always be determined in any given case from the circumstances of the problem. The physical interpretation of Ψ is, that it is the *electric tension* at each point of space.

The physical meaning of the terms in the expression for the electromotive force depending on the motion of the body, may be made simpler by supposing the field of magnetic force uniformly magnetized with intensity a in the direction of the axis of x. Then if l, m, n be the direction-cosines of any portion of a linear conductor, and S its length, the electromotive force resolved in the direction of the conductor will be

$$e = S\left(Pl + Qm + Rn\right) \dots\dots\dots\dots (78),$$

or

$$e = S\mu a \left(m\frac{dz}{dt} - n\frac{dy}{dt}\right) \dots\dots\dots (79),$$

that is, the product of μa, the quantity of magnetic induction over unit of area multiplied by $S\left(m\dfrac{dz}{dt} - n\dfrac{dy}{dt}\right)$, the area swept out by the conductor S in unit of time, resolved perpendicular to the direction of the magnetic force.

The electromotive force in any part of a conductor due to its motion is therefore measured by the *number* of lines of magnetic force which it crosses in unit of time; and the total electromotive force in a closed conductor is measured by the change of the number of lines of force which pass through it; and this is true whether the change be produced by the motion of the conductor or by any external cause.

In order to understand the mechanism by which the motion of a conductor across lines of magnetic force generates an electromotive force in that conductor, we must remember that in Prop. X. we have proved that the change of form of a portion of the medium containing vortices produces a change of the velocity of those vortices; and in particular that an extension of the medium in the direction of the axes of the vortices, combined with a contraction in all directions perpendicular to this, produces an increase of velocity of the vortices; while a shortening of the axis and bulging of the sides produces a diminution of the velocity of the vortices.

This change of the velocity of the vortices arises from the internal effects of change of form, and is independent of that produced by external electromotive forces. If, therefore, the change of velocity be prevented or checked, electromotive forces will arise, because each vortex will press on the surrounding particles in the direction in which it tends to alter its motion.

Let A, fig. 4, p. 488, represent the section of a vertical wire moving in the direction of the arrow from west to east, across a system of lines of magnetic force

61—2

running north and south. The curved lines in fig. 4 represent the lines of fluid motion about the wire, the wire being regarded as stationary, and the fluid as having a motion relative to it. It is evident that, from this figure, we can trace the variations of form of an element of the fluid, as the form of the element depends, not on the absolute motion of the whole system, but on the relative motion of its parts.

In front of the wire, that is, on its east side, it will be seen that as the wire approaches each portion of the medium, that portion is more and more compressed in the direction from east to west, and extended in the direction from north to south; and since the axes of the vortices lie in the north and south direction, their velocity will continually tend to increase by Prop. X., unless prevented or checked by electromotive forces acting on the circumference of each vortex.

We shall consider an electromotive force as positive when the vortices tend to move the interjacent particles *upwards* perpendicularly to the plane of the paper.

The vortices appear to revolve as the hands of a watch when we look at them from south to north; so that each vortex moves upwards on its west side, and downwards on its east side. In front of the wire, therefore, where each vortex is striving to increase its velocity, the electromotive force upwards must be greater on its west than on its east side. There will therefore be a continual increase of upward electromotive force from the remote east, where it is zero, to the front of the moving wire, where the upward force will be strongest.

Behind the wire a different action takes place. As the wire moves away from each successive portion of the medium, that portion is extended from east to west, and compressed from north to south, so as to tend to diminish the velocity of the vortices, and therefore to make the upward electromotive force greater on the east than on the west side of each vortex. The upward electromotive force will therefore increase continually from the remote west, where it is zero, to the back of the moving wire, where it will be strongest.

It appears, therefore, that a vertical wire moving eastwards will experience an electromotive force tending to produce in it an upward current. If there is no conducting circuit in connexion with the ends of the wire, no current will be formed, and the magnetic forces will not be altered; but if such a circuit exists, there will be a current, and the lines of magnetic force and the velocity of the vortices will be altered from their state previous to the motion of the wire. The change in the lines of force is shewn in fig. 5. The vortices in front of the wire, instead of merely producing pressures, actually increase in velocity, while those behind have their velocity diminished, and those at the sides of the wire have the direction of their axes altered; so that the final effect is to produce a force acting on the wire as a resistance to its motion. We may now recapitulate the assumptions we have made, and the results we have obtained.

(1) Magneto-electric phenomena are due to the existence of matter under certain conditions of motion or of pressure in every part of the magnetic field, and not to direct action at a distance between the magnets or currents. The substance producing these effects may be a certain part of ordinary matter, or it may be an æther associated with matter. Its density is greatest in iron, and least in diamagnetic substances; but it must be in all cases, except that of iron, very rare, since no other substance has a large ratio of magnetic capacity to what we call a vacuum.

(2) The condition of any part of the field, through which lines of magnetic force pass, is one of unequal pressure in different directions, the direction of the lines of force being that of least pressure, so that the lines of force may be considered lines of tension.

(3) This inequality of pressure is produced by the existence in the medium of vortices or eddies, having their axes in the direction of the lines of force, and having their direction of rotation determined by that of the lines of force.

We have supposed that the direction was that of a watch to a spectator looking from south to north. We might with equal propriety have chosen the reverse direction, as far as known facts are concerned, by supposing resinous electricity instead of vitreous to be positive. The effect of these vortices depends on their density, and on their velocity at the circumference, and is independent of their diameter. The density must be proportional to the capacity of the substance for magnetic induction, that of the vortices in air being 1. The velocity must be very great, in order to produce so powerful effects in so rare a medium.

The size of the vortices is indeterminate, but is probably very small as compared with that of a complete molecule of ordinary matter*.

* The angular momentum of the system of vortices depends on their average diameter; so that if the diameter were sensible, we might expect that a magnet would behave as if it contained a revolving body

(4) The vortices are separated from each other by a single layer of round particles, so that a system of cells is formed, the partitions being these layers of particles, and the substance of each cell being capable of rotating as a vortex.

(5) The particles forming the layer are in *rolling contact* with both the vortices which they separate, but do not rub against each other. They are perfectly free to roll between the vortices and so to change their place, provided they keep within one *complete molecule* of the substance; but in passing from one molecule to another they experience resistance, and generate irregular motions, which constitute heat. These particles, in our theory, play the part of electricity. Their motion of translation constitutes an electric current, their rotation serves to transmit the motion of the vortices from one part of the field to another, and the tangential pressures thus called into play constitute electromotive force. The conception of a particle having its motion connected with that of a vortex by perfect rolling contact may appear somewhat awkward. I do not bring it forward as a mode of connexion existing in nature, or even as that which I would willingly assent to as an electrical hypothesis. It is, however, a mode of connexion which is mechanically conceivable, and easily investigated, and it serves to bring out the actual mechanical connexions between the known electro-magnetic phenomena; so that I venture to say that any one who understands the provisional and temporary character of this hypothesis, will find himself rather helped than hindered by it in his search after the true interpretation of the phenomena.

The action between the vortices and the layers of particles is in part tangential; so that if there were any slipping or differential motion between the parts in contact, there would be a loss of the energy belonging to the lines of force, and a gradual transformation of that energy into heat. Now we know that the lines of force about a magnet are maintained for an indefinite time without any expenditure of energy; so that we must conclude that wherever there is tangential action between different parts of the medium, there is no motion of slipping between those parts. We must therefore conceive that the vortices and particles roll together without slipping; and that the interior strata of each vortex receive their proper velocities from the exterior stratum without slipping, that is, the angular velocity must be the same throughout each vortex.

within it, and that the existence of this rotation might be detected by experiments on the free rotation of a magnet. I have made experiments to investigate this question, but have not yet fully tried the apparatus.

The only process in which electro-magnetic energy is lost and transformed into heat, is in the passage of electricity from one molecule to another. In all other cases the energy of the vortices can only be diminished when an equivalent quantity of mechanical work is done by magnetic action.

(6) The effect of an electric current upon the surrounding medium is to make the vortices in contact with the current revolve so that the parts next to the current move in the same direction as the current. The parts furthest from the current will move in the opposite direction; and if the medium is a conductor of electricity, so that the particles are free to move in any direction, the particles touching the outside of these vortices will be moved in a direction contrary to that of the current, so that there will be an induced current in the opposite direction to the primary one.

If there were no resistance to the motion of the particles, the induced current would be equal and opposite to the primary one, and would continue as long as the primary current lasted, so that it would prevent all action of the primary current at a distance. If there is a resistance to the induced current, its particles act upon the vortices beyond them, and transmit the motion of rotation to them, till at last all the vortices in the medium are set in motion with such velocities of rotation that the particles between them have no motion except that of rotation, and do not produce currents.

In the transmission of the motion from one vortex to another, there arises a force between the particles and the vortices, by which the particles are pressed in one direction and the vortices in the opposite direction. We call the force acting on the particles the electromotive force. The reaction on the vortices is equal and opposite, so that the electromotive force cannot move any part of the medium as a whole, it can only produce currents. When the primary current is stopped, the electromotive forces all act in the opposite direction.

(7) When an electric current or a magnet is moved in presence of a conductor, the velocity of rotation of the vortices in any part of the field is altered by that motion. The force by which the proper amount of rotation is transmitted to each vortex, constitutes in this case also an electromotive force, and, if permitted, will produce currents.

(8) When a conductor is moved in a field of magnetic force, the vortices in it and in its neighbourhood are moved out of their places, and are changed in form. The force arising from these changes constitutes the electromotive

force on a moving conductor, and is found by calculation to correspond with that determined by experiment.

We have now shewn in what way electro-magnetic phenomena may be imitated by an imaginary system of molecular vortices. Those who have been already inclined to adopt an hypothesis of this kind, will find here the conditions which must be fulfilled in order to give it mathematical coherence, and a comparison, so far satisfactory, between its necessary results and known facts. Those who look in a different direction for the explanation of the facts, may be able to compare this theory with that of the existence of currents flowing freely through bodies, and with that which supposes electricity to act at a distance with a force depending on its velocity, and therefore not subject to the law of conservation of energy.

The facts of electro-magnetism are so complicated and various, that the explanation of any number of them by several different hypotheses must be interesting, not only to physicists, but to all who desire to understand how much evidence the explanation of phenomena lends to the credibility of a theory, or how far we ought to regard a coincidence in the mathematical expression of two sets of phenomena as an indication that these phenomena are of the same kind. We know that partial coincidences of this kind have been discovered; and the fact that they are only partial is proved by the divergence of the laws of the two sets of phenomena in other respects. We may chance to find, in the higher parts of physics, instances of more complete coincidence, which may require much investigation to detect their ultimate divergence.

NOTE.

Since the first part of this paper was written, I have seen in Crelle's *Journal* for 1859, a paper by Prof. Helmholtz on Fluid Motion, in which he has pointed out that the lines of fluid motion are arranged according to the same laws as the lines of magnetic force, the path of an electric current corresponding to a line of axes of those particles of the fluid which are in a state of rotation. This is an additional instance of a *physical analogy*, the investigation of which may illustrate both electro-magnetism and hydrodynamics.

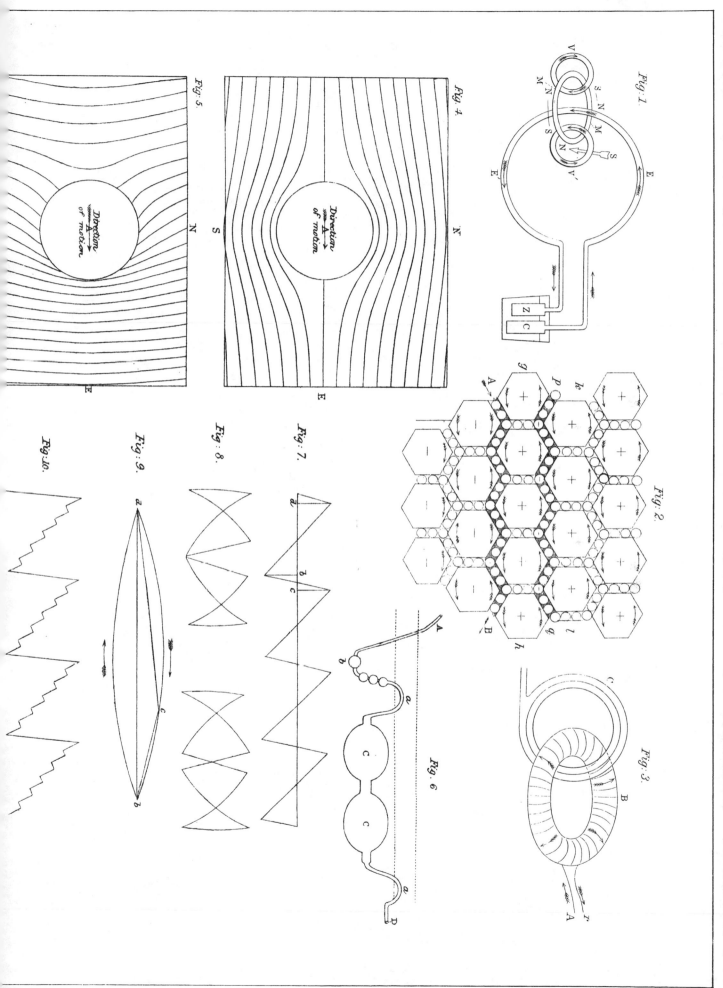

[From the *Philosophical Magazine* for January and February, 1862.]

PART III.

THE THEORY OF MOLECULAR VORTICES APPLIED TO STATICAL ELECTRICITY.

In the first part of this paper* I have shewn how the forces acting between magnets, electric currents, and matter capable of magnetic induction may be accounted for on the hypothesis of the magnetic field being occupied with innumerable vortices of revolving matter, their axes coinciding with the direction of the magnetic force at every point of the field.

The centrifugal force of these vortices produces pressures distributed in such a way that the final effect is a force identical in direction and magnitude with that which we observe.

In the second part† I described the mechanism by which these rotations may be made to coexist, and to be distributed according to the known laws of magnetic lines of force.

I conceived the rotating matter to be the substance of certain cells, divided from each other by cell-walls composed of particles which are very small compared with the cells, and that it is by the motions of these particles, and their tangential action on the substance in the cells, that the rotation is communicated from one cell to another.

I have not attempted to explain this tangential action, but it is necessary to suppose, in order to account for the transmission of rotation from the exterior to the interior parts of each cell, that the substance in the cells possesses elasticity of figure, similar in kind, though different in degree, to that observed in solid bodies. The undulatory theory of light requires us to admit this kind of elasticity in the luminiferous medium, in order to account for transverse vibrations. We need not then be surprised if the magneto-electric medium possesses the same property.

* *Phil. Mag.* March, 1861 [pp. 451—466 of this vol.].
† *Phil. Mag.* April and May, 1861 [pp. 467—488 of this vol.].

According to our theory, the particles which form the partitions between the cells constitute the matter of electricity. The motion of these particles constitutes an electric current; the tangential force with which the particles are pressed by the matter of the cells is electromotive force, and the pressure of the particles on each other corresponds to the tension or potential of the electricity.

If we can now explain the condition of a body with respect to the surrounding medium when it is said to be "charged" with electricity, and account for the forces acting between electrified bodies, we shall have established a connexion between all the principal phenomena of electrical science.

We know by experiment that electric tension is the same thing, whether observed in statical or in current electricity; so that an electromotive force produced by magnetism may be made to charge a Leyden jar, as is done by the coil machine.

When a difference of tension exists in different parts of any body, the electricity passes, or tends to pass, from places of greater to places of smaller tension. If the body is a conductor, an actual passage of electricity takes place; and if the difference of tensions is kept up, the current continues to flow with a velocity proportional inversely to the resistance, or directly to the conductivity of the body.

The electric resistance has a very wide range of values, that of the metals being the smallest, and that of glass being so great that a charge of electricity has been preserved* in a glass vessel for years without penetrating the thickness of the glass.

Bodies which do not permit a current of electricity to flow through them are called insulators. But though electricity does not flow through them, the electrical effects are propagated through them, and the amount of these effects differs according to the nature of the body; so that equally good insulators may act differently as dielectrics†.

Here then we have two independent qualities of bodies, one by which they allow of the passage of electricity through them, and the other by which they allow of electrical action being transmitted through them without any electricity being allowed to pass. A conducting body may be compared to a porous membrane which opposes more or less resistance to the passage of a fluid,

---

* By Professor W. Thomson.   † Faraday, *Experimental Researches*, Series XI.

while a dielectric is like an elastic membrane which may be impervious to the fluid, but transmits the pressure of the fluid on one side to that on the other.

As long as electromotive force acts on a conductor, it produces a current which, as it meets with resistance, occasions a continual transformation of electrical energy into heat, which is incapable of being restored again as electrical energy by any reversion of the process.

Electromotive force acting on a dielectric produces a state of polarization of its parts similar in distribution to the polarity of the particles of iron under the influence of a magnet*, and, like the magnetic polarization, capable of being described as a state in which every particle has its poles in opposite conditions.

In a dielectric under induction, we may conceive that the electricity in each molecule is so displaced that one side is rendered positively, and the other negatively electrical, but that the electricity remains entirely connected with the molecule, and does not pass from one molecule to another.

The effect of this action on the whole dielectric mass is to produce a general displacement of the electricity in a certain direction. This displacement does not amount to a current, because when it has attained a certain value it remains constant, but it is the commencement of a current, and its variations constitute currents in the positive or negative direction, according as the displacement is increasing or diminishing. The amount of the displacement depends on the nature of the body, and on the electromotive force; so that if h is the displacement, R the electromotive force, and E a coefficient depending on the nature of the dielectric,

$$R = - 4\pi E^2 h \,;$$

and if r is the value of the electric current due to displacement,

$$r = \frac{dh}{dt} \,.$$

These relations are independent of any theory about the internal mechanism of dielectrics; but when we find electromotive force producing electric displacement in a dielectric, and when we find the dielectric recovering from its state of electric displacement with an equal electromotive force, we cannot help

* See Prof. Mossotti, "Discussione Analitica," *Memorie della Soc. Italiana* (Modena), Vol. XXIV. Part 2, p. 49.

regarding the phenomena as those of an elastic body, yielding to a pressure, and recovering its form when the pressure is removed.

According to our hypothesis, the magnetic medium is divided into cells, separated by partitions formed of a stratum of particles which play the part of electricity. When the electric particles are urged in any direction, they will, by their tangential action on the elastic substance of the cells, distort each cell, and call into play an equal and opposite force arising from the elasticity of the cells. When the force is removed, the cells will recover their form, and the electricity will return to its former position.

In the following investigation I have considered the relation between the displacement and the force producing it, on the supposition that the cells are spherical. The actual form of the cells probably does not differ from that of a sphere sufficiently to make much difference in the numerical result.

I have deduced from this result the relation between the statical and dynamical measures of electricity, and have shewn, by a comparison of the electro-magnetic experiments of MM. Kohlrausch and Weber with the velocity of light as found by M. Fizeau, that the elasticity of the magnetic medium in air is the same as that of the luminiferous medium, if these two coexistent, coextensive, and equally elastic media are not rather one medium.

It appears also from Prop. XV. that the attraction between two electrified bodies depends on the value of E^2, and that therefore it would be less in turpentine than in air, if the quantity of electricity in each body remains the same. If, however, the *potentials* of the two bodies were given, the attraction between them would vary inversely as E^2, and would be greater in turpentine than in air.

PROP. XII. To find the conditions of equilibrium of an elastic sphere whose surface is exposed to normal and tangential forces, the tangential forces being proportional to the sine of the distance from a given point on the sphere.

Let the axis of z be the axis of spherical co-ordinates.

Let ξ, η, ζ be the displacements of any particle of the sphere in the directions of x, y, and z.

Let p_{xx}, p_{yy}, p_{zz} be the stresses normal to planes perpendicular to the three axes, and let p_{yz}, p_{zx}, p_{xy} be the stresses of distortion in the planes yz, zx, and xy.

Let μ be the coefficient of cubic elasticity, so that if

$$p_{xx} = p_{yy} = p_{zz} = p,$$

$$p = \mu \left(\frac{d\xi}{dx} + \frac{d\eta}{dy} + \frac{d\zeta}{dz} \right) \quad\dots\dots\dots\dots\dots\dots (80).$$

Let m be the coefficient of rigidity, so that

$$p_{xx} - p_{yy} = m \left(\frac{d\xi}{dx} - \frac{d\eta}{dy} \right), \ \&c. \quad\dots\dots\dots\dots (81).$$

Then we have the following equations of elasticity in an isotropic medium,

$$p_{xx} = (\mu - \tfrac{1}{3}m) \left(\frac{d\xi}{dx} + \frac{d\eta}{dy} + \frac{d\zeta}{dz} \right) + m \frac{d\xi}{dx} \quad\dots\dots\dots\dots (82);$$

with similar equations in y and z, and also

$$p_{yz} = \frac{m}{2} \left(\frac{d\eta}{dz} + \frac{d\zeta}{dy} \right), \ \&c. \quad\dots\dots\dots\dots\dots (83).$$

In the case of the sphere, let us assume the radius $= a$, and

$$\xi = exz, \quad \eta = ezy, \quad \zeta = f(x^2 + y^2) + gz^2 + d \dots\dots\dots\dots (84).$$

Then

$$\left. \begin{aligned}
p_{xx} &= 2 (\mu - \tfrac{1}{3}m)(e+g) z + mez = p_{yy} \\[2mm]
p_{zz} &= 2 (\mu - \tfrac{1}{3}m)(e+g) z + 2mgz \\[2mm]
p_{yz} &= \frac{m}{2} (e + 2f) y \\[2mm]
p_{zx} &= \frac{m}{2} (e + 2f) z \\[2mm]
p_{xy} &= 0
\end{aligned} \right\} \quad\dots\dots\dots\dots (85).$$

The equation of internal equilibrium with respect to z is

$$\frac{d}{dx} p_{zx} + \frac{d}{dy} p_{yz} + \frac{d}{dz} p_{zz} = 0 \dots\dots\dots\dots\dots\dots (86),$$

which is satisfied in this case if

$$m (e + 2f + 2g) + 2 (\mu - \tfrac{1}{3}m)(e + g) = 0 \quad\dots\dots\dots\dots (87).$$

The tangential stress on the surface of the sphere, whose radius is a at an angular distance θ from the axis in plane xz,

$$T = (p_{xx} - p_{zz}) \sin\theta\cos\theta + p_{xz} (\cos^2\theta - \sin^2\theta) \dots\dots\dots\dots (88)$$

$$= 2m (e + f - g) a \sin\theta\cos^2\theta - \frac{ma}{2} (e + 2f) \sin\theta \dots\dots\dots\dots (89).$$

In order that T may be proportional to $\sin\theta$, the first term must vanish, and therefore

$$g = e + f \dots\dots\dots (90),$$

$$T = -\frac{ma}{2}(e + 2f)\sin\theta \dots\dots (91).$$

The normal stress on the surface at any point is

$$N = p_{xx}\sin^2\theta + p_{yy}\cos^2\theta + 2p_{xz}\sin\theta\cos\theta$$

$$= 2\left(\mu - \tfrac{1}{3}m\right)(e+g)\,a\cos\theta + 2ma\cos\theta\{(e+f)\sin^2\theta + g\cos^2\theta\}\dots\dots(92);$$

or by (87) and (90), $\qquad N = -ma(e+2f)\cos\theta \dots\dots (93).$

The tangential displacement of any point is

$$t = \xi\cos\theta - \zeta\sin\theta = -(a^2f + d)\sin\theta \dots\dots(94).$$

The normal displacement is

$$n = \xi\sin\theta + \zeta\cos\theta = \{a^2(e+f) + d\}\cos\theta \dots\dots(95).$$

If we make $\qquad a^2(e+f) + d = 0 \dots\dots(96),$

there will be no normal displacement, and the displacement will be entirely tangential, and we shall have

$$t = a^2 e\sin\theta \dots\dots (97).$$

The whole work done by the superficial forces is

$$U = \tfrac{1}{2}\Sigma\,(Tt)\,dS,$$

the summation being extended over the surface of the sphere.

The energy of elasticity in the substance of the sphere is

$$U = \tfrac{1}{2}\Sigma\left\{\frac{d\xi}{dx}\,p_{xx} + \frac{d\eta}{dy}\,p_{yy} + \frac{d\zeta}{dz}\,p_{zz} + \left(\frac{d\eta}{dz} + \frac{d\zeta}{dy}\right)p_{yz} + \left(\frac{d\zeta}{dx} + \frac{d\xi}{dz}\right)p_{zx} + \left(\frac{d\xi}{dy} + \frac{d\eta}{dx}\right)p_{xy}\right\}dV,$$

the summation being extended to the whole contents of the sphere.

We find, as we ought, that these quantities have the same value, namely

$$U = -\tfrac{2}{3}\pi a^5 me\,(e + 2f) \dots\dots(98).$$

We may now suppose that the tangential action on the surface arises from a layer of particles in contact with it, the particles being acted on by their own mutual pressure, and acting on the surfaces of the two cells with which they are in contact.

We assume the axis of z to be in the direction of maximum variation of the pressure among the particles, and we have to determine the relation between an electromotive force R acting on the particles in that direction, and the electric displacement h which accompanies it.

PROP. XIII.—To find the relation between electromotive force and electric displacement when a uniform electromotive force R acts parallel to the axis of z.

Take any element δS of the surface, covered with a stratum whose density is ρ, and having its normal inclined θ to the axis of z; then the tangential force upon it will be

$$\rho R\delta S \sin \theta = 2\, T\delta S \dots\dots\dots\dots\dots\dots\dots\dots(99),$$

T being, as before, the tangential force on each side of the surface. Putting $\rho = \dfrac{1}{2\pi}$ as in equation (34)*, we find

$$R = -2\pi ma\,(e+2f)\dots\dots\dots\dots\dots\dots(100).$$

The displacement of electricity due to the distortion of the sphere is

$$\Sigma\delta S\tfrac{1}{2}\rho l \sin \theta \text{ taken over the whole surface}\dots\dots\dots\dots(101);$$

and if h is the electric displacement per unit of volume, we shall have

$$\tfrac{4}{3}\pi a^3 h = \tfrac{2}{3}a^4 e \dots\dots\dots\dots\dots\dots\dots (102),$$

or

$$h = \frac{1}{2\pi}\, ae \dots\dots\dots\dots\dots\dots\dots (103);$$

so that

$$R = 4\pi^2 m\, \frac{e+2f}{e}\, h \dots\dots\dots\dots\dots\dots (104),$$

or we may write

$$R = -4\pi E^2 h \dots\dots\dots\dots\dots\dots\dots(105),$$

provided we assume

$$E^2 = -\pi m\, \frac{e+2f}{e} \dots\dots\dots\dots\dots\dots (106).$$

Finding e and f from (87) and (90), we get

$$E^2 = \pi m\, \frac{3}{1+\dfrac{5}{3}\dfrac{m}{\mu}} \dots\dots\dots\dots\dots\dots(107).$$

The ratio of m to μ varies in different substances; but in a medium whose elasticity depends entirely upon forces acting between pairs of particles, this ratio is that of 6 to 5, and in this case

$$E^2 = \pi m \dots\dots\dots\dots\dots\dots\dots\dots\dots\dots(108).$$

* *Phil. Mag.* April, 1861 [p. 471 of this vol.].

When the resistance to compression is infinitely greater than the resistance to distortion, as in a liquid rendered slightly elastic by gum or jelly,

$$E^2 = 3\pi m. \quad\dotfill (109).$$

The value of E^2 must lie between these limits. It is probable that the substance of our cells is of the former kind, and that we must use the first value of E^2, which is that belonging to a hypothetically "perfect" solid*, in which

$$5m = 6\mu \quad\dotfill (110),$$

so that we must use equation (108).

PROP. XIV.—To correct the equations (9)† of electric currents for the effect due to the elasticity of the medium.

We have seen that electromotive force and electric displacement are connected by equation (105). Differentiating this equation with respect to t, we find

$$\frac{dR}{dt} = -4\pi E^2 \frac{dh}{dt} \quad\dotfill (111),$$

shewing that when the electromotive force varies, the electric displacement also varies. But a variation of displacement is equivalent to a current, and this current must be taken into account in equations (9) and added to r. The three equations then become

$$\left. \begin{aligned} p &= \frac{1}{4\pi}\left(\frac{d\gamma}{dy} - \frac{d\beta}{dz} - \frac{1}{E^2}\frac{dP}{dt}\right) \\ q &= \frac{1}{4\pi}\left(\frac{d\alpha}{dy} - \frac{d\gamma}{dx} - \frac{1'}{E^2}\frac{dQ}{dt}\right) \\ r &= \frac{1}{4\pi}\left(\frac{d\beta}{dx} - \frac{d\alpha}{dy} - \frac{1}{E^2}\frac{dR}{dt}\right) \end{aligned} \right\} \quad\dotfill (112),$$

where p, q, r are the electric currents in the directions of x, y, and z; α, β, γ are the components of magnetic intensity; and P, Q, R are the electromotive forces. Now if e be the quantity of free electricity in unit of volume, then the equation of continuity will be

$$\frac{dp}{dx} + \frac{dq}{dy} + \frac{dr}{dz} + \frac{de}{dt} = 0 \quad\dotfill (113).$$

---

* See Rankine "On Elasticity," *Camb. and Dub. Math. Journ.* 1851.
† *Phil. Mag.* March, 1861 [p. 462 of this vol.].

Differentiating (112) with respect to x, y, and z respectively, and substituting, we find

$$\frac{de}{dt} = \frac{1}{4\pi E^2} \frac{d}{dt} \left(\frac{dP}{dx} + \frac{dQ}{dy} + \frac{dR}{dz} \right) \dots \dots \dots \dots (114);$$

whence

$$e = \frac{1}{4\pi E^2} \left(\frac{dP}{dx} + \frac{dQ}{dy} + \frac{dR}{dz} \right) \dots \dots \dots \dots (115),$$

the constant being omitted, because $e = 0$ when there are no electromotive forces.

PROP. XV.—To find the force acting between two electrified bodies.

The energy in the medium arising from the electric displacements is

$$U = - \Sigma \tfrac{1}{2} (Pf + Qg + Rh)\, \delta V \dots \dots \dots \dots (116),$$

where P, Q, R are the forces, and f, g, h the displacements. Now when there is no motion of the bodies or alteration of forces, it appears from equations (77)* that

$$P = -\frac{d\Psi}{dx}, \quad Q = -\frac{d\Psi}{dy}, \quad R = -\frac{d\Psi}{dz} \dots \dots \dots \dots (118);$$

and we know by (105) that

$$P = -4\pi E^2 f, \quad Q = -4\pi E^2 g, \quad R = -4\pi E^2 h \dots \dots \dots \dots (119);$$

whence

$$U = \frac{1}{8\pi E^2} \Sigma \left(\overline{\frac{d\Psi}{dx}}\Big|^2 + \overline{\frac{d\Psi}{dy}}\Big|^2 + \overline{\frac{d\Psi}{dz}}\Big|^2 \right) \delta V \dots \dots \dots \dots (120).$$

Integrating by parts throughout all space, and remembering that Ψ vanishes at an infinite distance,

$$U = -\frac{1}{8\pi E^2} \Sigma \Psi \left(\frac{d^2\Psi}{dx^2} + \frac{d^2\Psi}{dy^2} + \frac{d^2\Psi}{dz^2} \right) \delta V \dots \dots \dots \dots (121);$$

or by (115),

$$U = \tfrac{1}{2} \Sigma (\Psi e)\, \delta V \dots \dots \dots \dots (122).$$

Now let there be two electrified bodies, and let e_1 be the distribution of electricity in the first, and Ψ_1 the electric tension due to it, and let

$$e_1 = \frac{1}{4\pi E^2} \left(\frac{d^2\Psi_1}{dx^2} + \frac{d^2\Psi_1}{dy^2} + \frac{d^2\Psi_1}{dz^2} \right) \dots \dots \dots \dots (123).$$

Let e_2 be the distribution of electricity in the second body, and Ψ_2 the tension due to it; then the whole tension at any point will be $\Psi_1 + \Psi_2$, and the expansion for U will become

$$U = \tfrac{1}{2} \Sigma (\Psi_1 e_1 + \Psi_2 e_2 + \Psi_1 e_2 + \Psi_2 e_1)\, \delta V \dots \dots \dots \dots (124).$$

* *Phil. Mag.* May, 1861 [p. 482 of this vol.].

Let the body whose electricity is e_1 be moved in any way, the electricity moving along with the body, then since the distribution of tension Ψ_1 moves with the body, the value of $\Psi_1 e_1$ remains the same.

$\Psi_2 e_2$ also remains the same; and Green has shewn (Essay on Electricity, p. 10) that $\Psi_1 e_2 = \Psi_2 e_1$, so that the work done by moving the body against electric forces

$$W = \delta U = \delta \Sigma \, (\Psi_2 e_1) \, \delta V \dots\dots\dots\dots\dots (125).$$

And if e_1 is confined to a small body,

$$W = e_1 \delta \Psi_2,$$

or

$$F dr = e_1 \frac{d\Psi_2}{dr} dr \dots\dots\dots\dots\dots\dots (126),$$

where F is the resistance and dr the motion.

If the body e_2 be small, then if r is the distance from e_2, equation (123) gives

$$\Psi_2 = E^2 \frac{e_2}{r} \, ;$$

whence

$$F = - E^2 \frac{e_1 e_2}{r^2} \dots\dots\dots\dots\dots\dots (127) ;$$

or the force is a repulsion varying inversely as the square of the distance.

Now let η_1 and η_2 be the same quantities of electricity measured statically, then we know by definition of electrical quantity

$$F = - \frac{\eta_1 \eta_2}{r^2} \dots\dots\dots\dots\dots\dots (128) ;$$

and this will be satisfied provided

$$\eta_1 = E e_1 \text{ and } \eta_2 = E e_2 \dots\dots\dots\dots\dots\dots (129) ;$$

so that the quantity E previously determined in Prop. XIII. is the number by which the electrodynamic measure of any quantity of electricity must be multiplied to obtain its electrostatic measure.

That electric current which, circulating round a ring whose area is unity, produces the same effect on a distant magnet as a magnet would produce whose strength is unity and length unity placed perpendicularly to the plane of the ring, is a unit current; and E units of electricity, measured statically, traverse the section of this current in one second,—these units being such that any two of them, placed at unit of distance, repel each other with unit of force.

We may suppose either that E units of positive electricity move in the positive direction through the wire, or that E units of negative electricity move in the negative direction, or, thirdly, that $\frac{1}{2}E$ units of positive electricity move in the positive direction, while $\frac{1}{2}E$ units of negative electricity move in the negative direction at the same time.

The last is the supposition on which MM. Weber and Kohlrausch* proceed, who have found
$$\tfrac{1}{2}E = 155{,}370{,}000{,}000 \ldots\ldots\ldots\ldots (130),$$
the unit of length being the millimetre, and that of time being one second, whence
$$E = 310{,}740{,}000{,}000 \ldots\ldots\ldots\ldots (131).$$

PROP. XVI.—To find the rate of propagation of transverse vibrations through the elastic medium of which the cells are composed, on the supposition that its elasticity is due entirely to forces acting between pairs of particles.

By the ordinary method of investigation we know that
$$V = \sqrt{\frac{m}{\rho}} \ldots\ldots\ldots\ldots (132),$$
where m is the coefficient of transverse elasticity, and ρ is the density. By referring to the equations of Part I., it will be seen that if ρ is the density of the matter of the vortices, and μ is the "coefficient of magnetic induction,"
$$\mu = \pi\rho \ldots\ldots\ldots\ldots (133);$$
whence
$$\pi m = V^2\mu \ldots\ldots\ldots\ldots (134);$$
and by (108),
$$E = V\sqrt{\mu} \ldots\ldots\ldots\ldots (135).$$
In air or vacuum $\mu = 1$, and therefore
$$\left.\begin{array}{l} V = E \\ = 310{,}740{,}000{,}000 \text{ millimetres per second} \\ = 193{,}088 \text{ miles per second} \end{array}\right\} \ldots\ldots\ldots (136).$$

* *Abhandlungen der König. Sächsischen Gesellschaft,* Vol. III. (1857); p. 260.

The velocity of light in air, as determined by M. Fizeau[*], is 70,843 leagues per second (25 leagues to a degree) which gives

$$V = 314{,}858{,}000{,}000 \text{ millimetres}$$

$$= 195{,}647 \text{ miles per second} \dots\dots\dots\dots\dots (137).$$

The velocity of transverse undulations in our hypothetical medium, calculated from the electro-magnetic experiments of MM. Kohlrausch and Weber, agrees so exactly with the velocity of light calculated from the optical experiments of M. Fizeau, that we can scarcely avoid the inference that *light consists in the transverse undulations of the same medium which is the cause of electric and magnetic phenomena.*

PROP. XVII.—To find the electric capacity of a Leyden jar composed of any given dielectric placed between two conducting surfaces.

Let the electric tensions or potentials of the two surfaces be Ψ_1 and Ψ_2. Let S be the area of each surface, and θ the distance between them, and let e and $-e$ be the quantities of electricity on each surface; then the capacity

$$C = \frac{e}{\Psi_1 - \Psi_2} \dots\dots\dots\dots\dots\dots\dots\dots\dots\dots (138).$$

Within the dielectric we have the variation of Ψ perpendicular to the surface

$$= \frac{\Psi_1 - \Psi_2}{\theta}.$$

Beyond either surface this variation is zero.

Hence by (115) applied at the surface, the electricity on unit of area is

$$\frac{\Psi_1 - \Psi_2}{4\pi E^2 \theta} \dots\dots\dots\dots\dots\dots\dots\dots\dots\dots (139);$$

and we deduce the whole capacity of the apparatus,

$$C = \frac{S}{4\pi E^2 \theta} \dots\dots\dots\dots\dots\dots\dots\dots\dots\dots (140);$$

so that the quantity of electricity required to bring the one surface to a

---

[*] *Comptes Rendus,* Vol. XXIX. (1849), p. 90. In Galbraith and Haughton's *Manual of Astronomy,* M. Fizeau's result is stated at 169,944 geographical miles of 1000 fathoms, which gives 193,118 statute miles; the value deduced from aberration is 192,000 miles.

given tension varies directly as the surface, inversely as the thickness, and inversely as the square of E.

Now the coefficient of induction of dielectrics is deduced from the capacity of induction-apparatus formed of them; so that if D is that coefficient, D varies inversely as E^2, and is unity for air. Hence

$$D = \frac{V^2}{V_1^2 \mu} \quad \dots \dots \dots \dots \dots \dots \dots \dots (141),$$

where V and V_1 are the velocities of light in air and in the medium. Now if i is the index of refraction, $\frac{V}{V_1} = i$, and

$$D = \frac{i^2}{\mu} \quad \dots \dots \dots \dots \dots \dots \dots \dots (142);$$

so that the inductive power of a dielectric varies directly as the square of the index of refraction, and inversely as the magnetic inductive power.

In dense media, however, the optical, electric, and magnetic phenomena may be modified in different degrees by the particles of gross matter; and their mode of arrangement may influence these phenomena differently in different directions. The axes of optical, electric, and magnetic properties will probably coincide; but on account of the unknown and probably complicated nature of the reactions of the heavy particles on the ætherial medium, it may be impossible to discover any general numerical relations between the optical, electric, and magnetic ratios of these axes.

It seems probable, however, that the value of E, for any given axis, depends upon the velocity of light whose vibrations are parallel to that axis, or whose plane of polarization is perpendicular to that axis.

In a uniaxal crystal, the axial value of E will depend on the velocity of the extraordinary ray, and the equatorial value will depend on that of the ordinary ray.

In "positive" crystals, the axial value of E will be the least and in negative the greatest.

The value of D_1, which varies inversely as E^2, will, *cæteris paribus*, be greatest for the axial direction in positive crystals, and for the equatorial direction in negative crystals, such as Iceland spar. If a spherical portion of a crystal, radius $= a$, be suspended in a field of electric force which would act on unit of electricity with force $= 1$, and if D_1 and D_2 be the coefficients of dielectric induction along the two axes in the plane of rotation, then if θ be the inclination of the axis to the electric force, the moment tending to turn the sphere will be

$$\tfrac{3}{2}\,\frac{(D_1 - D_2)}{(2D_1 + 1)(2D_2 + 1)}\,I^2 a^3 \sin 2\theta \,\ldots\ldots\ldots\ldots\,(143),$$

and the axis of greatest dielectric induction (D_1) will tend to become parallel to the lines of electric force.

PART IV.

THE THEORY OF MOLECULAR VORTICES APPLIED TO THE ACTION OF MAGNETISM ON POLARIZED LIGHT.

The connexion between the distribution of lines of magnetic force and that of electric currents may be completely expressed by saying that the work done on a unit of imaginary magnetic matter, when carried round any closed curve, is proportional to the quantity of electricity which passes through the closed curve. The mathematical form of this law may be expressed as in equations (9)*, which I here repeat, where a, β, γ are the rectangular components of magnetic intensity, and p, q, r are the rectangular components of steady electric currents,

$$\left.\begin{aligned}
p &= \frac{1}{4\pi}\left(\frac{d\gamma}{dy} - \frac{d\beta}{dz}\right)\\[4pt]
q &= \frac{1}{4\pi}\left(\frac{da}{dz} - \frac{d\gamma}{dx}\right)\\[4pt]
r &= \frac{1}{4\pi}\left(\frac{d\beta}{dx} - \frac{da}{dy}\right)
\end{aligned}\right\}\,\ldots\ldots\ldots\ldots\ldots\ldots\,(9).$$

The same mathematical connexion is found between other sets of phenomena in physical science.

(1) If a, β, γ represent displacements, velocities, or forces, then p, q, r will be rotatory displacements, velocities of rotation, or moments of couples producing rotation, in the elementary portions of the mass.

---

* *Phil. Mag.* March, 1861 [p. 462 of this vol.].

(2) If α, β, γ represent rotatory displacements in a uniform and continuous substance, then p, q, r represent the *relative* linear displacement of a particle with respect to those in its immediate neighbourhood. See a paper by Prof. W. Thomson "On a Mechanical Representation of Electric, Magnetic, and Galvanic Forces," *Camb. and Dublin Math. Journal*, Jan. 1847.

(3) If α, β, γ represent the rotatory velocities of vortices whose centres are fixed, then p, q, r represent the velocities with which loose particles placed between them would be carried along. See the second part of this paper (*Phil. Mag.* April, 1861) [p. 469].

It appears from all these instances that the connexion between magnetism and electricity has the same mathematical form as that between certain pairs of phenomena, of which one has a *linear* and the other a *rotatory* character. Professor Challis* conceives magnetism to consist in currents of a fluid whose direction corresponds with that of the lines of magnetic force; and electric currents, on this theory, are accompanied by, if not dependent on, a rotatory motion of the fluid about the axis of the current. Professor Helmholtz† has investigated the motion of an incompressible fluid, and has conceived lines drawn so as to correspond at every point with the instantaneous axis of rotation of the fluid there. He has pointed out that the lines of fluid motion are arranged according to the same laws with respect to the lines of rotation, as those by which the lines of magnetic force are arranged with respect to electric currents. On the other hand, in this paper I have regarded magnetism as a phenomenon of rotation, and electric currents as consisting of the actual translation of particles, thus assuming the inverse of the relation between the two sets of phenomena.

Now it seems natural to suppose that all the direct effects of any cause which is itself of a longitudinal character, must be themselves longitudinal, and that the direct effects of a rotatory cause must be themselves rotatory. A motion of translation along an axis cannot produce a rotation about that axis unless it meets with some special mechanism, like that of a screw, which connects a motion in a given direction along the axis with a rotation in a given direction round it; and a motion of rotation, though it may produce tension along the axis, cannot of itself produce a current in one direction along the axis rather than the other.

* *Phil. Mag.* December, 1860, January and February, 1861.
† Crelle, *Journal*, Vol. LV. (1858), p. 25.

Electric currents are known to produce effects of transference in the direction of the current. They transfer the electrical state from one body to another, and they transfer the elements of electrolytes in opposite directions, but they do not* cause the plane of polarization of light to rotate when the light traverses the axis of the current.

On the other hand, the magnetic state is not characterized by any strictly longitudinal phenomenon. The north and south poles differ only in their names, and these names might be exchanged without altering the statement of any magnetic phenomenon; whereas the positive and negative poles of a battery are completely distinguished by the different elements of water which are evolved there. The magnetic state, however, is characterized by a well-marked rotatory phenomenon discovered by Faraday†—the rotation of the plane of polarized light when transmitted along the lines of magnetic force.

When a transparent diamagnetic substance has a ray of plane-polarized light passed through it, and if lines of magnetic force are then produced in the substance by the action of a magnet or of an electric current, the plane of polarization of the transmitted light is found to be changed, and to be turned through an angle depending on the intensity of the magnetizing force within the substance.

The direction of this rotation in diamagnetic substances is the same as that in which positive electricity must circulate round the substance in order to produce the actual magnetizing force within it; or if we suppose the horizontal part of terrestrial magnetism to be the magnetizing force acting on the substance, the plane of polarization would be turned in the direction of the earth's true rotation, that is, from west upwards to east.

In paramagnetic substances, M. Verdet‡ has found that the plane of polarization is turned in the opposite direction, that is, in the direction in which negative electricity would flow if the magnetization were effected by a helix surrounding the substance.

In both cases the absolute direction of the rotation is the same, whether the light passes from north to south or from south to north,—a fact which distinguishes this phenomenon from the rotation produced by quartz, turpentine, &c.,

* Faraday, *Experimental Researches*, 951—954, and 2216—2220.
† Ibid., Series XIX.
‡ *Comptes Rendus*, Vol. XLIII. p. 529; Vol. XLIV. p. 1209.

in which the absolute direction of rotation is reversed when that of the light is reversed. The rotation in the latter case, whether related to an axis, as in quartz, or not so related, as in fluids, indicates a relation between the direction of the ray and the direction of rotation, which is similar in its formal expression to that between the longitudinal and rotatory motions of a right-handed or a left-handed screw; and it indicates some property of the substance the mathematical form of which exhibits right-handed or left-handed relations, such as are known to appear in the external forms of crystals having these properties. In the magnetic rotation no such relation appears, but the direction of rotation is directly connected with that of the magnetic lines, in a way which seems to indicate that magnetism is really a phenomenon of rotation.

The transference of electrolytes in fixed directions by the electric current, and the rotation of polarized light in fixed directions by magnetic force, are the facts the consideration of which has induced me to regard magnetism as a phenomenon of rotation, and electric currents as phenomena of translation, instead of following out the analogy pointed out by Helmholtz, or adopting the theory propounded by Professor Challis.

The theory that electric currents are linear, and magnetic forces rotatory phenomena, agrees so far with that of Ampère and Weber; and the hypothesis that the magnetic rotations exist wherever magnetic force extends, that the centrifugal force of these rotations accounts for magnetic attractions, and that the inertia of the vortices accounts for induced currents, is supported by the opinion of Professor W. Thomson*. In fact the whole theory of molecular vortices developed in this paper has been suggested to me by observing the direction in which those investigators who study the action of media are looking for the explanation of electro-magnetic phenomena.

Professor Thomson has pointed out that the cause of the magnetic action on light must be a real rotation going on in the magnetic field. A *right-handed* circularly polarized ray of light is found to travel with a different velocity according as it passes from north to south, or from south to north, along a line of magnetic force. Now, whatever theory we adopt about the direction of vibrations in plane-polarized light, the geometrical arrangement of the parts of the medium during the passage of a right-handed circularly polarized ray is exactly the same whether the ray is moving north or south. The only difference

* See Nichol's *Cyclopædia*, art. "Magnetism, Dynamical Relations of," edition 1860; *Proceedings of Royal Society*, June 1856 and June 1861; and *Phil. Mag.* 1857.

is, that the particles describe their circles in opposite directions. Since, therefore, the *configuration* is the same in the two cases, the forces acting between particles must be the same in both, and the motions due to these forces must be equal in velocity if the medium was originally at rest; but if the medium be in a state of rotation, either as a whole or in molecular vortices, the circular vibrations of light may differ in velocity according as their direction is similar or contrary to that of the vortices.

We have now to investigate whether the hypothesis developed in this paper—that magnetic force is due to the centrifugal force of small vortices, and that these vortices consist of the same matter the vibrations of which constitute light—leads to any conclusions as to the effect of magnetism on polarized light. We suppose transverse vibrations to be transmitted through a magnetized medium. How will the propagation of these vibrations be affected by the circumstance that portions of that medium are in a state of rotation?

In the following investigation, I have found that the only effect which the rotation of the vortices will have on the light will be to make the plane of polarization rotate in the *same* direction as the vortices, through an angle proportional—

(A) to the thickness of the substance,

(B) to the resolved part of the magnetic force parallel to the ray,

(C) to the index of refraction of the ray,

(D) inversely to the square of the wave-length in air,

(E) to the *mean radius* of the vortices,

(F) to the capacity for magnetic induction.

A and B have been fully investigated by M. Verdet*, who has shewn that the rotation is strictly proportional to the thickness and to the magnetizing force, and that, when the ray is inclined to the magnetizing force, the rotation is as the cosine of that inclination. D has been supposed to give the true relation between the rotation of different rays; but it is probable that C must be taken into account in an accurate statement of the phenomena. The rotation varies, not exactly inversely as the square of the wave length, but a little faster; so that for the highly refrangible rays the rotation is greater than that given by this law, but more nearly as the index of refraction divided by the square of the wave-length.

* *Annales de Chimie et de Physique*, sér. 3, Vol. XLI. p. 370; Vol. XLIII. p. 37.

The relation (E) between the amount of rotation and the size of the vortices shews that different substances may differ in rotating power independently of any observable difference in other respects. We know nothing of the absolute size of the vortices; and on our hypothesis the optical phenomena are probably the only data for determining their relative size in different substances.

On our theory, the direction of the rotation of the plane of polarization depends on that of the mean moment of momenta, or *angular momentum*, of the molecular vortices; and since M. Verdet has discovered that magnetic substances have an effect on light opposite to that of diamagnetic substances, it follows that the molecular rotation must be opposite in the two classes of substances.

We can no longer, therefore, consider diamagnetic bodies as being those whose coefficient of magnetic induction is less than that of space empty of gross matter. We must admit the diamagnetic state to be the *opposite* of the paramagnetic; and that the vortices, or at least the influential majority of them, in diamagnetic substances, revolve in the direction in which positive electricity revolves in the magnetizing bobbin, while in paramagnetic substances they revolve in the opposite direction.

This result agrees so far with that part of the theory of M. Weber* which refers to the paramagnetic and diamagnetic conditions. M. Weber supposes the electricity in paramagnetic bodies to revolve the same way as the surrounding helix, while in diamagnetic bodies it revolves the opposite way. Now if we regard negative or resinous electricity as a substance the absence of which constitutes positive or vitreous electricity, the results will be those actually observed. This will be true independently of any other hypothesis than that of M. Weber about magnetism and diamagnetism, and does not require us to admit either M. Weber's theory of the mutual action of electric particles in motion, or our theory of cells and cell-walls.

I am inclined to believe that iron differs from other substances in the manner of its action as well as in the intensity of its magnetism; and I think its behaviour may be explained on our hypothesis of molecular vortices, by supposing that the particles of the *iron itself* are set in rotation by the tangential action of the vortices, in an opposite direction to their own. These large heavy particles would thus be revolving exactly as we have supposed the

* Taylor's *Scientific Memoirs*, Vol. v. p. 477.

64—2

infinitely small particles constituting electricity to revolve, but without being free like them to change their place and form currents.

The whole *energy* of rotation of the magnetized field would thus be greatly increased, as we know it to be; but the *angular momentum* of the iron particles would be opposite to that of the æthereal cells and immensely greater, so that the total angular momentum of the substance will be in the direction of rotation of the iron, or the reverse of that of the vortices. Since, however, the angular momentum depends on the absolute size of the revolving portions of the substance, it may depend on the state of aggregation or chemical arrangement of the elements, as well as on the ultimate nature of the components of the substance. Other phenomena in nature seem to lead to the conclusion that all substances are made up of a number of parts, finite in size, the particles composing these parts being themselves capable of internal motion.

PROP. XVIII.—To find the angular momentum of a vortex.

The angular momentum of any material system about an axis is the sum of the products of the mass, dm, of each particle multiplied by twice the area it describes about that axis in unit of time; or if A is the angular momentum about the axis of x,

$$A = \Sigma dm \left(y \frac{dz}{dt} - z \frac{dy}{dt} \right).$$

As we do not know the distribution of density within the vortex, we shall determine the relation between the angular momentum and the energy of the vortex which was found in Prop. VI.

Since the time of revolution is the same throughout the vortex, the mean angular velocity ω will be uniform and $= \dfrac{a}{r}$, where a is the velocity at the circumference, and r the radius. Then

$$A = \Sigma dm r^2 \omega,$$

and the energy

$$E = \tfrac{1}{2} \Sigma dm r^2 \omega^2 = \tfrac{1}{2} A \omega,$$

$$= \frac{1}{8\pi} \mu a^2 V \text{ by Prop. VI.}^*$$

whence

$$A = \frac{1}{4\pi} \mu r a V \quad \dotfill \quad (144)$$

* *Phil. Mag.* April 1861 [p. 472 of this vol.].

for the axis of x, with similar expressions for the other axes, V being the volume, and r the radius of the vortex.

PROP. XIX.—To determine the conditions of undulatory motion in a medium containing vortices, the vibrations being perpendicular to the direction of propagation.

Let the waves be plane-waves propagated in the direction of z, and let the axis of x and y be taken in the directions of greatest and least elasticity in the plane xy. Let x and y represent the displacement parallel to these axes, which will be the same throughout the same wave-surface, and therefore we shall have x and y functions of z and t only.

Let X be the tangential stress on unit of area parallel to xy, tending to move the part next the origin in the direction of x.

Let Y be the corresponding tangential stress in the direction of y.

Let k_1 and k_2 be the coefficients of elasticity with respect to these two kinds of tangential stress; then, if the medium is at rest,

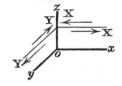

$$X = k_1 \frac{dx}{dz}, \quad Y = k_2 \frac{dy}{dz}.$$

Now let us suppose vortices in the medium whose velocities are represented as usual by the symbols a, β, γ, and let us suppose that the value of a is increasing at the rate $\frac{da}{dt}$, on account of the action of the tangential stresses alone, there being no electromotive force in the field. The angular momentum in the stratum whose area is unity, and thickness dz, is therefore increasing at the rate $\frac{1}{4\pi} \mu r \frac{da}{dt} dz$; and if the part of the force Y which produces this effect is Y', then the moment of Y' is $-Y'dz$, so that $Y' = -\frac{1}{4\pi} \mu r \frac{da}{dt}$.

The complete value of Y when the vortices are in a state of varied motion is

$$\left. \begin{array}{l} Y = k_2 \dfrac{dy}{dz} - \dfrac{1}{4\pi} \mu r \dfrac{da}{dt} \\[2ex] X = k_1 \dfrac{dx}{dz} + \dfrac{1}{4\pi} \mu r \dfrac{d\beta}{dt} \end{array} \right\} \dots\dots\dots\dots\dots\dots\dots (145).$$

Similarly,

The whole force acting upon a stratum whose thickness is dz and area unity, is $\dfrac{dX}{dz}\,dz$ in the direction of x, and $\dfrac{dY}{dz}\,dz$ in direction of y. The mass of the stratum is ρdz, so that we have as the equations of motion,

$$\left.\begin{aligned}\rho\,\frac{d^2x}{dt^2}&=\frac{dX}{dz}=k_1\frac{d^2x}{dz^2}+\frac{d}{dz}\frac{1}{4\pi}\mu r\frac{d\beta}{dt}\\[1mm]\rho\,\frac{d^2y}{dt^2}&=\frac{dY}{dz}=k_2\frac{d^2y}{dz^2}-\frac{d}{dz}\frac{1}{4\pi}\mu r\frac{d\alpha}{dt}\end{aligned}\right\}\ \ldots\ldots\ldots\ldots\ (146).$$

Now the changes of velocity $\dfrac{d\alpha}{dt}$ and $\dfrac{d\beta}{dt}$ are produced by the motion of the medium containing the vortices, which distorts and twists every element of its mass; so that we must refer to Prop. X.* to determine these quantities in terms of the motion. We find there at equation (68),

$$d\alpha=\alpha\frac{d}{dx}\,\delta x+\beta\frac{d}{dy}\,\delta x+\gamma\frac{d}{dz}\,\delta x\,\ldots\ldots\ldots\ldots\ldots\ldots\ (68).$$

Since δx and δy are functions of z and t only, we may write this equation

$$\left.\begin{aligned}\frac{d\alpha}{dt}&=\gamma\frac{d^2x}{dz\,dt}\\[1mm]\frac{d\beta}{dt}&=\gamma\frac{d^2y}{dz\,dt}\end{aligned}\right\}\ \ldots\ldots\ldots\ldots\ldots\ldots\ldots\ldots\ (147),$$

and in like manner,

so that if we now put $k_1=a^2\rho$, $k_2=b^2\rho$, and $\dfrac{1}{4\pi}\dfrac{\mu r}{\rho}\gamma=c^2$, we may write the equations of motion

$$\left.\begin{aligned}\frac{d^2x}{dt^2}&=a^2\frac{d^2x}{dz^2}+c^2\frac{d^3y}{dz^2dt}\\[1mm]\frac{d^2y}{dt^2}&=b^2\frac{d^2y}{dz^2}-c^2\frac{d^3x}{dz^2dt}\end{aligned}\right\}\ \ldots\ldots\ldots\ldots\ (148).$$

These equations may be satisfied by the values

$$\left.\begin{aligned}x&=A\cos(nt-mz+\alpha)\\y&=B\sin(nt-mz+\alpha)\end{aligned}\right\}\ \ldots\ldots\ldots\ldots\ (149),$$

provided

$$\left.\begin{aligned}(n^2-m^2a^2)\,A&=m^2nc^2B\\(n^2-m^2b^2)\,B&=m^2nc^2A\end{aligned}\right\}\ \ldots\ldots\ldots\ldots\ (150).$$

and

* *Phil. Mag.* May 1861 [p. 481 of this vol.].

Multiplying the last two equations together, we find

$$(n^2 - m^2a^2)\,(n^2 - m^2b^2) = m^4n^2c^4 \dots\dots\dots\dots\dots(151)$$

an equation quadratic with respect to m^2, the solution of which is

$$m^2 = \frac{2n^2}{a^2 + b^2 \mp \sqrt{(a^2 - b^2)^2 + 4n^2c^4}} \dots\dots\dots\dots(152).$$

These values of m^2 being put in the equations (150) will each give a ratio of A and B,

$$\frac{A}{B} = \frac{a^2 - b^2 \mp \sqrt{(a^2 - b^2)^2 + 4n^2c^4}}{2nc^2},$$

which being substituted in equations (149), will satisfy the original equations (148). The most general undulation of such a medium is therefore compounded of two elliptic undulations of different eccentricities travelling with different velocities and rotating in opposite directions. The results may be more easily explained in the case in which $a = b$; then

$$m^2 = \frac{n^2}{a^2 \mp nc^2} \quad \text{and} \quad A = \mp B \dots\dots\dots\dots(153).$$

Let us suppose that the value of A is unity for both vibrations, then we shall have

$$\left.\begin{aligned}
x &= \cos\left(nt - \frac{nz}{\sqrt{a^2 - nc^2}}\right) + \cos\left(nt - \frac{nz}{\sqrt{a^2 + nc^2}}\right) \\
y &= -\sin\left(nt - \frac{nz}{\sqrt{a^2 - nc^2}}\right) + \sin\left(nt - \frac{nz}{\sqrt{a^2 + nc^2}}\right)
\end{aligned}\right\} \dots\dots\dots(154).$$

The first terms of x and y represent a circular vibration in the negative direction, and the second term a circular vibration in the positive direction, the positive having the greatest velocity of propagation. Combining the terms, we may write

$$\left.\begin{aligned}
x &= 2\cos(nt - pz)\cos qz \\
y &= 2\cos(nt - pz)\sin qz
\end{aligned}\right\} \dots\dots\dots\dots\dots(155),$$

where

$$p = \frac{n}{2\sqrt{a^2 - nc^2}} + \frac{n}{2\sqrt{a^2 + nc^2}} \Bigg\} \dots\dots\dots\dots(156).$$

and

$$q = \frac{n}{2\sqrt{a^2 - nc^2}} - \frac{n}{2\sqrt{a^2 + nc^2}} \Bigg\}$$

These are the equations of an undulation consisting of a plane vibration whose periodic time is $\dfrac{2\pi}{n}$, and wave-length $\dfrac{2\pi}{p}=\lambda$, propagated in the direction of z with a velocity $\dfrac{n}{p}=v$, while the plane of the vibration revolves about the axis of z in the positive direction so as to complete a revolution when $z=\dfrac{2\pi}{q}$.

Now let us suppose c^2 small, then we may write

$$p=\frac{n}{a} \quad \text{and} \quad q=\frac{n^2 c^2}{2a^3} \quad\dots\dots\dots\dots\dots\dots (157);$$

and remembering that $c^2=\dfrac{1}{4\pi}\dfrac{r}{\rho}\mu\gamma$, we find

$$q=\frac{\pi}{2}\frac{r}{\rho}\frac{\mu\gamma}{\lambda^2 v} \quad\dots\dots\dots\dots\dots\dots\dots\dots (158).$$

Here r is the radius of the vortices, an unknown quantity. ρ is the density of the luminiferous medium in the body, which is also unknown; but if we adopt the theory of Fresnel, and make s the density in space devoid of gross matter, then

$$\rho=si^2 \quad\dots\dots\dots\dots\dots\dots\dots\dots\dots\dots (159),$$

where i is the index of refraction.

On the theory of MacCullagh and Neumann,

$$\rho=s \quad\dots\dots\dots\dots\dots\dots\dots\dots\dots (160)$$

in all bodies.

μ is the coefficient of magnetic induction, which is unity in empty space or in air.

γ is the velocity of the vortices at their circumference estimated in the ordinary units. Its value is unknown, but it is proportional to the intensity of the magnetic force.

Let Z be the magnetic intensity of the field, measured as in the case of terrestrial magnetism, then the intrinsic energy in air per unit of volume is

$$\frac{1}{8\pi}Z^2=\frac{1}{8\pi}\pi s\gamma^2,$$

where s is the density of the magnetic medium in air, which we have reason to believe the same as that of the luminiferous medium. We therefore put

$$\gamma=\frac{1}{\sqrt{\pi s}}Z \quad\dots\dots\dots\dots\dots\dots\dots\dots (161),$$

λ is the wave-length of the undulation in the substance. Now if Λ be the wave-length for the same ray in air, and i the index of refraction of that ray in the body,

$$\lambda = \frac{\Lambda}{i} \quad\text{.....................................} (162).$$

Also v, the velocity of light in the substance, is related to V, the velocity of light in air, by the equation

$$v = \frac{V}{i} \quad\text{....................................} (163).$$

Hence if z be the thickness of the substance through which the ray passes, the angle through which the plane of polarization will be turned will be in degrees,

$$\theta = \frac{180}{\pi} qz \quad\text{...............................} (164);$$

or, by what we have now calculated,

$$\theta = 90^{\circ} \frac{1}{\sqrt{\pi}} \cdot \frac{r}{s^{\frac{3}{2}}} \frac{\mu i Z z}{\Lambda^{2} V} \quad\text{............................} (165).$$

In this expression all the quantities are known by experiment except r, the radius of the vortices in the body, and s, the density of the luminiferous medium in air.

The experiments of M. Verdet* supply all that is wanted except the determination of Z in absolute measure; and this would also be known for all his experiments, if the value of the galvanometer deflection for a semi-rotation of the testing bobbin in a known magnetic field, such as that due to terrestrial magnetism at Paris, were once for all determined.

* *Annales de Chimie et de Physique*, sér. 3, Vol. XLI. p. 370.

[From the *London, Edinburgh, and Dublin Philosophical Magazine and Journal of Science.* Vol. XXVII. Fourth Series.]

XXIV. *On Reciprocal Figures and Diagrams of Forces.*

RECIPROCAL figures are such that the properties of the first relative to the second are the same as those of the second relative to the first. Thus inverse figures and polar reciprocals are instances of two different kinds of reciprocity.

The kind of reciprocity which we have here to do with has reference to figures consisting of straight lines joining a system of points, and forming closed rectilinear figures; and it consists in the directions of all lines in the one figure having a constant relation to those of the lines in the other figure which correspond to them.

In plane figures, corresponding lines may be either parallel, perpendicular, or at any constant angle. Lines meeting in a point in one figure form a closed polygon in the other.

In figures in space, the lines in one figure are perpendicular to planes in the other, and the planes corresponding to lines which meet in a point form a closed polyhedron.

The conditions of reciprocity may be considered from a purely geometrical point of view; but their chief importance arises from the fact that either of the figures being considered as a system of points acted on by forces along the lines of connexion, the other figure is a diagram of forces, in which these forces are represented in plane figures by lines, and in solid figures by the areas of planes.

The properties of the "triangle" and "polygon" of forces have been long known, and the "diagram" of forces has been used in the case of the funicular polygon; but I am not aware of any more general statement of the method of drawing diagrams of forces before Professor Rankine applied it to frames, roofs, &c. in his *Applied Mechanics*, p. 137, &c. The "polyhedron of forces," or the equilibrium of forces perpendicular and proportional to the areas of the faces of a polyhedron, has, I believe, been enunciated independently at various times; but the application to a "frame" is given by Professor Rankine in the *Philosophical Magazine*, February, 1864.

I propose to treat the question geometrically, as reciprocal figures are subject to certain conditions besides those belonging to diagrams of forces.

On Reciprocal Plane Figures.

Definition.—Two plane figures are reciprocal when they consist of an equal number of lines, so that corresponding lines in the two figures are parallel, and corresponding lines which converge to a point in one figure form a closed polygon in the other.

Note.—If corresponding lines in the two figures, instead of being parallel are at right angles or any other angle, they may be made parallel by turning one of the figures round in its own plane.

Since every polygon in one figure has three or more sides, every point in the other figure must have three or more lines converging to it; and since every line in the one figure has two and only two extremities to which lines converge, every line in the other figure must belong to two, and only two closed polygons. The simplest plane figure fulfilling these conditions is that formed by the six lines which join four points in pairs. The reciprocal figure consists of six lines parallel respectively to these, the points in the one figure corresponding to triangles in the other.

General Relation between the Numbers of Points, Lines, and Polygons in Reciprocal Figures.

The effect of drawing a line, one of whose extremities is a point connected with the system of lines already drawn, is either to introduce one new point into the system, or to complete one new polygon, or to divide a polygon into two parts, according as it is drawn to an isolated point, or a point already connected with the system. Hence the sum of points and polygons in the

65—2

system is increased by one for every new line. But the simplest figure consists of four points, four polygons, and six lines. Hence the sum of the points and polygons must always exceed the number of lines by two.

Note.—This is the same relation which connects the numbers of summits, faces, and edges of polyhedra.

Conditions of indeterminateness and impossibility in drawing reciprocal Diagrams.

Taking any line parallel to one of the lines of the figure for a base, every new point is to be determined by the intersection of two new lines. Calling s the number of points or summits, e the number of lines or edges, and f the number of polygons or faces, the assumption of the first line determines two points, and the remaining $s-2$ points are determined by $2(s-2)$ lines. Hence if

$$e = 2s - 3,$$

every point may be determined. If e be less, the form of the figure will be in some respects indeterminate; and if e be greater, the construction of the figure will be impossible, unless certain conditions among the directions of the lines are fulfilled.

These are the conditions of drawing any diagram in which the directions of the lines are arbitrarily given; but when one diagram is already drawn in which e is greater than $2s-3$, the directions of the lines will not be altogether arbitrary, but will be subject to $e-(2s-3)$ conditions.

Now if e', s', f' be the values of e, s, and f in the reciprocal diagram

$$e = e', \quad s = f', \quad f = s',$$
$$e = s + f - 2, \quad e' = s' + f' - 2.$$

Hence if $s = f$, $e = 2i - 2$; and there will be one condition connecting the directions of the lines of the original diagram, and this condition will ensure the possibility of constructing the reciprocal diagram. If

$$s > f, \quad e > 2s - 2, \quad \text{and} \quad e' < 2s' - 2;$$

so that the construction of the reciprocal diagram will be possible, but indeterminate to the extent of $s - f$ variables.

If $s < f$, the construction of the reciprocal diagram will be impossible unless $(s - f)$ conditions be fulfilled in the original diagram.

If any number of the points of the figure are so connected among themselves as to form an equal number of closed polygons, the conditions of constructing the reciprocal figure must be found by considering these points separately, and then examining their connexion with the rest.

Let us now consider a few cases of reciprocal figures in detail. The simplest case is that of the figure formed by the six lines connecting four points in a plane. If we now draw the six lines connecting the centres of the four circles which pass through three out of the four points, we shall have a reciprocal figure, the corresponding lines in the two figures being at right angles.

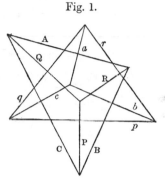

Fig. 1.

The reciprocal figure formed in this way is definite in size and position; but any figure similar to it and placed in any position is still reciprocal to the original figure. If the reciprocal figures are lettered as in fig. 1, we shall have the relation

$$\frac{AP}{ap} = \frac{BQ}{bq} = \frac{CR}{cr}.$$

In figures 2 and II. we have a pair of reciprocal figures in which the lines are more numerous, but the construction very easy. There are seven points in each figure corresponding to seven polygons in the other.

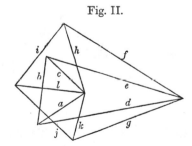

Fig. 2.   Fig. II.

The four points of triple concourse of lines *ABC*, *BDE*, *HIL*, *LJK* correspond to four triangles, *abc*, *bde*, *hil*, *ljk*.

The three points of quadruple concourse *ADFH*, *CEGK*, *IFGJ* correspond to three quadrilaterals, *adfh*, *cegk*, *ifgj*.

The five triangles *ADB*, *EBC*, *GJK*, *IJL*, *HIF* correspond to five points of triple concourse, *adb*, *ebc*, *gjk*, *ijl*, *hif*.

The quadrilateral *DEGF* corresponds to the point of quadruple concourse *degf*.

The pentagon *ACKLH* corresponds to the meeting of the five lines *acklh*.

In drawing the reciprocal of fig. 2, it is best to begin with a point of triple concourse. The reciprocal triangle of this point being drawn, determines three lines of the new figure. If the other extremities of any of the lines meeting in this point are points of triple concourse, we may in the same way determine more lines, two at a time. In drawing these lines, we have only to remember that those lines which in the first figure form a polygon, start from one point in the reciprocal figure. In this way we may proceed as long as we can always determine all the lines except two of each successive polygon.

The case represented in figs. 3 and III. is an instance of a pair of reciprocal figures fulfilling the conditions of possibility and determinateness, but

Fig. III.

Fig. 3.

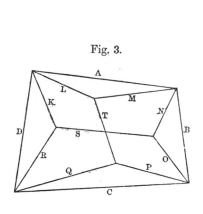

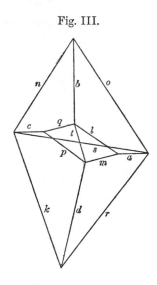

presenting a slight difficulty in drawing by the foregoing rule. Each figure has here eight points and eight polygons; but after we have drawn the lines *s*, *n*, *o*, *k*, *r*, we cannot proceed with the figure simply by drawing the last two lines of polygons, because the next polygons to be drawn are quadrilaterals, and we have only one side of each given. The easiest way to proceed is to produce *abcd* till they form a quadrilateral, then to draw a subsidiary figure similar to *tlmpq*, with *abcd* similarly situated, and then to reduce the latter figure to such a scale and position that *a*, *b*, *c*, *d* coincide in both figures.

In figures 4 and IV. the condition that the number of polygons is equal to the number of points is not fulfilled. In fig. 4 there are five points and

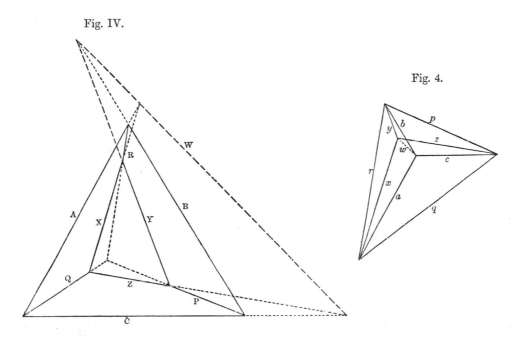

Fig. IV.

Fig. 4.

six triangles; in fig. IV. there are six points, two triangles, and three quadrilaterals. Hence if fig. 4 is given, fig. IV. is indeterminate to the extent of one variable, besides the elements of scale and position. In fact when we have drawn ABC and indicated the directions of P, Q, R, we may fix on any point of P as one of the angles of XYZ and complete the triangle XYZ. The size of XYZ is therefore indeterminate. Conversely, if fig. IV. is given, fig. 4 cannot be constructed unless one condition be fulfilled. That condition is that P, Q, and R meet in a point. When this is fulfilled, it follows by geometry that the points of concourse of A and X, B and Y, and C and Z lie in one straight line W, which is parallel to w in fig. 4. The condition may also be expressed by saying that fig. IV. must be a perspective projection of a polyhedron whose quadrilateral faces are planes. The planes of these faces intersect at the concourse of P, Q, R, and those of the triangular faces intersect in the line W.

Figs. 5 and V. represent another case of the same kind. In fig. 5 we have six points and eight triangles; fig. V. is therefore capable of two degrees of variability, and is subject to two conditions.

The conditions are that the four intersections of corresponding sides of opposite quadrilaterals in fig. V. shall lie in one straight line, parallel to the

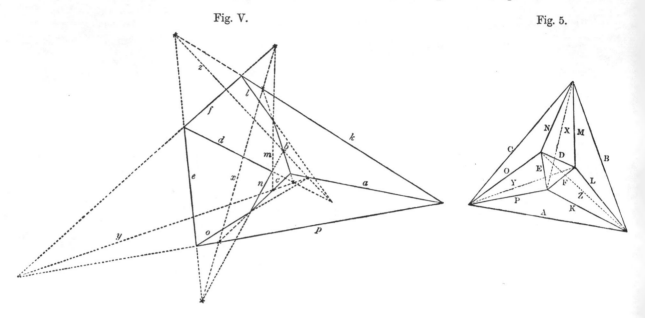

Fig. V.   Fig. 5.

line joining the opposite points of fig. 5 which correspond to these quadrilaterals. There are three such lines marked x, y, z, and four points of intersection lie on each line.

We may express this condition also by saying that fig. V. must be a perspective projection of a plane-sided polyhedron, the intersections of opposite planes being the lines x, y, z.

In fig. 6, let $ABCDE$ be a portion of a polygon bounded by other polygons of which the edges are $PQRST$, one or more of these edges meeting each angle of the polygon.

In fig. VI., let $abcde$ be lines parallel to $ABCDE$ and meeting in a point, and let these be terminated by the lines $pqrst$ parallel to $PQRST$, one or more of these lines completing each sector of fig. VI.

In fig. 6 draw Y through the intersections of AC and PQ, and in fig. VI. draw y through the intersections of a, p and c, q. Then the figures of six lines $ABCPQY$ and $abcpqy$ will be reciprocal, and y will be parallel to Y. Draw X parallel to x, and through the intersections of TX and CE draw Z, and in fig. VI. draw z through the intersections of cx and et; then $CDETXZ$

and *cdetxz* will be reciprocal, and Z will be parallel to z. Then through the intersections of AE and YZ draw W, and through those of *ay* and *ez* draw w; and since $ACEYZW$ and *aceyzw* are reciprocal, W will be parallel to w.

Fig. 6.

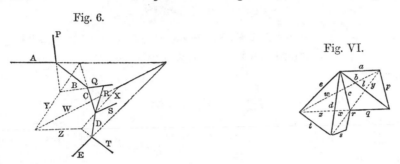

Fig. VI.

By going round the remaining sides of the polygon $ABCDE$ in the same way, we should find by the intersections of lines another point, the line joining which with the intersection of AE would be parallel to w, and therefore we should have three points in one line; namely, the intersection of Y and Z, the point determined by a similar process carried on on the other part of the circumference of the polygon, and the intersection of A and E; and we should find similar conditions for every pair of sides of every polygon.

Now the conditions of the figure 6 being a perspective projection of a plane-sided polyhedron are exactly the same. For A being the intersection of the faces AP and AB, and C that of BC and QC, the intersection AC will be a point in the intersection of the faces AP and CQ.

Similarly the intersection PQ will be another point in it, so that Y is the line of intersection of the faces AP and CQ.

In the same way Z is the intersection of ET and CQ, so that the intersection of Y and Z is a point in the intersection of AP and ET.

Another such point can be determined by going round the remaining sides of the polygon; and these two points, together with the intersections of the lines AE, must all be in one straight line, namely, the intersection of the faces AP and ET.

Hence the conditions of the possibility of reciprocity in plane figures are the same as those of each figure being the perspective projection of a plane-sided polyhedron. When the number of points is in every part of the figure equal to or less than the number of polygons, this condition is fulfilled of itself. When the number of points exceeds the number of polygons, there will be an impossible case, unless certain conditions are fulfilled so that certain sets of intersections lie in straight lines.

Application to Statics.

The doctrine of reciprocal figures may be treated in a purely geometrical manner, but it may be much more clearly understood by considering it as a method of calculating the forces among a system of points in equilibrium; for,

If forces represented in magnitude by the lines of a figure be made to act between the extremities of the corresponding lines of the reciprocal figure, then the points of the reciprocal figure will all be in equilibrium under the action of these forces.

For the forces which meet in any point are parallel and proportional to the sides of a polygon in the other figure.

If the points between which the forces are to act are known, the problem of determining the relations among the magnitudes of the forces so as to produce equilibrium will be indeterminate, determinate, or impossible, according as the construction of the reciprocal figure is so.

Reciprocal figures are mechanically reciprocal; that is, either may be taken as representing a system of points, and the other as representing the magnitudes of the forces acting between them.

In figures like 1, 2 and II., 3 and III., in which the equation

$$e = 2s - 2$$

is true, the forces are determinate in their ratios; so that one being given, the rest may be found.

When $e > 2s - 2$, as in figs. 4 and 5, the forces are indeterminate, so that more than one must be known to determine the rest, or else certain relations among them must be given, such as those arising from the elasticity of the parts of a frame.

When $e < 2s - 2$, the determination of the forces is impossible except under certain conditions. Unless these be fulfilled, as in figs. IV. and V., no forces along the lines of the figure can keep its points in equilibrium, and the figure, considered as a frame, may be said to be loose.

When the conditions are fulfilled, the pieces of the frame can support forces, but in such a way that a small disfigurement of the frame may produce infinitely great forces in some of the pieces, or may throw the frame into a loose condition at once.

The conditions, however, of the possibility of determining the ratios of the forces in a frame are not coextensive with those of finding a figure perfectly reciprocal to the frame.   The condition of determinate forces is

$$e = 2s - 2 \; ;$$

the condition of reciprocal figures is that every line belongs to two polygons only, and

$$e = s + f - 2.$$

In fig. 7 we have six points connected by ten lines in such a way that the forces are all determinate; but since the line L is a side of three triangles, we cannot draw a reciprocal figure, for we should have to draw a straight line l with three ends.

If we attempt to draw the reciprocal figure as in fig. VII., we shall find that, in order to represent the reciprocals of all the lines of fig. 7 and fix their relations, we must repeat two of them, as h and e by h' and e, so as to form a parallelogram. Fig. VII. is then a complete representation of the relations of the force which would produce equilibrium in fig. 7; but it is redundant by the repetition of h and e, and the two figures are not reciprocal.

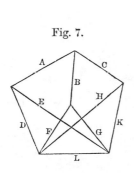

Fig. 7.

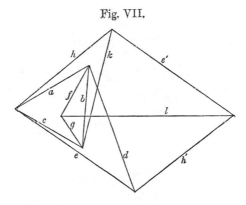

Fig. VII.

On Reciprocal Figures in three dimensions.

Definition.—Figures in three dimensions are reciprocal when they can be so placed that every line in the one figure is perpendicular to a plane face of the other, and every point of concourse of lines in the one figure is represented by a closed polyhedron with plane faces.

66—2

The simplest case is that of five points in space with their ten connecting lines, forming ten triangular faces enclosing five tetrahedrons. By joining the five points which are the centres of the spheres circumscribing these five tetrahedrons, we have a reciprocal figure of the kind described by Professor Rankine in the *Philosophical Magazine*, February 1864; and forces proportional to the areas of the triangles of one figure, if applied along the corresponding lines of connexion of the other figure, will keep its points in equilibrium.

In order to have perfect reciprocity between two figures, each figure must be made up of a number of closed polyhedra having plane faces of separation, and such that each face belongs to two and only two polyhedra, corresponding to the extremities of the reciprocal line in the other figure. Every line in the figure is the intersection of three or more plane faces, because the plane face in the reciprocal figure is bounded by three or more straight lines.

Let s be the number of points or summits, e the number of lines or edges, f the number of faces, and c the number of polyhedra or cells. Then if about one of the summits in which polyhedra meet, and σ edges and η faces, we describe a polyhedral cell, it will have ϕ faces and σ summits and η edges, and we shall have

$$\eta = \phi + \sigma - 2 ;$$

s, the number of summits, will be decreased by one and increased by σ;

c, the number of cells, will be increased by one;

f, the number of faces, will be increased by ϕ;

e, the number of edges, will be increased by η;

so that $e + c - (s + f)$ will be increased by $\eta + 1 - (\sigma + \phi - 1)$, which is zero, or this quantity is constant. Now in the figure of five points already discussed, $e = 10$, $c = 5$, $s = 5$, $f = 10$; so that generally

$$e + c = s + f,$$

in figures made up of cells in the way described.

The condition of a reciprocal figure being indeterminate, determinate, or impossible except in particular cases, is

$$e \gtreqless 3s - 5.$$

This condition is sufficient to determine the possibility of finding a system of forces along the edges which will keep the summits in equilibrium; but it is manifest that the mechanical problem may be solved, though the reciprocal figure cannot be constructed owing to the condition of all the sides of a face lying in a plane not being fulfilled, or owing to a face belonging to more than two cells. Hence the mechanical interest of reciprocal figures in space rapidly diminishes with their complexity.

Diagrams of forces in which the forces are represented by lines may be always constructed in space as well as in a plane, but in general some of the lines must be repeated.

Thus in the figure of five points, each point is the meeting place of four lines. The forces in these lines may be represented by five gauche quadrilaterals (that is, quadrilaterals not in one plane); and one of these being chosen, the other four may be applied to its sides and to each other so as to form five sides of a gauche hexahedron. The sixth side, that opposite the original quadrilateral, will be a parallelogram, the opposite sides of which are repetitions of the same line.

We have thus a complete but redundant diagram of forces consisting of eight points joined by twelve lines, two pairs of the lines being repetitions. This is a more convenient though less elegant construction of a diagram of forces, and it never becomes geometrically impossible as long as the problem is mechanically possible, however complicated the original figure may be.

[From the *Royal Society Transactions*, Vol. CLV.]

XXV. *A Dynamical Theory of the Electromagnetic Field.*

(Received October 27,—Read December 8, 1864.)

PART I.

INTRODUCTORY.

(1) THE most obvious mechanical phenomenon in electrical and magnetical experiments is the mutual action by which bodies in certain states set each other in motion while still at a sensible distance from each other. The first step, therefore, in reducing these phenomena into scientific form, is to ascertain the magnitude and direction of the force acting between the bodies, and when it is found that this force depends in a certain way upon the relative position of the bodies and on their electric or magnetic condition, it seems at first sight natural to explain the facts by assuming the existence of something either at rest or in motion in each body, constituting its electric or magnetic state, and capable of acting at a distance according to mathematical laws.

In this way mathematical theories of statical electricity, of magnetism, of the mechanical action between conductors carrying currents, and of the induction of currents have been formed. In these theories the force acting between the two bodies is treated with reference only to the condition of the bodies and their relative position, and without any express consideration of the surrounding medium.

These theories assume, more or less explicitly, the existence of substances the particles of which have the property of acting on one another at a distance by attraction or repulsion. The most complete development of a theory of this kind is that of M. W. Weber*, who has made the same theory include electrostatic and electromagnetic phenomena.

In doing so, however, he has found it necessary to assume that the force between two electric particles depends on their relative velocity, as well as on their distance.

This theory, as developed by MM. W. Weber and C. Neumann†, is exceedingly ingenious, and wonderfully comprehensive in its application to the phenomena of statical electricity, electromagnetic attractions, induction of currents and diamagnetic phenomena; and it comes to us with the more authority, as it has served to guide the speculations of one who has made so great an advance in the practical part of electric science, both by introducing a consistent system of units in electrical measurement, and by actually determining electrical quantities with an accuracy hitherto unknown.

(2) The mechanical difficulties, however, which are involved in the assumption of particles acting at a distance with forces which depend on their velocities are such as to prevent me from considering this theory as an ultimate one, though it may have been, and may yet be useful in leading to the coordination of phenomena.

I have therefore preferred to seek an explanation of the fact in another direction, by supposing them to be produced by actions which go on in the surrounding medium as well as in the excited bodies, and endeavouring to explain the action between distant bodies without assuming the existence of forces capable of acting directly at sensible distances.

(3) The theory I propose may therefore be called a theory of the *Electromagnetic Field*, because it has to do with the space in the neighbourhood of the electric or magnetic bodies, and it may be called a *Dynamical* Theory, because it assumes that in that space there is matter in motion, by which the observed electromagnetic phenomena are produced.

(4) The electromagnetic field is that part of space which contains and surrounds bodies in electric or magnetic conditions.

* "Electrodynamische Maassbestimmungen." *Leipzic Trans.* Vol. i. 1849, and Taylor's *Scientific Memoirs*, Vol. v. art. xiv.

† *Explicare tentatur quomodo fiat ut lucis planum polarizationis per vires electricas vel magneticas declinetur.*—Halis Saxonum, 1858.

It may be filled with any kind of matter, or we may endeavour to render it empty of all gross matter, as in the case of Geissler's tubes and other so-called vacua.

There is always, however, enough of matter left to receive and transmit the undulations of light and heat, and it is because the transmission of these radiations is not greatly altered when transparent bodies of measurable density are substituted for the so-called vacuum, that we are obliged to admit that the undulations are those of an æthereal substance, and not of the gross matter, the presence of which merely modifies in some way the motion of the æther.

We have therefore some reason to believe, from the phenomena of light and heat, that there is an æthereal medium filling space and permeating bodies, capable of being set in motion and of transmitting that motion from one part to another, and of communicating that motion to gross matter so as to heat it and affect it in various ways.

(5) Now the energy communicated to the body in heating it must have formerly existed in the moving medium, for the undulations had left the source of heat some time before they reached the body, and during that time the energy must have been half in the form of motion of the medium and half in the form of elastic resilience. From these considerations Professor W. Thomson has argued*, that the medium must have a density capable of comparison with that of gross matter, and has even assigned an inferior limit to that density.

(6) We may therefore receive, as a datum derived from a branch of science independent of that with which we have to deal, the existence of a pervading medium, of small but real density, capable of being set in motion, and of transmitting motion from one part to another with great, but not infinite, velocity.

Hence the parts of this medium must be so connected that the motion of one part depends in some way on the motion of the rest; and at the same time these connexions must be capable of a certain kind of elastic yielding, since the communication of motion is not instantaneous, but occupies time.

The medium is therefore capable of receiving and storing up two kinds of energy, namely, the "actual" energy depending on the motions of its parts, and "potential" energy, consisting of the work which the medium will do in recovering from displacement in virtue of its elasticity.

---

* "On the Possible Density of the Luminiferous Medium, and on the Mechanical Value of a Cubic Mile of Sunlight," *Transactions of the Royal Society of Edinburgh* (1854), p. 57.

The propagation of undulations consists in the continual transformation of one of these forms of energy into the other alternately, and at any instant the amount of energy in the whole medium is equally divided, so that half is energy of motion, and half is elastic resilience.

(7) A medium having such a constitution may be capable of other kinds of motion and displacement than those which produce the phenomena of light and heat, and some of these may be of such a kind that they may be evidenced to our senses by the phenomena they produce.

(8) Now we know that the luminiferous medium is in certain cases acted on by magnetism; for Faraday* discovered that when a plane polarized ray traverses a transparent diamagnetic medium in the direction of the lines of magnetic force produced by magnets or currents in the neighbourhood, the plane of polarization is caused to rotate.

This rotation is always in the direction in which positive electricity must be carried round the diamagnetic body in order to produce the actual magnetization of the field.

M. Verdet† has since discovered that if a paramagnetic body, such as solution of perchloride of iron in ether, be substituted for the diamagnetic body, the rotation is in the opposite direction.

Now Professor W. Thomson‡ has pointed out that no distribution of forces acting between the parts of a medium whose only motion is that of the luminous vibrations, is sufficient to account for the phenomena, but that we must admit the existence of a motion in the medium depending on the magnetization, in addition to the vibratory motion which constitutes light.

It is true that the rotation by magnetism of the plane of polarization has been observed only in media of considerable density; but the properties of the magnetic field are not so much altered by the substitution of one medium for another, or for a vacuum, as to allow us to suppose that the dense medium does anything more than merely modify the motion of the ether. We have therefore warrantable grounds for inquiring whether there may not be a motion of the ethereal medium going on wherever magnetic effects are observed, and

---

* *Experimental Researches*, Series XIX.

† *Comptes Rendus* (1856, second half year, p. 529, and 1857, first half year, p. 1209).

‡ *Proceedings of the Royal Society*, June 1856 and June 1861.

we have some reason to suppose that this motion is one of rotation, having the direction of the magnetic force as its axis.

(9) We may now consider another phenomenon observed in the electro-magnetic field. When a body is moved across the lines of magnetic force it experiences what is called an electromotive force; the two extremities of the body tend to become oppositely electrified, and an electric current tends to flow through the body. When the electromotive force is sufficiently powerful, and is made to act on certain compound bodies, it decomposes them, and causes one of their components to pass towards one extremity of the body, and the other in the opposite direction.

Here we have evidence of a force causing an electric current in spite of resistance; electrifying the extremities of a body in opposite ways, a condition which is sustained only by the action of the electromotive force, and which, as soon as that force is removed, tends, with an equal and opposite force, to produce a counter current through the body and to restore the original electrical state of the body; and finally, if strong enough, tearing to pieces chemical compounds and carrying their components in opposite directions, while their natural tendency is to combine, and to combine with a force which can generate an electromotive force in the reverse direction.

This, then, is a force acting on a body caused by its motion through the electromagnetic field, or by changes occurring in that field itself; and the effect of the force is either to produce a current and heat the body, or to decompose the body, or, when it can do neither, to put the body in a state of electric polarization,—a state of constraint in which opposite extremities are oppositely electrified, and from which the body tends to relieve itself as soon as the disturbing force is removed.

(10) According to the theory which I propose to explain, this "electro-motive force" is the force called into play during the communication of motion from one part of the medium to another, and it is by means of this force that the motion of one part causes motion in another part. When electromotive force acts on a conducting circuit, it produces a current, which, as it meets with resistance, occasions a continual transformation of electrical energy into heat, which is incapable of being restored again to the form of electrical energy by any reversal of the process.

(11) But when electromotive force acts on a dielectric it produces a state of polarization of its parts similar in distribution to the polarity of the parts of a mass of iron under the influence of a magnet, and like the magnetic polarization, capable of being described as a state in which every particle has its opposite poles in opposite conditions*.

In a dielectric under the action of electromotive force, we may conceive that the electricity in each molecule is so displaced that one side is rendered positively and the other negatively electrical, but that the electricity remains entirely connected with the molecule, and does not pass from one molecule to another. The effect of this action on the whole dielectric mass is to produce a general displacement of electricity in a certain direction. This displacement does not amount to a current, because when it has attained to a certain value it remains constant, but it is the commencement of a current, and its variations constitute currents in the positive or the negative direction according as the displacement is increasing or decreasing. In the interior of the dielectric there is no indication of electrification, because the electrification of the surface of any molecule is neutralized by the opposite electrification of the surface of the molecules in contact with it; but at the bounding surface of the dielectric, where the electrification is not neutralized, we find the phenomena which indicate positive or negative electrification.

The relation between the electromotive force and the amount of electric displacement it produces depends on the nature of the dielectric, the same electromotive force producing generally a greater electric displacement in solid dielectrics, such as glass or sulphur, than in air.

(12) Here, then, we perceive another effect of electromotive force, namely, electric displacement, which according to our theory is a kind of elastic yielding to the action of the force, similar to that which takes place in structures and machines owing to the want of perfect rigidity of the connexions.

(13) The practical investigation of the inductive capacity of dielectrics is rendered difficult on account of two disturbing phenomena. The first is the conductivity of the dielectric, which, though in many cases exceedingly small, is not altogether insensible. The second is the phenomenon called electric absorp-

---

* Faraday, *Experimental Researches*, Series XI.; Mossotti, *Mem. della Soc. Italiana* (Modena), Vol. XXIV. Part 2, p. 49.

tion *, in virtue of which, when the dielectric is exposed to electromotive force, the electric displacement gradually increases, and when the electromotive force is removed, the dielectric does not instantly return to its primitive state, but only discharges a portion of its electrification, and when left to itself gradually acquires electrification on its surface, as the interior gradually becomes depolarized. Almost all solid dielectrics exhibit this phenomenon, which gives rise to the residual charge in the Leyden jar, and to several phenomena of electric cables described by Mr F. Jenkin †.

(14) We have here two other kinds of yielding besides the yielding of the perfect dielectric, which we have compared to a perfectly elastic body. The yielding due to conductivity may be compared to that of a viscous fluid (that is to say, a fluid having great internal friction), or a soft solid on which the smallest force produces a permanent alteration of figure increasing with the time during which the force acts. The yielding due to electric absorption may be compared to that of a cellular elastic body containing a thick fluid in its cavities. Such a body, when subjected to pressure, is compressed by degrees on account of the gradual yielding of the thick fluid; and when the pressure is removed it does not at once recover its figure, because the elasticity of the substance of the body has gradually to overcome the tenacity of the fluid before it can regain complete equilibrium.

Several solid bodies in which no such structure as we have supposed can be found, seem to possess a mechanical property of this kind ‡; and it seems probable that the same substances, if dielectrics, may possess the analogous electrical property, and if magnetic, may have corresponding properties relating to the acquisition, retention, and loss of magnetic polarity.

(15) It appears therefore that certain phenomena in electricity and magnetism lead to the same conclusion as those of optics, namely, that there is an æthereal medium pervading all bodies, and modified only in degree by their presence; that the parts of this medium are capable of being set in motion by electric currents and magnets; that this motion is communicated from one

---

* Faraday, *Experimental Researches*, 1233—1250.

† *Reports of British Association*, 1859, p. 248; and *Report of Committee of Board of Trade on Submarine Cables*, pp. 136 & 464.

‡ As, for instance, the composition of glue, treacle, &c., of which small plastic figures are made, which after being distorted gradually recover their shape.

part of the medium to another by forces arising from the connexions of those parts; that under the action of these forces there is a certain yielding depending on the elasticity of these connexions; and that therefore energy in two different forms may exist in the medium, the one form being the actual energy of motion of its parts, and the other being the potential energy stored up in the connexions, in virtue of their elasticity.

(16) Thus, then, we are led to the conception of a complicated mechanism capable of a vast variety of motion, but at the same time so connected that the motion of one part depends, according to definite relations, on the motion of other parts, these motions being communicated by forces arising from the relative displacement of the connected parts, in virtue of their elasticity. Such a mechanism must be subject to the general laws of Dynamics, and we ought to be able to work out all the consequences of its motion, provided we know the form of the relation between the motions of the parts.

(17) We know that when an electric current is established in a conducting circuit, the neighbouring part of the field is characterized by certain magnetic properties, and that if two circuits are in the field, the magnetic properties of the field due to the two currents are combined. Thus each part of the field is in connexion with both currents, and the two currents are put in connexion with each other in virtue of their connexion with the magnetization of the field. The first result of this connexion that I propose to examine, is the induction of one current by another, and by the motion of conductors in the field.

The second result, which is deduced from this, is the mechanical action between conductors carrying currents. The phenomenon of the induction of currents has been deduced from their mechanical action by Helmholtz * and Thomson †. I have followed the reverse order, and deduced the mechanical action from the laws of induction. I have then described experimental methods of determining the quantities L, M, N, on which these phenomena depend.

(18) I then apply the phenomena of induction and attraction of currents to the exploration of the electromagnetic field, and the laying down systems of lines of magnetic force which indicate its magnetic properties. By exploring

---

* "Conservation of Force," *Physical Society of Berlin*, 1847; and Taylor's *Scientific Memoirs*, 1853, p. 114.

† *Reports of the British Association*, 1848; *Philosophical Magazine*, Dec. 1851.

the same field with a magnet, I shew the distribution of its equipotential magnetic surfaces, cutting the lines of force at right angles.

In order to bring these results within the power of symbolical calculation, I then express them in the form of the General Equations of the Electromagnetic Field. These equations express—

(A)   The relation between electric displacement, true conduction, and the total current, compounded of both.

(B)   The relation between the lines of magnetic force and the inductive coefficients of a circuit, as already deduced from the laws of induction.

(C)   The relation between the strength of a current and its magnetic effects, according to the electromagnetic system of measurement.

(D)   The value of the electromotive force in a body, as arising from the motion of the body in the field, the alteration of the field itself, and the variation of electric potential from one part of the field to another.

(E)   The relation between electric displacement, and the electromotive force which produces it.

(F)   The relation between an electric current, and the electromotive force which produces it.

(G)   The relation between the amount of free electricity at any point, and the electric displacements in the neighbourhood.

(H)   The relation between the increase or diminution of free electricity and the electric currents in the neighbourhood.

There are twenty of these equations in all, involving twenty variable quantities.

(19) I then express in terms of these quantities the intrinsic energy of the Electromagnetic Field as depending partly on its magnetic and partly on its electric polarization at every point.

From this I determine the mechanical force acting, 1st, on a moveable conductor carrying an electric current; 2ndly, on a magnetic pole; 3rdly, on an electrified body.

The last result, namely, the mechanical force acting on an electrified body, gives rise to an independent method of electrical measurement founded on its electrostatic effects.   The relation between the units employed in the two methods is shewn to depend on what I have called the "electric elasticity" of the medium, and to be a velocity, which has been experimentally determined by MM. Weber and Kohlrausch.

I then shew how to calculate the electrostatic capacity of a condenser, and the specific inductive capacity of a dielectric.

The case of a condenser composed of parallel layers of substances of different electric resistances and inductive capacities is next examined, and it is shewn that the phenomenon called electric absorption will generally occur, that is, the condenser, when suddenly discharged, will after a short time shew signs of a *residual* charge.

(20)   The general equations are next applied to the case of a magnetic disturbance propagated through a non-conducting field, and it is shewn that the only disturbances which can be so propagated are those which are transverse to the direction of propagation, and that the velocity of propagation is the velocity v, found from experiments such as those of Weber, which expresses the number of electrostatic units of electricity which are contained in one electro-magnetic unit.

This velocity is so nearly that of light, that it seems we have strong reason to conclude that light itself (including radiant heat, and other radiations if any) is an electromagnetic disturbance in the form of waves propagated through the electromagnetic field according to electromagnetic laws.   If so, the agreement between the elasticity of the medium as calculated from the rapid alternations of luminous vibrations, and as found by the slow processes of electrical experiments, shews how perfect and regular the elastic properties of the medium must be when not encumbered with any matter denser than air.   If the same character of the elasticity is retained in dense transparent bodies, it appears that the square of the index of refraction is equal to the product of the specific dielectric capacity and the specific magnetic capacity.   Conducting media are shewn to absorb such radiations rapidly, and therefore to be generally opaque.

The conception of the propagation of transverse magnetic disturbances to the exclusion of normal ones is distinctly set forth by Professor Faraday[*] in his "Thoughts on Ray Vibrations."   The electromagnetic theory of light, as

---

[*] *Philosophical Magazine*, May 1846, or *Experimental Researches*, III. p. 447.

proposed by him, is the same in substance as that which I have begun to develope in this paper, except that in 1846 there were no data to calculate the velocity of propagation.

(21) The general equations are then applied to the calculation of the coefficients of mutual induction of two circular currents and the coefficient of self-induction in a coil. The want of uniformity of the current in the different parts of the section of a wire at the commencement of the current is investigated, I believe for the first time, and the consequent correction of the coefficient of self-induction is found.

These results are applied to the calculation of the self-induction of the coil used in the experiments of the Committee of the British Association on Standards of Electric Resistance, and the value compared with that deduced from the experiments.

PART II.

ON ELECTROMAGNETIC INDUCTION.

Electromagnetic Momentum of a Current.

(22) We may begin by considering the state of the field in the neighbourhood of an electric current. We know that magnetic forces are excited in the field, their direction and magnitude depending according to known laws upon the form of the conductor carrying the current. When the strength of the current is increased, all the magnetic effects are increased in the same proportion. Now, if the magnetic state of the field depends on motions of the medium, a certain force must be exerted in order to increase or diminish these motions, and when the motions are excited they continue, so that the effect of the connexion between the current and the electromagnetic field surrounding it, is to endow the current with a kind of momentum, just as the connexion between the driving-point of a machine and a fly-wheel endows the driving-point with an additional momentum, which may be called the momentum of the fly-wheel reduced to the driving-point. The unbalanced force acting on the driving-point increases this momentum, and is measured by the rate of its increase.

In the case of electric currents, the resistance to sudden increase or diminution of strength produces effects exactly like those of momentum, but the amount of this momentum depends on the shape of the conductor and the relative position of its different parts.

Mutual Action of two Currents.

(23) If there are two electric currents in the field, the magnetic force at any point is that compounded of the forces due to each current separately, and since the two currents are in connexion with every point of the field, they will be in connexion with each other, so that any increase or diminution of the one will produce a force acting with or contrary to the other.

Dynamical Illustration of Reduced Momentum.

(24) As a dynamical illustration, let us suppose a body C so connected with two independent driving-points A and B that its velocity is p times that of A together with q times that of B. Let u be the velocity of A, v that of B, and w that of C, and let δx, δy, δz be their simultaneous displacements, then by the general equation of dynamics*,

$$C \frac{dw}{dt} \delta z = X \delta x + Y \delta y,$$

where X and Y are the forces acting at A and B.

But
$$\frac{dw}{dt} = p \frac{du}{dt} + q \frac{dv}{dt},$$

and
$$\delta z = p \delta x + q \delta y.$$

Substituting, and remembering that δx and δy are independent,

$$\left. \begin{array}{l} X = \dfrac{d}{dt} \left(Cp^2 u + Cpqv \right) \\ Y = \dfrac{d}{dt} \left(Cpqu + Cq^2 v \right) \end{array} \right\} \dots\dots\dots\dots\dots\dots\dots (1).$$

We may call $Cp^2 u + Cpqv$ the momentum of C referred to A, and $Cpqu + Cq^2 v$ its momentum referred to B; then we may say that the effect of the force X is to increase the momentum of C referred to A, and that of Y to increase its momentum referred to B.

* Lagrange, *Méc. Anal.* II. 2, § 5.

If there are many bodies connected with A and B in a similar way but with different values of p and q, we may treat the question in the same way by assuming

$$L = \Sigma\,(Cp^2), \quad M = \Sigma\,(Cpq), \quad \text{and} \quad N = \Sigma\,(Cq^2),$$

where the summation is extended to all the bodies with their proper values of C, p, and q. Then the momentum of the system referred to A is

$$Lu + Mv,$$

and referred to B,

$$Mu + Nv,$$

and we shall have

$$\left.\begin{aligned} X &= \frac{d}{dt}\,(Lu + Mv) \\[2mm] Y &= \frac{d}{dt}\,(Mu + Nv) \end{aligned}\right\} \quad \dots\dots\dots\dots\dots\dots (2),$$

where X and Y are the external forces acting on A and B.

(25) To make the illustration more complete we have only to suppose that the motion of A is resisted by a force proportional to its velocity, which we may call Ru, and that of B by a similar force, which we may call Sv, R and S being coefficients of resistance. Then if ξ and η are the forces on A and B,

$$\left.\begin{aligned} \xi &= X + Ru = Ru + \frac{d}{dt}\,(Lu + Mv) \\[2mm] \eta &= Y + Sv = Sv + \frac{d}{dt}\,(Mu + Nv) \end{aligned}\right\} \quad \dots\dots\dots\dots\dots (3).$$

If the velocity of A be increased at the rate $\dfrac{du}{dt}$, then in order to prevent B from moving a force, $\eta = \dfrac{d}{dt}\,(Mu)$ must be applied to it.

This effect on B, due to an increase of the velocity of A, corresponds to the electromotive force on one circuit arising from an increase in the strength of a neighbouring circuit.

This dynamical illustration is to be considered merely as assisting the reader to understand what is meant in mechanics by Reduced Momentum. The facts of the induction of currents as depending on the variations of the quantity called Electromagnetic Momentum, or Electrotonic State, rest on the experiments of Faraday[*], Felici[†], &c.

* *Experimental Researches*, Series i., ix.   † *Annales de Chimie*, sér. 3, xxxiv. (1852), p. 64.

Coefficients of Induction for Two Circuits.

(26) In the electromagnetic field the values of L, M, N depend on the distribution of the magnetic effects due to the two circuits, and this distribution depends only on the form and relative position of the circuits. Hence L, M, N are quantities depending on the form and relative position of the circuits, and are subject to variation with the motion of the conductors. It will be presently seen that L, M, N are geometrical quantities of the nature of lines, that is, of one dimension in space; L depends on the form of the first conductor, which we shall call A, N on that of the second, which we shall call B, and M on the relative position of A and B.

(27) Let ξ be the electromotive force acting on A, x the strength of the current, and R the resistance, then Rx will be the resisting force. In steady currents the electromotive force just balances the resisting force, but in variable currents the resultant force $\xi - Rx$ is expended in increasing the "electromagnetic momentum," using the word momentum merely to express that which is generated by a force acting during a time, that is, a velocity existing in a body.

In the case of electric currents, the force in action is not ordinary mechanical force, at least we are not as yet able to measure it as common force, but we call it electromotive force, and the body moved is not merely the electricity in the conductor, but something outside the conductor, and capable of being affected by other conductors in the neighbourhood carrying currents. In this it resembles rather the reduced momentum of a driving-point of a machine as influenced by its mechanical connexions, than that of a simple moving body like a cannon ball, or water in a tube.

Electromagnetic Relations of two Conducting Circuits.

(28) In the case of two conducting circuits, A and B, we shall assume that the electromagnetic momentum belonging to A is

$$Lx + My,$$

and that belonging to B, $\qquad Mx + Ny,$

where L, M, N correspond to the same quantities in the dynamical illustration, except that they are supposed to be capable of variation when the conductors A or B are moved.

68—2

Then the equation of the current x in A will be

$$\xi = Rx + \frac{d}{dt}(Lx + My) \dots\dots\dots\dots\dots (4),$$

and that of y in B

$$\eta = Sy + \frac{d}{dt}(Mx + Ny) \dots\dots\dots\dots\dots (5),$$

where ξ and η are the electromotive forces, x and y the currents, and R and S the resistances in A and B respectively.

Induction of one Current by another.

(29) Case 1st. Let there be no electromotive force on B, except that which arises from the action of A, and let the current of A increase from 0 to the value x, then

$$Sy + \frac{d}{dt}(Mx + Ny) = 0,$$

whence

$$Y = \int_0^t y\,dt = -\frac{M}{S}x, \dots\dots\dots\dots\dots (6)$$

that is, a quantity of electricity Y, being the total induced current, will flow through B when x rises from 0 to x. This is induction by variation of the current in the primary conductor. When M is positive, the induced current due to increase of the primary current is negative.

Induction by Motion of Conductor.

(30) Case 2nd. Let x remain constant, and let M change from M to M', then

$$Y = -\frac{M' - M}{S}x; \dots\dots\dots\dots\dots (7)$$

so that if M is increased, which it will be by the primary and secondary circuits approaching each other, there will be a negative induced current, the total quantity of electricity passed through B being Y.

This is induction by the relative motion of the primary and secondary conductors.

Equation of Work and Energy.

(31) To form the equation between work done and energy produced, multiply (1) by x and (2) by y, and add

$$\xi x + \eta y = Rx^2 + Sy^2 + x \frac{d}{dt}(Lx + My) + y \frac{d}{dt}(Mx + Ny) \ldots\ldots\ldots(8).$$

Here ξx is the work done in unit of time by the electromotive force ξ acting on the current x and maintaining it, and ηy is the work done by the electromotive force η. Hence the left-hand side of the equation represents the work done by the electromotive forces in unit of time.

Heat produced by the Current.

(32) On the other side of the equation we have, first,

$$Rx^2 + Sy^2 = H \ldots\ldots\ldots\ldots\ldots\ldots\ldots\ldots(9),$$

which represents the work done in overcoming the resistance of the circuits in unit of time. This is converted into heat. The remaining terms represent work not converted into heat. They may be written

$$\tfrac{1}{2} \frac{d}{dt}(Lx^2 + 2Mxy + Ny^2) + \tfrac{1}{2} \frac{dL}{dt} x^2 + \frac{dM}{dt} xy + \tfrac{1}{2} \frac{dN}{dt} y^2.$$

Intrinsic Energy of the Currents.

(33) If L, M, N are constant, the whole work of the electromotive forces which is not spent against resistance will be devoted to the development of the currents. The whole intrinsic energy of the currents is therefore

$$\tfrac{1}{2}Lx^2 + Mxy + \tfrac{1}{2}Ny^2 = E \ldots\ldots\ldots\ldots\ldots\ldots (10).$$

This energy exists in a form imperceptible to our senses, probably as actual motion, the seat of this motion being not merely the conducting circuits, but the space surrounding them.

Mechanical Action between Conductors.

(34) The remaining terms,

$$\tfrac{1}{2} \frac{dL}{dt} x^2 + \frac{dM}{dt} xy + \tfrac{1}{2} \frac{dN}{dt} y^2 = W \ldots\ldots\ldots\ldots (11),$$

represent the work done in unit of time arising from the variations of L, M, and N, or, what is the same thing, alterations in the form and position of the conducting circuits A and B.

Now if work is done when a body is moved, it must arise from ordinary mechanical force acting on the body while it is moved. Hence this part of the expression shews that there is a mechanical force urging every part of the conductors themselves in that direction in which L, M, and N will be most increased.

The existence of the electromagnetic force between conductors carrying currents is therefore a direct consequence of the joint and independent action of each current on the electromagnetic field. If A and B are allowed to approach a distance ds, so as to increase M from M to M' while the currents are x and y, then the work done will be

$$(M' - M)\, xy,$$

and the force in the direction of ds will be

$$\frac{dM}{ds}\, xy \dots\dots\dots(12),$$

and this will be an attraction if x and y are of the same sign, and if M is increased as A and B approach.

It appears, therefore, that if we admit that the unresisted part of electromotive force goes on as long as it acts, generating a self-persistent state of the current, which we may call (from mechanical analogy) its electromagnetic momentum, and that this momentum depends on circumstances external to the conductor, then both induction of currents and electromagnetic attractions may be proved by mechanical reasoning.

What I have called electromagnetic momentum is the same quantity which is called by Faraday * the electrotonic state of the circuit, every change of which involves the action of an electromotive force, just as change of momentum involves the action of mechanical force.

If, therefore, the phenomena described by Faraday in the Ninth Series of his *Experimental Researches* were the only known facts about electric currents, the laws of Ampère relating to the attraction of conductors carrying currents,

* *Experimental Researches*, Series i. 60, &c.

as well as those of Faraday about the mutual induction of currents, might be deduced by mechanical reasoning.

In order to bring these results within the range of experimental verification, I shall next investigate the case of a single current, of two currents, and of the six currents in the electric balance, so as to enable the experimenter to determine the values of L, M, N.

Case of a single Circuit.

(35) The equation of the current x in a circuit whose resistance is R, and whose coefficient of self-induction is L, acted on by an external electromotive force ξ, is

$$\xi - Rx = \frac{d}{dt} Lx \quad\dots\dots\dots\dots\dots\dots\dots\dots (13).$$

When ξ is constant, the solution is of the form

$$x = b + (a - b)\, e^{-\frac{R}{L}t},$$

where a is the value of the current at the commencement, and b is its final value.

The total quantity of electricity which passes in time t, where t is great, is

$$\int_0^t x\, dt = bt + (a - b)\frac{L}{R} \quad\dots\dots\dots\dots\dots\dots (14).$$

The value of the integral of x^2 with respect to the time is

$$\int_0^t x^2 dt = b^2 t + (a - b)\frac{L}{R}\left(\frac{3b + a}{2}\right) \quad\dots\dots\dots\dots (15).$$

The actual current changes gradually from the initial value a to the final value b, but the values of the integrals of x and x^2 are the same as if a steady current of intensity $\frac{1}{2}(a + b)$ were to flow for a time $2\frac{L}{R}$, and were then succeeded by the steady current b. The time $2\frac{L}{R}$ is generally so minute a fraction of a second, that the effects on the galvanometer and dynamometer may be calculated as if the impulse were instantaneous.

If the circuit consists of a battery and a coil, then, when the circuit is first completed, the effects are the same as if the current had only half its final strength during the time $2\dfrac{L}{R}$. This diminution of the current, due to induction, is sometimes called the counter-current.

(36) If an additional resistance r is suddenly thrown into the circuit, as by breaking contact, so as to force the current to pass through a thin wire of resistance r, then the original current is $a=\dfrac{\xi}{R}$, and the final current is $b=\dfrac{\xi}{R+r}$.

The current of induction is then $\tfrac{1}{2}\xi\dfrac{2R+r}{R(R+r)}$, and continues for a time $2\dfrac{L}{R+r}$. This current is greater than that which the battery can maintain in the two wires R and r, and may be sufficient to ignite the thin wire r.

When contact is broken by separating the wires in air, this additional resistance is given by the interposed air, and since the electromotive force across the new resistance is very great, a spark will be forced across.

If the electromotive force is of the form $E\sin pt$, as in the case of a coil revolving in the magnetic field, then

$$x=\frac{E}{\rho}\sin{(pt-a)},$$

where $\rho^{2}=R^{2}+L^{2}p^{2}$, and $\tan a=\dfrac{Lp}{R}$.

Case of two Circuits.

(37) Let R be the primary circuit and S the secondary circuit, then· we have a case similar to that of the induction coil.

The equations of currents are those marked A and B, and we may here assume L, M, N as constant because there is no motion of the conductors. The equations then become

$$\left.\begin{aligned} Rx+L\,\frac{dx}{dt}+M\,\frac{dy}{dt}&=\xi\\ Sy+M\,\frac{dx}{dt}+N\,\frac{dy}{dt}&=0 \end{aligned}\right\}\dotsb(13^{*}).$$

To find the total quantity of electricity which passes, we have only to integrate these equations with respect to t; then if x_0, y_0 be the strengths of the currents at time 0, and x_1, y_1 at time t, and if X, Y be the quantities of electricity passed through each circuit during time t,

$$\left.\begin{array}{l} X = \dfrac{1}{R}\left\{\xi t + L\left(x_0 - x_1\right) + M\left(y_0 - y_1\right)\right\} \\[2mm] Y = \dfrac{1}{S}\left\{M\left(x_0 - x_1\right) + N\left(y_0 - y_1\right)\right\} \end{array}\right\} \ \dots\dots\dots\dots (14^*).$$

When the circuit R is completed, then the total currents up to time t, when t is great, are found by making

$$x_0 = 0, \quad x_1 = \frac{\xi}{R}, \quad y_0 = 0, \quad y_1 = 0 ;$$

then
$$X = x_1\left(t - \frac{L}{R}\right), \quad Y = -\frac{M}{S} x_1 \dots\dots\dots\dots (15^*).$$

The value of the total counter-current in R is therefore independent of the secondary circuit, and the induction current in the secondary circuit depends only on M, the coefficient of induction between the coils, S the resistance of the secondary coil, and x_1 the final strength of the current in R.

When the electromotive force ξ ceases to act, there is an extra current in the primary circuit, and a positive induced current in the secondary circuit, whose values are equal and opposite to those produced on making contact.

(38)   All questions relating to the total quantity of transient currents, as measured by the impulse given to the magnet of the galvanometer, may be solved in this way without the necessity of a complete solution of the equations. The heating effect of the current, and the impulse it gives to the suspended coil of Weber's dynamometer, depend on the square of the current at every instant during the short time it lasts. Hence we must obtain the solution of the equations, and from the solution we may find the effects both on the galvanometer and dynamometer; and we may then make use of the method of Weber for estimating the intensity and duration of a current uniform while it lasts which would produce the same effects.

(39)   Let n_1, n_2 be the roots of the equation

$$(LN-M^2)\,n^2+(RN+LS)\,n+RS=0 \ldots\ldots\ldots\ldots\ldots(16),$$

and let the primary coil be acted on by a constant electromotive force Rc, so that c is the constant current it could maintain; then the complete solution of the equations for making contact is

$$x=\frac{c}{S}\,\frac{n_1 n_2}{n_1-n_2}\left\{\left(\frac{S}{n_1}+N\right)e^{n_1 t}-\left(\frac{S}{n_2}+N\right)e^{n_2 t}+S\frac{n_1-n_2}{n_1 n_2}\right\}\ldots\ldots\ldots(17),$$

$$y=\frac{cM}{S}\,\frac{n_1 n_2}{n_1-n_2}\left\{e^{n_1 t}-e^{n_2 t}\right\}\ldots\ldots\ldots\ldots\ldots\ldots\ldots\ldots\ldots\ldots (18).$$

From these we obtain for calculating the impulse on the dynamometer,

$$\int x^2 dt=c^2\left\{t-\tfrac{3}{2}\frac{L}{R}-\tfrac{1}{2}\frac{M^2}{RN+LS}\right\}\ldots\ldots\ldots\ldots (19),$$

$$\int y^2 dt=c^2\tfrac{1}{2}\frac{M^2 R}{S(RN+LS)}\ldots\ldots\ldots\ldots\ldots\ldots\ldots\ldots (20).$$

The effects of the current in the secondary coil on the galvanometer and dynamometer are the same as those of a uniform current

$$-\tfrac{1}{2}c\,\frac{MR}{RN+LS}$$

for a time

$$2\left(\frac{L}{R}+\frac{N}{S}\right).$$

(40)   The equation between work and energy may be easily verified. The work done by the electromotive force is

$$\xi\int x\,dt=c^2\,(Rt-L).$$

Work done in overcoming resistance and producing heat,

$$R\int x^2 dt+S\int y^2 dt=c^2\,(Rt-\tfrac{3}{2}L).$$

Energy remaining in the system, $=\tfrac{1}{2}c^2 L.$

(41)   If the circuit R is suddenly and completely interrupted while carrying a current c, then the equation of the current in the secondary coil would be

$$y=c\,\frac{M}{N}\,e^{-\frac{S}{N}t}.$$

This current begins with a value $c\dfrac{M}{N}$, and gradually disappears.

The total quantity of electricity is $c\dfrac{M}{S}$, and the value of $\int y^2 dt$ is $c^2\dfrac{M^2}{2SN}$.

The effects on the galvanometer and dynamometer are equal to those of a uniform current $\frac{1}{2}c\dfrac{M}{N}$ for a time $2\dfrac{N}{S}$.

The heating effect is therefore greater than that of the current on making contact.

(42) If an electromotive force of the form $\xi = E\cos pt$ acts on the circuit R, then if the circuit S is removed, the value of x will be

$$x = \frac{E}{A}\sin\left(pt - a\right),$$

where

$$A^2 = R^2 + L^2 p^2,$$

and

$$\tan a = \frac{Lp}{R}.$$

The effect of the presence of the circuit S in the neighbourhood is to alter the value of A and a, to that which they would be if R became

$$R + p^2\frac{MS}{S^2 + p^2 N^2},$$

and L became

$$L - p^2\frac{MN}{S^2 + p^2 N^2}.$$

Hence the effect of the presence of the circuit S is to increase the apparent resistance and diminish the apparent self-induction of the circuit R.

On the Determination of Coefficients of Induction by the Electric Balance.

(43) The electric balance consists of six conductors joining four points, A, C, D, E, two and two. One pair, AC, of these points is connected through the battery B. The opposite pair, DE, is connected through the galvanometer G. Then if the resistances of the four remaining conductors are represented by P, Q, R, S, and the currents in them by x, $x-z$, y, and $y+z$,

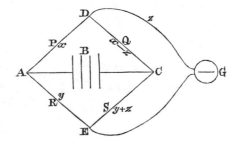

the current through G will be z. Let the potentials at the four points be A, C, D, E. Then the conditions of steady currents may be found from the equations

$$\left.\begin{aligned} Px &= A - D, & Q\,(x-z) &= D - C \\ Ry &= A - E, & S\,(y+z) &= E - C \\ Gz &= D - E, & B\,(x+y) &= -A + C + F \end{aligned}\right\} \dots\dots\dots\dots (21).$$

Solving these equations for z, we find

$$z\left\{\frac{1}{P}+\frac{1}{Q}+\frac{1}{R}+\frac{1}{S}+B\left(\frac{1}{P}+\frac{1}{R}\right)\left(\frac{1}{Q}+\frac{1}{S}\right)+G\left(\frac{1}{P}+\frac{1}{Q}\right)\left(\frac{1}{R}+\frac{1}{S}\right)\right.$$
$$\left.+\frac{BG}{PQRS}\,(P+Q+R+S)\right\}=F\left(\frac{1}{PS}-\frac{1}{QR}\right)\dots\dots(22).$$

In this expression F is the electromotive force of the battery, z the current through the galvanometer when it has become steady. P, Q, R, S the resistances in the four arms. B that of the battery and electrodes, and G that of the galvanometer.

(44) If $PS = QR$, then $z = 0$, and there will be no steady current, but a transient current through the galvanometer may be produced on making or breaking circuit on account of induction, and the indications of the galvanometer may be used to determine the coefficients of induction, provided we understand the actions which take place.

We shall suppose $PS = QR$, so that the current z vanishes when sufficient time is allowed, and

$$x\,(P+Q) = y\,(R+S) = \frac{F\,(P+Q)\,(R+S)}{(P+Q)\,(R+S)+B\,(P+Q)\,(R+S)} \dots\dots (23).$$

Let the induction coefficients between P, Q, R, S be given by the following Table, the coefficient of induction of P on itself being p, between P and Q, h, and so on.

Let g be the coefficient of induction of the galvanometer on itself, and let it be out of the reach of the inductive influence of P, Q, R, S (as it must be in order to avoid direct action of P, Q, R, S on the needle) Let X, Y, Z be the integrals of x, y, z with respect to t. At

	P	Q	R	S
P	p	h	k	l
Q	h	q	m	n
R	k	m	r	o
S	l	n	o	s

making contact x, y, z are zero. After a time z disappears, and x and y reach constant values. The equations for each conductor will therefore be

$$\left.\begin{aligned}
PX &+ (p+h)\,x + (k+l)\,y = \int A\,dt - \int D\,dt \\
Q(X-Z) &+ (h+q)\,x + (m+n)\,y = \int D\,dt - \int C\,dt \\
RY &+ (k+m)\,x + (r+o)\,y = \int A\,dt - \int E\,dt \\
S(Y+Z) &+ (l+n)\,x + (o+s)\,y = \int E\,dt - \int C\,dt \\
GZ &= \int D\,dt - \int E\,dt.
\end{aligned}\right\} \quad \ldots\ldots\ldots\ldots (24).$$

Solving these equations for Z, we find

$$\left.\begin{aligned}
Z\left\{ \frac{1}{P} + \frac{1}{Q} + \frac{1}{R} + \frac{1}{S} + B\left(\frac{1}{P}+\frac{1}{R}\right)\left(\frac{1}{Q}+\frac{1}{S}\right) + G\left(\frac{1}{P}+\frac{1}{Q}\right)\left(\frac{1}{R}+\frac{1}{S}\right) \right. \\
\left. + \frac{BG}{PQRS}(P+Q+R+S) \right\} \\
= -F\frac{1}{PS}\left\{ \frac{p}{P} - \frac{q}{Q} - \frac{r}{R} + \frac{s}{S} + h\left(\frac{1}{P}-\frac{1}{Q}\right) + k\left(\frac{1}{R}-\frac{1}{P}\right) + l\left(\frac{1}{R}+\frac{1}{Q}\right) \right. \\
\left. - m\left(\frac{1}{P}+\frac{1}{S}\right) + n\left(\frac{1}{Q}-\frac{1}{S}\right) + o\left(\frac{1}{S}-\frac{1}{R}\right) \right\}
\end{aligned}\right\} \ldots(25).$$

(45) Now let the deflection of the galvanometer by the instantaneous current whose intensity is Z be a.

*Let the permanent deflection produced by making the ratio of PS to QR, ρ instead of unity, be θ.

Also let the time of vibration of the galvanometer needle from rest to rest be T.

Then calling the quantity

$$\frac{p}{P} - \frac{q}{Q} - \frac{r}{R} + \frac{s}{S} + h\left(\frac{1}{P}-\frac{1}{Q}\right) + k\left(\frac{1}{R}-\frac{1}{P}\right) + l\left(\frac{1}{R}+\frac{1}{Q}\right) - m\left(\frac{1}{P}+\frac{1}{S}\right)$$
$$+ n\left(\frac{1}{Q}-\frac{1}{S}\right) + o\left(\frac{1}{S}-\frac{1}{R}\right) = \tau \ldots\ldots (26),$$

we find

$$\frac{Z}{z} = \frac{2\sin\frac{1}{2}a}{\tan\theta}\frac{T}{\pi} = \frac{\tau}{1-\rho} \quad \ldots\ldots\ldots\ldots\ldots\ldots (27).$$

* [In those circumstances the values of x and y found in Art. 44 require modification before being inserted in equation (24). This has been pointed out by Lord Rayleigh, who employed the method described in the text in his second determination of the British unit of resistance in absolute measure. See the *Philosophical Transactions*, 1882, Part II. pp. 677, 678.]

In determining τ by experiment, it is best to make the alteration of resistance in one of the arms by means of the arrangement described by Mr Jenkin in the Report of the British Association for 1863, by which any value of ρ from 1 to 1·01 can be accurately measured.

We observe (a) the greatest deflection due to the impulse of induction when the galvanometer is in circuit, when the connexions are made, and when the resistances are so adjusted as to give no permanent current.

We then observe (β) the greatest deflection produced by the permanent current when the resistance of one of the arms is increased in the ratio of 1 to ρ, the galvanometer not being in circuit till a little while after the connexion is made with the battery.

In order to eliminate the effects of resistance of the air, it is best to vary ρ till $\beta = 2a$ nearly; then

$$\tau = T \frac{1}{\pi} (1 - \rho) \frac{2 \sin \frac{1}{2} a}{\tan \frac{1}{2} \beta} \dots\dots\dots\dots\dots\dots\dots (28).$$

If all the arms of the balance except P consist of resistance coils of very fine wire of no great length and doubled before being coiled, the induction coefficients belonging to these coils will be insensible, and τ will be reduced to $\frac{p}{P}$. The electric balance therefore affords the means of measuring the self-induction of any circuit whose resistance is known.

(46) It may also be used to determine the coefficient of induction between two circuits, as for instance, that between P and S which we have called m; but it would be more convenient to measure this by directly measuring the current, as in (37), without using the balance. We may also ascertain the equality of $\frac{p}{P}$ and $\frac{q}{Q}$ by there being no current of induction, and thus, when we know the value of p, we may determine that of q by a more perfect method than the comparison of deflections.

Exploration of the Electromagnetic Field.

(47) Let us now suppose the primary circuit A to be of invariable form, and let us explore the electromagnetic field by means of the secondary circuit B, which we shall suppose to be variable in form and position.

We may begin by supposing B to consist of a short straight conductor with its extremities sliding on two parallel conducting rails, which are put in connexion at some distance from the sliding-piece.

Then, if sliding the moveable conductor in a given direction increases the value of M, a negative electromotive force will act in the circuit B, tending to produce a negative current in B during the motion of the sliding-piece.

If a current be kept up in the circuit B, then the sliding-piece will itself tend to move in that direction, which causes M to increase. At every point of the field there will always be a certain direction such that a conductor moved in that direction does not experience any electromotive force in whatever direction its extremities are turned. A conductor carrying a current will experience no mechanical force urging it in that direction or the opposite.

This direction is called the direction of the line of magnetic force through that point.

Motion of a conductor across such a line produces electromotive force in a direction perpendicular to the line and to the direction of motion, and a conductor carrying a current is urged in a direction perpendicular to the line and to the direction of the current.

(48) We may next suppose B to consist of a very small plane circuit capable of being placed in any position and of having its plane turned in any direction. The value of M will be greatest when the plane of the circuit is perpendicular to the line of magnetic force. Hence if a current is maintained in B it will tend to set itself in this position, and will of itself indicate, like a magnet, the direction of the magnetic force.

On Lines of Magnetic Force.

(49) Let any surface be drawn, cutting the lines of magnetic force, and on this surface let any system of lines be drawn at small intervals, so as to lie side by side without cutting each other. Next, let any line be drawn on the surface cutting all these lines, and let a second line be drawn near it, its distance from the first being such that the value of M for each of the small spaces enclosed between these two lines and the lines of the first system is equal to unity.

In this way let more lines be drawn so as to form a second system, so that the value of M for every reticulation formed by the intersection of the two systems of lines is unity.

Finally, from every point of intersection of these reticulations let a line be drawn through the field, always coinciding in direction with the direction of magnetic force.

(50) In this way the whole field will be filled with lines of magnetic force at regular intervals, and the properties of the electromagnetic field will be completely expressed by them.

For, 1st, If any closed curve be drawn in the field, the value of M for that curve will be expressed by the *number* of lines of force which *pass through* that closed curve.

2ndly. If this curve be a conducting circuit and be moved through the field, an electromotive force will act in it, represented by the rate of decrease of the number of lines passing through the curve.

3rdly. If a current be maintained in the circuit, the conductor will be acted on by forces tending to move it so as to increase the number of lines passing through it, and the amount of work done by these forces is equal to the current in the circuit multiplied by the number of additional lines.

4thly. If a small plane circuit be placed in the field, and be free to turn, it will place its plane perpendicular to the lines of force. A small magnet will place itself with its axis in the direction of the lines of force.

5thly. If a long uniformly magnetized bar is placed in the field, each pole will be acted on by a force in the direction of the lines of force. The number of lines of force passing through unit of area is equal to the force acting on a unit pole multiplied by a coefficient depending on the magnetic nature of the medium, and called the coefficient of magnetic induction.

In fluids and isotropoic solids the value of this coefficient μ is the same in whatever direction the lines of force pass through the substance, but in crystallized, strained, and organized solids the value of μ may depend on the direction of the lines of force with respect to the axes of crystallization, strain, or growth.

In all bodies μ is affected by temperature, and in iron it appears to diminish as the intensity of the magnetization increases.

On Magnetic Equipotential Surfaces.

(51) If we explore the field with a uniformly magnetized bar, so long that one of its poles is in a very weak part of the magnetic field, then the magnetic forces will perform work on the other pole as it moves about the field.

If we start from a given point, and move this pole from it to any other point, the work performed will be independent of the path of the pole between the two points; provided that no electric current passes between the different paths pursued by the pole.

Hence, when there are no electric currents but only magnets in the field, we may draw a series of surfaces such that the work done in passing from one to another shall be constant whatever be the path pursued between them. Such surfaces are called Equipotential Surfaces, and in ordinary cases are perpendicular to the Lines of magnetic force.

If these surfaces are so drawn that, when a unit pole passes from any one to the next in order, unity of work is done, then the work done in any motion of a magnetic pole will be measured by the strength of the pole multiplied by the number of surfaces which it has passed through in the positive direction.

(52) If there are circuits carrying electric currents in the field, then there will still be equipotential surfaces in the parts of the field external to the conductors carrying the currents, but the work done on a unit pole in passing from one to another will depend on the number of times which the path of the pole circulates round any of these currents. Hence the potential in each surface will have a series of values in arithmetical progression, differing by the work done in passing completely round one of the currents in the field.

The equipotential surfaces will not be continuous closed surfaces, but some of them will be limited sheets, terminating in the electric circuit as their common edge or boundary. The number of these will be equal to the amount of work done on a unit pole in going round the current, and this by the ordinary measurement $= 4\pi\gamma$, where γ is the value of the current.

These surfaces, therefore, are connected with the electric current as soap-bubbles are connected with a ring in M. Plateau's experiments. Every current γ has $4\pi\gamma$ surfaces attached to it. These surfaces have the current for their common edge, and meet it at equal angles. The form of the surfaces in other parts depends on the presence of other currents and magnets, as well as on the shape of the circuit to which they belong.

PART III.

GENERAL EQUATIONS OF THE ELECTROMAGNETIC FIELD.

(53) Let us assume three rectangular directions in space as the axes of x, y, and z, and let all quantities having direction be expressed by their components in these three directions.

Electrical Currents (p, q, r).

(54) An electrical current consists in the transmission of electricity from one part of a body to another. Let the quantity of electricity transmitted in unit of time across unit of area perpendicular to the axis of x be called p, then p is the component of the current at that place in the direction of x.

We shall use the letters p, q, r to denote the components of the current per unit of area in the directions of x, y, z.

Electrical Displacements (f, g, h).

(55) Electrical displacement consists in the opposite electrification of the sides of a molecule or particle of a body which may or may not be accompanied with transmission through the body. Let the quantity of electricity which would appear on the faces $dy \cdot dz$ of an element dx, dy, dz cut from the body be $f \cdot dy \cdot dz$, then f is the component of electric displacement parallel to x. We shall use f, g, h to denote the electric displacements parallel to x, y, z respectively.

The variations of the electrical displacement must be added to the currents p, q, r to get the total motion of electricity, which we may call p', q', r', so that

$$\left.\begin{aligned} p' &= p + \frac{df}{dt} \\ q' &= q + \frac{dg}{dt} \\ r' &= r + \frac{dh}{dt} \end{aligned}\right\} \quad \dots\dots\dots\dots\dots\dots\dots\dots\dots\dots (A).$$

Electromotive Force (P, Q, R).

(56) Let P, Q, R represent the components of the electromotive force at any point. Then P represents the difference of potential per unit of length in a conductor placed in the direction of x at the given point. We may suppose an indefinitely short wire placed parallel to x at a given point and touched, during the action of the force P, by two small conductors, which are then insulated and removed from the influence of the electromotive force. The value of P might then be ascertained by measuring the charge of the conductors.

Thus if l be the length of the wire, the difference of potential at its ends will be Pl, and if C be the capacity of each of the small conductors the charge on each will be $\frac{1}{2}CPl$. Since the capacities of moderately large conductors, measured on the electromagnetic system, are exceedingly small, ordinary electromotive forces arising from electromagnetic actions could hardly be measured in this way. In practice such measurements are always made with long conductors, forming closed or nearly closed circuits.

Electromagnetic Momentum (F, G, H).

(57) Let F, G, H represent the components of electromagnetic momentum at any point of the field, due to any system of magnets or currents.

Then F is the total impulse of the electromotive force in the direction of x that would be generated by the removal of these magnets or currents from the field, that is, if P be the electromotive force at any instant during the removal of the system

$$F = \int P dt.$$

Hence the part of the electromotive force which depends on the motion of magnets or currents in the field, or their alteration of intensity, is

$$P = -\frac{dF}{dt}, \quad Q = -\frac{dG}{dt}, \quad R = -\frac{dH}{dt} \dots\dots\dots\dots\dots\dots (29).$$

Electromagnetic Momentum of a Circuit.

(58) Let s be the length of the circuit, then if we integrate

$$\int \left(F\frac{dx}{ds} + G\frac{dy}{ds} + H\frac{dz}{ds} \right) ds \dots\dots\dots\dots\dots\dots\dots (30)$$

70—2

round the circuit, we shall get the total electromagnetic momentum of the circuit, or the number of lines of magnetic force which pass through it, the variations of which measure the total electromotive force in the circuit. This electromagnetic momentum is the same thing to which Professor Faraday has applied the name of the Electrotonic State.

If the circuit be the boundary of the elementary area $dy\,dz$, then its electromagnetic momentum is

$$\left(\frac{dH}{dy} - \frac{dG}{dz}\right) dy\,dz,$$

and this is the number of lines of magnetic force which pass through the area $dy\,dz$.

Magnetic Force (a, β, γ).

(59) Let a, β, γ represent the force acting on a unit magnetic pole placed at the given point resolved in the directions of x, y, and z.

Coefficient of Magnetic Induction (μ).

(60) Let μ be the ratio of the magnetic induction in a given medium to that in air under an equal magnetizing force, then the number of lines of force in unit of area perpendicular to x will be μa (μ is a quantity depending on the nature of the medium, its temperature, the amount of magnetization already produced, and in crystalline bodies varying with the direction).

(61) Expressing the electric momentum of small circuits perpendicular to the three axes in this notation, we obtain the following

Equations of Magnetic Force,

$$\left.\begin{aligned}
\mu a &= \frac{dH}{dy} - \frac{dG}{dz} \\[2mm]
\mu \beta &= \frac{dF}{dz} - \frac{dH}{dx} \\[2mm]
\mu \gamma &= \frac{dG}{dx} - \frac{dF}{dy}
\end{aligned}\right\} \quad \dots\dots\dots\dots\dots\dots\dots\dots\text{(B)}.$$

Equations of Currents.

(62)   It is known from experiment that the motion of a magnetic pole in the electromagnetic field in a closed circuit cannot generate work unless the circuit which the pole describes passes round an electric current.   Hence, except in the space occupied by the electric currents,

$$a dx + \beta dy + \gamma dz = d\phi \ \dots\dots\dots\dots\dots\dots \ (31)$$

a complete differential of ϕ, the magnetic potential.

The quantity ϕ may be susceptible of an indefinite number of distinct values, according to the number of times that the exploring point passes round electric currents in its course, the difference between successive values of ϕ corresponding to a passage completely round a current of strength c being $4\pi c$.

Hence if there is no electric current,

$$\frac{d\gamma}{dy} - \frac{d\beta}{dz} = 0 ;$$

but if there is a current p',

$$\frac{d\gamma}{dy} - \frac{d\beta}{dz} = 4\pi p'$$

Similarly,

$$\frac{da}{dz} - \frac{d\gamma}{dx} = 4\pi q' \ \left.\right\} \ \dots\dots\dots\dots\dots\dots\dots \ (C).$$

$$\frac{d\beta}{dx} - \frac{da}{dy} = 4\pi r'$$

We may call these the Equations of Currents.

Electromotive Force in a Circuit.

(63)   Let ξ be the electromotive force acting round the circuit A, then

$$\xi = \int \left(P \frac{dx}{ds} + Q \frac{dy}{ds} + R \frac{dz}{ds} \right) ds \ \dots\dots\dots\dots\dots \ (32),$$

where ds is the element of length, and the integration is performed round the circuit.

Let the forces in the field be those due to the circuits A and B, then the electromagnetic momentum of A is

$$\int \left(F\frac{dx}{ds} + G\frac{dy}{ds} + H\frac{dz}{ds} \right) ds = Lu + Mv \ldots\ldots\ldots\ldots (33),$$

where u and v are the currents in A and B, and

$$\xi = -\frac{d}{dt}(Lu + Mv)\ldots\ldots\ldots\ldots\ldots\ldots (34).$$

Hence, if there is no motion of the circuit A,

$$\left.
\begin{aligned}
P &= -\frac{dF}{dt} - \frac{d\psi}{dx} \\[2mm]
Q &= -\frac{dG}{dt} - \frac{d\psi}{dy} \\[2mm]
R &= -\frac{dH}{dt} - \frac{d\psi}{dz}
\end{aligned}
\right\} \ldots\ldots\ldots\ldots\ldots\ldots (35),$$

where ψ is a function of x, y, z, and t, which is indeterminate as far as regards the solution of the above equations, because the terms depending on it will disappear on integrating round the circuit. The quantity ψ can always, however, be determined in any particular case when we know the actual conditions of the question. The physical interpretation of ψ is, that it represents the *electric potential* at each point of space.

Electromotive Force on a Moving Conductor.

(64) Let a short straight conductor of length a, parallel to the axis of x, move with a velocity whose components are $\dfrac{dx}{dt}$, $\dfrac{dy}{dt}$, $\dfrac{dz}{dt}$, and let its extremities slide along two parallel conductors with a velocity $\dfrac{ds}{dt}$. Let us find the alteration of the electromagnetic momentum of the circuit of which this arrangement forms a part.

In unit of time the moving conductor has travelled distances $\dfrac{dx}{dt}$, $\dfrac{dy}{dt}$, $\dfrac{dz}{dt}$ along the directions of the three axes, and at the same time the lengths of the parallel conductors included in the circuit have each been increased by $\dfrac{ds}{dt}$.

Hence the quantity

$$\int \left(F\frac{dx}{ds} + G\frac{dy}{ds} + H\frac{dz}{ds} \right) ds$$

will be increased by the following increments,

$$a\left(\frac{dF}{dx}\frac{dx}{dt} + \frac{dF}{dy}\frac{dy}{dt} + \frac{dF}{dz}\frac{dz}{dt} \right), \text{ due to motion of conductor,}$$

$$-a\frac{ds}{dt}\left(\frac{dF}{dx}\frac{dx}{ds} + \frac{dG}{dx}\frac{dy}{ds} + \frac{dH}{dx}\frac{dz}{ds} \right), \text{ due to lengthening of circuit.}$$

The total increment will therefore be

$$a\left(\frac{dF}{dy} - \frac{dG}{dx} \right)\frac{dy}{dt} - a\left(\frac{dH}{dx} - \frac{dF}{dz} \right)\frac{dz}{dt} ;$$

or, by the equations of Magnetic Force (8),

$$-a\left(\mu\gamma\frac{dy}{dt} - \mu\beta\frac{dz}{dt} \right).$$

If P is the electromotive force in the moving conductor parallel to x referred to unit of length, then the actual electromotive force is Pa; and since this is measured by the decrement of the electromagnetic momentum of the circuit, the electromotive force due to motion will be

$$P = \mu\gamma\frac{dy}{dt} - \mu\beta\frac{dz}{dt} \quad\ldots\ldots\ldots\ldots\ldots\ldots\ldots\ldots\ldots(36).$$

(65)   The complete equations of electromotive force on a moving conductor may now be written as follows:—

Equations of Electromotive Force.

$$\left.\begin{aligned}
P &= \mu\left(\gamma\frac{dy}{dt} - \beta\frac{dz}{dt} \right) - \frac{dF}{dt} - \frac{d\psi}{dx} \\
Q &= \mu\left(a\frac{dz}{dt} - \gamma\frac{dx}{dt} \right) - \frac{dG}{dt} - \frac{d\psi}{dy} \\
R &= \mu\left(\beta\frac{dx}{dt} - a\frac{dy}{dt} \right) - \frac{dH}{dt} - \frac{d\psi}{dz}
\end{aligned}\right\} \ldots\ldots\ldots\ldots\ldots\ldots(D).$$

The first term on the right-hand side of each equation represents the electromotive force arising from the motion of the conductor itself. This electromotive force is perpendicular to the direction of motion and to the lines of magnetic force ; and if a parallelogram be drawn whose sides represent in direction and magnitude the velocity of the conductor and the magnetic induction at that point of the field, then the area of the parallelogram will represent the electromotive force due to the motion of the conductor, and the direction of the force is perpendicular to the plane of the parallelogram.

The second term in each equation indicates the effect of changes in the position or strength of magnets or currents in the field.

The third term shews the effect of the electric potential ψ. It has no effect in causing a circulating current in a closed circuit. It indicates the existence of a force urging the electricity to or from certain definite points in the field.

Electric Elasticity.

(66) When an electromotive force acts on a dielectric, it puts every part of the dielectric into a polarized condition, in which its opposite sides are oppositely electrified. The amount of this electrification depends on the electromotive force and on the nature of the substance, and, in solids having a structure defined by axes, on the direction of the electromotive force with respect to these axes. In isotropic substances, if k is the ratio of the electromotive force to the electric displacement, we may write the

Equations of Electric Elasticity,

$$\left.\begin{array}{l} P = kf \\ Q = kg \\ R = kh \end{array}\right\} \dots\dots\dots\dots\dots\dots\dots\dots\dots (E).$$

Electric Resistance.

(67) When an electromotive force acts on a conductor it produces a current of electricity through it. This effect is additional to the electric displacement already considered. In solids of complex structure, the relation between the electromotive force and the current depends on their direction through the solid.

In isotropic substances, which alone we shall here consider, if ρ is the specific resistance referred to unit of volume, we may write the

Equations of Electric Resistance,

$$\left.\begin{aligned} P &= -\rho p \\ Q &= -\rho q \\ R &= -\rho r \end{aligned}\right\} \dots\dots\dots\dots\dots\dots\dots\dots\text{(F)}.$$

Electric Quantity.

(68) Let e represent the quantity of free positive electricity contained in unit of volume at any part of the field, then, since this arises from the electrification of the different parts of the field not neutralizing each other, we may write the

Equation of Free Electricity,

$$e + \frac{df}{dx} + \frac{dg}{dy} + \frac{dh}{dz} = 0 \dots\dots\dots\dots\dots\dots\text{(G)}.$$

(69) If the medium conducts electricity, then we shall have another condition, which may be called, as in hydrodynamics, the

Equation of Continuity,

$$\frac{de}{dt} + \frac{dp}{dx} + \frac{dq}{dy} + \frac{dr}{dz} = 0 \dots\dots\dots\dots\dots\dots\text{(H)}.$$

(70) In these equations of the electromagnetic field we have assumed twenty variable quantities, namely,

For Electromagnetic Momentum	F	G	H
„ Magnetic Intensity	a	β	γ
„ Electromotive Force	P	Q	R
„ Current due to true Conduction	p	q	r
„ Electric Displacement	f	g	h
„ Total Current (including variation of displacement)	p'	q'	r'
„ Quantity of Free Electricity	e		
„ Electric Potential	Ψ		

Between these twenty quantities we have found twenty equations, viz.

Three equations of Magnetic Force .... (B)

  „   Electric   Currents .... (C)

  „   Electromotive Force .... (D)

  „   Electric Elasticity .... (E)

  „   Electric Resistance .... (F)

  „   Total Currents .... (A)

One equation of Free Electricity .... (G)

  „   Continuity... (H)

These equations are therefore sufficient to determine all the quantities which occur in them, provided we know the conditions of the problem. In many questions, however, only a few of the equations are required.

Intrinsic Energy of the Electromagnetic Field.

(71) We have seen (33) that the intrinsic energy of any system of currents is found by multiplying half the current in each circuit into its electromagnetic momentum. This is equivalent to finding the integral

$$E = \tfrac{1}{2}\Sigma\,(Fp' + Gq' + Hr')\,dV \dots\dots\dots\dots\dots\dots(37)$$

over all the space occupied by currents, where p, q, r are the components of currents, and F, G, H the components of electromagnetic momentum.

Substituting the values of p', q', r' from the equations of Currents (C), this becomes

$$\frac{1}{8\pi}\Sigma\left\{F\left(\frac{d\gamma}{dy} - \frac{d\beta}{dz}\right) + G\left(\frac{d\alpha}{dz} - \frac{d\gamma}{dx}\right) + H\left(\frac{d\beta}{dx} - \frac{d\alpha}{dy}\right)\right\}dV.$$

Integrating by parts, and remembering that α, β, γ vanish at an infinite distance, the expression becomes

$$\frac{1}{8\pi}\Sigma\left\{\alpha\left(\frac{dH}{dy} - \frac{dG}{dz}\right) + \beta\left(\frac{dF}{dz} - \frac{dH}{dx}\right) + \gamma\left(\frac{dG}{dx} - \frac{dF}{dy}\right)\right\}dV,$$

where the integration is to be extended over all space. Referring to the equations of Magnetic Force (B), p. 556, this becomes

$$E = \frac{1}{8\pi}\Sigma\left\{\alpha\,.\,\mu\alpha + \beta\,.\,\mu\beta + \gamma\,.\,\mu\gamma\right\}dV \dots\dots\dots\dots(38),$$

where a, β, γ are the components of magnetic intensity or the force on a unit magnetic pole, and μa, $\mu\beta$, $\mu\gamma$ are the components of the quantity of magnetic induction, or the number of lines of force in unit of area.

In isotropic media the value of μ is the same in all directions, and we may express the result more simply by saying that the intrinsic energy of any part of the magnetic field arising from its magnetization is

$$\frac{\mu}{8\pi}I^2$$

per unit of volume, where I is the magnetic intensity.

(72) Energy may be stored up in the field in a different way, namely, by the action of electromotive force in producing electric displacement. The work done by a variable electromotive force, P, in producing a variable displacement, f, is got by integrating

$$\int P df$$

from $P = 0$ to the given value of P.

Since $P = kf$, equation (E), this quantity becomes

$$\int kf df = \tfrac{1}{2}kf^2 = \tfrac{1}{2}Pf.$$

Hence the intrinsic energy of any part of the field, as existing in the form of electric displacement, is

$$\tfrac{1}{2}\Sigma(Pf + Qg + Rh)dV.$$

The total energy existing in the field is therefore

$$E = \Sigma\left\{\frac{1}{8\pi}(a\mu a + \beta\mu\beta + \gamma\mu\gamma) + \tfrac{1}{2}(Pf + Qg + Rh)\right\}dV \dots\dots\dots(\mathrm{I}).$$

The first term of this expression depends on the magnetization of the field, and is explained on our theory by actual motion of some kind. The second term depends on the electric polarization of the field, and is explained on our theory by strain of some kind in an elastic medium.

(73) I have on a former occasion * attempted to describe a particular kind of motion and a particular kind of strain, so arranged as to account for the phenomena. In the present paper I avoid any hypothesis of this kind; and in

* "On Physical Lines of Force," *Philosophical Magazine*, 1861—62. (In this vol. p. 451.)

71—2

using such words as electric momentum and electric elasticity in reference to the known phenomena of the induction of currents and the polarization of dielectrics, I wish merely to direct the mind of the reader to mechanical phenomena which will assist him in understanding the electrical ones. All such phrases in the present paper are to be considered as illustrative, not as explanatory.

(74) In speaking of the Energy of the field, however, I wish to be understood literally. All energy is the same as mechanical energy, whether it exists in the form of motion or in that of elasticity, or in any other form. The energy in electromagnetic phenomena is mechanical energy. The only question is, Where does it reside? On the old theories it resides in the electrified bodies, conducting circuits, and magnets, in the form of an unknown quality called potential energy, or the power of producing certain effects at a distance. On our theory it resides in the electromagnetic field, in the space surrounding the electrified and magnetic bodies, as well as in those bodies themselves, and is in two different forms, which may be described without hypothesis as magnetic polarization and electric polarization, or, according to a very probable hypothesis, as the motion and the strain of one and the same medium.

(75) The conclusions arrived at in the present paper are independent of this hypothesis, being deduced from experimental facts of three kinds:

1. The induction of electric currents by the increase or diminution of neighbouring currents according to the changes in the lines of force passing through the circuit.

2. The distribution of magnetic intensity according to the variations of a magnetic potential.

3. The induction (or influence) of statical electricity through dielectrics.

We may now proceed to demonstrate from these principles the existence and laws of the mechanical forces which act upon electric currents, magnets, and electrified bodies placed in the electromagnetic field.

PART IV.

MECHANICAL ACTIONS IN THE FIELD.

Mechanical Force on a Moveable Conductor.

(76) We have shewn (§§ 34 & 35) that the work done by the electro-magnetic forces in aiding the motion of a conductor is equal to the product of the current in the conductor multiplied by the increment of the electromagnetic momentum due to the motion.

Let a short straight conductor of length a move parallel to itself in the direction of x, with its extremities on two parallel conductors. Then the increment of the electromagnetic momentum due to the motion of a will be

$$a\left(\frac{dF}{dx}\frac{dx}{ds} + \frac{dG}{dx}\frac{dy}{ds} + \frac{dH}{dx}\frac{dz}{ds}\right)\delta x.$$

That due to the lengthening of the circuit by increasing the length of the parallel conductors will be

$$-a\left(\frac{dF}{dx}\frac{dx}{ds} + \frac{dF}{dy}\frac{dy}{ds} + \frac{dF}{dz}\frac{dz}{ds}\right)\delta x.$$

The total increment is

$$a\delta x\left\{\frac{dy}{ds}\left(\frac{dG}{dx} - \frac{dF}{dy}\right) - \frac{dz}{ds}\left(\frac{dF}{dz} - \frac{dH}{dx}\right)\right\},$$

which is by the equations of Magnetic Force (B), p. 556,

$$a\delta x\left(\frac{dy}{ds}\mu\gamma - \frac{dz}{ds}\mu\beta\right).$$

Let X be the force acting along the direction of x per unit of length of the conductor, then the work done is $Xa\delta x$.

Let C be the current in the conductor, and let p', q', r' be its components, then

$$Xa\delta x = Ca\delta x\left(\frac{dy}{ds}\mu\gamma - \frac{dz}{ds}\mu\beta\right),$$

or

Similarly,

$$
\left.\begin{aligned}
X &= \mu\gamma q' - \mu\beta r' \\
Y &= \mu a r' - \mu\gamma p' \\
Z &= \mu\beta p' - \mu a q'
\end{aligned}\right\} \quad \dots\dots\dots\dots\dots\dots\dots\dots\dots\text{(J)}.
$$

These are the equations which determine the mechanical force acting on a conductor carrying a current. The force is perpendicular to the current and to the lines of force, and is measured by the area of the parallelogram formed by lines parallel to the current and lines of force, and proportional to their intensities.

Mechanical Force on a Magnet.

(77) In any part of the field not traversed by electric currents the distribution of magnetic intensity may be represented by the differential coefficients of a function which may be called the magnetic potential. When there are no currents in the field, this quantity has a single value for each point. When there are currents, the potential has a series of values at each point, but its differential coefficients have only one value, namely,

$$
\frac{d\phi}{dx} = a, \quad \frac{d\phi}{dy} = \beta, \quad \frac{d\phi}{dz} = \gamma.
$$

Substituting these values of a, β, γ in the expression (equation 38) for the intrinsic energy of the field, and integrating by parts, it becomes

$$
-\Sigma \left\{ \phi \, \frac{1}{8\pi} \left(\frac{d\mu a}{dx} + \frac{d\mu\beta}{dy} + \frac{d\mu\gamma}{dz} \right) \right\} dV.
$$

The expression

$$
\Sigma \left(\frac{d\mu a}{dx} + \frac{d\mu\beta}{dy} + \frac{d\mu\gamma}{dz} \right) dV = \Sigma m \, dV \dots\dots\dots\dots\dots \text{(39)}
$$

indicates the number of lines of magnetic force which have their origin within the space V. Now a magnetic pole is known to us only as the origin or termination of lines of magnetic force, and a unit pole is one which has 4π lines belonging to it, since it produces unit of magnetic intensity at unit of distance over a sphere whose surface is 4π.

Hence if m is the amount of free positive magnetism in unit of volume, the above expression may be written $4\pi m$, and the expression for the energy of the field becomes

$$
E = -\Sigma \left(\tfrac{1}{2}\phi m \right) dV \dots\dots\dots\dots\dots\dots\dots\text{(40)}.
$$

If there are two magnetic poles m_1 and m_2 producing potentials ϕ_1 and ϕ_2 in the field, then if m_2 is moved a distance dx, and is urged in that direction by a force X, then the work done is Xdx, and the decrease of energy in the field is

$$d\left\{\tfrac{1}{2}\left(\phi_1+\phi_2\right)\left(m_1+m_2\right)\right\},$$

and these must be equal by the principle of Conservation of Energy.

Since the distribution ϕ_1 is determined by m_1, and ϕ_2 by m_2, the quantities $\phi_1 m_1$ and $\phi_2 m_2$ will remain constant.

It can be shewn also, as Green has proved (Essay, p. 10), that

$$m_1\phi_2 = m_2\phi_1,$$

so that we get
$$Xdx = d\left(m_2\phi_1\right),$$

or
$$X = m_2\frac{d\phi_1}{dx} = m_2 a_1,$$

where a_1 represents the magnetic intensity due to m_1. $\left.\begin{array}{l}\\ \\ \\ \\ \\ \\ \\ \end{array}\right\}$ .... (K).

Similarly,
$$Y = m_2\beta_1,$$
$$Z = m_2\gamma_1.$$

So that a magnetic pole is urged in the direction of the lines of magnetic force with a force equal to the product of the strength of the pole and the magnetic intensity.

(78) If a single magnetic pole, that is, one pole of a very long magnet, be placed in the field, the only solution of ϕ is

$$\phi_1 = -\frac{m_1}{\mu}\frac{1}{r}\dots\dots\dots\dots\dots\dots\dots(41),$$

where m_1 is the strength of the pole, and r the distance from it.

The repulsion between two poles of strength m_1 and m_2 is

$$m_2\frac{d\phi_1}{dr} = \frac{m_1 m_2}{\mu r^2}\dots\dots\dots\dots\dots\dots\dots(42).$$

In air or any medium in which $\mu = 1$ this is simply $\frac{m_1 m_2}{r^2}$, but in other media the force acting between two given magnetic poles is inversely proportional to the coefficient of magnetic induction for the medium. This may be explained by the magnetization of the medium induced by the action of the poles.

Mechanical Force on an Electrified Body.

(79) If there is no motion or change of strength of currents or magnets in the field, the electromotive force is entirely due to variation of electric potential, and we shall have (§ 65)

$$P = -\frac{d\Psi}{dx}, \quad Q = -\frac{d\Psi}{dy}, \quad R = -\frac{d\Psi}{dz}.$$

Integrating by parts the expression (I) for the energy due to electric displacement, and remembering that P, Q, R vanish at an infinite distance, it becomes

$$\tfrac{1}{2}\Sigma \left\{ \Psi \left(\frac{df}{dx} + \frac{dg}{dy} + \frac{dh}{dz} \right) dV, \right.$$

or by the equation of Free Electricity (G), p. 561,

$$-\tfrac{1}{2}\Sigma (\Psi e) \, dV.$$

By the same demonstration as was used in the case of the mechanical action on a magnet, it may be shewn that the mechanical force on a small body containing a quantity e_2 of free electricity placed in a field whose potential arising from other electrified bodies is Ψ_1, has for components

$$\left. \begin{aligned} X &= e_2 \frac{d\Psi_1}{dx} = -P_1 e_2 \\[1em] Y &= e_2 \frac{d\Psi_1}{dy} = -Q_1 e_2 \\[1em] Z &= e_3 \frac{d\Psi_1}{dz} = -R_1 e_2 \end{aligned} \right\} \quad \dots\dots\dots\dots\dots\dots\dots\dots \text{(D)}.$$

So that an electrified body is urged in the direction of the electromotive force with a force equal to the product of the quantity of free electricity and the electromotive force.

If the electrification of the field arises from the presence of a small electrified body containing e_1 of free electricity, the only solution of Ψ_1 is

$$\Psi_1 = \frac{k}{4\pi} \frac{e_1}{r} \quad \dots\dots\dots\dots\dots\dots\dots\dots\dots \text{(43)},$$

where r is the distance from the electrified body.

The repulsion between two electrified bodies e_1, e_2 is therefore

$$e_2 \frac{d\Psi_1}{dr} = \frac{k}{4\pi} \frac{e_1 e_2}{r^2} \quad \dots\dots\dots\dots\dots\dots\dots\dots \text{(44)}.$$

Measurement of Electrostatic Effects.

(80) The quantities with which we have had to do have been hitherto expressed in terms of the Electromagnetic System of measurement, which is founded on the mechanical action between currents. The electrostatic system of measurement is founded on the mechanical action between electrified bodies, and is independent of, and incompatible with, the electromagnetic system; so that the units of the different kinds of quantity have different values according to the system we adopt, and to pass from the one system to the other, a reduction of all the quantities is required.

According to the electrostatic system, the repulsion between two small bodies charged with quantities η_1, η_2 of electricity is

$$\frac{\eta_1 \eta_2}{r^2},$$

where r is the distance between them.

Let the relation of the two systems be such that one electromagnetic unit of electricity contains v electrostatic units; then $\eta_1 = v e_1$ and $\eta_2 = v e_2$, and this repulsion becomes

$$v^2 \frac{e_1 e_2}{r^2} = \frac{k}{4\pi} \frac{e_1 e_2}{r^2} \text{ by equation (44)} \dots\dots\dots\dots (45),$$

whence k, the coefficient of "electric elasticity" in the medium in which the experiments are made, *i.e.* common air, is related to v, the number of electrostatic units in one electromagnetic unit, by the equation

$$k = 4\pi v^2 \dots\dots\dots\dots\dots\dots\dots\dots\dots\dots (46).$$

The quantity v may be determined by experiment in several ways. According to the experiments of MM. Weber and Kohlrausch,

$$v = 310,740,000 \text{ metres per second.}$$

(81) It appears from this investigation, that if we assume that the medium which constitutes the electromagnetic field is, when dielectric, capable of receiving in every part of it an electric polarization, in which the opposite sides of every element into which we may conceive the medium divided are oppositely electrified, and if we also assume that this polarization or electric displacement is proportional to the electromotive force which produces or maintains it, then we

can shew that electrified bodies in a dielectric medium will act on one another with forces obeying the same laws as are established by experiment.

The energy, by the expenditure of which electrical attractions and repulsions are produced, we suppose to be stored up in the dielectric medium which surrounds the electrified bodies, and not on the surface of those bodies themselves, which on our theory are merely the bounding surfaces of the air or other dielectric in which the true springs of action are to be sought.

Note on the Attraction of Gravitation.

(82) After tracing to the action of the surrounding medium both the magnetic and the electric attractions and repulsions, and finding them to depend on the inverse square of the distance, we are naturally led to inquire whether the attraction of gravitation, which follows the same law of the distance, is not also traceable to the action of a surrounding medium.

Gravitation differs from magnetism and electricity in this; that the bodies concerned are all of the same kind, instead of being of opposite signs, like magnetic poles and electrified bodies, and that the force between these bodies is an attraction and not a repulsion, as is the case between like electric and magnetic bodies.

The lines of gravitating force near two dense bodies are exactly of the same form as the lines of magnetic force near two poles of the same name; but whereas the poles are repelled, the bodies are attracted. Let E be the intrinsic energy of the field surrounding two gravitating bodies M_1, M_2, and let E' be the intrinsic energy of the field surrounding two magnetic poles, m_1, m_2 equal in numerical value to M_1, M_2, and let X be the gravitating force acting during the displacement δx, and X' the magnetic force,

$$X\delta x = \delta E, \qquad X'\delta x = \delta E';$$

now X and X' are equal in numerical value, but of opposite signs; so that

$$\delta E = -\delta E',$$

or

$$E = C - E'$$

$$= C - \Sigma \frac{1}{8\pi} \left(\alpha^2 + \beta^2 + \gamma^2 \right) dV,$$

where α, β, γ are the components of magnetic intensity. If R be the resultant gravitating force, and R' the resultant magnetic force at a corresponding part of the field,

$$R = -R', \text{ and } \alpha^2 + \beta^2 + \gamma^2 = R^2 = R'^2.$$

Hence

$$E = C - \Sigma \frac{1}{8\pi} R^2 dV \dots\dots\dots\dots\dots\dots\dots\dots\dots (47).$$

The intrinsic energy of the field of gravitation must therefore be less wherever there is a resultant gravitating force.

As energy is essentially positive, it is impossible for any part of space to have negative intrinsic energy. Hence those parts of space in which there is no resultant force, such as the points of equilibrium in the space between the different bodies of a system, and within the substance of each body, must have an intrinsic energy per unit of volume greater than

$$\frac{1}{8\pi} R^2,$$

where R is the greatest possible value of the intensity of gravitating force in any part of the universe.

The assumption, therefore, that gravitation arises from the action of the surrounding medium in the way pointed out, leads to the conclusion that every part of this medium possesses, when undisturbed, an enormous intrinsic energy, and that the presence of dense bodies influences the medium so as to diminish this energy wherever there is a resultant attraction.

As I am unable to understand in what way a medium can possess such properties, I cannot go any further in this direction in searching for the cause of gravitation.

PART V.

THEORY OF CONDENSERS.

Capacity of a Condenser.

(83) The simplest form of condenser consists of a uniform layer of insulating matter bounded by two conducting surfaces, and its capacity is measured by the quantity of electricity on either surface when the difference of potentials is unity.

Let S be the area of either surface, a the thickness of the dielectric, and k its coefficient of electric elasticity; then on one side of the condenser the potential is Ψ_1, and on the other side $\Psi_1 + 1$, and within its substance

$$\frac{d\Psi}{dx} = \frac{1}{a} = kf \dots\dots\dots\dots\dots\dots\dots\dots (48).$$

Since $\dfrac{d\Psi}{dx}$ and therefore f is zero outside the condenser, the quantity of electricity on its first surface $= -Sf$, and on the second $+Sf$. The capacity of the condenser is therefore $Sf = \dfrac{S}{ak}$ in electromagnetic measure.

Specific Capacity of Electric Induction (D).

(84) If the dielectric of the condenser be air, then its capacity in electrostatic measure is $\dfrac{S}{4\pi a}$ (neglecting corrections arising from the conditions to be fulfilled at the edges). If the dielectric have a capacity whose ratio to that of air is D, then the capacity of the condenser will be $\dfrac{DS}{4\pi a}$.

Hence
$$D = \frac{k_0}{k} \dots\dots\dots\dots\dots\dots\dots\dots\dots (49),$$

where k_0 is the value of k in air, which is taken for unity.

Electric Absorption.

(85) When the dielectric of which the condenser is formed is not a perfect insulator, the phenomena of conduction are combined with those of electric displacement. The condenser, when left charged, gradually loses its charge, and in some cases, after being discharged completely, it gradually acquires a new charge of the same sign as the original charge, and this finally disappears. These phenomena have been described by Professor Faraday (*Experimental Researches,* Series XI.) and by Mr F. Jenkin (*Report of Committee of Board of Trade on Submarine Cables*), and may be classed under the name of "Electric Absorption."

(86) We shall take the case of a condenser composed of any number of parallel layers of different materials. If a constant difference of potentials between its extreme surfaces is kept up for a sufficient time till a condition of permanent steady flow of electricity is established, then each bounding surface will have a charge of electricity depending on the nature of the substances on each side of it. If the extreme surfaces be now discharged, these internal charges will gradually be dissipated, and a certain charge may reappear on the extreme surfaces if they are insulated, or, if they are connected by a conductor, a certain quantity of electricity may be urged through the conductor during the re-establishment of equilibrium.

Let the thickness of the several layers of the condenser be a_1, a_2, &c.

Let the values of k for these layers be respectively k_1, k_2, k_3, and let

$$a_1 k_2 + a_2 k_2 + \&c. = ak \dots\dots\dots\dots\dots\dots\dots (50),$$

where k is the "electric elasticity" of air, and a is the thickness of an equivalent condenser of air.

Let the resistances of the layers be respectively r_1, r_2, &c., and let $r_1 + r_2 + \&c. = r$ be the resistance of the whole condenser, to a steady current through it per unit of surface.

Let the electric displacement in each layer be f_1, f_2, &c.

Let the electric current in each layer be p_1, p_2, &c.

Let the potential on the first surface be Ψ_1, and the electricity per unit of surface e_1.

Let the corresponding quantities at the boundary of the first and second surface be Ψ_2 and e_2, and so on. Then by equations (G) and (H),

$$\left.\begin{array}{ll} e_1 = -f_1, & \dfrac{de_1}{dt} = -p_1, \\[2ex] e_2 = f_1 - f_2, & \dfrac{de_2}{dt} = p_1 - p_2, \\[2ex] \&c. & \&c. \end{array}\right\} \quad\ldots\ldots\ldots\ldots\ldots\ldots(51),$$

But by equations (E) and (F),

$$\left.\begin{array}{l} \Psi_1 - \Psi_2 = a_1 k_1 f_1 = -r_1 p_1 \\[1ex] \Psi_2 - \Psi_3 = a_2 k_2 f_2 = -r_2 p_2, \\[1ex] \&c. \quad\&c. \quad\&c. \end{array}\right\} \quad\ldots\ldots\ldots\ldots\ldots (52).$$

After the electromotive force has been kept up for a sufficient time the current becomes the same in each layer, and

$$p_1 = p_2 = \&c. = p = \frac{\Psi}{r},$$

where Ψ is the total difference of potentials between the extreme layers. We have then

and
$$\left.\begin{array}{ll} f_1 = -\dfrac{\Psi}{r}\dfrac{r_1}{a_1 k_1}, & f_2 = -\dfrac{\Psi}{r}\dfrac{r_2}{a_2 k_2}, \quad\&c. \\[3ex] e_1 = \dfrac{\Psi}{r}\dfrac{r_1}{a_1 k_1}, & e_2 = \dfrac{\Psi}{r}\left(\dfrac{r_2}{a_2 k_2} - \dfrac{r_1}{a k_1}\right), \quad\&c. \end{array}\right\} \quad\ldots\ldots\ldots (53).$$

These are the quantities of electricity on the different surfaces.

(87) Now let the condenser be discharged by connecting the extreme surfaces through a perfect conductor so that their potentials are instantly rendered equal, then the electricity on the extreme surfaces will be altered, but that on the internal surfaces will not have time to escape. The total difference of potentials is now

$$\Psi' = a_1 k_1 e'_1 + a_2 k_2 (e'_1 + e_2) + a_3 k_3 (e'_1 + e_2 + e_3), \quad\&c. = 0 \ldots\ldots\ldots\ldots(54),$$

whence if e'_1 is what e_1 becomes at the instant of discharge,

$$e'_1 = \frac{\Psi}{r}\frac{r_1}{a_1 k_1} - \frac{\Psi}{ak} = e_1 - \frac{\Psi}{ak}\ldots\ldots\ldots\ldots\ldots\ldots (55).$$

The instantaneous discharge is therefore $\dfrac{\Psi}{ak}$, or the quantity which would be discharged by a condenser of air of the equivalent thickness a, and it is unaffected by the want of perfect insulation.

(88) Now let us suppose the connexion between the extreme surfaces broken, and the condenser left to itself, and let us consider the gradual dissipation of the internal charges. Let Ψ' be the difference of potential of the extreme surfaces at any time t; then

$$\Psi' = a_1 k_1 f_1 + a_2 k_2 f_2 + \&\text{c.} \dots\dots\dots\dots\dots\dots (56);$$

but

$$a_1 k_1 f_1 = -r_1 \frac{df_1}{dt},$$

$$a_2 k_2 f_2 = -r_2 \frac{df_2}{dt}.$$

Hence $f_1 = A_1 e^{-\frac{a_1 k_1}{r_1} t}$, $f_2 = A_2 e^{-\frac{a_2 k_2}{r_2} t}$, &c.; and by referring to the values of e'_1, e_2, &c., we find

$$
\left.
\begin{aligned}
A_1 &= \frac{\Psi}{r}\frac{r_1}{a_1 k_1} - \frac{\Psi}{ak} \\[2mm]
A_2 &= \frac{\Psi}{r}\frac{r_2}{a_2 k_2} - \frac{\Psi}{ak} \\[2mm]
&\text{\&c.}
\end{aligned}
\right\} \dots\dots\dots\dots\dots\dots (57),
$$

so that we find for the difference of extreme potentials at any time,

$$\Psi' = \Psi \left\{ \left(\frac{r_1}{r} - \frac{a_1 k_1}{ak}\right) e^{-\frac{a_1 k_1}{r_1} t} + \left(\frac{r_2}{r} - \frac{a_2 k_2}{ak}\right) e^{-\frac{a_2 k_2}{r_2} t} + \&\text{c.} \right\} \dots\dots\dots (58).$$

(89) It appears from this result that if all the layers are made of the same substance, Ψ' will be zero always. If they are of different substances, the order in which they are placed is indifferent, and the effect will be the same whether each substance consists of one layer, or is divided into any number of thin layers and arranged in any order among thin layers of the other substances. Any substance, therefore, the parts of which are not mathematically homogeneous, though they may be apparently so, may exhibit phenomena of absorption. Also, since the order of magnitude of the coefficients is the same as that of the indices, the value of Ψ' can never change sign, but must start from zero, become positive, and finally disappear.

(90) Let us next consider the total amount of electricity which would pass from the first surface to the second, if the condenser, after being thoroughly saturated by the current and then discharged, has its extreme surfaces connected by a conductor of resistance R. Let p be the current in this conductor; then, during the discharge,

$$\Psi' = p_1 r_1 + p_2 r_2 + \&c. = pR \dots\dots\dots (59).$$

Integrating with respect to the time, and calling q_1, q_2, q the quantities of electricity which traverse the different conductors,

$$q_1 r_1 + q_2 r_2 + \&c. = qR \dots\dots\dots (60).$$

The quantities of electricity on the several surfaces will be

$$e'_1 - q - q_1,$$
$$e_2 + q_1 - q_2,$$
$$\&c. ;$$

and since at last all these quantities vanish, we find

$$q_1 = e'_1 - q,$$
$$q_2 = e'_1 + e_2 - q ;$$

whence

$$qR = \frac{\Psi}{r}\left(\frac{r_1^2}{a_1 k_1} + \frac{r_2^2}{a_2 k_2} + \&c.\right) - \frac{\Psi r}{ak},$$

or

$$q = \frac{\Psi}{akrR}\left\{ a_1 k_1 a_2 k_2 \left(\frac{r_1}{a_1 k_1} - \frac{r_2}{a_2 k_2}\right)^2 + a_2 k_2 a_3 k_3 \left(\frac{r_2}{a_2 k_2} - \frac{r_3}{a_3 k_3}\right)^2 + \&c.\right\}\dots\dots(61),$$

a quantity essentially positive; so that, when the primary electrification is in one direction, the secondary discharge is always in the same direction as the primary discharge *.

* Since this paper was communicated to the Royal Society, I have seen a paper by M. Gaugain in the *Annales de Chimie* for 1864, in which he has deduced the phenomena of electric absorption and secondary discharge from the theory of compound condensers.

PART VI.

ELECTROMAGNETIC THEORY OF LIGHT.

(91) At the commencement of this paper we made use of the optical hypothesis of an elastic medium through which the vibrations of light are propagated, in order to shew that we have warrantable grounds for seeking, in the same medium, the cause of other phenomena as well as those of light. We then examined electromagnetic phenomena, seeking for their explanation in the properties of the field which surrounds the electrified or magnetic bodies. In this way we arrived at certain equations expressing certain properties of the electromagnetic field. We now proceed to investigate whether these properties of that which constitutes the electromagnetic field, deduced from electromagnetic phenomena alone, are sufficient to explain the propagation of light through the same substance.

(92) Let us suppose that a plane wave whose direction cosines are l, m, n is propagated through the field with a velocity V. Then all the electromagnetic functions will be functions of

$$w = lx + my + nz - Vt.$$

The equations of Magnetic Force (B), p. 556, will become

$$\mu a = m \frac{dH}{dw} - n \frac{dG}{dw},$$

$$\mu \beta = n \frac{dF}{dw} - l \frac{dH}{dw},$$

$$\mu \gamma = l \frac{dG}{dw} - m \frac{dF}{dw}.$$

If we multiply these equations respectively by l, m, n, and add, we find

$$l\mu a + m\mu\beta + n\mu\gamma = 0 \quad \ldots \ldots \ldots \ldots \ldots \ldots \ldots \ldots \ldots (62),$$

which shews that the direction of the magnetization must be in the plane of the wave.

(93) If we combine the equations of Magnetic Force (B) with those of Electric Currents (C), and put for brevity

$$\frac{dF}{dx} + \frac{dG}{dy} + \frac{dH}{dz} = J, \text{ and } \frac{d^2}{dx^2} + \frac{d^2}{dy^2} + \frac{d^2}{dz^2} = \nabla^2 \quad \dots\dots\dots\dots \text{(63)},$$

$$\left. \begin{aligned} 4\pi\mu p' &= \frac{dJ}{dx} - \nabla^2 F \\[4pt] 4\pi\mu q' &= \frac{dJ}{dy} - \nabla^2 G \\[4pt] 4\pi\mu r' &= \frac{dJ}{dz} - \nabla^2 H \end{aligned} \right\} \quad \dots\dots\dots\dots\dots\dots\dots \text{(64)}.$$

If the medium in the field is a perfect dielectric there is no true conduction, and the currents p', q', r' are only variations in the electric displacement, or, by the equations of Total Currents (A),

$$p' = \frac{df}{dt}, \qquad q' = \frac{dg}{dt}, \qquad r' = \frac{dh}{dt} \quad \dots\dots\dots\dots\dots \text{(65)}.$$

But these electric displacements are caused by electromotive forces, and by the equations of Electric Elasticity (E),

$$P = kf, \qquad Q = kg, \qquad R = kh \quad \dots\dots\dots\dots\dots\dots \text{(66)}.$$

These electromotive forces are due to the variations either of the electro-magnetic or the electrostatic functions, as there is no motion of conductors in the field; so that the equations of electromotive force (D) are

$$\left. \begin{aligned} P &= -\frac{dF}{dt} - \frac{d\Psi}{dx} \\[4pt] Q &= -\frac{dG}{dt} - \frac{d\Psi}{dy} \\[4pt] R &= -\frac{dH}{dt} - \frac{d\Psi}{dz} \end{aligned} \right\} \quad \dots\dots\dots\dots\dots\dots \text{(67)}.$$

(94) Combining these equations, we obtain the following:—

$$\left. \begin{aligned} k\left(\frac{dJ}{dx} - \nabla^2 F\right) + 4\pi\mu\left(\frac{d^2 F}{dt^2} + \frac{d^2\Psi}{dx\,dt}\right) &= 0 \\[4pt] k\left(\frac{dJ}{dy} - \nabla^2 G\right) + 4\pi\mu\left(\frac{d^2 G}{dt^2} + \frac{d^2\Psi}{dy\,dt}\right) &= 0 \\[4pt] k\left(\frac{dJ}{dz} - \nabla^2 H\right) + 4\pi\mu\left(\frac{d^2 H}{dt^2} + \frac{d^2\Psi}{dz\,dt}\right) &= 0 \end{aligned} \right\} \quad \dots\dots\dots \text{(68)}.$$

If we differentiate the third of these equations with respect to y, and the second with respect to z, and subtract, J and Ψ disappear, and by remembering the equations (B) of magnetic force, the results may be written

$$\left. \begin{array}{l} k\nabla^2\mu a = 4\pi\mu \dfrac{d^2}{dt^2}\mu a \\[2mm] k\nabla^2\mu\beta = 4\pi\mu \dfrac{d^2}{dt^2}\mu\beta \\[2mm] k\nabla^2\mu\gamma = 4\pi\mu \dfrac{d^2}{dt^2}\mu\gamma \end{array} \right\} \dots\dots\dots\dots\dots\dots (69).$$

(95) If we assume that a, β, γ are functions of $lx + my + nz - Vt = w$, the first equation becomes

$$k\mu \frac{d^2a}{dw^2} = 4\pi\mu^2 V^2 \frac{d^2a}{dw^2} \dots\dots\dots\dots\dots\dots (70),$$

or

$$V = \pm\sqrt{\frac{k}{4\pi\mu}} \dots\dots\dots\dots\dots\dots (71).$$

The other equations give the same value for V, so that the wave is propagated in either direction with a velocity V.

This wave consists entirely of magnetic disturbances, the direction of magnetization being in the plane of the wave. No magnetic disturbance whose direction of magnetization is not in the plane of the wave can be propagated as a plane wave at all.

Hence magnetic disturbances propagated through the electromagnetic field agree with light in this, that the disturbance at any point is transverse to the direction of propagation, and such waves may have all the properties of polarized light.

(96) The only medium in which experiments have been made to determine the value of k is air, in which $\mu = 1$, and therefore, by equation (46),

$$V = v \dots\dots\dots\dots\dots\dots\dots\dots (72).$$

By the electromagnetic experiments of MM. Weber and Kohlrausch [*],

$$v = 310{,}740{,}000 \text{ metres per second}$$

---

[*] *Leipzig Transactions*, Vol. v. (1857), p. 260, or Poggendorff's *Annalen*, Aug. 1856, p. 10.

is the number of electrostatic units in one electromagnetic unit of electricity, and this, according to our result, should be equal to the velocity of light in air or vacuum.

The velocity of light in air, by M. Fizeau's * experiments, is

$$V = 314,858,000 \,;$$

according to the more accurate experiments of M. Foucault †,

$$V = 298,000,000.$$

The velocity of light in the space surrounding the earth, deduced from the coefficient of aberration and the received value of the radius of the earth's orbit, is

$$V = 308,000,000.$$

(97) Hence the velocity of light deduced from experiment agrees sufficiently well with the value of v deduced from the only set of experiments we as yet possess. The value of v was determined by measuring the electromotive force with which a condenser of known capacity was charged, and then discharging the condenser through a galvanometer, so as to measure the quantity of electricity in it in electromagnetic measure. The only use made of light in the experiment was to see the instruments. The value of V found by M. Foucault was obtained by determining the angle through which a revolving mirror turned, while the light reflected from it went and returned along a measured course. No use whatever was made of electricity or magnetism.

The agreement of the results seems to shew that light and magnetism are affections of the same substance, and that light is an electromagnetic disturbance propagated through the field according to electromagnetic laws.

(98) Let us now go back upon the equations in (94), in which the quantities J and Ψ occur, to see whether any other kind of disturbance can be propagated through the medium depending on these quantities which disappeared from the final equations.

* *Comptes Rendus*, Vol. xxix. (1849), p. 90.
† Ibid. Vol. lv. (1862), pp. 501, 792.

If we determine χ from the equation

$$\nabla^2\chi = \frac{d^2\chi}{dx^2} + \frac{d^2\chi}{dy^2} + \frac{d^2\chi}{dz^2} = J \ldots\ldots\ldots\ldots\ldots\ldots (73),$$

and F', G', H' from the equations

$$F' = F - \frac{d\chi}{dx}, \quad G' = G - \frac{d\chi}{dy}, \quad H' = H - \frac{d\chi}{dz} \ldots\ldots\ldots\ldots (74),$$

then

$$\frac{dF'}{dx} + \frac{dG'}{dy} + \frac{dH'}{dz} = 0 \ldots\ldots\ldots\ldots\ldots\ldots\ldots (75),$$

and the equations in (94) become of the form

$$k\nabla^2 F' = 4\pi\mu \left\{ \frac{d^2F'}{dt^2} + \frac{d}{dxdt}\left(\Psi + \frac{d\chi}{dt}\right) \right\} \ldots\ldots\ldots\ldots\ldots (76).$$

Differentiating the three equations with respect to x, y, and z, and adding, we find that

$$\Psi = -\frac{d\chi}{dt} + \phi(x, y, z) \ldots\ldots\ldots\ldots\ldots\ldots\ldots (77),$$

and that

$$\left. \begin{array}{l} k\nabla^2 F' = 4\pi\mu \dfrac{d^2F'}{dt^2} \\[2mm] k\nabla^2 G' = 4\pi\mu \dfrac{d^2G'}{dt^2} \\[2mm] k\nabla^2 H' = 4\pi\mu \dfrac{d^2H'}{dt^2} \end{array} \right\} \ldots\ldots\ldots\ldots\ldots\ldots (78).$$

Hence the disturbances indicated by F', G', H' are propagated with the velocity $V = \sqrt{\dfrac{k}{4\pi\mu}}$ through the field; and since

$$\frac{dF'}{dx} + \frac{dG'}{dy} + \frac{dH'}{dz} = 0,$$

the resultant of these disturbances is in the plane of the wave.

(99) The remaining part of the total disturbances F, G, H being the part depending on χ, is subject to no condition except that expressed in the equation

$$\frac{d\Psi}{dt} + \frac{d^2\chi}{dt^2} = 0.$$

If we perform the operation ∇^2 on this equation, it becomes

$$ke = \frac{dJ}{dt} - k\nabla^2 \phi \, (x, \; y, \; z)\dots\dots\dots\dots\dots(79).$$

Since the medium is a perfect insulator, e, the free electricity, is immoveable, and therefore $\frac{dJ}{dt}$ is a function of x, y, z, and the value of J is either constant or zero, or uniformly increasing or diminishing with the time; so that no disturbance depending on J can be propagated as a wave.

(100) The equations of the electromagnetic field, deduced from purely experimental evidence, shew that transversal vibrations only can be propagated. If we were to go beyond our experimental knowledge and to assign a definite density to a substance which we should call the electric fluid, and select either vitreous or resinous electricity as the representative of that fluid, then we might have normal vibrations propagated with a velocity depending on this density. We have, however, no evidence as to the density of electricity, as we do not even know whether to consider vitreous electricity as a substance or as the absence of a substance.

Hence electromagnetic science leads to exactly the same conclusions as optical science with respect to the direction of the disturbances which can be propagated through the field; both affirm the propagation of transverse vibrations, and both give the same velocity of propagation. On the other hand, both sciences are at a loss when called on to affirm or deny the existence of normal vibrations.

Relation between the Index of Refraction and the Electromagnetic Character of the substance.

(101) The velocity of light in a medium, according to the Undulatory Theory, is

$$\frac{1}{i} V_0,$$

where i is the index of refraction and V_0 is the velocity in vacuum. The velocity, according to the Electromagnetic Theory, is

$$\sqrt{\frac{k}{4\pi\mu}},$$

where, by equations (49) and (71), $k = \frac{1}{D} k_0$, and $k_0 = 4\pi V_0^2$.

Hence $$D = \frac{i^2}{\mu} \dots\dots\dots\dots\dots\dots\dots\dots\dots\dots\dots\dots (80),$$

or the Specific Inductive Capacity is equal to the square of the index of refraction divided by the coefficient of magnetic induction.

Propagation of Electromagnetic Disturbances in a Crystallized Medium.

(102) Let us now calculate the conditions of propagation of a plane wave in a medium for which the values of k and μ are different in different directions. As we do not propose to give a complete investigation of the question in the present imperfect state of the theory as extended to disturbances of short period, we shall assume that the axes of magnetic induction coincide in direction with those of electric elasticity.

(103) Let the values of the magnetic coefficient for the three axes be λ, μ, ν, then the equations of magnetic force (B) become

$$\left. \begin{aligned} \lambda a &= \frac{dH}{dy} - \frac{dG}{dz} \\[4pt] \mu \beta &= \frac{dF}{dz} - \frac{dH}{dx} \\[4pt] \nu \gamma &= \frac{dG}{dx} - \frac{dF}{dy} \end{aligned} \right\} \dots\dots\dots\dots\dots\dots (81).$$

The equations of electric currents (C) remain as before.

The equations of electric elasticity (E) will be

$$\left. \begin{aligned} P &= 4\pi a^2 f \\ Q &= 4\pi b^2 g \\ R &= 4\pi c^2 h \end{aligned} \right\} \dots\dots\dots\dots\dots\dots\dots (82),$$

where $4\pi a^2$, $4\pi b^2$, and $4\pi c^2$ are the values of k for the axes of x, y, z.

Combining these equations with (A) and (D), we get equations of the form

$$\frac{1}{\mu\nu}\left(\lambda\frac{d^2F}{dx^2} + \mu\frac{d^2F}{dy^2} + \nu\frac{d^2F}{dz^2}\right) - \frac{1}{\mu\nu}\frac{d}{dx}\left(\lambda\frac{dF}{dx} + \mu\frac{dG}{dy} + \nu\frac{dH}{dz}\right) = \frac{1}{a^2}\left(\frac{d^2F}{dt^2} + \frac{d^2\Psi}{dxdt}\right)\dots(83).$$

(104) If l, m, n are the direction-cosines of the wave, and V its velocity, and if

$$lx + my + nz - Vt = w \dots\dots\dots\dots\dots\dots (84),$$

then F, G, H, and Ψ will be functions of w; and if we put F', G', H', Ψ' for the second differentials of these quantities with respect to w, the equations will be

$$\left.\begin{aligned}
\left\{V^2 - a^2\left(\frac{m^2}{\nu} + \frac{n^2}{\mu}\right)\right\} F' + \frac{a^2 lm}{\nu} G' + \frac{a^2 ln}{\mu} H' - lV\Psi' = 0 \\
\left\{V^2 - b^2\left(\frac{n^2}{\lambda} + \frac{l^2}{\nu}\right)\right\} G' + \frac{b^2 mn}{\lambda} H' + \frac{b^2 ml}{\nu} F' - mV\Psi' = 0 \\
\left\{V^2 - c^2\left(\frac{l^2}{\mu} + \frac{m^2}{\lambda}\right)\right\} H' + \frac{c^2 nl}{\mu} F' + \frac{c^2 nm}{\lambda} G' - nV\Psi' = 0
\end{aligned}\right\} \dots\dots (85).$$

If we now put

$$\left.\begin{aligned}
V^4 - V^2 \frac{1}{\lambda\mu\nu}\left\{l^2\lambda(b^2\mu + c^2\nu) + m^2\mu(c^2\nu + a^2\lambda) + n^2\nu(a^2\lambda + b^2\mu)\right\} \\
+ \frac{a^2 b^2 c^2}{\lambda\mu\nu}\left(\frac{l^2}{a^2} + \frac{m^2}{b^2} + \frac{n^2}{c^2}\right)(l^2\lambda + m^2\mu + n^2\nu) = U
\end{aligned}\right\} \dots (86),$$

we shall find

$$F' V^2 U - l\Psi' VU = 0 \dots\dots\dots\dots (87),$$

with two similar equations for G' and H'. Hence either

$$V = 0 \dots\dots\dots\dots\dots\dots\dots\dots (88),$$
$$U = 0 \dots\dots\dots\dots\dots\dots\dots\dots (89),$$

or

$$VF' = l\Psi', \quad VG' = m\Psi' \text{ and } VH' = n\Psi' \dots\dots (90).$$

The third supposition indicates that the resultant of F', G', H' is in the direction normal to the plane of the wave; but the equations do not indicate that such a disturbance, if possible, could be propagated, as we have no other relation between Ψ' and F', G', H'.

The solution $V = 0$ refers to a case in which there is no propagation.

* The solution $U = 0$ gives two values for V^2 corresponding to values of F', G', H', which are given by the equations

$$\frac{l}{a^2} F' + \frac{m}{b^2} G' + \frac{n}{c^2} H' = 0 \dots\dots\dots\dots\dots\dots (91),$$

$$\frac{a^2 l\lambda}{F'}(b^2\mu - c^2\nu) + \frac{b^2 m\mu}{G'}(c^2\nu - a^2\lambda) + \frac{c^2 n\nu}{H'}(a^2\lambda - b^2\mu) = 0 \dots\dots (92).$$

---

* [Although it is not expressly stated in the text it should be noticed that in finding equations (91) and (92) the quantity Ψ' is put equal to zero. See § 98 and also the corresponding treatment of this subject in the Electricity and Magnetism ii. § 796. It may be observed that the

(105) The velocities along the axes are as follows:—

Direction of propagation		x	y	z
Direction of the electric displacements	x		$\dfrac{a^2}{\nu}$	$\dfrac{a^2}{\mu}$
	y	$\dfrac{b^2}{\nu}$		$\dfrac{b^2}{\lambda}$
	z	$\dfrac{c^2}{\mu}$	$\dfrac{c^2}{\lambda}$	

Now we know that in each principal plane of a crystal the ray polarized in that plane obeys the ordinary law of refraction, and therefore its velocity is the same in whatever direction in that plane it is propagated.

If polarized light consists of electromagnetic disturbances in which the electric displacement is in the plane of polarization, then

$$a^2 = b^2 = c^2 \dots\dots\dots\dots\dots\dots\dots\dots\dots (93).$$

If, on the contrary, the electric displacements are perpendicular to the plane of polarization,

$$\lambda = \mu = \nu \dots\dots\dots\dots\dots\dots\dots\dots\dots (94).$$

We know, from the magnetic experiments of Faraday, Plücker, &c., that in many crystals λ, μ, ν are unequal.

equations referred to and the table given in § 105 may perhaps be more readily understood from a different mode of elimination. If we write

$$\lambda l^2 + \mu m^2 + \nu n^2 = P\lambda\mu\nu \text{ and } \lambda l F' + \mu m G' + \nu n H' = Q\lambda\mu\nu,$$

it is readily seen that

$$F' = l \frac{V\Psi' - a^2\lambda Q}{V^2 - a^2\lambda P},$$

with similar expressions for G', H'. From these we readily obtain by reasoning similar to that in § 104, the equation corresponding to (86), viz.:

$$\frac{l^2\lambda}{V^2 - a^2\lambda P} + \frac{m^2\mu}{V^2 - b^2\mu P} + \frac{n^2\nu}{V^2 - c^2\nu P} = 0.$$

This form of the equation agrees with that given in the Electricity and Magnetism ii. § 797.

By means of this equation the equations (91) and (92) readily follow when $\Psi' = 0$. The ratios of F' : G' : H' for any direction of propagation may also be determined.]

The experiments of Knoblauch* on electric induction through crystals seem to shew that a, b and c may be different.

The inequality, however, of λ, μ, ν is so small that great magnetic forces are required to indicate their difference, and the differences do not seem of sufficient magnitude to account for the double refraction of the crystals.

On the other hand, experiments on electric induction are liable to error on account of minute flaws, or portions of conducting matter in the crystal.

Further experiments on the magnetic and dielectric properties of crystals are required before we can decide whether the relation of these bodies to magnetic and electric forces is the same, when these forces are permanent as when they are alternating with the rapidity of the vibrations of light.

Relation between Electric Resistance and Transparency.

(106) If the medium, instead of being a perfect insulator, is a conductor whose resistance per unit of volume is ρ, then there will be not only electric displacements, but true currents of conduction in which electrical energy is transformed into heat, and the undulation is thereby weakened. To determine the coefficient of absorption, let us investigate the propagation along the axis of x of the transverse disturbance G.

By the former equations

$$\frac{d^2G}{dx^2} = -4\pi\mu(q')$$

$$= -4\pi\mu\left(\frac{df}{dt} + q\right) \text{ by (A)},$$

$$\frac{d^2G}{dx^2} = +4\pi\mu\left(\frac{1}{k}\frac{d^2G}{dt^2} - \frac{1}{\rho}\frac{dG}{dt}\right) \text{ by (E) and (F)} \dots\dots\dots\dots (95).$$

If G is of the form

$$G = e^{-px}\cos(qx + nt)\dots\dots\dots\dots\dots\dots\dots\dots(96),$$

we find that

$$p = \frac{2\pi\mu}{\rho}\frac{n}{q} = \frac{2\pi\mu}{\rho}\frac{V}{i}\dots\dots\dots\dots\dots\dots\dots (97),$$

where V is the velocity of light in air, and i is the index of refraction. The proportion of incident light transmitted through the thickness x is

$$e^{-2px}\dots\dots\dots\dots\dots\dots\dots\dots\dots\dots (98).$$

* *Philosophical Magazine*, 1852.

Let R be the resistance in electromagnetic measure of a plate of the substance whose thickness is x, breadth b, and length l, then

$$R = \frac{l\rho}{bx},$$

$$2px = 4\pi\mu \frac{V}{i} \frac{l}{bR} \dots\dots\dots\dots\dots\dots\dots (99).$$

(107) Most transparent solid bodies are good insulators, whereas all good conductors are very opaque.

Electrolytes allow a current to pass easily and yet are often very transparent. We may suppose, however, that in the rapidly alternating vibrations of light, the electromotive forces act for so short a time that they are unable to effect a complete separation between the particles in combination, so that when the force is reversed the particles oscillate into their former position without loss of energy.

Gold, silver, and platinum are good conductors, and yet when reduced to sufficiently thin plates they allow light to pass through them. If the resistance of gold is the same for electromotive forces of short period as for those with which we make experiments, the amount of light which passes through a piece of gold-leaf, of which the resistance was determined by Mr C. Hockin, would be only 10^{-50} of the incident light, a totally imperceptible quantity. I find that between $\frac{1}{500}$ and $\frac{1}{1000}$ of green light gets through such gold-leaf. Much of this is transmitted through holes and cracks; there is enough, however, transmitted through the gold itself to give a strong green hue to the transmitted light. This result cannot be reconciled with the electromagnetic theory of light, unless we suppose that there is less loss of energy when the electromotive forces are reversed with the rapidity of the vibrations of light than when they act for sensible times, as in our experiments.

Absolute Values of the Electromotive and Magnetic Forces called into play in the Propagation of Light.

(108) If the equation of propagation of light is

$$F = A \cos \frac{2\pi}{\lambda}(z - Vt),$$

74—2

the electromotive force will be

$$P = -A\frac{2\pi}{\lambda}V\sin\frac{2\pi}{\lambda}(z - Vt);$$

and the energy per unit of volume will be

$$\frac{P^2}{8\pi\mu V^2},$$

where P represents the greatest value of the electromotive force. Half of this consists of magnetic and half of electric energy.

The energy passing through a unit of area is

$$W = \frac{P^2}{8\pi\mu V};$$

so that

$$P = \sqrt{8\pi\mu VW},$$

where V is the velocity of light, and W is the energy communicated to unit of area by the light in a second.

According to Pouillet's data, as calculated by Professor W. Thomson *, the mechanical value of direct sunlight at the Earth is

83·4 foot-pounds per second per square foot.

This gives the maximum value of P in direct sunlight at the Earth's distance from the Sun,

$$P = 60,000,000,$$

or about 600 Daniell's cells per metre.

At the Sun's surface the value of P would be about

13,000 Daniell's cells per metre.

At the Earth the maximum magnetic force would be ·193 †.

At the Sun it would be 4·13.

These electromotive and magnetic forces must be conceived to be reversed twice in every vibration of light; that is, more than a thousand million million times in a second.

---

* *Transactions of the Royal Society of Edinburgh*, 1854 ("Mechanical Energies of the Solar System").

† The horizontal magnetic force at Kew is about 1·76 in metrical units.

PART VII.

CALCULATION OF THE COEFFICIENTS OF ELECTROMAGNETIC INDUCTION.

General Methods.

(109) The electromagnetic relations between two conducting circuits, A and B, depend upon a function M of their form and relative position, as has been already shewn.

M may be calculated in several different ways, which must of course all lead to the same result.

First Method. M is the electromagnetic momentum of the circuit B when A carries a unit current, or

$$M = \int \left(F \frac{dx}{ds'} + G \frac{dy}{ds'} + H \frac{dz}{ds'} \right) ds',$$

where F, G, H are the components of electromagnetic momentum due to a unit current in A, and ds' is an element of length of B, and the integration is performed round the circuit of B.

To find F, G, H, we observe that by (B) and (C)

$$\frac{d^2 F}{dx^2} + \frac{d^2 F}{dy^2} + \frac{d^2 F}{dz^2} = -4\pi\mu p',$$

with corresponding equations for G and H, p', q', and r' being the components of the current in A.

Now if we consider only a single element ds of A, we shall have

$$p' = \frac{dx}{ds} ds, \qquad q' = \frac{dy}{ds} ds, \qquad r' = \frac{dz}{ds} ds,$$

and the solution of the equation gives

$$F = \frac{\mu}{\rho} \frac{dx}{ds} ds, \qquad G = \frac{\mu}{\rho} \frac{dy}{ds} ds, \qquad H = \frac{\mu}{\rho} \frac{dz}{ds} ds,$$

where ρ is the distance of any point from ds. Hence

$$M = \iint \frac{\mu}{\rho} \left(\frac{dx}{ds}\frac{dx}{ds'} + \frac{dy}{ds}\frac{dy}{ds'} + \frac{dz}{ds}\frac{dz}{ds'} \right) ds\,ds'$$

$$= \iint \frac{\mu}{\rho} \cos \theta \, ds\,ds',$$

where θ is the angle between the directions of the two elements ds, ds', and ρ is the distance between them, and the integration is performed round both circuits.

In this method we confine our attention during integration to the two linear circuits alone.

(110) Second Method. M is the number of lines of magnetic force which pass through the circuit B when A carries a unit current, or

$$M = \Sigma \left(\mu a l + \mu \beta m + \mu \gamma n \right) dS',$$

where μa, $\mu \beta$, $\mu \gamma$ are the components of magnetic induction due to unit current in A, S' is a surface bounded by the current B, and l, m, n are the direction-cosines of the normal to the surface, the integration being extended over the surface.

We may express this in the form

$$M = \mu \Sigma \frac{1}{\rho^2} \sin \theta \sin \theta' \sin \phi \, dS' ds,$$

where dS' is an element of the surface bounded by B, ds is an element of the circuit A, ρ is the distance between them, θ and θ' are the angles between ρ and ds and between ρ and the normal to dS' respectively, and ϕ is the angle between the planes in which θ and θ' are measured. The integration is performed round the circuit A and over the surface bounded by B.

This method is most convenient in the case of circuits lying in one plane, in which case $\sin \theta = 1$, and $\sin \phi = 1$.

(111) Third Method. M is that part of the intrinsic magnetic energy of the whole field which depends on the product of the currents in the two circuits, each current being unity.

Let a, β, γ be the components of magnetic intensity at any point due to the first circuit, a', β', γ' the same for the second circuit; then the intrinsic energy of the element of volume dV of the field is

$$\frac{\mu}{8\pi}\{(a+a')^2+(\beta+\beta')^2+(\gamma+\gamma')^2\}dV.$$

The part which depends on the product of the currents is

$$\frac{\mu}{4\pi}(aa'+\beta\beta'+\gamma\gamma')dV.$$

Hence if we know the magnetic intensities I and I' due to the unit current in each circuit, we may obtain M by integrating

$$\frac{\mu}{4\pi}\Sigma\mu I I'\cos\theta dV$$

over all space, where θ is the angle between the directions of I and I'.

Application to a Coil.

(112) To find the coefficient (M) of mutual induction between two circular linear conductors in parallel planes, the distance between the curves being everywhere the same, and small compared with the radius of either.

If r be the distance between the curves, and a the radius of either, then when r is very small compared with a, we find by the second method, as a first approximation,

$$M=4\pi a\left(\log_e\frac{8a}{r}-2\right).$$

To approximate more closely to the value of M, let a and a_1 be the radii of the circles, and b the distance between their planes; then

$$r^2=(a-a_1)^2+b^2.$$

We obtain M by considering the following conditions:—

1st. M must fulfil the differential equation

$$\frac{d^2M}{da^2}+\frac{d^2M}{db^2}+\frac{1}{a}\frac{dM}{da}=0.$$

This equation being true for any magnetic field symmetrical with respect to the common axis of the circles, cannot of itself lead to the determination of M as a function of a, a_1, and b. We therefore make use of other conditions.

2ndly.   The value of M must remain the same when a and a_1 are exchanged.

3rdly.   The first two terms of M must be the same as those given above.

M may thus be expanded in the following series :—

$$M = 4\pi a \log \frac{8a}{r} \left\{ 1 + \frac{1}{2}\frac{a-a_1}{a} + \frac{1}{16}\frac{3b^2+(a_1-a)^2}{a^2} - \frac{1}{32}\frac{\{3b^2+(a-a_1)^2\}(a-a_1)}{a^3} + \&c. \right\}$$

$$- 4\pi a \left[2 + \frac{1}{2}\frac{a-a_1}{a} + \frac{1}{16}\frac{b^2-3(a-a_1)^2}{a^2} - \frac{1}{48}\frac{\{6b^2-(a-a_1)^2\}(a-a_1)}{a^3} + \&c. \right].$$

(113)   We may apply this result to find the coefficient of self-induction (L) of a circular coil of wire whose section is small compared with the radius of the circle.

Let the section of the coil be a rectangle, the breadth in the plane of the circle being c, and the depth perpendicular to the plane of the circle being b.

Let the mean radius of the coil be a, and the number of windings n ; then we find, by integrating,

$$L = \frac{n^2}{b^2 c^2} \iiiint M(xy\,x'y')\, dx\, dy\, dx'\, dy',$$

where $M(xy\,x'y')$ means the value of M for the two windings whose coordinates are xy and $x'y'$ respectively; and the integration is performed first with respect to x and y over the rectangular section, and then with respect to x' and y' over the same space.

$$L = 4\pi n^2 a \left\{ \log_\epsilon \frac{8a}{r} + \frac{1}{12} - \frac{4}{3}\left(\theta - \frac{\pi}{4}\right)\cot 2\theta - \frac{\pi}{3}\cos 2\theta - \frac{1}{6}\cot^2\theta \log\cos\theta - \frac{1}{6}\tan^2\theta \log\sin\theta \right\}$$

$$+ \frac{\pi n^2 r^2}{24a}\left\{ \log\frac{8a}{r}(2\sin^2\theta + 1) + 3\cdot45 + 27\cdot475\cos^2\theta - 3\cdot2\left(\frac{\pi}{2} - \theta\right)\frac{\sin^3\theta}{\cos\theta} + \frac{1}{5}\frac{\cos^4\theta}{\sin^2\theta}\log\cos\theta \right.$$

$$\left. + \frac{13}{3}\frac{\sin^4\theta}{\cos^2\theta}\log\sin\theta \right\} + \&c.$$

Here $a =$ mean radius of the coil.

,, $r =$ diagonal of the rectangular section $= \sqrt{b^2 + c^2}$.

,, $\theta =$ angle between r and the plane of the circle.

,, $n =$ number of windings.

The logarithms are Napierian, and the angles are in circular measure.

In the experiments made by the Committee of the British Association for determining a standard of Electrical Resistance, a double coil was used, consisting of two nearly equal coils of rectangular section, placed parallel to each other, with a small interval between them.

The value of L for this coil was found in the following way.

The value of L was calculated by the preceding formula for six different cases, in which the rectangular section considered has always the same breadth, while the depth was

$$A, \ B, \ C, \ A+B, \ B+C, \ A+B+C,$$

and $n=1$ in each case.

Calling the results $\quad L(A), \quad L(B), \quad L(C),$ &c.,

we calculate the coefficient of mutual induction $M(AC)$ of the two coils thus,

$$2ACM(AC) = (A+B+C)^2 L(A+B+C) - (A+B)^2 L(A+B)$$
$$- (B+C)^2 L(B+C) + B^2 L(B).$$

Then if n_1 is the number of windings in the coil A and n_2 in the coil C, the coefficient of self-induction of the two coils together is

$$L = n_1^2 L(A) + 2n_1 n_2 M(AC) + n_2^2 L(C).$$

(114) These values of L are calculated on the supposition that the windings of the wire are evenly distributed so as to fill up exactly the whole section. This, however, is not the case, as the wire is generally circular and covered with insulating material. Hence the current in the wire is more concentrated than it would have been if it had been distributed uniformly over the section, and the currents in the neighbouring wires do not act on it exactly as such a uniform current would do.

The corrections arising from these considerations may be expressed as numerical quantities, by which we must multiply the length of the wire, and they are the same whatever be the form of the coil.

Let the distance between each wire and the next, on the supposition that they are arranged in square order, be D, and let the diameter of the wire be d, then the correction for diameter of wire is

$$+ 2\left(\log \frac{D}{d} + \tfrac{4}{3}\log 2 + \frac{\pi}{3} - \tfrac{11}{6}\right).$$

The correction for the eight nearest wires is

$$+0{\cdot}0236.$$

For the sixteen in the next row $+0{\cdot}00083.$

These corrections being multiplied by the length of wire and added to the former result, give the true value of L, considered as the measure of the potential of the coil on itself for unit current in the wire when that current has been established for some time, and is uniformly distributed through the section of the wire.

(115) But at the commencement of a current and during its variation the current is not uniform throughout the section of the wire, because the inductive action between different portions of the current tends to make the current stronger at one part of the section than at another. When a uniform electromotive force P arising from any cause acts on a cylindrical wire of specific resistance ρ, we have

$$p\rho = P - \frac{dF}{dt},$$

where F is got from the equation

$$\frac{d^2F}{dr^2} + \frac{1}{r}\frac{dF}{dr} = -4\pi\mu p,$$

r being the distance from the axis of the cylinder.

Let one term of the value of F be of the form Tr^n, where T is a function of the time, then the term of p which produced it is of the form

$$-\frac{1}{4\pi\mu}\, n^2 T r^{n-2}.$$

Hence if we write

$$F = T + \frac{\mu\pi}{\rho}\left(-P + \frac{dT}{dt}\right)r^2 + \overline{\frac{\mu\pi}{\rho}}\Big|^2 \frac{1}{1^2 \cdot 2^2}\frac{d^2T}{dt^2}r^4 + \&c.$$

$$p\rho = \left(P - \frac{dT}{dt}\right) - \frac{\mu\pi}{\rho}\frac{d^2T}{dt^2}r^2 - \overline{\frac{\mu\pi}{\rho}}\Big|^2 \frac{1}{1^2 \cdot 2^2}\frac{d^3T}{dt^3}r^4 - \&c.$$

The total counter current of self-induction at any point is

$$\int\left(\frac{P}{\rho} - p\right)dt = \frac{1}{\rho}\,T + \frac{\mu\pi}{\rho^2}\frac{dT}{dt}r^2 + \frac{\mu^2\pi^2}{\rho^3}\frac{1}{1^2 2^2}\frac{d^2T}{dt^2}r^4 + \&c.$$

from $t=0$ to $t=\infty$.

When $t = 0$, $p = 0$, $\quad \therefore \left(\dfrac{dT}{dt}\right) = P$, $\left(\dfrac{d^2T}{dt^2}\right)_0 = 0$, &c.

When $t = \infty$, $p = \dfrac{P}{\rho}$, $\quad \therefore \left(\dfrac{dT}{dt}\right)_\infty = 0$, $\left(\dfrac{d^2T}{dt^2}\right)_\infty = 0$, &c.

$$\int_0^\infty \int_0^r 2\pi \left(\dfrac{P}{\rho} - p\right) r\,dr\,dt = \dfrac{1}{\rho}\, T\pi r^2 + \tfrac{1}{2}\dfrac{\mu\pi^2}{\rho^2}\dfrac{dT}{dt}\, r^4 + \dfrac{\mu^2\pi^3}{\rho^3}\dfrac{1}{1^2 . 2^2 . 3}\dfrac{d^2T}{dt^2}\, r^6 + \&\text{c.}$$

from $t = 0$ to $= \infty$.

When $t = 0$, $p = 0$ throughout the section, $\therefore \left(\dfrac{dT}{dt}\right)_0 = P$, $\left(\dfrac{d^2T}{dt^2}\right)_0 = 0$, &c.

When $t = \infty$, $p = 0$ $\qquad$,, $\qquad$,, $\qquad$,, $\qquad \therefore \left(\dfrac{dT}{dt}\right)_\infty = 0$, $\left(\dfrac{d^2T}{dt^2}\right)_\infty = 0$, &c.

Also if l be the length of the wire, and R its resistance,

$$R = \dfrac{\rho l}{\pi r^2};$$

and if C be the current when established in the wire, $C = \dfrac{Pl}{R}$.

The total counter current may be written

$$\dfrac{l}{R}(T_\infty - T_0) - \tfrac{1}{2}\mu\,\dfrac{l}{R}\,C = -\dfrac{LC}{R} \text{ by § (35).}$$

Now if the current instead of being variable from the centre to the circumference of the section of the wire had been the same throughout, the value of F would have been

$$F = T + \mu\gamma \left(1 - \dfrac{r^2}{r_0^2}\right),$$

where γ is the current in the wire at any instant, and the total counter current would have been

$$\int_0^\infty \int_0^r \dfrac{1}{\rho}\dfrac{dF}{dt}\, 2\pi r\,dr = \dfrac{l}{R}(T_\infty - T_0) - \tfrac{3}{4}\mu\,\dfrac{l}{R}\,C = -\dfrac{L'C}{R}, \text{ say.}$$

Hence $\qquad\qquad\qquad L = L' - \tfrac{1}{4}\mu l,$

or the value of L which must be used in calculating the self-induction of a wire for variable currents is less than that which is deduced from the supposition of the current being constant throughout the section of the wire by $\tfrac{1}{4}\mu l$, where l is the length of the wire, and μ is the coefficient of magnetic induction for the substance of the wire.

(116) The dimensions of the coil used by the Committee of the British Association in their experiments at King's College in 1864 were as follows :—

metre.

$$\text{Mean radius} \dots\dots\dots\dots\dots = a = \cdot 158194$$
$$\text{Depth of each coil} \dots\dots\dots = b = \cdot 01608$$
$$\text{Breadth of each coil} \dots\dots\dots = c = \cdot 01841$$
$$\text{Distance between the coils} \dots\dots = \cdot 02010$$
$$\text{Number of windings} \dots\dots\dots n = 313$$
$$\text{Diameter of wire} \dots\dots\dots\dots = \cdot 00126$$

The value · of L derived from the first term of the expression is 437440 metres.

The correction depending on the radius not being infinitely great compared with the section of the coil as found from the second term is -7345 metres.

The correction depending on the diameter of the wire is per unit of length	$+ \cdot 44997$
Correction of eight neighbouring wires	$+ \cdot 0236$
For sixteen wires next to these	$+ \cdot 0008$
Correction for variation of current in different parts of section	$- \cdot 2500$
Total correction per unit of length	$\cdot 22437$
Length	311·236 metres.
Sum of corrections of this kind	70 ,,
Final value of L by calculation	430165 ,,

This value of L was employed in reducing the observations, according to the method explained in the Report of the Committee*. The correction depending on L varies as the square of the velocity. The results of sixteen experiments to which this correction had been applied, and in which the velocity varied from 100 revolutions in seventeen seconds to 100 in seventy-seven seconds,

---

* *British Association Reports*, 1863, p. 169.

were compared by the method of least squares to determine what further correction depending on the square of the velocity should be applied to make the outstanding errors a minimum.

The result of this examination shewed that the calculated value of L should be multiplied by 1·0618 to obtain the value of L, which would give the most consistent results.

We have therefore L by calculation................................. 430165 metres.

Probable value of L by method of least squares 456748   ,,

Result of rough experiment with the Electric Balance (see § 46) 410000   ,,

The value of L calculated from the dimensions of the coil is probably much more accurate than either of the other determinations.

[From the *Philosophical Magazine*, Vol. XXVII.]

* XXVI. *On the Calculation of the Equilibrium and Stiffness of Frames.*

THE theory of the equilibrium and deflections of frameworks subjected to the action of forces is sometimes considered as more complicated than it really is, especially in cases in which the framework is not simply stiff, but is strengthened (or weakened as it may be) by additional connecting pieces.

I have therefore stated a general method of solving all such questions in the least complicated manner. The method is derived from the principle of Conservation of Energy, and is referred to in Lamé's *Leçons sur l'Elasticité*, Leçon 7me, as Clapeyron's Theorem ; but I have not yet seen any detailed application of it.

If such questions were attempted, especially in cases of three dimensions, by the regular method of equations of forces, every point would have three equations to determine its equilibrium, so as to give $3s$ equations between e unknown quantities, if s be the number of points and e the number of connexions. There are, however, six equations of equilibrium of the system which must be fulfilled necessarily by the forces, on account of the equality of action and reaction in each piece. Hence if

$$e = 3s - 6,$$

the effect of any external force will be definite in producing tensions or pressures in the different pieces ; but if $e > 3s - 6$, these forces will be indeterminate. This indeterminateness is got rid of by the introduction of a system of e equations of elasticity connecting the force in each piece with the change in its length. In order, however, to know the changes of length, we require to assume $3s$ displacements of the s points ; 6 of these displacements, however, are equivalent to the motion of a rigid body so that we have $3s - 6$ displacements of points, e extensions and e forces to determine from $3s - 6$ equations of forces, e

---

* [Owing to an oversight this paper is out of its proper place ; it should have been immediately before the memoir on "The Electro-magnetic Field." (No. XXV.)]

equations of extensions, and e equations of elasticity; so that the solution is always determinate.

The following method enables us to avoid unnecessary complexity by treating separately all pieces which are additional to those required for making the frame stiff, and by proving the identity in form between the equations of forces and those of extensions by means of the principle of work.

On the Stiffness of Frames.

Geometrical definition of a Frame. A frame is a system of lines connecting a number of points.

A stiff frame is one in which the distance between any two points cannot be altered without altering the length of one or more of the connecting lines of the frame.

A frame of s points in space requires *in general* $3s - 6$ connecting lines to render it stiff. In those cases in which stiffness can be produced with a smaller number of lines, certain conditions must be fulfilled, rendering the case one of a maximum or minimum value of one or more of its lines. The stiffness of such frames is of an inferior order, as a small disturbing force may produce a displacement infinite in comparison with itself.

A frame of s points in a plane requires in general $2s - 3$ connecting lines to render it stiff.

A frame of s points in a line requires $s - 1$ connecting lines.

A frame may be either simply stiff, or it may be self-strained by the introduction of additional connecting lines having tensions or pressures along them.

In a frame which is simply stiff, the forces in each connecting line arising from the application of a force of pressure or tension between any two points of the frame may be calculated either by equations of forces, or by drawing diagrams of forces according to known methods.

In general, the lines of connexion in one part of the frame may be affected by the action of this force, while those in other parts of the frame may not be so affected.

Elasticity and Extensibility of a connecting piece.

Let e be the extension produced in a piece by tension-unity acting in it, then e may be called its extensibility. Its elasticity, that is, the force required to produce extension-unity, will be $\dfrac{1}{e}$. We shall suppose that the effect of pressure in producing compression of the piece is equal to that of tension in producing extension, and we shall use e indifferently for extensibility and compressibility.

Work done against Elasticity.

Since the extension is proportional to the force, the whole work done will be the product of the extension and the mean value of the force; or if x is the extension and F the force,

$$x = eF,$$

$$\text{work} = \tfrac{1}{2}Fx = \tfrac{1}{2}eF^2 = \tfrac{1}{2}\dfrac{1}{e}x^2.$$

When the piece is inextensible, or $e = 0$, then all the work applied at one end is transmitted to the other, and the frame may be regarded as a machine whose efficiency is perfect. Hence the following

THEOREM. If p be the tension of the piece A due to a tension-unity between the points B and C, then an extension-unity taking place in A will bring B and C nearer by a distance p.

For let X be the tension and x the extension of A, Y the tension and y the extension of the line BC; then supposing all the other pieces inextensible, no work will be done except in stretching A, or

$$\tfrac{1}{2}Xx + \tfrac{1}{2}Yy = 0.$$

But $X = pY$, therefore $y = -px$, which was to be proved.

PROBLEM I. A tension F is applied between the points B and C of a frame which is simply stiff; to find the extension of the line joining D and E, all the pieces except A being inextensible, the extensibility of A being e.

Determine the tension in each piece due to unit tension between B and C, and let p be the tension in A due to this cause.

Determine also the tension in each piece due to unit tension between D and E, and let y be the tension in the piece A due to this cause.

Then the actual tension of A is Fp, and its extension is eFp, and the extension of the line DE due to this cause is $-Fepq$ by the last theorem.

Cor. If the other pieces of the frame are extensible, the complete value of the extension in DE due to a tension F in BC is

$$- F\Sigma(epq),$$

where $\Sigma(epq)$ means the sum of the products of epq, which are to be found for each piece in the same way as they were found for A.

The extension of the line BC due to a tension F in BC itself will be

$$- F\Sigma(ep^2),$$

$\Sigma(ep^2)$ may therefore be called the resultant extensibility along BC.

PROBLEM II. A tension F is applied between B and C; to find the extension between D and E when the frame is not simply stiff, but has additional pieces R, S, T, &c. whose elasticities are known.

Let p and q, as before, be the tensions in the piece A due to unit tensions in BC and DE, and let r, s, t, &c. be the tensions in A due to unit tension in R, S, T, &c.; also let R, S, T be the tensions of R, S, T, and ρ, σ, τ their extensibilities. Then the tension A

$$= Fp + Rr + Ss + Tt + \&c.;$$

the extension of A

$$= e(Fp + Rr + Ss + Tt + \&c.);$$

the extension of R

$$= - F\Sigma(epr) - R\Sigma er^2 - S\Sigma ers - T\Sigma ert + \&c. = R\rho \,;$$

extension of S

$$= - F\Sigma(eps) - R\Sigma(ers) - S\Sigma es^2 - T\Sigma(est) = S\sigma \,;$$

extension of T

$$= - F\Sigma(ept) - R\Sigma(ert) - S\Sigma(est) - T\Sigma(et^2) = T\tau \,;$$

also extension of DE

$$= - F\Sigma(epq) - R\Sigma(eqr) - S\Sigma(eqs) - T\Sigma(eqt) = x,$$

the extension required. Here we have as many equations to determine R, S, T, &c. as there are of these unknown quantities, and by the last equation we determine x the extension of DE from F the tension in BC.

Thus, if there is only one additional connexion R, we find

$$R = - F\,\frac{\Sigma(epr)}{\Sigma(er^2) + \rho},$$

and

$$x = - F\left\{\Sigma(epq) + \frac{\Sigma(epr)\Sigma(eqr)}{\Sigma(er^2) + \rho}\right\}.$$

If there are two additional connexions R and S, with elasticities ρ and σ,

$$x = \frac{-F}{\Sigma e(r^2+\rho)\Sigma e(s^2+\sigma) - (\Sigma(ers))^2} \left\{ \begin{array}{l} \Sigma(epr)\Sigma(ers)\Sigma(eqs) + \Sigma(eps)\Sigma(eqr)\Sigma(ers) + \Sigma(epq)\Sigma e(r^2+\rho)\Sigma e(s^2+\sigma) \\ -\Sigma(epr)\Sigma(eqr)\Sigma e(s^2+\sigma) - \Sigma(eps)\Sigma(eqs)\Sigma e(r^2+\rho) - \Sigma(epq)(\Sigma(ers))^2 \end{array} \right\}.$$

The expressions for the extensibility, when there are many additional pieces, are of course very complicated.

It will be observed, however, that p and q always enter into the equations in the same way, so that we may establish the following general

THEOREM. The extension in BC, due to unity of tension along DE, is always equal to the tension in DE due to unity of tension in BC. Hence we have the following method of determining the displacement produced at any joint of a frame due to forces applied at other joints.

1st. Select as many pieces of the frame as are sufficient to render all its points stiff. Call the remaining pieces R, S, T, &c.

2nd. Find the tension on each piece due to unit of tension in the direction of the force proposed to be applied. Call this the value of p for each piece.

3rd. Find the tension on each piece due to unit of tension in the direction of the displacement to be determined. Call this the value of q for each piece.

4th. Find the tension on each piece due to unit of tension along R, S, T, &c., the additional pieces of the frame. Call these the values of r, s, t, &c. for each piece.

5th. Find the extensibility of each piece and call it e, those of the additional pieces being ρ, σ, τ, &c.

6th. R, S, T, &c. are to be determined from the equations

$$R\rho + R\Sigma(er^2) + S(ers) + T\Sigma(ert) + F\Sigma(epr) = 0,$$
$$S\sigma + R\Sigma(ers) + S(es^2) + T\Sigma(est) + F\Sigma(eps) = 0,$$
$$T\tau + R\Sigma(ert) + S(est) + T\Sigma(et^2) + F\Sigma(ept) = 0,$$

as many equations as there are quantities to be found.

7th. x, the extension required, is then found from the equation

$$x = -F\Sigma(epq) - R\Sigma(erq) - S\Sigma(eqs) - T\Sigma(eqt).$$

In structures acted on by weights in which we wish to determine the deflection at any point, we may regard the points of support as the extremities of pieces connecting the structure with the centre of the earth; and if the supports are capable of resisting a horizontal thrust, we must suppose them connected by a piece of equivalent elasticity. The deflection is then the shortening of a piece extending from the given point to the centre of the earth.

EXAMPLE. Thus in a triangular or Warren girder of length l, depth d, with a load W placed at a distance a from one end, 0; to find the deflection at a point distant b from the same end, due to the yielding of a piece of the boom whose extensibility is e, distant x from the same end.

The pressure of the support at $0 = W \dfrac{l-a}{l}$; and if x is less than a, the force at x will be $\dfrac{W}{dl} x(l-a)$, or

$$p = \frac{x(l-a)}{dl}.$$

If x is greater than a,

$$p = \frac{a(l-x)}{dl}.$$

Similarly, if x is less than b,

$$q = \frac{x(l-b)}{dl};$$

but if x is greater than b,

$$q = \frac{b(l-x)}{dl}.$$

The deflection due to x is therefore $Wepq$, where the proper values of p and q must be taken according to the relative position of a, b, and x.

If a, b, l, x represent the *number* of the respective pieces, reckoning from the beginning and calling the first joint 0, the second joint and the piece opposite 1, &c., and if L be the length of each piece, and the extensibility of each piece $= e$, then the deflection of b due to W at a will be, by summation of series,

$$= \frac{1}{6} WeL^2 . \frac{a(l-b)}{d^2 l} \{2b(l-a) - (b-a)^2 + 1\}.$$

76—2

This is the deflection due to the yielding of all the horizontal pieces. The greater the number of pieces, the less is the importance of the last term.

Let the inclination of the pieces of the web be α, then the force on a piece between 0 and a is $W \dfrac{l-a}{l \sin \alpha}$, or

$$p' = \frac{l-a}{l \sin \alpha} \text{ when } x < a,$$

and

$$p' = \frac{a}{l \sin \alpha} \text{ when } x > a.$$

Also

$$q' = \frac{l-b}{l \sin \alpha} \text{ when } x < b,$$

$$q' = \frac{b}{l \sin \alpha} \text{ when } x > b.$$

If e' be the extensibility of a piece of the web, we have to sum $W \Sigma e' p' q'$ to get the deflection due to the yielding of the web,

$$= \frac{W e'}{l^2 \sin^2 \alpha} a(l-b)\{l + 2(b-a)\}.$$

INDEX TO VOL. I.

CAMBRIDGE : PRINTED BY C. J. CLAY, M.A. AND SONS, AT THE UNIVERSITY PRESS.

Printed in the United States
By Bookmasters